S0-DXN-541

Digital Computing and Numerical Methods

(with FORTRAN-IV, WATFOR, and WATFIV Programming)

Digital Computing and Numerical Methods

(with FORTRAN-IV, WATFOR, and WATFIV Programming)

Brice Carnahan and James O. Wilkes

Professors of Chemical Engineering
The University of Michigan

JOHN WILEY and SONS, Inc.
New York · London · Sydney · Toronto

Library of Congress Cataloging in Publication Data

Carnahan, Brice.
Digital computing and numerical methods.
Includes bibliographical references.
1. Electronic data processing—Engineering.
2. FORTRAN (Computer program language) 3. Numerical calculations. I. Wilkes, James O. joint author.
II. Title.
TA345.C38 001.6'4 72-13010
ISBN 0-471-13500-3

Printed in the United States of America

10 9 8 7 6 5 4 3 2 1

Preface

We hope that this introductory text will appeal to those who are interested in learning about the elements of digital-computer organization, FORTRAN programming, the formulation of algorithms, and the theory and applications of numerical methods. The level is suitable for the early years at a university, and the orientation is toward engineering and applied mathematics.

Chapter 1 introduces the beginner to the functional organization of digital computers, including material on the internal representation of numbers and characters, which is relevant to the later study of FORTRAN. Chapter 2 deals with the formulation of computational procedures or algorithms, with emphasis on the flow-diagram approach. Chapter 3 continues with a synopsis of the FORTRAN-IV language, together with important variations that are available in the WATFOR and WATFIV dialects; several illustrative complete programs are included. Operating systems are discussed in a general manner in Chapter 4. Each reader should have available additional specific information concerning the particular installation that he is using.

The remaining chapters (5 to 10) are devoted to an introduction to numerical methods, including the solution of single and simultaneous equations, numerical approximation and integration, the solution of ordinary differential equations, and optimization techniques. Algorithms for the various methods are developed in the text and are also illustrated by completely documented computer programs, enabling the reader to gain an appreciation of what to expect during the implementation of particular numerical techniques on a computer. Certain important general-purpose programs, such as those involving the Lagrange interpolating polynomial and the Gauss-Jordan solution of an arbitrary set of simultaneous equations, have intentionally been omitted; the student should benefit from writing his own programs in these areas. The reader who wishes to study numerical methods (excluding optimization techniques) in greater depth may be interested in our other text, coauthored with H. A. Luther: *Applied Numerical Methods*, Wiley, 1969.

There is a set of unworked problems at the end of each chapter, except for Chapter 4. Some of these problems involve the derivation of formulas or proofs; others involve hand calculations; and the rest are concerned with the computer solution of a variety of problems, many of which are drawn from various branches of engineering and applied mathematics.

We thank our friends and colleagues who have expressed interest in our work, have made constructive criticisms, and have suggested some of the problems discussed herein.

BRICE CARNAHAN
JAMES O. WILKES

Contents

APPENDICES

Digital Computing and Numerical Methods

(with FORTRAN-IV, WATFOR, and WATFIV Programming)

CHAPTER 1
An Introduction to Digital Computers

1.1 Introduction

Although digital computers are often viewed as fast desk calculators or super slide rules, they are, in fact, rather general devices for manipulating *symbolic* information. In most applications the symbols being manipulated are numbers or digits (hence the name *digital computer*), and the operations being performed on the symbols are the standard arithmetical operations such as addition and subtraction. However, the symbols might just as easily have nonnumerical values and the operations be nonnumerical in nature. For example, the symbols might be the characters such as letters and punctuation marks in an English sentence, and the operations might result in the parsing of the sentence for its word and phrase structure. In fact, a *general-purpose* digital computer can be instructed to accept, to store, to manipulate, and to display virtually *any* kind of properly encoded symbolic information.

The term general-purpose digital computer applies to those digital computing machines that are designed to solve essentially any "computable" problem. Although *computability* has a rigorous mathematical or logical meaning, an intuitive understanding of the word is adequate for most computer users. A computable problem is one that can be stated unambiguously in mathematical or symbolic form and for which a terminating *solution procedure* or *algorithm* can be outlined, step-by-step, as described in Chapter 2. Often we prepare a solution procedure in a graphical form, called a *flow diagram*. If the procedure is in the form of a list of orders or commands that can be interpreted directly by a computer, it is called a *program*. Thus the terms procedure, algorithm, flow diagram, and program have closely related meanings; each is a representation of *how* to solve a problem. A *programmer* develops an algorithm to solve his problem and then writes a program to implement the algorithm on a computer.

In passing, we should mention that other kinds of electronic computers are also manufactured. *Analog computers* consist of electronic elements such as resistors, capacitors, and amplifiers that can be used to develop an electronic circuit analog of a physical system or of a system of equations. Here, electrical characteristics of the circuit (for example, voltage and current) are used to represent physical quantities in the real system being modeled (for example, pressure and flow rate in a piping system). A *hybrid* computer consists of a general-purpose digital computer *and* an analog computer, plus special equipment to interface the two different machines. In addition, a variety of *special-purpose* computers, both digital and analog, is available for solving very special problems, such as controlling non-skid brakes in automobiles or carrying out aircraft or missile guidance calculations. In this text, we shall be concerned only with general-purpose digital computers.

While we usually view a digital computer as a single problem-solving machine, every computer is in fact a collection of a large number of interconnected electronic and electromechanical devices, all directed by a central control unit. At the detailed *hardware* level, involving transistors, circuits, etc., computers are very complicated indeed. Fortunately, an understanding of computer operation and the ability to make effective use of a computer does *not* require knowledge either of electronics or of the hardware. An overall view of the computer's organization with emphasis on function rather than on electrical or mechanical details is sufficient. In this respect, computers are similar to many of the machines produced by our technology (automobiles, television sets, airplanes) that can be used very effectively by those without extensive technical training (in combustion chemistry, electronics, or aerodynamics, for example). The essential questions are:

1. What can the machine do?
2. How do we get the machine to do what it is supposed to be able to do? That is, how do we communicate with the machine, and vice versa?

1.2 Digital Computer Organization

Viewed functionally, all equipment items associated with a digital computer can be grouped into four general categories:

1. Memory
2. Input/output
3. Arithmetic
4. Control

The machine shown in Fig. 1.1 is a hypothetical one, but is typical of all currently available digital computers. Specific operating details for each of the many computers now in use will differ from those for the machine described here.

1.3 The Memory

The memory or *store* is the heart of the digital computer. The principal *central* or *fast* memory of most existing computers is a collection of electronic devices made of a ferromagnetic material that can be permanently magnetized by appropriate electrical impulses. Nearly all such storage elements are stable in only one of two states; the two-state polarity of the magnetic field produced in an individual storage element can be used to represent any *binary* or *two-state* information.

A small ferromagnetic toroid, called a *magnetic core*, is the principal storage element of most computers now on the market. Each donut-shaped core, typically about 0.020 in. in diameter, has associated with it a set of wires. (Plate 1 shows a highly magnified view of several magnetic cores with their associated wires.) By sending a particular combination of pulses through the wires, the core can be made into a permanent magnet with one of two different polar orientations for the magnetic field, analogous to creating a bar magnet with either the north-south or the south-north orientation for the magnetic field. Once the orientation has been established, it is possible to sense the current state of the magnet or to reverse the polarity of the field, again

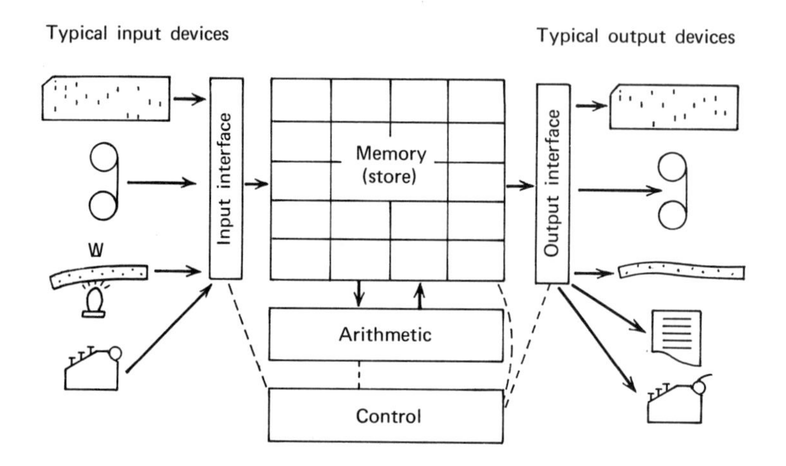

Figure 1.1 Overall digital computer organization.

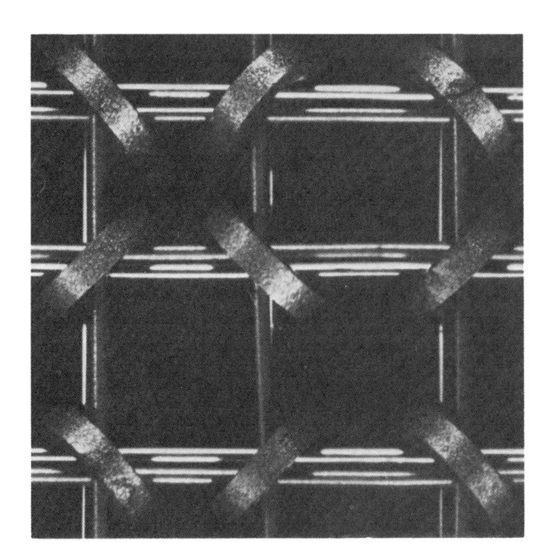

Plate 1. Magnetic core storage elements. This magnified view shows several cores (viewed on edge) with their associated wires. (Courtesy IBM Corporation.)

by sending particular combinations of pulses through the wires associated with the core. Thus the "permanent" magnet should be viewed as only *temporary*, since the magnetic field can be switched at will.

On first glance, the simple two-state device appears to be severely limited as an information storage medium. However, given a large collection of such devices, virtually any kind of symbolic information can be stored. (An array of many magnetic core storage elements arranged in a *core plane* is shown in Plate 2.) Consider the problem of storing the answers to 100 true/false examination questions, for example. Let the two magnetic-field orientations be denoted as state 1 and state 0. Now associate the value *true* with state 1 and the value *false* with state 0. Given 100 magnetic cores numbered 1, 2, 3, . . . , 100, let the magnetic field of the ith core represent the answer given for the ith question on the examination; thus, if the answers given for the first five examination questions were true, true, false, true, and false, the magnetic fields of the first five cores would be set to states 1, 1, 0, 1, and 0, respectively. Since the cores are permanent magnets, we can at any later time retrieve the stored values by sensing the polarity of the magnetic field associated with each core. At some other time (when we no longer need to save the examination answers) we might use the same 100 magnetic cores to save some other kind of two-state information, for example, the signs + or − of 100 numbers; in this case, state 0 might be associated with the plus sign and state 1 with the minus sign.

Of course, the obvious question is: "How can a decimal number, such as 3.14159, be saved?" Clearly, it is impossible to store unique representations of the ten decimal digits in a single two-state device such as a magnetic core. However, either of the two digits in the binary number system (see Section 1.13) could be stored in a single two-state device, by associating the digit 0 with state 0 and the digit 1 with state 1. Therefore, given enough magnetic cores (30 to 40) it should be possible to represent, say, a ten-digit *decimal* number in *binary* form in the computer's memory. Subsequent arithmetical operations performed on such numbers would have to be carried out using binary arithmetic, of course. Hence, most digital computers, particularly those intended for scientific calculations, are called *binary* computers. Details of number systems and of base 2, base 8, and base 16 arithmetic are covered later, in Sections 1.13, 1.14, and 1.15.

Other symbolic information, such as letters and punctuation marks, can be stored similarly by encoding each character as a unique string of ones and zeros and then by setting the magnetic states of several magnetic cores accordingly (see Section 1.17 for the scheme used in the IBM 360/370 computers). Fortunately, the essential conversions from decimal and character form to internal binary form and *vice versa* are handled automatically by the newer machines. The user need not know about binary notation to make effective use of a digital computer.

For illustrative purposes we shall describe a *hypothetical* computer with a memory consisting of *ten*-state rather than *two*-state devices. Let each of the ten

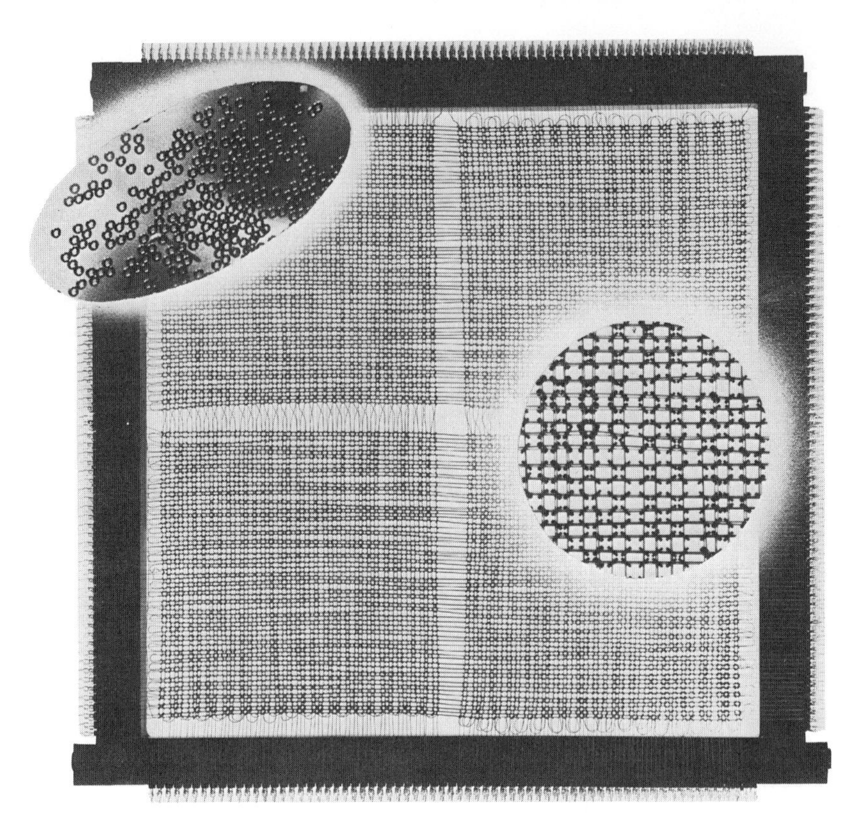

Plate 2. Magnetic core storage plane. This planar arrangement contains over 15,000 individually strung magnetic cores on a 4 in. by 4 in. frame. The inserts show a magnified collection of some unstrung cores and a detail of a portion of the core plane (see also Plate 1). (Courtesy IBM Corporation.)

states represent one of the ten decimal digits, 0, 1, 2, . . . , 9; then one storage element can be used to store any of the ten decimal digits. (The only purpose in inventing this hypothetical storage element is to avoid the confusion of binary arithmetic.) Let the entire memory of this hypothetical computer consist of 10,000 such ten-state devices, so that a total of 10,000 digits may be stored in the memory at any one time.

To simplify the problem of locating any sequence of digits in the memory, the overall collection of storage elements in most scientific computers is divided into smaller collections containing just a few digits; these collections of digits are called *cells*, *words*, *memory* or *machine words*, *storage*, *memory*, or *machine locations*, among others. A machine in which the number of digits in each word is not variable is said to have a *fixed word length*. Each word in the memory is assigned a numerical *address*, usually in sequential order starting with address 0. In our 10,000 decimal digit memory, let the word length be fixed and equal to ten digits, so that the memory contains a total of 1000 words (this memory would normally be termed a "1K store" in computer parlance). Let the memory be enlarged slightly, so that there is a sign associated with each word. This could be accomplished by adding 1000 magnetic cores to the memory, one to represent the sign of each word. If we assign to the first signed collection of ten digits (that is, to the first word) the address 000, to the second word the address 001, to the third the address 002, etc., and to the last or 1000th word the address 999, just three decimal digits are needed to describe the address of any of the 1000 machine words (see Fig. 1.2).

It is very important to distinguish be-

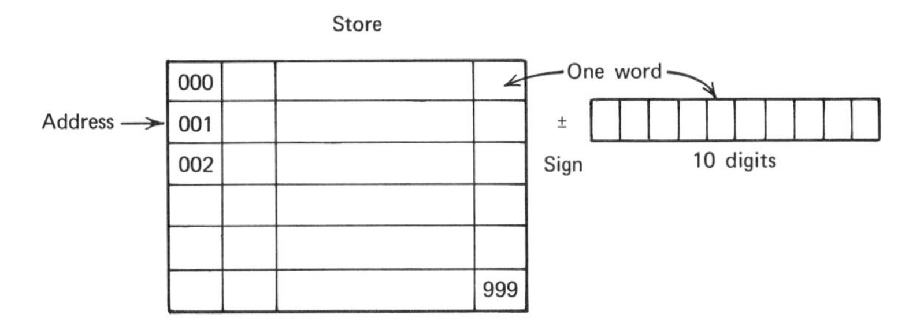

Figure 1.2 Division of the memory into words.

tween the *address* of a memory word and the *content* of the memory word. The three-digit address, specifies which word in the memory is to be examined. The *content* of that address is the signed ten-digit number stored in the memory elements of that particular word in the store.

The stores of most digital computers are built so that the content of a memory word may be retrieved or *read* without destroying it (*nondestructive read-out*); on the other hand, when a new number is stored or *written* into a memory word, the previous content of that word is lost (*destructive read-in*). This is completely analogous to the operation of a magnetic-tape recorder. Recorded information may be played back without destroying it; when a new signal is recorded over previous information, the earlier recording is destroyed or erased in the process. Note that in our machine only a finite set of numbers ($2 \times 10^{10} - 1$ altogether) can be represented, namely all numbers (ignoring any placing of a decimal point) from -9999999999 to $+9999999999$. Thus it is not possible to represent irrational numbers (or for that matter any number with more than ten significant digits) in our computer's memory. For this reason, information that is essentially *continuous* (for example, a voltage from a temperature indicator) must be transformed into *discrete digital form* before processing on a digital machine.

The *access time* (the time required to retrieve the contents of a memory location) for the *fast* or *core store* depends on the size and cost of the computer, and ranges from about 50 nsec (nanoseconds or billionths of a second) to a few microseconds. Contents of the fast memory are available in *random-access* fashion; that is, the access time is the same for all memory locations, regardless of the address.

The fast store is normally the most expensive part of the computer. Because of the cost, those who lease or buy computers can afford only a limited amount of such storage. Even the largest machines seldom have more than 500,000 words of fast memory. To back up the fast store, there is a variety of other secondary storage devices with longer access times, but with much larger capacities and lower initial costs. The most important of these are the magnetic drum, the disk file, the data cell, and magnetic tape.

Magnetic drums are rapidly rotating cylinders, coated with a ferromagnetic material similar to the oxide coating on magnetic tape. Information encoded in magnetic form, in a fashion similar to that used in the fast memory, is written on or read from cylindrical tracks by read/write heads much like those found in a magnetic tape recorder. Storage capacities vary from hundreds to millions of words: Access times for the highest-speed drums are of the order of 5 msec (milliseconds) (several orders of magnitude longer than the access time for information in the fast memory). Magnetic drums are not strictly random-access devices, since the desired storage location may not be positioned exactly at the reading or writing station at the time its content is desired; the drum

***Plate 3.** Magnetic drum unit. Each set of vertical bars (about 10 in. long) and wires shown is associated with one read/write station located on the periphery of the drum (the drum itself is contained inside the visible housing). This drum has five such read/write stations located every 72° of angular displacement around the drum. (Courtesy IBM Corporation.)*

must rotate until the storage location is positioned at the read/write head for the track. The access time can be reduced by stationing several heads at different angular locations around the drum for each track. Plate 3 shows an external view of a magnetic drum unit; the wiring in view is associated with the several read/write heads.

Disk files contain one or more plates (usually several inches in diameter, with a thin magnetic film on each surface) rigidly attached to a rotating spindle. As the plates rotate in the manner of phonograph records, reading and writing heads (analogous to the pickup cartridge in a phonograph) can either play back old information or record new information in circular tracks on the surface. The capacities of such files vary from a few thousand to tens of millions of words. Most large computing systems (see Chapter 4, for example) would have several drives in operation at the same time. [A disk unit with nine drives (eight active drives and one spare) is shown in Plate 4.] In many cases, the disk drive mechanism is designed to accept replaceable *disk packs.* [A disk pack with several surfaces (plates) is shown in Plate 5.] This allows the user to attach different files of disk storage to the computer at different times, similar to the changing of records on the turntable of a home phonograph. Access times again depend upon the position of the desired information on the surface of the disk with respect to the position of the reading and writing heads, and are of the order of 30 msec for the faster units.

In addition to serving as an input and output device (see below) the *magnetic-tape unit* can serve as a read/write station for the magnetic-tape storage medium. The total capacity is enormous, as reels of tape can be mounted and removed from particular drives as necessary. Depending on how the information is stored on the tape, the access time for a stored item may vary

Plate 4. Disk file containing nine drive units. Individual disk packs (see Plate 5) are mounted on each drive spindle. (Courtesy IBM Corporation.)

Plate 5. Replaceable disk pack. This pack contains 11 plates (approximately 12 in. in diameter) with a total of 20 surfaces for magnetic storage (the outer surfaces of the end plates are not used). (Courtesy IBM Corporation.)

from a few milliseconds (if the tape is already in proper position) to minutes (if the tape must be wound or rewound over hundreds or thousands of feet to position it properly). Plate 6 shows a typical tape drive. The mounting and take-up reels are at the top and the drive capstans and read/write heads (inside the small protective covers) are at the bottom of the photograph.

The *data cell* is a device that uses thin magnetic strips as the storage medium. The data cell can retrieve the appropriate strip, and position it for reading or writing with access times of the order of 100 msec. The data cell may be viewed as a magnetic tape unit in which the tape is cut into a large number of short strips. Capacities are normally measured in the hundreds of millions of words. Plate 7 shows a data cell unit with a capacity of 400 million characters (over 3.2 billion binary digits). The magnetic strips are positioned inside the cylindrical canister (about 15 in. high) shown in the center of the picture. When a particular strip is to be read, the canister is rotated to position it under a fixed read/write station.

In most computing systems, information on magnetic drums, magnetic tape, and data cell storage is saved in sequential rather than random order. This means that when the device is positioned properly to retrieve the first piece of stored information, many immediately adjacent data items can be retrieved in rapid sequence; for example, all of the information in one track of a magnetic drum could be read in one revolution of the drum. The same comment applies to the writing of information onto these devices.

Plate 6. Magnetic tape drive. This nine-track tape unit can read/write 1600 bytes of information (see Section 1.17) per linear inch of the 1/2 in. wide tape, at a transfer rate of 160,000 bytes per second. (Courtesy IBM Corporation.)

Plate 7. Data cell. Magnetic strips containing stored information are positioned inside one rotatable canister; total storage capacity is over 3.2 billion binary digits (see Section 1.17) of information. (Courtesy IBM Corporation.)

1.4 Input and Output (I/O) Equipment

The function of the input/output equipment is to allow communication between the machine user and the store. There is a large number of such devices in use. Some of the more common input devices are: (1) punched-card readers, (2) punched paper tape readers, (3) typewriters, (4) teletypewriters, and (5) magnetic-tape units. Less common are cathode ray tubes with light-pen input, which allow the computer user to draw pictures on the face of a cathode ray (television) tube or to point to images displayed on the tube. In fact, almost any signal-generating device (for example, an electrocardiograph machine) can be adapted as a suitable input device for a computer. In some of the larger computer systems it is now becoming fairly common to use a small computer as an input terminal device for a larger computer (see Chapter 4).

Without doubt, the most common input medium for today's computers is the punched card. The punched card (see Fig. 1.3) now favored and usually called an IBM card (although nearly all computer manufacturers use it) contains 80 columns with 12 row positions in each column. The cards are punched on a typewriterlike machine called a *keypunch*. When a particular key is depressed on the keypunch, a unique combination of holes is punched in one of the columns on the card; thus at most 80 different symbols may be punched on one card. The card reader is equipped with light sources on one side of the card to be read and sensing photocells on the other side. As the punched card passes through the read station, the pattern of light and dark corresponding to the "holes" and "no holes" are intercepted and appropriate photoelectric signals are generated.

Between these predominantly mechanical input devices and the computer's store, which operates completely electronically, there are conversion devices that we shall label the *input interface*. The function of this buffering equipment is to accept impulses sent by the card reader, tape reader, typewriter, or any other input device, and then to convert the impulses into appropriate internal form and store the accepted information in the memory. For example, a card might have holes punched in the first 12 columns for a ten-digit number with a decimal point and a plus or minus sign. The card reader senses the locations of the holes on the card and sends some signals to the buffering equipment. Subsequently, the signed ten-digit number would be stored in some memory word. The particular memory location to be used is determined by the program and will be described later.

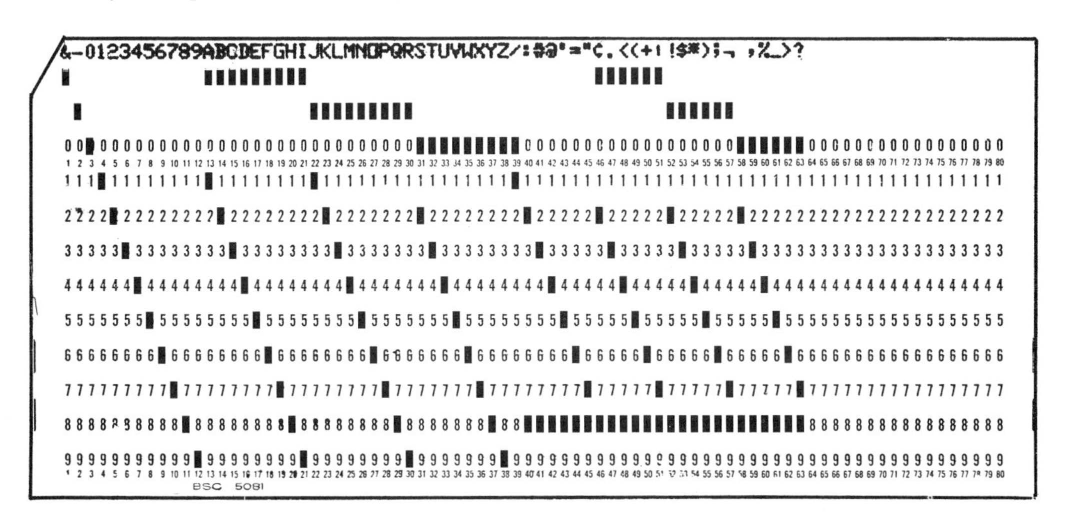

Figure 1.3 A punched card showing the hole codes in black.

The output devices are generally quite similar to the input devices; for example, a card punch, a magnetic-tape unit, a paper-tape punch, a typewriter, a teletypewriter, or a line printer. Similar input/output devices are often housed in a single unit; for example, a card or paper tape read/punch unit or a magnetic-tape unit that can both read and write (that is, play back and record). Between the memory and the output devices there is again some interfacing equipment.

At this point our hypothetical machine consists of input devices, a memory, and output devices (with suitable electronic equipment at the interfaces). The machine can accept information from its environment, store the information in digital form in the memory, and display information present in the memory (see Fig. 1.4).

***Figure 1.4** Input devices read information into the store. Output devices display information retrieved from the store.*

1.5 The Arithmetic Unit

The ability to save and retrieve information is useful in itself. However, to solve a problem, we would like to *read* information or *data* into some words of the memory, *operate* on this information in some meaningful way to produce *results*, which could be stored in other words of the memory, and finally to *display* the data and results on some output device.

In order to do operations on information in the memory (for the moment these operations may be assumed to be arithmetical in nature), calculating equipment analogous to the gears and cogs of a mechanical desk calculator is needed. In a digital computer, such operations are performed by strictly electronic devices. The *arithmetic unit* of a computer contains all the necessary circuitry to carry out the standard arithmetic operations on the contents of memory words (on the numbers stored in the memory) and can perform many other manipulations such as the shifting or digitwise examination of numbers, the comparison of numbers for sign, relative magnitude, and so forth (see Fig. 1.5).

Each digital computer has a fixed number of distinctly different *operations*, called *machine instructions*, which the arithmetic unit is capable of executing. (In addition, some other machine instructions are used for controlling the reading and writing operations of the input/output devices.) Most large computers have 200 or more such operations in their *instruction repertoire*. The instruction repertoire for each different model of computer is usually different from that of all other machines.

The arithmetic unit usually contains several *registers* (similar to the sets of dials on a desk calculator) that have immediate two-way access to any word in the memory. Most computers have at least two such registers; the most important of these are usually called the *accumulator* (AC), and the *multiplier/quotient unit* (MQ). Operations involving addition and subtraction are done in the accumulator, which corresponds to the accumulating register (one of the sets of dials) in a desk calculator. For example, the ten digits from location 000 could be put into the accumulator, the ten digits from location 001 could be added to the contents of the accumulator, and finally the resulting contents of the accumulator, namely the sum of the two ten-digit numbers, could be stored back into the memory in location 002. For multiplication and division operations, both the accumulator and the multiplier/quotient unit would normally be used. Some of the instructions involve other registers in the arithmetic unit. Many machines, for example, have a set of very useful counting and address modification registers called *index*

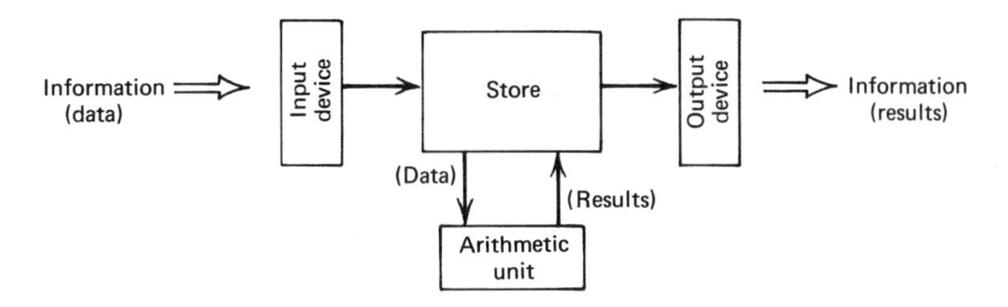

Figure 1.5 Flow of information in a digital computer.

registers. Some of the new machines have several general-purpose registers that can function either as arithmetic or index registers.

With the addition of the arithmetic unit, the digital computer now begins to assume a meaningful form. The machine can read data from its environment and enter them into the memory. The contents of various memory words can be manipulated in the arithmetic unit by means of the operations that the computer is designed to perform. The results of these operations can be stored in the memory along with the original data and can subsequently be retrieved for display on the output equipment. The sequence of events (see also Fig. 1.5) is:

1. Read data into the memory via the input equipment.
2. Operate (in the arithmetic unit) on the data stored in the memory.
3. Store (in the memory) the results of operations in the arithmetic unit.
4. Retrieve the results from the memory for display on the output equipment.

1.6 The Control Unit

Obviously, in order to produce useful results (to process data in a meaningful way) the computer must have associated with it a controlling device that supervises the sequence of activities taking place in *all* parts of the machine. This control equipment must decide: (1) when (and with which input device) to bring information into the memory, (2) where to place the information in the memory, (3) what sequence of operations or manipulations on information in the memory is to be done in the arithmetic unit, (4) where intermediate or final results of operations in the arithmetic unit are to be saved in the memory, and (5) when (and on which output device) data and results are to be displayed.

With the addition of the control unit (see Fig. 1.1) we now have a machine that can be directed to accept data, operate on the data to produce results, and display the results for the machine user, that is, a machine that is capable of solving suitably stated and defined problems, given a list of commands to be performed. Often, the combination of arithmetic and control units is called the *central processing unit* (CPU) or the *main frame*.

How does the machine user indicate what the machine is to do to solve his problem? First he must examine his problem and then outline a step-by-step *procedure*, sometimes called an *algorithm*, for its solution. Then he makes a list of commands from the machine's instruction repertoire, called a *program*, that he wants the machine to execute to implement the algorithm. The instructions must be ordered in the proper sequence; *only* those instructions the machine is designed to execute, namely those in the instruction repertoire, may appear in the program.

When one uses a desk calculator, the available instructions consist of addition, subtraction, multiplication, division, shifting, clearing registers, entering the contents of the keyboard into the registers, and so forth. Unless the calculator is designed to take square roots automatically, one cannot command the calculator to compute a square root. Instead, some numerical procedure (see Chapter 5) that uses only the available operations is required. Likewise, a digital computer can be instructed to carry out only those instructions that have been incorporated into its design.

1.7 The Notion of a Program—A Desk-Calculator Analogy

To illustrate the development of an algorithm and of a program to implement it on a machine, consider the following simple situation. You are a researcher who has collected hundreds, perhaps thousands, of sets of data in the laboratory. The data have been tabulated as shown in Fig. 1.6 under column headings X, Y, and Z; a line number has been given to each row to identify the various data sets.

	X column 1	Y column 2	Z column 3	U column 4
1	1.56	0.72	−1.5	
2	16.42	0.59	4.6	
3	1.72	9.3	0.45	
4	16.97	5.62	4.31	
5				
⋮				

Figure 1.6 Tabulated data sets.

Suppose that for each data set you wish to have the number in column 1 added to the number in column 2, the sum divided by the number in column 3, and the final result entered in column 4 under the heading U. You wish to have a co-worker compute these results for you. Verbal instructions such as: "For each row, take the number in the first column, add it to the number in the second column, divide the result by the number in the third column, and put the answer in the fourth column," would probably be adequate to convey your intentions.

However, let us assume that the computations are to be done on a mechanical desk calculator, and that we wish to be very explicit about the algorithm to be followed in solving the problem. Let X, Y, Z, and U be *symbolic names* standing for column 1, column 2, column 3, and column 4, respectively. The *name*, X, and the *place*, column 1, are interchangeable; thus X may be viewed as a symbolic address, and will be used as such later when we shall use names to refer to memory locations in a computer store. Let $\overline{X}$, $\overline{Y}$, $\overline{Z}$, and $\overline{U}$ indicate the numerical *values* to be found in the places X, Y, Z, and U for a particular row; that is, the numbers in columns 1, 2, 3, and 4 in the row. That is, $\overline{X}$ is the *content* of the place X. Then we could write the above instructions rather tersely for each row as follows:

$$\overline{U} \leftarrow \frac{(\overline{X}+\overline{Y})}{\overline{Z}}$$

Here, the left arrow is intended to indicate *substitution* or *assignment*. The instructions might be verbalized as follows: for the row of interest take the content of X, add the content of Y, divide by the content of Z, and make the result the content of U by writing the numerical result in the fourth column.

A typical mechanical desk calculator (see Fig. 1.7) consists of: (1) a keyboard that can serve to hold or store one number temporarily, (2) an accumulator (AC) register, involved in addition and subtraction operations, (3) a multiplier/quotient (MQ) register, involved in multiplication and division operations, (4) a set of operator buttons to initiate operations, and (5) appropriate equipment (mechanical linkages) to implement the specified operations. Typical operations available on desk calculators are: clear(CL), add(+), subtract (−), enter multiplier(EM), multiply(×), enter dividend(ED), divide(÷), shift left(SL), and shift right(SR). If we use the symbolic

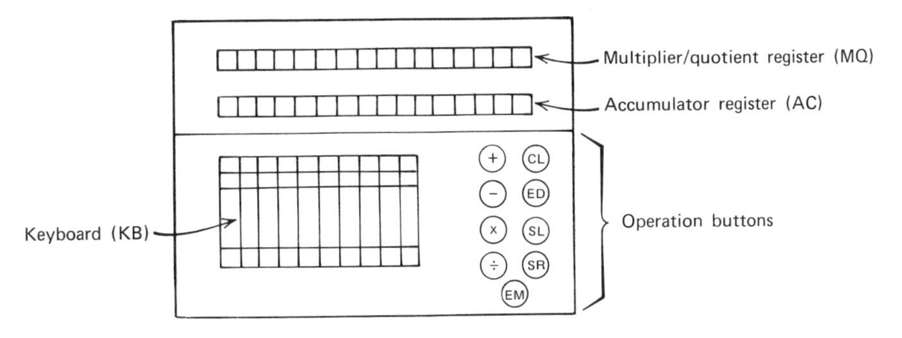

Figure 1.7 A typical mechanical desk calculator.

names KB, AC, and MQ to denote the keyboard, accumulator register, and multiplier/quotient register, respectively, then the first seven of the encircled symbols above will cause the machine to take the actions shown in Table 1.1.

Table 1.1 Desk-Calculator Operations

Operation	Action
(CL)	$\overline{AC} \leftarrow 0$, $\overline{MQ} \leftarrow 0$, $\overline{KB} \leftarrow 0$
(+)	$\overline{AC} \leftarrow \overline{AC} + \overline{KB}$
(−)	$\overline{AC} \leftarrow \overline{AC} - \overline{KB}$
(×)	$\overline{AC} \leftarrow \overline{MQ} \times \overline{KB}$
(ED)[a]	$\overline{AC} \leftarrow \overline{KB}$
(÷)	$\overline{MQ} \leftarrow \overline{AC}/\overline{KB}$
(EM)	$\overline{MQ} \leftarrow \overline{KB}$

[a]The dividend is entered into the left half of the accumulator of most desk calculators.

Now, let I be a symbolic address whose content $\overline{I}$ is the identifying row number for the data set with values $\overline{X}$, $\overline{Y}$, and $\overline{Z}$. Then a more detailed description of the algorithm to solve the stated problem is shown in Table 1.2. The presence of an operator symbol indicates that the desk calculator operator should push the corresponding operation button.

The algorithm outlined in Table 1.2 may be viewed as a *program* for the desk calculator operator, who serves as the *control unit* for the calculating machine. In addition, the operator serves as an *input device* by entering data into the one-word store (the keyboard) of the calculator, and as the *output device* by writing the result on the data sheet. Note that the operator need know nothing about the nature of the operations that he initiates by pushing the appropriate buttons; the only requirement is that he carry out each step of the program in the proper sequence. Note, in addition, that *no data values* appear anywhere in the *program.* Operands for operations are always referenced by their *locations*, X, Y, Z, U, KB, AC, MQ. Because of this, this simple program could be used to process *any* number of data sets in sequence.

1.8 Machine-Language Instructions

When using a desk calculator, the sequence of instructions to be executed is determined by the machine user. The user functions as the control unit in deciding which number or operation is to be used next. With the very high internal operating speeds of a digital computer (on the larger machines, one million or more individual instructions may be executed *per second*), it is impractical to have the machine user stand before the console pushing buttons in sequence as he does at the keyboard of a desk calculator. Consequently, some other approach is necessary to allow very rapid processing of machine instructions. Because direct communication between the machine and

Table 1.2 Desk Calculator Program

Step	Instruction	Comment
1	$I \leftarrow 1$	initialize the row counter
2	go to row $\bar{I}$ on the data sheet[a]	pick data set
3	(CL)	calculate $\bar{X}+\bar{Y}$
4	$\overline{KB} \leftarrow \bar{X}$	
5	(+)	
6	$\overline{KB} \leftarrow \bar{Y}$	
7	(+)	
8	$\overline{KB} \leftarrow \overline{AC}$[b]	calculate $(\bar{X}+\bar{Y})/\bar{Z}$
9	(ED)	
10	$\overline{KB} \leftarrow \bar{Z}$	
11	(÷)	
12	$\bar{U} \leftarrow \overline{MQ}$	write answer $\bar{U}$ on data sheet
13	(CL)	calculate $\bar{I} \leftarrow \bar{I}+1$ (row identifier for the next data set)
14	$\overline{KB} \leftarrow \bar{I}$	
15	(+)	
16	$KB \leftarrow 1$	
17	(+)	
18	$\bar{I} \leftarrow \overline{AC}$	
19	go to step 2 for next instruction	process next data set

[a]To allow for orderly termination of the program, step 2 could be amplified as follows: If row $\bar{I}$ exists on the data sheet, go to row $\bar{I}$ on the data sheet, and continue with step 3. If row $\bar{I}$ does not exist, stop, as all data have been processed.

[b]Although the value $\bar{X}+\bar{Y}$ already appears in the accumulator, the dividend must ordinarily be entered into the left half of the accumulator using the (ED) operation.

its environment involves the use of slow mechanical equipment, any approach that requires such contact continuously, such as pushing buttons, or even reading orders from cards, is impossibly slow. One solution to this problem was first suggested by Burks, Goldstine, and von Neumann [40]. Since a program is one kind of information, and a computer's memory can be viewed as a place to store *any* kind of information, the *program* can be *stored* in the memory along with the data and results. Virtually all of the general-purpose computers now available are organized in this way, and are known as *stored-program computers* or *von Neumann machines*. Since only digits may be stored in the machine's memory, the instructions must be encoded in digital form before being written into the store. The code used is called the *machine's language* and the coded program is known as a *machine-language program.*

How might we encode the sequence of instructions in a machine's repertoire? In our hypothetical machine each word in the memory contains ten digits with an associated sign. Therefore, for convenience, let us design our machine so that one coded machine instruction can be placed in one memory word, that is, let one instruction consist of ten digits and a sign. Divide the instruction word into four segments as shown in the Fig. 1.8. Let the sign and the first digit ($\pm\,\theta$) represent the *operation* that the machine is to carry out. Since any of the ten possible digits may appear in the digit position, the machine's instruction

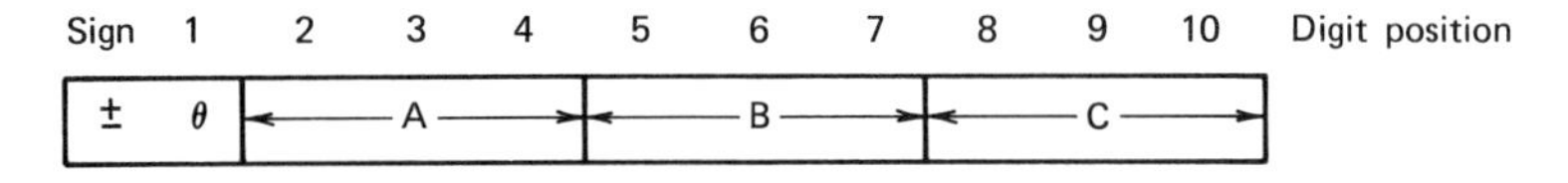

Figure 1.8 Format of a machine-language instruction for our hypothetical computer.

repertoire will be restricted to a total of 20 possible operations (−9, −8, . . . , −1, −0, +0, +1, . . . , +9). Divide the other nine digits into groups of three, such that the digits A in positions 2, 3, and 4 comprise the three-digit *address* of an operand, B in positions 5, 6, 7 the three-digit *address* of a second operand, and C in positions 8, 9, 10 the three-digit *address* of a third operand. Let the meaning of each instruction be as shown in Table 1.3 where $\bar{A}$ means the signed ten-digit *content* of address $\bar{A}$, $\bar{B}$ means the signed ten-digit content of address $\bar{B}$, and $\bar{C}$ means the signed ten-digit content of address C.

Assume that instructions are normally placed in sequence in the machine, that is, if the first instruction is stored in location 000, the second is in location 001, the third in location 002, etc. The machine will be designed to advance to the *next* location automatically for its next instruction. The scheme of placing instructions in order in sequential memory addresses is used in virtually all real machines. Frequently, however, it is essential to transfer out of the normal sequence either unconditionally or conditionally. The TRA instruction causes the control unit to go to the A rather than the next address for the next instruction. The TGT instruction allows for a two-way branch based on a conditional test involving two operand values.

A typical instruction for our machine might be

$$+4500501502$$

which, when interpreted as an instruction, would mean divide (+4) the ten-digit content $\bar{A}$ of location A (500) by the ten-digit content $\bar{B}$ of location B (501) and store the results as the ten-digit content $\bar{C}$ of location C (502). Such a coded instruction is called a *machine-language instruction.*

Table 1.3 Instruction Repertoire

Operation Code ±θ	Operation	Meaning
+0	READ	Read $\bar{A}$, $\bar{B}$, $\bar{C}$, from one data card.[a]
+1	ADD	$\bar{C} \leftarrow \bar{A} + \bar{B}$
+2	SUB	$\bar{C} \leftarrow \bar{A} - \bar{B}$
+3	MPY	$\bar{C} \leftarrow \bar{A} \times \bar{B}$
+4	DIV	$\bar{C} \leftarrow \bar{A}/\bar{B}$
+5	PRINT	Print $\bar{A}$, $\bar{B}$, $\bar{C}$ on one line of a printer[b]
+6	TRA	Transfer to address A for the next instruction.
+7	TGT	If $\bar{A} > \bar{B}$ then transfer to address C for the next instruction; otherwise go to the next address for the next instruction.
⋮	⋮	

[a]Let $\bar{A}$ be the signed ten-digit number in columns 1 to 12, $\bar{B}$ be the signed ten-digit number in columns 13 to 24, and $\bar{C}$ be the signed ten-digit number in columns 25 to 36 of a punched card.

[b]$\bar{A}$, $\bar{B}$, $\bar{C}$ will be printed in columns 1 to 15, 16 to 30, and 31 to 45, respectively, of the next line on the printer. This arrangement allows for each number to contain a sign, a decimal point, ten digits, and three blanks to separate the printed values for greater readability.

Note that we have not specified that certain parts of the memory could contain only instructions while others could contain only data or results. A coded instruction has the appearance of just another signed ten-digit number and cannot be distinguished *a priori* from some data item that might be stored in the memory. Such a storage scheme is termed *ambiguous* and gives the computer great power because the machine is able to treat instructions as data when appropriate; a program may be written to modify its own instructions as well as data while it is being executed by the computer.

1.9 A Machine-Language Program

Let us now consider the solution of the problem outlined in Section 1.7 on our hypothetical computer rather than on a desk calculator. Assume that the data sets of Fig. 1.6 have been punched on cards, such that each data card contains just one set, with $\overline{X}$, $\overline{Y}$, and $\overline{Z}$ punched in columns 1 to 12, 13 to 24, and 25 to 36, respectively. Assume that the first instruction in the program will be placed in location 000, the second in 001, and so forth, and that the values associated with the symbols X, Y, Z, and U, later to be called *variables*, are to be stored in the following memory locations:

Symbol	Address
X	501
Y	502
Z	503
U	504

Note that the letters X, Y, Z, U are simply *symbolic names* for the numerical *addresses* 501, 502, etc.; thus, X, Y, Z, and U are *symbolic addresses* for memory locations. A *flow diagram* (see Chapter 2) or graphical description of the algorithm for the calculation is shown in Fig. 1.9.

In the flow diagram, we have emphasized the distinction between a symbolic address, say X, and the content of the address, say $\overline{X}$, by using the overbar notation. However, it is common practice to draw flow diagrams for computer algorithms using the symbolic *address* when the *content* of the address is intended. For example, $\overline{U} \leftarrow (\overline{X}+\overline{Y})/\overline{Z}$, would normally be written simply as $U \leftarrow (X+Y)/Z$. In subsequent chapters, we shall follow convention and delete the overbar, blurring the distinction between the address and content of a memory location. However, it is crucial to a proper understanding of computer operation that this unfortunate ambiguity not lead to confusion on the part of the programmer.

A machine-language program for our hypothetical computer for this algorithm is shown in Table 1.4.

A new symbol T was introduced in the instruction in location 001, to save the results of an intermediate calculation, $\overline{X}+\overline{Y}$. $\overline{T}$ is assigned to location 505. The instruction in address 004 will cause the value $\overline{U}$ and the sequence +0501502503 +1501502505 to be printed. Note that this is an example of the use of an instruction as an *operand* in another instruction. (The printing of the contents of addresses 000 and 001 is not part of the algorithm of Fig. 1.9, and is included only to illustrate this important feature of machine language instructions. Since the contents of *three* memory locations will always be printed by the instruction in location 005, it would be preferable to change this instruction to say +5504006006, and then to enter a true zero, +0000000000, as the content of address 006.)

Having processed one data card and produced two printed lines for one set of values $\overline{X}$, $\overline{Y}$, $\overline{Z}$, the program returns to location 000, and reads another card with new values $\overline{X}$, $\overline{Y}$, and $\overline{Z}$, storing the values in the *same* locations used for the previous data set. As long as there are data cards in the card reader, the machine will continue to process them, one at a time.

We have not yet specified how the machine language program itself is entered into the memory or how the machine is directed to start at location 000 for the first instruction. This will be discussed later in Chapter 4.

Table 1.4 Machine-Language Program

Instruction Address	Operation $\pm\theta$	Addresses of Operands A	B	C	Action
000	+0	501	502	503	Read $\overline{X}, \overline{Y}, \overline{Z}$
001	+1	501	502	505	$\overline{T} \leftarrow \overline{X}+\overline{Y}$
002	+4	505	503	504	$\overline{U} \leftarrow \overline{T}/\overline{Z}$
003	+5	501	502	503	Print $\overline{X}, \overline{Y}, \overline{Z}$
004	+5	504	000	001	Print $\overline{U}, \overline{000}, \overline{001}$
005	+6	000	000	000	Go back to instruction address 000 for the next instruction

Our hypothetical computer would be called a *three-address machine* because three operand addresses appear in each instruction. Although such machines have been built, most machines now in use have just *one* operand address per instruction, and are termed *single-address* machines. In these computers, for binary operations such as addition, one operand is assumed to be present already in one of the registers of the arithmetic unit. For example, our single addition instruction $\overline{C} \leftarrow \overline{A}+\overline{B}$ might be equivalent to three one-address instructions such as the following:

$\overline{AC} \leftarrow \overline{A}$ Put the value of operand A into the accumulator (AC) register,

$\overline{AC} \leftarrow \overline{AC}+\overline{B}$ Add the value of operand B to the content of the accumulator register,

$\overline{C} \leftarrow \overline{AC}$ Store the result (left in the accumulator) in the memory location assigned to hold the value of operand C.

1.10 Symbolic Computer Languages—Assembly-Language Programming

As is obvious from the preceding example, the preparation of machine-language programs is very tedious. A complicated problem might require hundreds or even thousands of such machine instructions, making the writing of an error-free program virtually impossible. The changing of a single digit in a machine-language program will normally result in an error; either the machine will perform an improper command because the operation code in a command will not be the desired one, or an unintended operand will be used because an address will be incorrect. In addition, in order to write programs in the machine's language, one must be familiar with all or nearly all of the instructions in the instruction repertoire. In binary computers, that is, where only the digits 1 and 0 can be stored, the machine-language programming problem is even more complex since a machine instruction might consist of a sequence of 30 to 60 ones and zeros in various patterns.

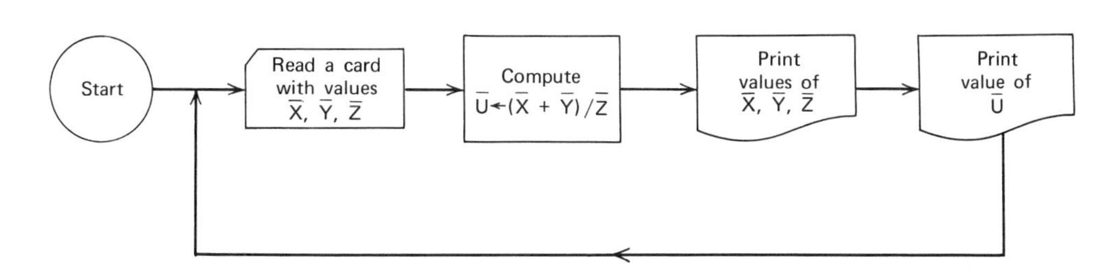

Figure 1.9 Flow diagram. Algorithm to compute $\overline{U} \leftarrow (\overline{X}+\overline{Y})/\overline{Z}$.

Obviously, better ways of communicating with the computer are required to achieve any kind of programming efficiency. The first approach to this problem was the development of *symbolic languages* for describing algorithms. Symbols rather than sequences of digits are used to represent the operation codes and the memory locations involved. For example, we might write the program for the algorithm of Fig. 1.9 as shown in Table 1.5.

The operation SYMB is not actually one of the machine operations, and so is termed a *pseudooperation*; the only function of SYMB is to relate the symbolic names and their corresponding numerical addresses in the memory. Having written an algorithm or procedure in this symbolic form, it is a simple mechanical task to encode each of the symbols with appropriate digits to produce the machine-language program, that is, to translate the symbols of Table 1.5 to the digit strings of Table 1.4. Because the computer itself is especially well adapted for this sort of mechanical detail, machine-language *programs* have been written for most machines that automatically *translate* from the symbolic to digital code. Such programs are termed *assembly programs*.

The processing sequence is shown schematically in Fig. 1.10. First the *machine-language* version of the *assembly program* is brought into the memory, usually from the system *library* of programs stored on one of the secondary storage devices. Since the assembly program is already in the machine's language, the machine can execute its instructions without translation once it has been loaded. The program to be translated, such as the program of Table 1.5, is punched in the symbolic language on cards and is read as *data* by the assembly program. The letters in the symbolic-language program, automatically converted to digital codes by the input buffering equipment, are analyzed by the assembly program, that is, they are manipulated by the arithmetic unit according to the instructions in the assembly program, to produce the machine-language equivalent (for example, the program of Table 1.4).

The *machine-language* version of the *original symbolic program* produced by the translating program is then read into the machine's memory for subsequent execution. The symbolic version of the

Table 1.5 Symbolic Program Equivalent to the Machine Language Program of Table 1.4*

Symbolic Instruction Address	Operation	Symbolic Address of 1st Operand	Symbolic Address of 2nd Operand	Symbolic Address of 3rd Operand
START	READ	X	Y	Z
	ADD	X	Y	T
	DIV	T	Z	U
	PRINT	X	Y	Z
	PRINT	U	–	–
	TRA	START		
START	SYMB	000		
X	SYMB	501		
Y	SYMB	502		
Z	SYMB	503		
U	SYMB	504		
T	SYMB	505		

*The contents of addresses 000 and 001 will not be printed by the second PRINT instruction, as in the program of Table 1.4.

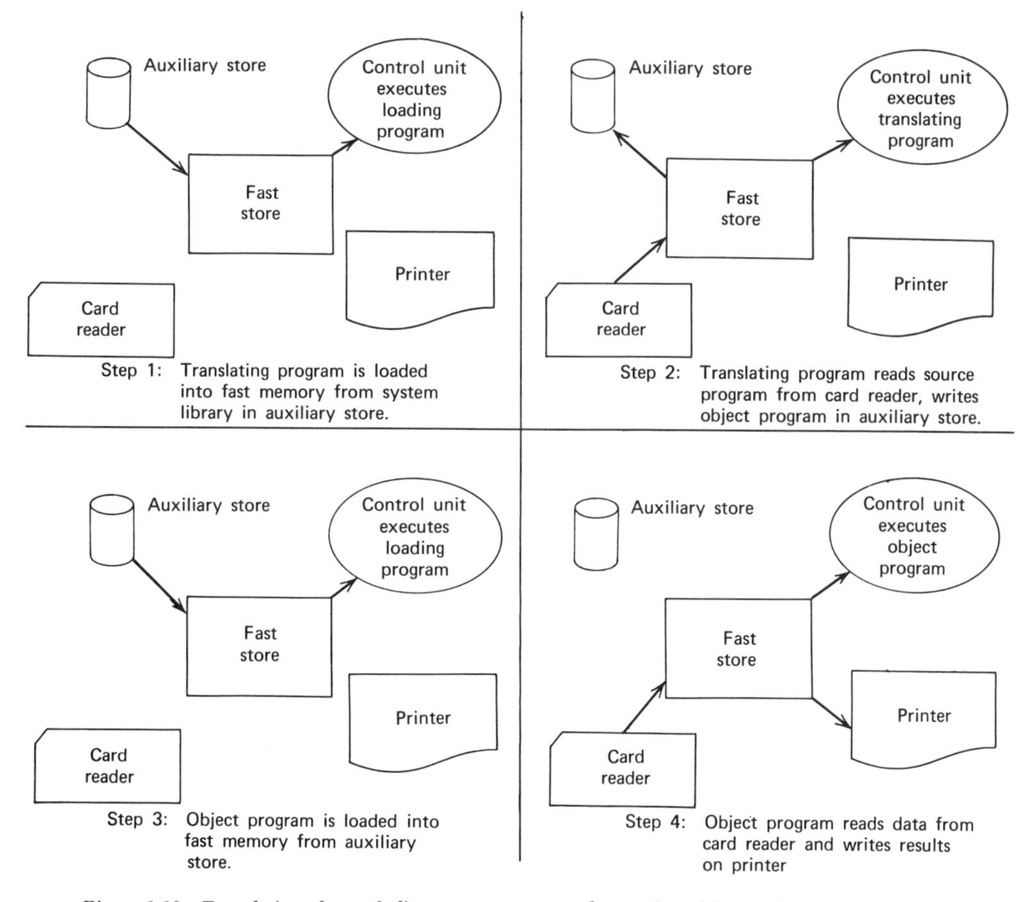

Figure 1.10 Translation of a symbolic source program and execution of the resulting object program.

program is usually called the *source program*; the machine-language equivalent of the source program is called the *object program.* Since the machine can do nothing without having a program in its memory or without receiving instructions one at a time from the operator's console, another program, usually called the *loader*, must be available to read both the translating program and the object program into the memory at the proper time. In turn, the loading program must *itself* be read into the memory. Details of the sequence are covered in Chapter 4.

1.11 Procedure-Oriented Languages

The development of assembly languages resulted in a very great saving of time and effort by computer programmers; however, such languages require detailed knowledge of the instruction repertoire, different for each model of computer; assembly languages by their nature are very *machine oriented.*

Casual users would prefer to have the ability to communicate with the machine in a more familiar symbolic form using algebraic notation, English words, etc., and in general to allow rather direct transcription from flow diagram to computer program. For example, writing the program of the preceding section in a form such as

```
START  READ X, Y, Z
       U = (X + Y)/Z
       PRINT X, Y, Z, U
       GO TO START
```

would considerably simplify the programming problem. Detailed knowledge of instructions in the repertoire of a particular machine would be unnecessary, and, since it is not oriented toward a particular machine, the language might hopefully be used to describe programs for more than one computer, making it possible to interchange programs with other computer users.

Computer languages similar to that illustrated above, and which allow rather straightforward description of an algorithm are called *algebraic*, *algorithmic*, or *procedure-oriented* languages. These languages, insofar as is possible, are *machine-independent*, that is, details of hardware organization or machine language are not apparent to the user of the language. Programs written in such languages cannot be executed directly by the computer; they must first be translated into equivalent machine-language programs. The programs that implement the translation are called *compilers* and operate very much like the assembly programs described in the previous section and illustrated in Fig. 1.10. A *source* program written in the procedure-oriented language is first translated into an *object* or machine-language program. The resulting object program is then stored in the machine's memory and executed by the computer. Thus the processing of a program written in a procedure-oriented language takes place in two steps: *translation* or *compilation*, and *execution*. In addition to translating the source program, compilers typically assign all addresses for both operands and instructions, eliminating the need for pseudooperations such as those shown in Table 1.5.

The most popular of the procedure-oriented languages is called FORTRAN (FORmula TRANslation) and will be described in detail in Chapter 3. FORTRAN is similar to other procedure-oriented languages such as BASIC, ALGOL, and PL/I. Since each different model of computer has a different machine language, there must be a different translating program or compiler for *each* procedure-oriented language for *each* different kind of computer. Fortunately, FORTRAN translators are available for virtually every machine on the commercial market, so that FORTRAN programs can be processed on almost any computer now in use. Translating programs for the assembly and procedure-oriented languages (assemblers and compilers) are usually supplied by the computer manufacturer as part of the *software* or programming support for the hardware.

1.12 Charles Babbage

The electronic digital computer is indeed one of the wonders of twentieth century science and engineering. Yet, the functional organization of digital machines described in preceding sections, was, with the single exception of the stored-program concept, completely formulated in the early nineteenth century by a remarkable Englishman named Charles Babbage. An heir to a banking fortune, Babbage was educated at Cambridge, and was for a time Lucasian Professor at Cambridge, the chair once held by Isaac Newton. He was fascinated by mathematics and engineering, and spent nearly all his adult life designing, promoting, and attempting to build mechanical calculating machines.

In the early 1820s, Babbage designed a machine that he called the *difference engine*, which would be able to calculate successive differences of tabular information and, in theory, produce very extensive and very accurate tables of functional values. Upon the recommendation of prominent members of the Royal Society and the Astronomical Society, Babbage convinced the Chancellor of the Exchequer to support the development of his engine. This was one of the first instances of major government support of a research and development project. Unfortunately, despite some major successes in designing new mechanisms and in improving the quality and accuracy of machining operations, Babbage and his chief engineer had a falling out (Babbage appears to have been an irascible genius), and after years of intermittent effort the project was abandoned in 1833. During the period 1823 to 1833, the British government advanced over £17,000 to Babbage, and he is thought to have spent a comparable amount from his personal fortune on the project; even today, this would represent an investment of over $80,000, and it must have been an astronomical sum at the time [41].

About a year after work on the difference engine had stopped, Babbage conceived a wholly new idea for a calculating machine, vastly more powerful than the difference engine, and functionally the mechanical equivalent of our modern electronic general-purpose digital computer. He called his machine the *analytical engine*. Briefly, his machine consisted of five functional units: a memory or *store*, an arithmetic unit or *mill*, an *input section*, an *output section*, and a part of the machine that functioned as a *control unit*. The machine was designed on a grand scale. For example, the store was to be able to hold 1000 50-digit numbers. The engine was to be designed so that any of the 1000 numbers in the store could be transmitted mechanically to the mill for processing. The mill would be capable of carrying out all of the common arithmetic operations and certain logical operations, such as comparing the value of one number with another for relative magnitude and algebraic sign. Results of computations in the mill could be transmitted mechanically to any location in the store. Babbage estimated that with a maximum velocity of 40 ft/min for any moving parts in his engine, it would be possible to perform 60 additions or subtractions per minute, and one multiplication or division per minute, all on numbers having 50 digits [43]!

Data were to be entered into the store from *punched cards*. Babbage got the idea for using punched cards as an input medium from the cards that were, and still are, used to control the weaving of patterns on the Jacquard loom. He envisioned that any numbers in the store could be retrieved mechanically and either: (1) printed on paper (two copies!), (2) cast directly into a stereotype mold suitable for printing tables, or (3) punched onto blank cards.

The machine was to be controlled by two sets of punched cards: the *operation* cards and the *variable* cards. The two sets of cards served as the program for the computer; the operation cards contained the operation codes, and the variable cards contained the addresses of the operands; that is, all the essential elements of a machine-language program were present in his design. He even envisaged the availability of prepunched decks to carry out common functional evaluations, anticipating the idea of a program library. The only essential organizational difference between Babbage's analytical engine and the digital computer described earlier is the stored-program concept.

Babbage spent much of the rest of his life developing his ideas about the analytical engine and searching for financial support to build his computer. He never published a complete description of his machine, since he considered an article written in 1842 by an Italian engineer named Menabrea to be the definitive reference. The article was translated into English and augmented with a set of notes [42] by a close friend of Babbage, Ada Augusta, Countess of Lovelace. Lady Lovelace, who was Lord Byron's daughter, was a mathematician of considerable ability, and wrote "... the analytical engine weaves algebraic patterns, just as the Jacquard loom weaves flowers and leaves ...."

In 1848, Babbage prepared a complete set of drawings for the construction of the analytical engine (he is credited with inventing much of the notation still used in mechanical drawings, particularly the notation used to describe the motion of machinery). In 1852 he made another attempt to secure government money to build his machine. In response, the Chancellor of the Exchequer wrote [43]:

"Mr. Babbage's projects appear to be so indefinitely expensive, the ultimate success so problematical and the expenditure certainly so large and so utterly incapable of being calculated, that the government would not be justified in taking upon itself any further liability."

In 1862, Babbage wrote [43]:

"It [the analytical engine] can not only

calculate the millions the ex-Chancellor of the Exchequer squandered, but it can deal with the smallest of quantities; nay it even feels for zeros [can find the roots of equations] . . . It may possibly enable him to unmuddle even his own accounts, and to—but as I have no wish to crucify him, I will leave his name in obscurity. The Herostratus of Science [who], if he escapes oblivion, will be linked to the destroyer of the Ephesian Temple."

He died in 1871, an embittered old man. Parts of the mill and printing mechanism were constructed from Babbage's drawings after his death, and are on view in the Science Museum, London (see Plate 8).

1.13 Number Systems*

Counting schemes probably antedate written history. It seems likely that the earliest efforts at quantitative measurement involved making a single mark for each item to be counted, for example, |, ||, |||, ||||, |||||, ||||||, . . . , etc. Later, more compact notation was developed by using different symbols to represent groups of marks. For example, the Romans used equivalents such as:

$$\begin{aligned} \mathrm{I} &\equiv | \\ \mathrm{V} &\equiv ||||| \\ \mathrm{X} &\equiv |||||||||| \equiv \mathrm{VV} \end{aligned}$$

It is a feature of the simplest counting schemes that the symbols retain their *values*, regardless of their *positions* in the symbol string. Thus, in Roman notation, the symbol X retains its value of ten in each of the following: X, XX, XXX, LX, LXX, LXXX, CX, CXX, ... Even in the Roman numbers XL ≡ XXXX, XC ≡ LXXXX, the X still retains its basic value; in these instances, however, the position of the X to the left of a symbol with a larger value indicates that ten is to be *subtracted* from the value of the symbol to its right. Curiously, the Romans had no symbol to indicate the absence of marks, that is, no symbol with the value zero.

With the introduction of the concept of zero, it was possible to create notations based on symbol position, such that each symbol in the symbol string becomes a multiplier for some *base* number.

In some cases, the base number changes from position to position, leading to cumbersome counting schemes. Most of the common English measures suffer from this characteristic. Lengths measured in yards, feet, and inches cannot be written compactly by juxtaposing the symbols. For example, 3 yards, 2 feet and 5 inches cannot be written unambiguously as length = 325 inches, because the base number for the feet position is 12 inches, while the base number for the yard position is 3 feet or 36 inches. Note that a similar measure in metric units, such as 3 meters, 3 decimeters, 2 centimeters, and 5 millimeters *can* be written unambiguously by juxtaposing the symbols, as in length = 3325 millimeters. This follows because the base for the centimeter position is 10 millimeters, the base for the decimeter position is 10 centimeters or 100 millimeters, and the base for the meter position is 10 decimeters or 1000 millimeters.

As the metric example of the preceding paragraph illustrates, a positional notation is much simpler to use if a single symbol in each position serves as a multiplier for a different power of the *same* number, known as the *base* or *radix* of the number system. For example, the base 10 or *decimal* system uses a set of *ten* basic symbols, the Arabic numerals, 1, 2, 3, 4, 5, 6, 7, 8, 9, 0, which, when standing alone, represent |, ||, |||, ||||, |||||, ||||||, |||||||, ||||||||, |||||||||, and *no* marks respectively. A *decimal number* consists of any string of the basic symbols or digits, a decimal point to separate the integral and fractional parts, and possibly a plus or minus sign preceding the first digit. Thus −245.02 is a valid decimal number. It may be

*An understanding of the material in the remaining sections of this chapter, although useful, is not essential to effective use of computers when programming in procedure-oriented languages, such as FORTRAN.

Plate 8. A part of the mill of Babbage's analytical engine, constructed after his death from original drawings. (British Crown Copyright. Science Museum, London.)

interpreted as

$$-(2\times 10^2+4\times 10^1+5\times 10^0 + 0\times 10^{-1}+2\times 10^{-2})$$

Note that each digit is a *multiplier* for a power of the base 10, the power being determined by the *position* of the digit with respect to the radix point (decimal point). Thus the same symbol, 2 in the example, represents different values by virtue of its different positions in the number. Note the difference between the positional approach and the Roman scheme; XX is twenty, not 100 plus 10 for example. In addition, the importance of the null character 0 should not be overlooked. Without the zero or some suitable substitute, it would not be possible to indicate that certain powers of the base do not contribute to the value of the number.

The essential features of a good positional notation are the following:

1. A single base or radix, b.
2. A set of unique symbols representing all integers from 1 to $b-1$.
3. A null symbol or zero.
4. A radix point to separate integral and fractional portions of the number.
5. A plus and minus sign.

A number in a positional notation with base b may then be written as

$$\pm d_n \cdots d_3 d_2 d_1 d_0 \,.\, d_{-1} d_{-2} d_{-3} \cdots d_{-m},$$

where the d_i are any of the b symbols representing the integers from 0 through $b-1$. The *value* associated with the number is

$$\pm d_n b^n + d_{n-1} b^{n-1} + \cdots + d_1 b^1 + d_0 + d_{-1} b^{-1} + \cdots + d_{-m} b^{-m}.$$

Viewed in this manner, it is clear that there is nothing unique about counting by tens, that is, about the decimal system. No doubt we count by tens because we have ten fingers. Had we only eight fingers (a possibility, presumably), then we probably would count by *eights* and use the *octal* system. In this case we have the following:

1. The base b, eight.
2. Eight integer symbols from 0 to 7.
3. An octal point.
4. A plus and minus sign.

Then, using a positional notation as before, the octal number $(215.23)_8$ would be equivalent in decimal notation to

$$2\times 8^2+1\times 8^1+5+2\times 8^{-1}+3\times 8^{-2}$$

or

$$(141\tfrac{19}{64})_{10}.$$

Clearly, if we choose to count by *pairs*, then we can create a consistent notation known as the *binary* system. We have

1. The base b, two.
2. The two integers 0, 1.
3. A binary point.
4. A plus and minus sign.

A binary number would then be any string of ones and zeros containing a point and possibly preceded by a sign, for example, $(1011.011)_2$. This binary number would be equivalent to the decimal number

$$1\times 2^3+0\times 2^2+1\times 2^1+1+0\times 2^{-1} +1\times 2^{-2}+1\times 2^{-3}$$

or

$$(11.375)_{10}.$$

What about counting by *dozens* or twelves? In this case we need to define unique symbols for the decimal numbers 10 and 11 since we must have $b=12$ unique symbols. Let $A \equiv (10)_{10}$ and $B \equiv (11)_{10}$; then the *duodecimal* or *base 12* system can be described in terms of:

1. The base b, twelve.
2. Twelve integers 0, 1, 2, 3, 4, 5, 6, 7, 8, 9, A, B.
3. A duodecimal point.
4. A plus and minus sign.

The base twelve number $(2B4.0A)_{12}$ would be equivalent to the decimal number

$$2\times 12^2+B\times 12^1+4+0\times 12^{-1} +A\times 12^{-2},$$

or

$$(2\times 144+11\times 12+4+10\times 12^{-2})_{10} = 424\tfrac{5}{72}.$$

The integral part, 2B4, could be viewed as 2 *gross* plus 11 *dozen* plus 4. This system has the attractive feature that the base is integrably divisible by several small integers, namely 1, 2, 3, 4, and 6; many common items (such as eggs, oranges, pencils, and bottles of soft drinks and beer) are still sold using essentially this counting scheme.

Another important system is the *hexadecimal* or *base 16 system.* If we let A, B, C, D, E, and F represent 10, 11, 12, 13, 14, and 15 in decimal notation, respectively, then the reader should verify that

$$(1B)_{16}=(23)_{12}=(33)_8=(11011)_2 =(27)_{10}.$$

The equivalents for the small integers in the binary, octal, decimal, duodecimal, and hexadecimal systems are shown in Table 1.6.

1.14 Conversion from One Number System to Another

Because computers normally use bistable devices for internal storage, the binary system and other systems whose base b is a power of 2 (for example, the octal and hexadecimal systems) play an important role in computer work near the hardware level. Note that conversions from the binary to octal or binary to hexadecimal systems and *vice versa* are particularly simple to implement. For example,

$$(1110111101.1011)_2 \leftrightarrow (1675.54)_8,$$

$$(1101111101.1011)_2 \leftrightarrow (37D.B)_{16}.$$

Table 1.6 Equivalents in Several Number Systems

Binary	Octal	Decimal	Duodecimal	Hexadecimal
0	0	0	0	0
1	1	1	1	1
10	2	2	2	2
11	3	3	3	3
100	4	4	4	4
101	5	5	5	5
110	6	6	6	6
111	7	7	7	7
1000	10	8	8	8
1001	11	9	9	9
1010	12	10	A	A
1011	13	11	B	B
1100	14	12	10	C
1101	15	13	11	D
1110	16	14	12	E
1111	17	15	13	F
10000	20	16	14	10

For conversions from binary to octal notation, collect the binary digits in groups of three (because 8 is 2^3) from the binary point, and use the binary-octal equivalents from Table 1.6. For conversions from binary to hexadecimal notation, collect the binary digits in groups of four (because 16 is 2^4) from the binary point, and use the appropriate binary-hexadecimal equivalents from Table 1.6, noting that $(1001)_2 = (9)_{16}$, $(1010)_2 = (A)_{16}$, and so on.

Conversions of numbers from base b to base 10 is rather straightforward, and has already been outlined in preceding examples. Conversion of decimal numbers to numbers in other bases is rather more tedious, though not particularly difficult. The integral and fractional parts of the number must be treated separately. Consider first the conversion of an integer in base 10 notation to an integer in base 8 notation. The key of course is to find out how many times various powers of 8 are contained in the starting decimal integer. Consider conversion of the decimal integers 6, 8, 21, 16, and 1973.

If we divide each number by 8, then the *integral* part of the quotient is the number of times 8 is contained in the number; the *remainder* will be a number smaller than 8, and hence must be the appropriate multiplier in the 8^0 position of the octal number equivalent. Thus:

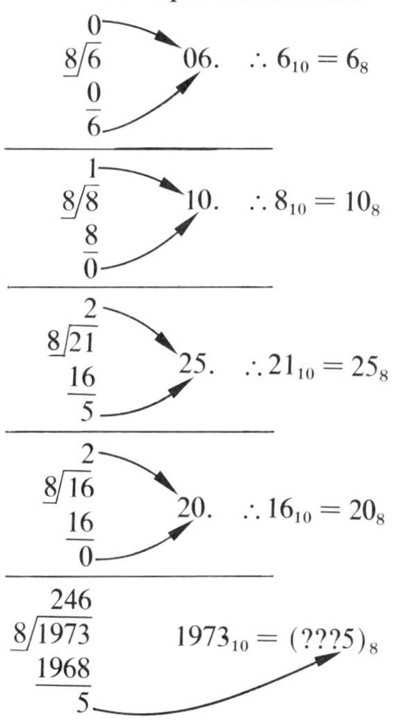

In the latter case, the digit nearest the octal point will be a 5. However, the presence of a quotient greater than 7 indicates that more than 7×8 from the original number is unaccounted for (to be precise there is $246 \times 8 = 1968$ unaccounted for). If we divide the quotient by 8 again we shall find the number of times $8^2 = 64$ can be factored from the remaining number. The remainder from this division will be the digit in the 8^1 position. Thus,

$$\begin{array}{r} 30 \\ 8\overline{)246} \\ \underline{240} \\ 6 \end{array} \qquad 1973_{10} = (??65)_8$$

Continuing in this manner,

$$\begin{array}{r} 3 \\ 8\overline{)30} \\ \underline{24} \\ 6 \end{array} \qquad 1973_{10} = 3665$$

As a check,

$$3 \times 8^3 + 6 \times 8^2 + 6 \times 8 + 5 = (1973)_{10}$$

Fractional parts of a decimal number can be converted by repeated multiplication of the new base (using the arithmetic of the old base). For example, consider conversion of the numbers 0.5, 0.625, and 0.1 from decimal to octal notation.

$$\begin{array}{r} 0.5 \\ \times 8 \\ \hline 4.0 \end{array} \qquad \therefore 0.5_{10} = 0.4_8$$

$$\begin{array}{r} 0.625 \\ \times \ 8 \\ \hline 5.000 \end{array} \qquad \therefore 0.625_{10} = 0.5$$

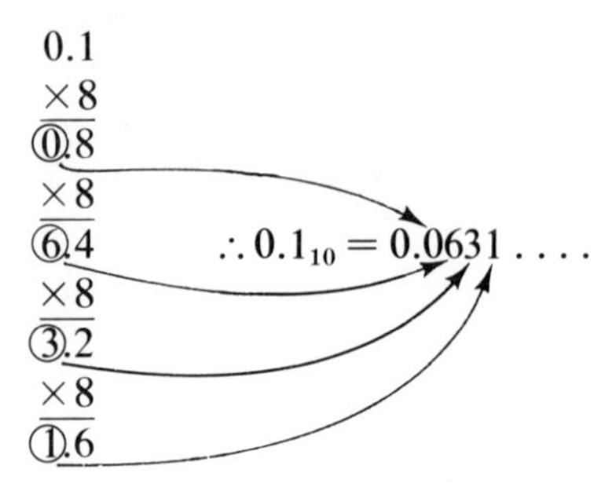

The *integral* parts of successive quotients become the digits, in order, in the converted fraction. Note that the process continues until the fractional part of the product vanishes. The last example illustrates the point that even simple rational numbers with a finite number of digits in one base may convert to unending digit strings in other bases.

As another example, convert $(3172.645)_{10}$ to octal and binary form.

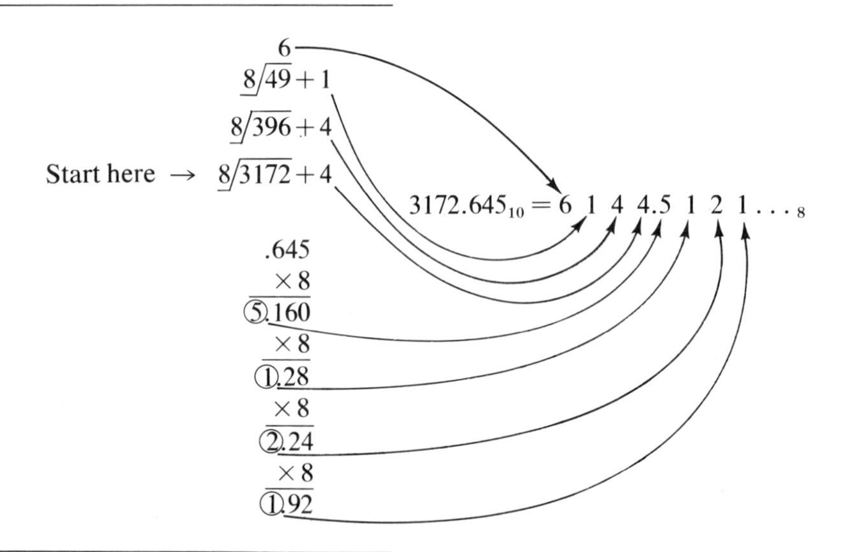

The binary equivalent may be found by converting directly from octal to binary,

$$(6144.5121\ldots)_8$$
$$= (110001100100.101001010001\ldots)_2$$

or by starting with the decimal number, as before (see top of page 27).

1.15 Binary and Octal Arithmetic

All the arithmetic operations used on decimal numbers may be implemented on

Remainders Integral parts of products

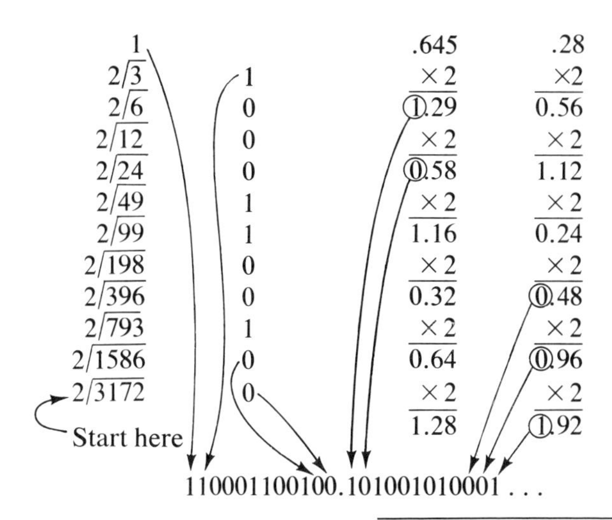

numbers in other bases as well. Of course the basic addition and multiplication tables *appear* to be different. For example, the addition and multiplication tables for binary arithmetic are:

Addition

First Operand \ Second Operand	0	1
0	0	1
1	1	10

Multiplication

First Operand \ Second Operand	0	1
0	0	0
1	0	1

The multiplication table for octal numbers is:

First Operand \ Second Operand	0	1	2	3	4	5	6	7
0	0	0	0	0	0	0	0	0
1	0	1	2	3	4	5	6	7
2	0	2	4	6	10	12	14	16
3	0	3	6	11	14	17	22	25
4	0	4	10	14	20	24	30	34
5	0	5	12	17	24	31	36	43
6	0	6	14	22	30	36	44	52
7	0	7	16	25	34	43	52	61

The following are examples of binary and octal arithmetic.

```
Binary   Decimal  |  Binary   Decimal
  101       5     |    101       5
 +111      +7     |   ×111      ×7
 1100      12     |    101      35
                  |   101
                  |  101
                  |  100011
```

```
          Octal                      Decimal

          52.3146                     42.4
  33.2)2203.3146              27.25)1155.400
        2102                        10900
         1013                         6540
          664                         5450
          1271                        10900
          1216                        10900
            534
            332
            2026
            1550
             2566
             2434
              132
```

1.16 Internal Number and Character Representation in a Digital Computer

In our hypothetical computer, we have decided to have a memory of 1000 words, each with hardware elements to store a sign and ten decimal digits. The format for the machine language instructions laid out in Fig. 1.8 and the operation codes assigned in Table 1.3 conform to this basic arrangement for the memory word. After the machine designer has decided on the general organization of the memory, and format and operation codes for the machine instructions, he must decide on the encoding scheme to be used for internally stored numbers, letters, truth values, etc.

What would be a reasonable scheme for storing *numerical information* in our memory? First, note that no provision has been made for the hardware representation of a decimal point in a memory word. One might consider the possibility of including a hardware element to represent the decimal point, but such an element would have no meaning in any word that contained the internal representation of fundamentally nonnumerical information such as an instruction. In addition, since the memory is organized to allow *any* numerically encoded information to be stored in *any* memory location, there is no rationale for including a decimal point in the hardware of some words but not others.

The absence of a decimal point in the hardware causes no serious problems, however. In most real computers the circuitry associated with the implementation of certain arithmetic operations is designed to assume that the decimal point for operands is located in a *fixed* position. Usually the fixed position is taken to be to the right of the rightmost digit, leading to numbers with only *integral* values. Thus a *data* word containing the digits

$$+4500501502$$

in the *integer* or *fixed point* representation would have the integral value 4,500,501,502 (compare this with the meaning of an *instruction* word with the *same* content shown in the machine-language program of Section 1.9). Note that for our hypothetical computer, this representation would allow the storage of all integers in the range −9999999999 to +9999999999, including zero.

The integer form has a certain appeal, since the sum, difference, and product of integers are also integers, and the hardware associated with these operations can effectively *ignore* the location of the point. There are some difficulties even here, however. For addition or subtraction, it is possible to generate an 11-digit result that could not be stored back into the memory in the integer form without losing the most significant digit. This phenomenon leads to an *overflow* condition, and ordinarily results in a program *interrupt*, a temporary or permanent suspension of computation. The programmer would normally be warned of the offense via an error message. Of course, if each of the two operands has nine or fewer digits, then no addition or subtraction overflow condition can occur.

Multiplication presents a somewhat more serious problem; two ten-digit operands produce a 20-digit product. In order to prevent loss of leading digits (multiplication overflow), the sum of the number of significant digits in the two operands must be ten or less.

The division operation introduces some interesting possibilities. Suppose the dividend is not an integral multiple of the divisor. Then the result would normally have a *fractional* part. But, by definition, a fraction cannot be stored into a memory word in integer representation. Therefore, the fractional part of a quotient resulting from the division operation with integer operands is lost. This is normally *not* considered to be an error in real machines, although the programmer might view things differently. Another possibility, more serious, is that the divisor might have a *zero* value. Real machines are usually designed to *ignore* division by zero and to give warning (an interrupt) that a zero divisor has been encountered.

Suppose that the problem to be solved on a machine has operands that are *not* integral in value. Then it is frequently possible to *scale* all values in the problem to achieve integer operands for all operations. For example, in a business calculation where all operands are measured in dollars, such as \$5.07 or \$22.57, the program could be written to carry out all calculations in integer form by converting all operands from units of dollars to units of cents, such as 507¢ or 2257¢.

As the preceding paragraphs indicate, the integer representation for numbers has major limitations, among the most important being:

1. Loss of fractional parts in division.
2. Possible loss by overflow of the most significant digits in addition, subtraction, and multiplication.
3. Inability to store fractions or fractional parts of numbers in the memory.
4. Limited range of numbers that can be stored in the memory.
5. Scaling problems encountered in practical problems.

Some of these difficulties might be overcome by having the arithmetic unit automatically locate the decimal point somewhere else in the string of digits retrieved from a storage word. However, it is effectively impossible to choose any *fixed* location that would satisfy all users of the computer. Some programmers might wish to use extremely large numbers, in which the decimal point would have to be positioned far to the right of the rightmost significant digit. For example, a chemist might wish to use Avogadro's number, 6.023×10^{23}, in his calculations. Still other programmers might want to use extremely small numbers. The chemist mentioned above might even need the cross-section of the nucleus, approximately 1×10^{-24} cm^2, in the same program where Avogadro's number is required.

In most practical calculations, the programmer would be willing to sacrifice some digits of *significance* if he were allowed a much larger number *range* and at the same time relieved of the *scaling* problem. For example, numbers used in engineering calculations are rarely known to more than four or five significant digits, but may vary over several orders of magnitude.

The computer designers' solution to this dilemma has been to develop an additional internal form for representing numerical information with a fractional part, or integral information too large to store in integer form. This representation is called the *floating-point* form, and is closely related to *normalized scientific notation.* In ordinary *scientific notation*, all numbers are represented as a decimal number, including a possible point (see Section 1.13), called the *mantissa*, followed by a *characteristic*, some power of the base that serves as a *scale factor* for the mantissa. For example, several different approximations of π, all in proper scientific notation, would be 3.14×10^0, 31.4×10^{-1}, and 0.000314×10^4. In a *normalized* scientific notation, the point in the mantissa is shifted until it immediately precedes the first significant digit in the number (never a zero unless the number is a true zero), so that the most significant digit immediately follows the decimal point. The exponent of the base in the scale factor is adjusted accordingly. The one and only three-digit approximation of π in normalized decimal scientific notation is 0.314×10^1.

In the floating-point representation, the computer storage word is typically partitioned into two parts, one to hold the most significant digits of the mantissa, and the other to hold the exponent. A possible design choice for our hypothetical computer would be to partition the word as shown in Fig. 1.11.

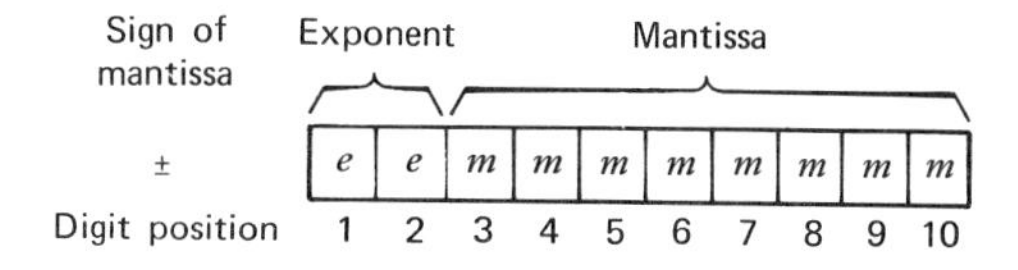

Figure 1.11 A possible format for internal representation of a floating-point number (m = mantissa, e = exponent). The number represented is $\pm 0.mmmmmmmm \times 10^{ee}$.

With this choice, the mantissa of the number could be represented with eight significant digits of accuracy, while two digits would be left for the exponent associated with the scale factor or characteristic. Let the *sign value* stored with the content of the memory location be interpreted as the *sign* of the *mantissa* of the stored floating-point word. Then negative and positive mantissas in the ranges

$$-.99999999 \text{ to } -.10000000$$
$$+.10000000 \text{ to } +.99999999$$

could be represented,* while the exponent would be in the range from 00 to 99. Thus the number 57.456×10^{10} would be represented in the internal floating-point form as

$$+1257456000 = 0.57456000 \times 10^{12}$$

while the number $-1.$ in floating point form would be represented by

$$-0110000000 = -0.10000000 \times 10^{1}.$$

Since our memory word only has *one* sign associated with its hardware elements, the scheme of the preceding paragraph has a serious drawback. We have no way of representing *negative exponents* in the scale factor; there would be no way to represent the constants $1. \times 10^{-24}$, 0.005, or -7.33×10^{-5}. A very simple solution to this problem is to save a *biased exponent* in the exponent portion of the word. For example, suppose that the *true exponent plus 50* were stored in the exponent position. Then the smallest storable exponent, 00, would correspond to the true exponent -50, while the largest exponent, 99, would correspond to the true exponent $+49$. Thus the allowable number range would still cover 100 orders of magnitude, but both negative and positive exponents in the scale factor could be represented. The format of this internal representation of a floating-point number is shown in Fig. 1.12, where m represents a decimal digit in the mantissa, and b a decimal digit in the biased exponent. Several constants are shown in equivalent external and internal form in Table 1.7.

*A true zero can be represented as well, in which case the mantissa is 0.00000000.

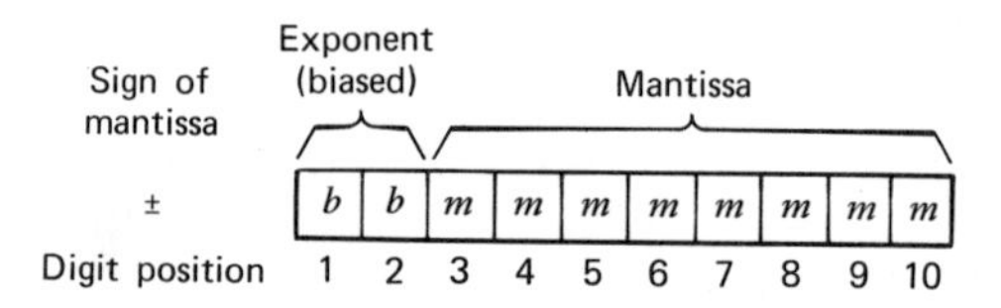

Figure 1.12 Floating-point format with a biased exponent (m = mantissa, b = biased exponent). The number represented is $\pm 10.mmmmmmmm \times 10^{(bb-50)}$.

Note that numbers having a magnitude larger than zero but smaller than 10^{-51} or greater than or equal to 10^{49} *cannot* be represented at all. In the floating-point form, it is possible to represent only a *discrete subset* of the rational numbers over a limited magnitude range, as shown in Fig. 1.13.

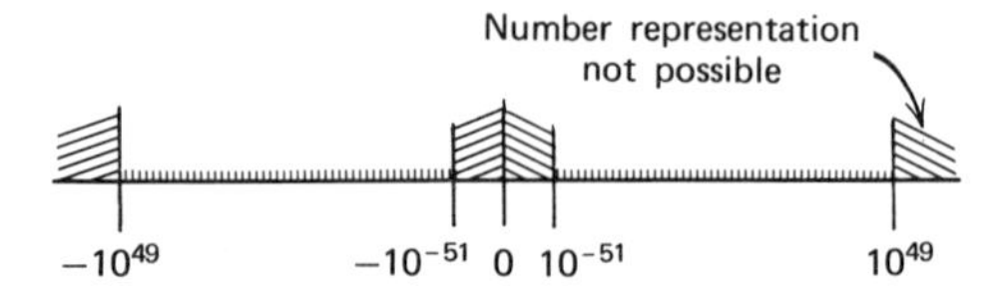

Figure 1.13 Numbers representable in the floating-point format of Fig. 1.12.

Integers with fewer than eight significant digits can be represented in *either* the integer or floating-point formats. For example, the integer -7563 could appear as:

Integer form:	-0000007563
Floating-point form:	-5475630000

This example illustrates the point that integer and floating-point representations are quite different in principle, and will therefore require *different* machine instructions for their manipulation. In real computers, in fact, there are usually at least *two* different machine instructions for each of the basic arithmetical operations. There will be one machine-language instruction to add operands in integer format (usually called "integer add") and another machine-language instruction to add operands in floating-point format ("floating add"). Ordinarily there are no instructions for adding operands of mixed format, one in integer and the other in floating-point representation.

Table 1.7 Internal Floating-point Number Representations for Our Hypothetical Computer

Number	Floating-point representation	Comment
$-0.99999999 \times 10^{49}$	−9999999999	smallest negative number
$-1.0000000 \times 10^{-51}$	−0010000000	largest negative number
0	±xx00000000	(xx may be any digits)
$+1.0000000 \times 10^{-51}$	+0010000000	smallest positive number
$+0.99999999 \times 10^{49}$	+9999999999	largest positive number
+2.0000000	+5120000000	
3.1415927	+5131415927	most accurate approximation of π
6.023×10^{23}	+7460230000	Avogadro's number to four significant digits

The format of the information content of a memory word is called the *mode* in computer parlance. We might say, for example, "the content of location 345 is of integer mode" or "the content of symbolic address X is of floating-point mode." In practice, a programmer would probably say "location 345 is integer," or "X is floating-point." But remember, it is not the *address* X but the *content* of the address X that is of floating-point mode.

We now have three different memory-word formats, one for machine-language instructions, one for integer-number representation, and one for floating-point number representation. Other kinds of information will require suitable formats for their internal representations as well. For example, suppose we wish to represent the English letters and punctuation marks in numerically coded form. Assuming that there are no more than 100 different symbols to be encoded, a two-decimal digit code might be assigned to each character as indicated in Table 1.8.

Table 1.8 Possible Internal Representation for Characters

Character	Code
A	21
B	22
C	23
D	24
E	25
⋮	
Y	45
Z	46
/	47
=	48
?	49
+	50
$	51
*	52
etc.	

Since each character requires two decimal digits for its representation, and since each word in our hypothetical machine has a signed ten-digit content, the code for *five* symbols could be stored in a *single* machine word. For example, the following would be equivalent.

External form: ACE**
Internal form: 2123255252

In this case, the sign associated with the storage word would have no particular meaning.

The significance of the ambiguous nature of memory should now be clear. Let a particular memory location have the content +4525503451. Then the content of the location might, among still other possibilities, mean *any* of the following:

An instruction: $\overline{451} \leftarrow \overline{525} \,/\, \overline{503}$
An integer number: +4,525,503,451

A floating-point number: $0.25503451 \times 10^{-5}$
A character string: YE+P$

When a program refers to a location in memory, the instructions making the reference must be appropriate for the format of the content of the location. It is the responsibility of the programmer to insure that all operands for all operations are of proper mode. He must know the mode associated with every memory location. No ambiguity is allowed in a program.

1.17 Number and Character Representation in the IBM 360/370 Computers

In the preceding section we have described possible representations of numbers and characters for the hypothetical decimal computer that is used, for purposes of illustration, throughout this chapter. In this section, we describe the representations of numbers and characters in a family of *real* machines manufactured by the International Business Machines Corporation, and marketed under the name "360" or "370" Series. Some of the other major manufacturers, though not all, use the same conventions.

The fast memory of every computer in the IBM 360 line consists of a large number of magnetic core storage elements, the number varying with the model and with customer requirements, as described in Section 1.3. Unlike our hypothetical computer, the IBM 360 memory is *not* subdivided directly into words of *fixed* length. Instead, the smallest segment of memory with a unique address consists of eight magnetic cores that can represent eight binary digits (0 or 1). The binary digits are usually called *binary bits*, and the addressable collection of eight binary bits is called a *byte*. Words in the IBM 360 or 370 memory can have a *length* of one, two, four, or eight bytes, known as *byte*, *half-word*, *full-word*, and *double-word*, respectively. (As an illustration of the proliferation of jargon, half a byte is called a "nibble" in some circles).

The eight bits in a byte can be interpreted as the representation of *two hexadecimal characters* (see Table 1.6); because of this, the 360 computers are often called *hexadecimal machines*. More fundamentally, of course, they are essentially binary machines, but the hexadecimal notation allows a considerably more compact external notation.

As was true for our hypothetical computer, each different mode or type of information to be saved in the memory has an appropriate word format for its internal representation. For FORTRAN programmers (see Section 3.27) the most important formats are those for the integer and floating-point number representations.

Words of *integer* format can contain either *two* or *four* bytes, as shown in Fig. 1.14. In each case, the decimal point is assumed to be to the right of the rightmost binary bit or hexadecimal character, and the leftmost bit is used to encode the *sign* of the stored integer, with 0 = plus and 1 = minus. If the integer is *positive*, then the remaining 15 bits in the two-byte format or 31 bits in the four-byte format are used to encode the *value* of the integer in straightforward binary form. If the integer is *negative*, then the remaining bits are used to encode its *magnitude* in *two's-complement* form. For the two- and four-byte forms, the two's-complement can be found by subtracting the magnitude from 2^{15} and 2^{31}, respectively.

Range of Two-Byte (Half-Word) Integer.
Binary:

$-(1000000000000000)_2$
to
$+(111111111111111)_2$,

Hexadecimal:

$-(8000)_{16}$ to $+(7FFF)_{16}$,

Decimal:

$-(2^{15})_{10} = -(32768)_{10}$ to
$+(2^{15}-1)_{10} = +(32767)_{10}$.

Range of Four-Byte (Full-Word) Integer.
Binary:

$-(10000000000000000000000000000000)_2$
to
$+(1111111111111111111111111111111)_2$,

Hexadecimal:

$-(80000000)_{16}$ to $+(7FFFFFFF)_{16}$,

Decimal:

$-(2^{31})_{10} = -(2{,}147{,}483{,}648)_{10}$ to
$+(2^{31}-1)_{10} = +(2{,}147{,}483{,}647)_{10}$.

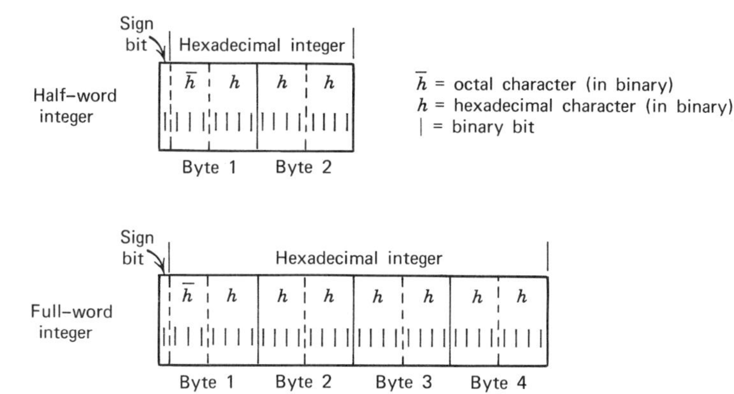

Figure 1.14 Formats for two- and four-byte integer representation in the IBM 360 computers.

Therefore, in terms of external decimal representation, the two- and four-byte integer formats allow somewhat more than four and nine digits of accuracy, respectively.

Words of *floating-point* format can contain four or eight bytes as shown in Fig. 1.15. In each of these forms, the number is effectively coded in *normalized hexadecimal scientific notation* with a biased exponent. A hexadecimal number in normalized scientific form consists of a fractional hexadecimal *mantissa* with the leading nonzero digit positioned immediately after the point, and a *scale factor* that is an integral power of 16. Examples are:

$$(0.11 \times 10^1)_{16} = (1.1)_{16} = (1 \times 16^0 + 1 \times 16^{-1})_{10} = (1.0625)_{10},$$
$$(0.24 \times 10^2)_{16} = (24)_{16} = (2 \times 16^1 + 4 \times 16^0)_{10} = (36)_{10},$$
$$(-0.\text{CF} \times 10^{-3})_{16} = (-0.000\text{CF})_{16} = -(12 \times 16^{-4} + 15 \times 16^{-5})_{10} = (207 \times 16^{-5})_{10}.$$

The four-byte (full-word or *single-precision*) form and the eight-byte (double-word or *double-precision*) form have a similar structure. The first byte contains the sign of the mantissa and a biased exponent of 16; the remaining bytes contain the most significant digits of the mantissa.

The eight bits of the first byte are partitioned as follows: the *first* bit is used to represent the *sign* of the mantissa with 0 = plus and 1 = minus. The remaining *seven* bits are used to encode a hexadecimal *exponent* of the number base, 16.

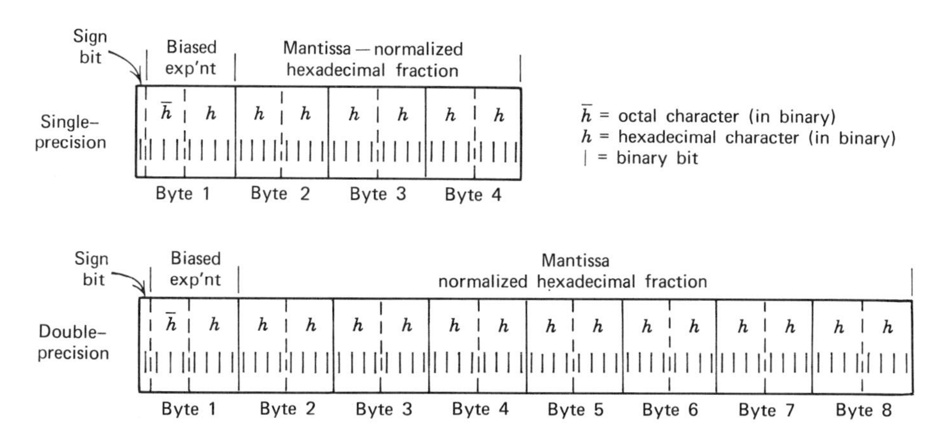

Figure 1.15 Formats for four-byte (single-precision) and eight-byte (double-precision) floating-point representation in the IBM 360 computers.

With seven bits it is possible to represent the hexadecimal integers 00 to 7F or, in decimal notation, 00 to 127. The biasing quantity is $(40)_{16}$ or $(64)_{10}$. Thus the scale factor or characteristic has the range 16^{-64} to 16^{63}.

The remaining three bytes in the single-precision or seven bytes in the double-precision format are allocated to storage of the mantissa in normalized hexadecimal form. The point is assumed to lie between the first and second bytes in both forms. Thus the number *range* in the single- and double-precision forms is identical, 16^{-65} to 16^{63}. But the number of significant digits of accuracy is considerably different, being six hexadecimal digits (the equivalent of about seven decimal digits) for single-precision and 14 hexadecimal digits (equivalent to about 17 decimal digits) for double-precision representation. Table 1.9 shows the internal representation of some floating-point numbers.

As is true of virtually all computers of significant size on the commercial market, most 360 systems have a full complement of integer operations and floating-point operations. In fact there may be several *different* arithmetic instructions (for addition, say) to account for both the *mode* and the *length* of the operands being considered. As described in Section 3.27 the normal word size assigned by the FORTRAN translator is four bytes for both integer and floating-point numbers. The floating-point numbers are normalized and the arithmetic operations performed on them yield normalized floating-point results. The programmer has the option of choosing two-byte integers and double-precision floating-point format for any or all operands. As we shall see in Chapter 3, it is the programmer's responsibility to ensure that the modes and lengths of all operands are proper for the intended operations.

In the 360 computers, letters, punctuation marks, and other symbols are encoded with two hexadecimal characters so that one *character* can be encoded in the memory element of *one byte* of storage. Table 1.10 shows the hexadecimal internal equivalents of the symbols in the FORTRAN character set (see Section 3.2). Notice that even the digits can be encoded as characters. The number of characters that can be encoded in a word of storage is exactly equivalent to the

Table 1.9 Internal Floating-point Number Representations in the IBM 360 Computers

Number (decimal)	Floating-point Representation	Type (precision)	Comment
16^{-65}	00100000	single	smallest positive number
$\sim 16^{63}$	7FFFFFFF	single	largest positive number
1.0625	4111000000000000	double	
36	42240000	single	
-207×16^{-5}	BDCF0000[a]	single	
$\sim -16^{63}$	FFFFFFFFFFFFFFFF[a]	double	smallest negative number
~ -0.1	C019999999999999[a]	double	

[a] If the leading hexadecimal character is greater than 7, then the mantissa is negative. Thus, (see Fig. 1.15) BDCF0000 could also be interpreted as $-$3DCF0000 and FFFFFFFFFFFFFFFF could be interpreted as $-$7FFFFFFFFFFFFFFF.

Table 1.10 Internal Character Representations in the IBM/360 Computers

Character	Hexadecimal Code	Character	Hexadecimal Code
A	C1	blank	40
B	C2	¢	4A
C	C3	.	4B
D	C4	<	4C
E	C5	(	4D
F	C6	+	4E
G	C7	|	4F
H	C8	&	50
I	C9	!	5A
J	D1	$	5B
K	D2	*	5C
L	D3	)	5D
M	D4	;	5E
N	D5	¬	5F
O	D6	-	60
P	D7	/	61
Q	D8	,	6B
R	D9	%	6C
S	E2	_	6D
T	E3	>	6E
U	E4	?	6F
V	E5	:	7A
W	E6	#	7B
X	E7	@	7C
Y	E8	'	7D
Z	E9	=	7E
0	F0	"	7F
1	F1		
2	F2		
3	F3		
4	F4		
5	F5		
6	F6		
7	F7		
8	F8		
9	F9		

number of bytes in that storage word; a full word has a capacity for encoding four characters, a double word capacity for eight characters, etc. The internal hexadecimal code of Table 1.10 is sometimes termed the EBCDIC code (Extended Binary Coded Decimal Interchange Code).

Problems

1.1 If the instruction in address 002 of the machine-language program of Table 1.4 were +4505505504, what results would be produced by the program during its execution on our hypothetical computer?

1.2 How would you modify the machine-language program of Table 1.4 to implement the calculation

$$\overline{\mathrm{U}} \leftarrow (\overline{\mathrm{X}}+\overline{\mathrm{Y}})/\overline{\mathrm{Z}}+4.1,$$

assuming that the machine instructions for the arithmetic operations in Table 1.3 are designed to employ floating-point operands with the internal representation shown in Fig. 1.12?

1.3 Using only the instructions listed in Table 1.3, write a machine-language program for our hypothetical computer that will implement the following computation:

$$\overline{\mathrm{U}} \leftarrow (|\overline{\mathrm{X}}|+|\overline{\mathrm{Y}}|)/\overline{\mathrm{Z}}+36.554.$$

1.4 Invent a new machine instruction with the format of Fig. 1.8 that would simplify the writing of the program of Problem 1.3.

1.5 Suppose that a new machine-language instruction named DTR (*d*ecrement and *t*ransfer) is available for the hypothetical machine described in Section 1.8. In the format of Table 1.3, let the instruction have the following characteristics.

Operation Code	Operation	Meaning
+8	DTR	If $\overline{\mathrm{A}} \geq 0$, decrement $\overline{\mathrm{A}}$ by $\overline{\mathrm{B}}$ (that is, $\overline{\mathrm{A}} \leftarrow \overline{\mathrm{A}}-\overline{\mathrm{B}}$) and transfer to address C for the next instruction; otherwise, simply go to the next address for the next instruction.

Of what value would the DTR instruction be? Suggest several uses.

1.6 Consider the problem of reading values for n examination scores $e_1, e_2, \ldots, e_n$ and calculating the average score. Suppose that the data are arranged so that n appears on the first data card, the first examination score appears on the second data card, the second score appears on the third data card, etc. Write a program for the hypothetical computer of Section 1.8 that makes use of the DTR instruction of Problem 1.5 and that reads the data, writes out the data, computes the average score, writes the average score, and then returns to read n and another set of scores.

1.7 Suppose that the data of Problem 1.6 were arranged with n on the first card, the first three scores on the second card, the next three scores on the third card, etc. Write a program for the hypothetical computer of Section 1.8 that reads the data for this case and computes and prints the average score.

1.8 Write a program for an extension of the algorithm developed in Problem 1.7 that sorts the scores into ascending sequence, and then prints the ordered scores, the range of the scores and the median of the scores (the median score has an equal number of scores above and below it in the sorted list; if n is even, then the average of the two most central scores should be used).

1.9 Convert $(101.11101)_2$ to equivalent octal and decimal notation.

1.10 Convert $(752.612)_{10}$ to equivalent binary, octal, and hexadecimal notation.

1.11 Use octal multiplication to find $(161.35)_8 \times (16.12)_8$.

1.12 Use binary division to evaluate $(11101.111)_2/(110.11)_2$.

1.13 Add $(1B.CDEF)_{16}$, $(23.DA47)_{16}$, and $(E23.16A1)_{16}$. What are the decimal equivalents?

1.14 Convert $(0.1)_{10}$ to binary, octal, and hexadecimal form.

1.15 Write the decimal number 753.602 in base 5 notation.

1.16 Convert the hexadecimal number AB15.06D to its duodecimal (base twelve) equivalent.

1.17 Write the decimal number 6.023×10^{23} in normalized hexadecimal scientific notation.

1.18 Could negative integers be used as bases for a positional number system? If so, find the base -3 equivalents of decimal integers between -10 and 10; are they unique?

1.19 Suppose that four new machine-language instructions, named SHR (*sh*ift *r*ight), SHL (*sh*ift *l*eft), READC (*read c*haracter), and PRINTC (*print c*haracter) are added to the instruction set for the hypothetical computer (see Table 1.3) along with the DTR instruction of Problem 1.5. Let these new instructions have the following characteristics.

For example, if address 507 contains the number -1234567890 and the instruction is $+9507003508$, the new content of address 508 will be -0001234567. Let the digital code for a blank be 60. Then, if locations 607, 456, and 345 contain the numbers $+5228253232$, $+3560243532$, and $+3245526060$, respectively, then the instruction -0607456345 will cause the characters *HELLO*b*DOLLY**bb* (*b* = blank) to be printed in the first 15 columns of a line on a printer.

Write a program, similar to that of Table 1.4, which will print a line with the characters "X, Y, Z, AND U =" before writing out the two lines with the results for each data set.

1.20* Using the instructions of Table 1.3 and Problems 1.5 and 1.19 write a machine-language program that reads three English words of five letters or less, starting in columns 1, 6, and 11 of a data card, sorts them into alphabetical order, and prints the words in alphabetical order on the printer.

Operation Code	Operation	Meaning
+9	SHR	$\overline{C} \leftarrow \overline{A}_{sr\ B}$ where $\overline{A}_{sr\ B}$ is the content of the A address shifted B places to the *right*; the sign of $\overline{A}$ is *not* shifted, the sign of $\overline{C}$ is the same as the sign of $\overline{A}$, any digits shifted off the end of the word are lost, and zeros are entered on the left as necessary to fill out the ten-digit word.
−9	SHL	$\overline{C} \leftarrow \overline{A}_{sl\ B}$ where $\overline{A}_{sl\ B}$ is the content of the A address shifted B places to the *left*; the sign of $\overline{A}$ is *not* shifted, the sign of $\overline{C}$ is the same as the sign of $\overline{A}$, any digits shifted off the end of the word are lost, and zeros are entered on the right as necessary to fill out the ten-digit word.
−0	READC	Reads the first 15 columns of a data card. The characters punched in the first five columns are coded in the digital code shown in Table 1.8, and the ten-digit code for the five characters becomes the content of the A address; the content of the second five columns is encoded and stored in address B; the content of the third five columns is similarly coded and stored in address C. The signs of $\overline{A}$, $\overline{B}$ and $\overline{C}$ will all be positive.
−5	PRINTC	The contents of addresses A, B, and C are assumed to be characters, digitally coded according to Table 1.8. The digital codes are converted to their character equivalents and printed in the first 15 columns of a new line on the printer.

*Difficult problem.

1.21* Write a machine-language program that reads n English words of 15 characters or less from n data cards, sorts them into alphabetical order, and then prints them out, one word to a line.

1.22 Let the hexadecimal representation of a four-byte IBM 360 word be

C8C5D3D6

(a) If this word were known to contain encoded characters, what is the external equivalent?
(b) If this word were known to be in floating-point format, what number is represented? Give both the hexadecimal and decimal equivalents.
(c) If this word were known to be in integer format, what number is represented? Give both hexadecimal and decimal equivalents.

1.23 Show the IBM 360 hexadecimal encoding of the number -8.7 in double-precision format.

1.24 Encode your name in IBM 360 character format. How many bytes and words are required?

1.25 Write the following decimal numbers in IBM 360 single-precision format in hexadecimal notation:

2.5
0.1
2.0
-5.5
16.
256.0625

1.26 Convert the following IBM 360 hexadecimal strings to external decimal form, assuming that they are written in (a) four-byte integer format and (b) four-byte floating-point format:

25346214
3AB6C100
D3EFA000
F1710000

*Difficult problem.

CHAPTER 2

Flow Diagrams and Algorithms

2.1 Introduction

The engineer or scientist frequently needs to perform certain calculations that will lead to the solution of a particular problem. The solution of such problems is greatly facilitated if we can specify the exact *sequence* of computational steps or operations that must be performed in order to achieve the desired result. The understanding is that once this procedure has been clearly established, then the subsequent process of substituting numbers and actually performing the calculations—whether by slide rule, by mental arithmetic, or by digital computer—is virtually automatic. Thus, laying out an appropriate and unambiguous computational procedure, or *algorithm*, is the key step to solving the problem. We shall emphasize the *flow diagram* as a convenient visual representation of a particular algorithm. A flow diagram consists of an unambiguous sequence of inscribed boxes, each having a characteristic shape that represents a particular type of operation, as explained below, and summarized in Appendix B.

Although the majority of the algorithms in this book are most conveniently implemented by digital computer, we shall endeavor to write algorithms and flow diagrams in terms of conventional algebra, independent of any particular computing language. Thus the flow-diagram form will represent a universal language, which can then readily be translated for implementation by specific computing devices. The emphasis in this text is on solving *numerical* problems; however, the flow diagram can serve equally well for representing algorithms for solving *nonnumerical* problems, in which the decisions made and actions taken do not necessarily involve numbers.

2.2 Variables, Storage, Accept and Display Operations

We start by taking a simple example, that of computing the density of a perfect gas from the formula

$$\rho = \frac{Mp}{RT}, \tag{2.1}$$

in which ρ = density (lb_m/cu ft), M = molecular weight of gas (lb_m/lb mole), p = absolute pressure (lb_f/sq in.), R = gas constant (10.73 lb_f/sq in. lb mole °R), and T = absolute temperature (°R, = °F + 460°). Here, typical engineering units have been indicated, although the formula holds for any consistent set of units. Although the computation of ρ may seem a trivial problem—and indeed it is—it is nevertheless instructive to examine the precise sequence of steps whereby it is accomplished.

First, we must obtain or be given definite values for the *variables* M, p, R, and T. Variables are so named because they can be assigned a whole range of different values (including the "constant" R, whose value depends on the units employed). This operation of obtaining values is represented in a flow diagram by the box

M, p, R, T

which means: *accept* or *read* specified *values* for the *variables* whose names appear in the box. Such values could be accepted in a variety of ways: by copying from a blackboard, looking up in tables, reading from magnetic tape, and so on. The actual shape used for designating the accept operation is that of a punched card, which represents a well-known way of communicating information to a digital computer.

Until we are ready to perform the calculations, the values for M, p, R, and T cannot "float in thin air," but must be recorded or stored somewhere—perhaps in our brain, on a piece of paper, in the registers of a desk calculator, in the memory of a digital computer, or by a particular setting on a slide rule. Thus, we can think in terms of four "pigeon-holes," memory locations, or simply boxes, labeled "M," "p," "R," and "T." Inside each box, the appropriate number is stored, replacing any existing values for the four variables that might happen to be in the boxes. In a computer, this feature of replacement is termed *destructive read-in.* (If the boxes really refer to storage locations in the memory of a computer, there will *always* be numbers stored in them—sometimes left there by the person who ran his program just before ours. Thus we must never assume that a variable has a particular value unless we have taken care to put it in the appropriate storage location.)

For example, if $M = 28.8$ (appropriate for air), $p = 14.7$ (atmospheric pressure), $R = 10.73$, and $T = 492°R$, then the contents of the boxes after the accept or read operation would be

M	p	R	T
28.8	14.7	10.73	492

These storage boxes will not actually appear in the flow diagram. However, we include them here in the discussion in order to emphasize that the mere mention of a variable *name* automatically presupposes the creation of a corresponding storage location for retaining the current *value* of that variable.

The question might be asked at this stage: "What are the current values of the variables M, p, R, and T? In answer, it would be appropriate to *display* their values, by writing them on a blackboard, by printing them on paper, by showing them on a television screen, and so on. Since the computer often communicates numerical values to us by typing or printing them, the following box, intended to represent a (very neatly) torn sheet of paper, is used for the display operation:

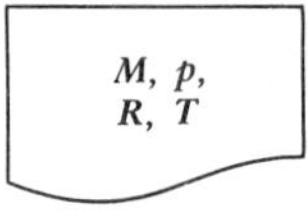

Note that the display operation leaves unchanged the values of M, p, R, and T that are currently in storage. In a computer, this feature is termed *nondestructive read-out.*

2.3 Arithmetic Operations and Substitution

We are now ready to compute the value for the density ρ, according to equation (2.1). We shall represent this operation (and all other computational steps) by a rectangular box, inscribed as follows:

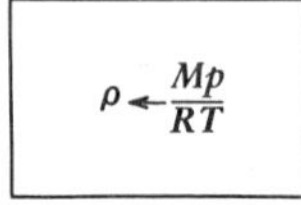

The interpretation is: take the values currently stored for M, p, R, and T, evaluate the expression $(M \times p) \div (R \times T)$, which has the approximate value 0.0801, and store the result in another box that is labeled "ρ." At this stage, we have five boxes or memory locations, containing the values

M	p	R	T
28.8	14.7	10.73	492

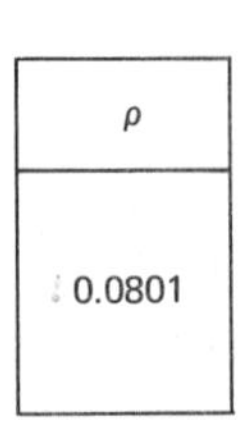

Note the following:

(a) Because of the nondestructive readout feature, although the values of M, p, R, and T have been supplied for use in evaluating Mp/RT, duplicate values for these variables still remain in memory.

(b) The exact sequence of multiplication and division operations in evaluating Mp/RT is assumed to be unimportant. However, if some special feature were particularly important (notably, that of integer division, in which the remainder is discarded in order to produce an integer result), then a note to that effect should be inscribed inside the box. It is also assumed that *temporary* or *scratch* storage is also provided as needed—for example, to hold the value of the numerator $M \times p$, while the denominator, $R \times T$, is being computed. A meaningful combination of constants, variables, arithmetic operators ($+, \times$, etc.), and parentheses is called an *arithmetic expression*; $(M \times p) \div (R \times T)$, or simply Mp/RT, is an example of such an expression.

(c) A left-pointing arrow is used instead of an equals sign in order to emphasize that *substitution* of a value into storage for ρ is intended, thereby *replacing* any previous value that might be there. (If the reader has any doubts on this point, let him consider which, if either, of the following makes sense: $i = i + 1$, or $i \leftarrow i + 1$.)

The final step in the sequence is to display the computed value for ρ:

ρ

At last, the above steps may be assembled into Fig. 2.1, which is a complete flow diagram for computing the density of a perfect gas!

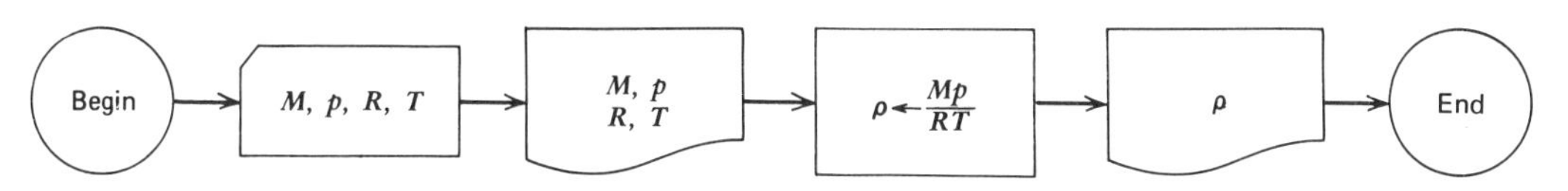

Figure 2.1 Flow diagram for computing density of a perfect gas.

Note that arrows are used for stepping from one operation to the next, and that circles inscribed with the words "Begin" and "End" are used for delineating the algorithm.

2.4 Decision Making

The flow diagram in Section 2.3 involved only a single path from "Begin" to "End." Now consider the problem of solving the quadratic equation

$$ax^2 + bx + c = 0, \tag{2.2}$$

which has two roots,

$$x_1 = \frac{-b + \sqrt{b^2 - 4ac}}{2a},$$

$$x_2 = \frac{-b - \sqrt{b^2 - 4ac}}{2a},$$

provided a is not zero. However, if a were zero, then there would only be a single root, $x = -c/b$. (Assume for the present that $b^2 - 4ac$ is nonnegative and there is no complication on that score.)

Clearly, at some stage in the flow diagram for the solution of (2.2), we must ask the question, "is the coefficient a zero?" Depending on whether the outcome is *yes* or *no* (*maybe* is unacceptable as a reply!), we shall follow one of *two* branching paths—one that computes the two roots x_1 and x_2, and the other that computes the single root x only.

This operation of testing and branching—making a decision, in effect—is conventionally represented in a flow diagram by placing the assertion to be tested inside a box resembling an oval:

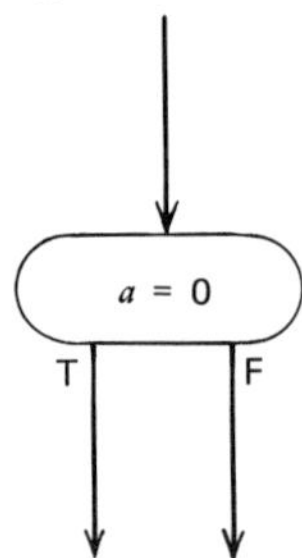

Depending on whether the statement "$a = 0$" is *true* (yes, a *is* zero) or *false* (no, a is *not* zero), the exit will be via the arrow marked T or via the one marked F. Note that the opposite statement "$a \neq 0$" could just as easily have been tested, as long as the subsequent actions were transposed accordingly.

The case of a being a very small number, such as -6.7×10^{-27} or 1.2×10^{-12}, might conceivably be encountered, in which event it would probably be appropriate to treat it as being zero for all practical purposes. Arbitrarily assume that we shall opt for the one-root case if $|a|$, the absolute value of a, is less than or equal to 10^{-10}, and for the two-root solution if it exceeds this value.

A possible complete flow diagram, involving a check of the input data, would be:

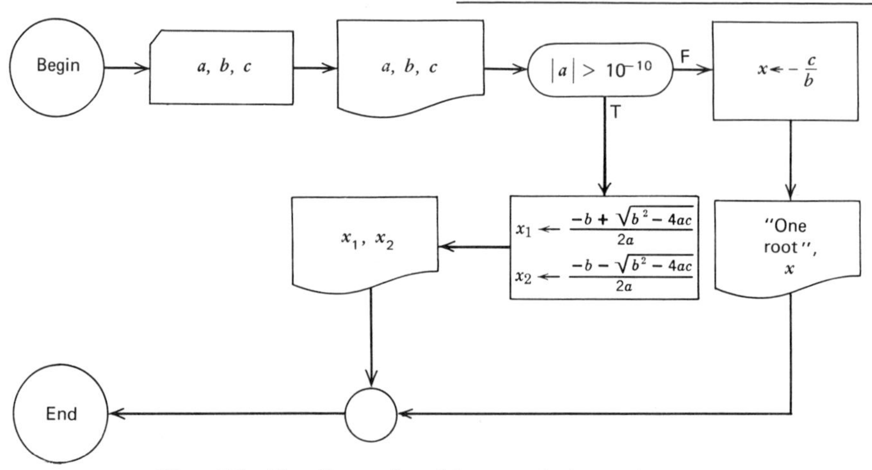

Figure 2.2 Flow diagram for solving a quadratic equation.

Referring to Fig. 2.2, note the following:

(a) The small circle indicates a *junction*, where two (or more) separate paths join together. Of course, only a single arrow may lead *out* of such a junction. The small circle may be inscribed with a number or letter if for some reason the junction requires particular identification (see Fig. 2.4, for example).

(b) An appropriate message has been added to the single-root case. In a display operation, a phrase in quotation marks is understood to be a *message*, not involving numerical values. Thus in the single-root case, the displayed value of x would be preceded by the comment: "One root."

EXAMPLE 2.1

Construct a flow diagram that will accept a value for n, compute the sum S of the arithmetic progression

$$S = 1 + 2 + \cdots + i + \cdots + n = \sum_{i=1}^{n} i, \quad (2.3)$$

and display the values for n and S. [This example is illustrative, and intentionally ignores the well-known formula, $S = n(n+1)/2$.]

A variable S, initially cleared to zero, is used for accumulating the sum. Successive values of i $(=1, 2, 3, \ldots)$ are then added to S. After each such addition, i is incremented and a test is made to see if i has exceeded its upper limit n. When this test is eventually passed, the results are displayed and the flow diagram terminates.

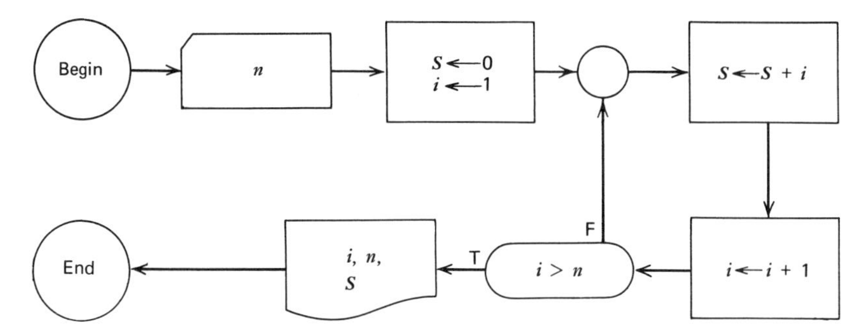

Figure 2.3 Summation of the numbers 1 through n.

Referring to Fig. 2.3, note the following:

(a) The final value of i is displayed along with n and S, and will be one greater than the upper limit n.

(b) The four boxes at the right-hand side of the flow diagram represent an *iterative loop*—that is, a set pattern of calculations that is repeated several times until a certain condition is eventually satisfied (see Section 2.7).

(c) The general format of the flow diagram is one that snakes backwards and forwards *across* the page. Although this pattern cannot be adhered to rigidly, experience has shown us that it does lead to a flow diagram that is fairly concise and easy to read.

2.5 Nested Conditionals

The decision-making process described in the previous section is also referred to as that of *testing a conditional.* In the examples in Section 2.4, we had asked only one question. However, if a decision is to be based on the outcome of the answers to *several* questions, we may be involved with *nested conditionals*, which are introduced in the following example.

EXAMPLE 2.2

Draw a flow diagram that will test to see whether x is negative, zero, or positive, and display an appropriate comment in each case.

Although there are *three* possibilities to be tested, a simple conditional applied once can only lead to *two* branches. Thus, at least two tests in sequence will be needed. The first test in Fig. 2.4 examines x to see if it is negative. If it is not, two possibilities remain: $x = 0$ or $x > 0$. The second test examines just one of these $(x = 0)$, and the other is automatically accounted for by the process of elimination. Clearly, there are several alternative ways of writing the flow diagram—for example, by first testing $x = 0$ and then testing $x < 0$.

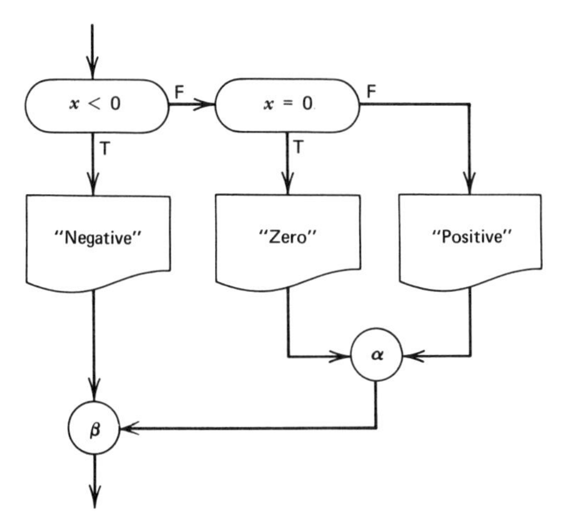

Figure 2.4 Testing for x negative, zero, or positive.

In Fig. 2.4, note that the two inner branches, starting with the conditional statement $x = 0$ and ending at the junction α, comprise one of the two outer branches, starting with the conditional statement $x < 0$ and ending at junction β. In other words, one of the conditionals is *nested* inside the other.

EXAMPLE 2.3

Repeat Fig. 2.2, the flow diagram for the solution of the quadratic equation (2.2),

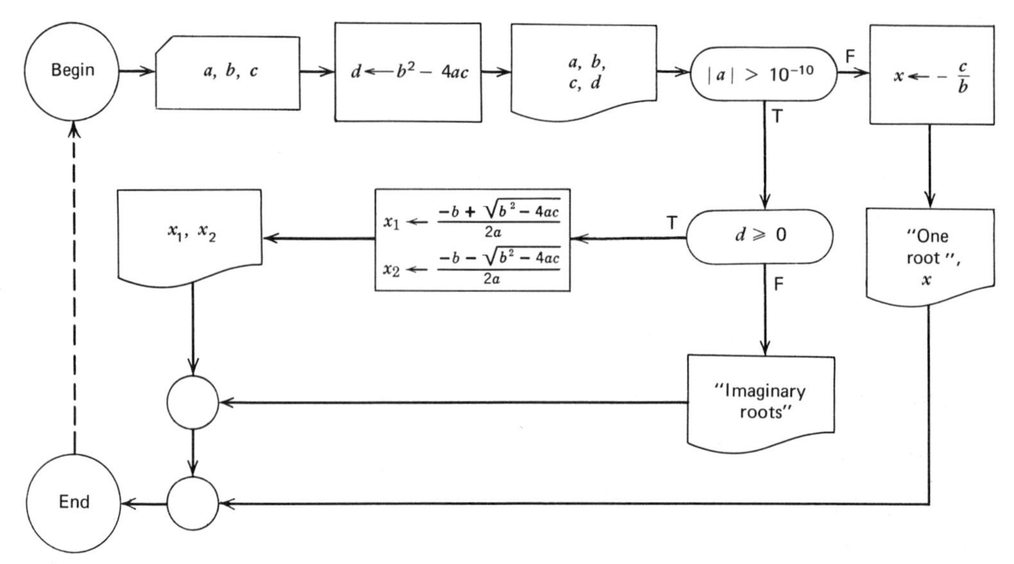

Figure 2.5 Modified flow diagram for solving a quadratic equation.

but now allow for the possibility that the values supplied for a, b, and c may lead to a discriminant, $d = b^2 - 4ac$, that is negative. In such event, merely display the message "imaginary roots" (without attempting to compute their real and imaginary parts).

An appropriate flow diagram is shown in Fig. 2.5. The dotted arrow leading from "End" to "Begin" suggests that the entire algorithm can be repeated for as many sets of values for a, b, and c as there are available. However, this feature is usually taken for granted, and such an arrow is almost always omitted.

2.6 Compound Conditionals

We may also investigate two or more conditional statements simultaneously, and ask a *compound question*, in which we are concerned with: (1) at least one statement being true, (2) all statements being true.

EXAMPLE 2.4

In a numerical simulation of a proposed experimental reactor, the temperature T and pressure p are being computed for many possible test conditions. If T exceeds an upper limit T_{max}, metallic corrosion is likely; if p exceeds an upper limit p_{max}, the safe working stress of the metal will be exceeded. In either event (or if both occur simultaneously), the reactor will probably explode after running for some time. Draw a flow diagram that will display a suitable warning message.

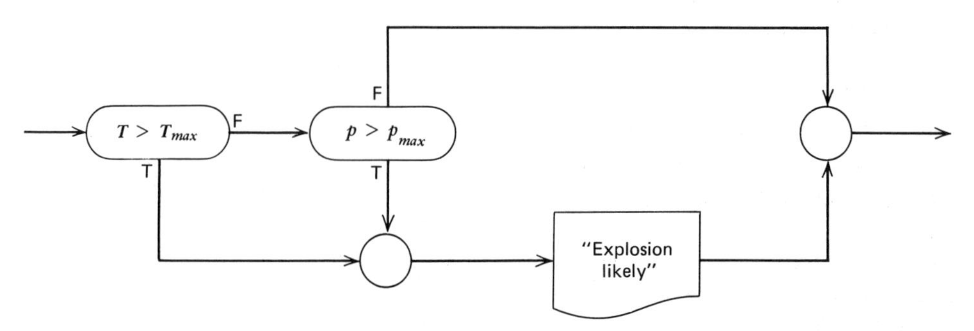

Figure 2.6 Test for explosion of reactor.

From Fig. 2.6, if *either* or *both* of the two conditional statements or *logical expressions* $T > T_{max}$, $p > p_{max}$ is true, the message is displayed. Otherwise, if both are false, no action is taken. With the aid of a *logical operator*, "or," the flow diagram may be written more simply, as shown in Fig. 2.7.

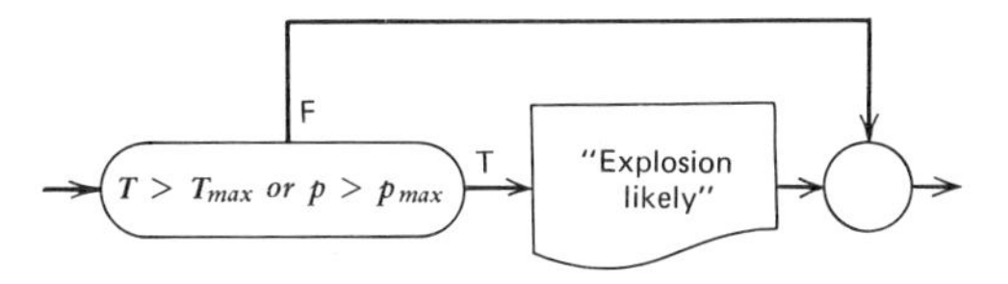

Figure 2.7 Modified test for explosion of reactor.

Based on what has just been said, we can construct a *truth table* for the *compound logical expression* Q_1 *or* Q_2, where Q_1 and Q_2 are individual logical expressions (T ≡ *true*, F ≡ *false*):

Q_1	Q_2	Q_1 *or* Q_2
T	T	T
T	F	T
F	T	T
F	F	F

Alternatively, we may only wish to take some action if *both* of two conditions are satisfied. In such cases, the logical operator "and" is appropriate. It has the following truth table:

Q_1	Q_2	Q_1 *and* Q_2
T	T	T
T	F	F
F	T	F
F	F	F

Observe that Q_1 *and* Q_2 is true only if both Q_1 and Q_2 are individually true.

EXAMPLE 2.5

Store the sum $(a+b)$ in x if y equals 2 *and at the same time* (z^2+4) is less than δ^2; otherwise, store the difference $(a-b)$ in x. Assuming that a, b, y, z, and δ have been assigned values previously, a suitable flow diagram is shown in Fig. 2.8.

2.7 Iteration

We are concerned here with calculations that are systematically repeated. For example, consider the flow diagram of Fig. 2.3, which represented the summation

$$S = 1+2+\cdots+i+\cdots+n = \sum_{i=1}^{n} i.$$

In slightly rearranged form, the portion that effects the summation is shown in Fig. 2.9.

Note that the operations inside the dashed hexagon of Fig. 2.9 involve:

(a) Setting a counter or index (i) to its initial value (1).
(b) Incrementing the counter by a fixed amount (1).
(c) Testing to see if the counter has exceeded an upper limit (n).

These combined operations of *initialization*, *incrementing*, and *testing* occur so

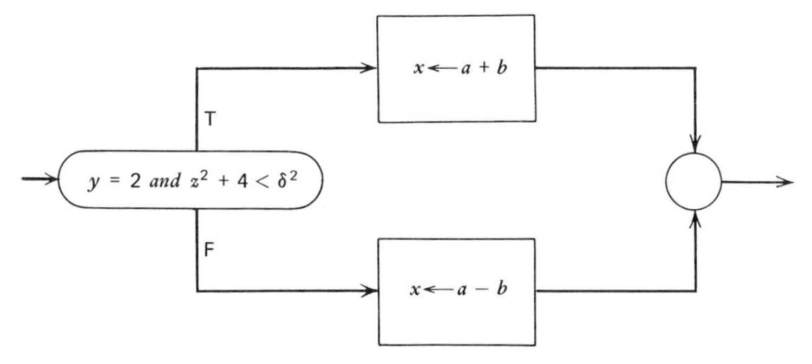

Figure 2.8 Flow diagram for Example 2.5.

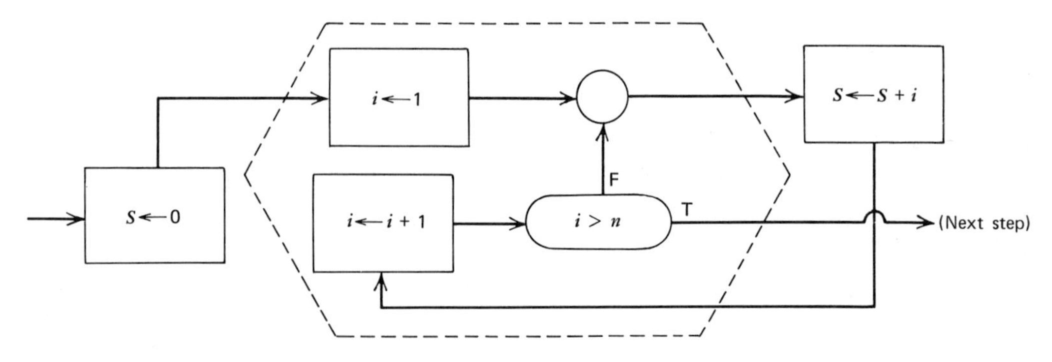

Figure 2.9 Summation of the numbers 1 through n.

frequently that we can incorporate them into a single box. A hexagonal shape is chosen to represent this *iteration* operation. The summation in Fig. 2.9 can now be written much more concisely, as shown in Fig. 2.10; here, the dotted line delineates the *scope* of the iteration—that is, just how far the computations proceed before we return to increment i and test it against the upper limit.

Note that the inscription in the hexagonal box of Fig. 2.10 follows the conventional agebraic notation: $i = 1, 2, \ldots, n$. The increment, which can be any specified value, equals the difference between the stated first and second values of the variable being incremented. Thus, $k = 2, 5, \ldots, m+1$ indicates that k is incremented in steps of 3 between 2 and $m+1$; $j = n, n-2, \ldots, m$ indicates that j is *decremented* in steps of 2 from n to m. In the latter case, j will be tested against the *lower* limit m, that is, the test $j < m$ will be made each time j is decremented.

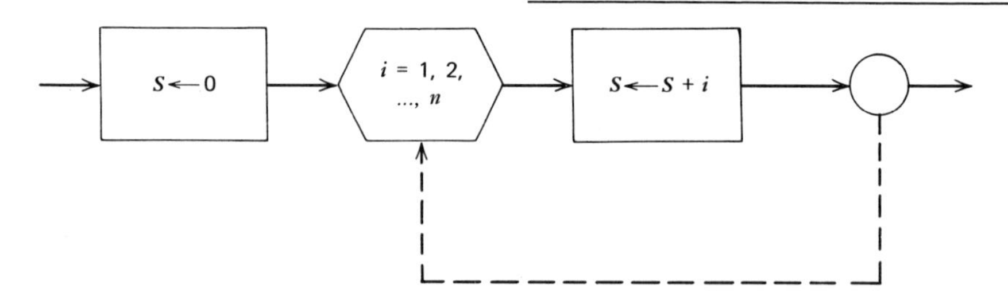

Figure 2.10 Summation of 1 through n using the iteration box.

EXAMPLE 2.6

Draw a flow diagram that will display a table of logarithms consisting of pairs of values of x and $\ln x$, with x running from specified values x_1 to x_2 in steps of Δx.

The flow diagram is shown in Fig. 2.11; if, for example, $x_1 = 0.1$, $x_2 = 10$, and $\Delta x = 0.1$, there would be 100 pairs of values displayed for x and $\ln x$ (that is, for $x = 0.1, 0.2, 0.3, \ldots, 9.9, 10.0$).

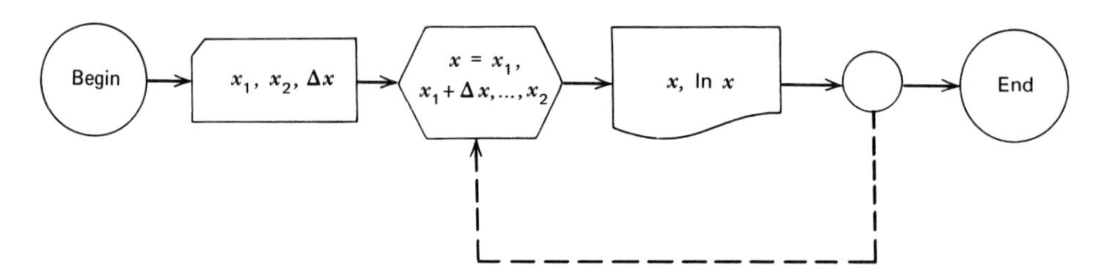

Figure 2.11 Flow diagram for tables of logarithms.

2.8 Subscripted Variables

When a group or *array* of variables has some characteristic in common, it is often convenient to assign them a common *array name* and to distinguish between the individual variables by appending one or more *subscripts* to that name. The situation is analogous to a family of people, each bearing the same surname, but having different first names or initials.

For example, if there are n students in a class, their scores on a particular examination might be stored in the variables x_1, $x_2, \ldots, x_n$; that is, the score of the first student is stored in x_1, the score of the second student in x_2, and so on, until we reach the score of the last student, which is stored in x_n. It is assumed that n already has a definite number stored in it, such as 15, 63, or whatever the class size is. (If *all* classes *always* contained exactly 20 students, for example, then we could refer to the scores more definitely as $x_1, x_2, \ldots, x_{20}$; however, the present arrangement, in which the class size is itself allowed to be a variable, permits more flexibility in practice.) In the present situation, with a single subscript appended to the common name x, the group x_1, $x_2, \ldots, x_n$, is called a *one-dimensional array* or *vector.*

Computations involving arrays can often be formulated very concisely by using the iteration box of Section 2.7, as illustrated in the following examples.

EXAMPLE 2.7

Draw a flow diagram that will accept values for n, the number of students in a class, their scores $x_1, x_2, \ldots, x_n$ on a particular test, and that will compute and display $\bar{x}$, the class average score. In the flow diagram, shown in Fig. 2.12, the iteration box is employed for accumulating the sum of the scores:

$$S = x_1 + x_2 + \cdots + x_i + \cdots + x_n = \sum_{i=1}^{n} x_i.$$

The average score is then simply $\bar{x} = S/n$.

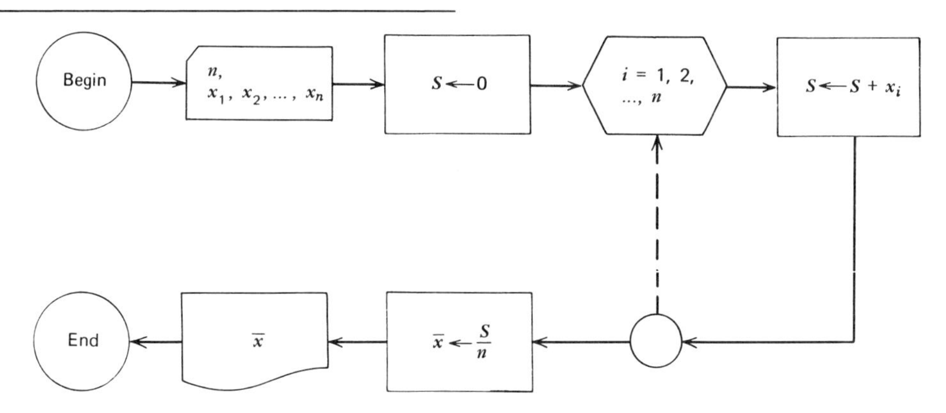

Figure 2.12 Computation of class average score.

Arrays may also be doubly subscripted, in which case they are called *two-dimensional arrays* or *matrices*. (Three or more subscripts can also be used, although the need seldom arises.) For example, suppose that a company owns n stores, and records the monthly profit for each store over a period of m months. Then, if we designate the profit made during the ith month by store j as p_{ij}, the profit records can be contained in the following matrix **P**:

$$\begin{matrix} p_{11} & p_{12} & \cdots & p_{1j} & \cdots & p_{1n} \\ p_{21} & p_{22} & \cdots & p_{2j} & \cdots & p_{2n} \\ \vdots & \vdots & & \vdots & & \vdots \\ p_{i1} & p_{i2} & \cdots & p_{ij} & \cdots & p_{in} \\ \vdots & \vdots & & \vdots & & \vdots \\ p_{m1} & p_{m2} & \cdots & p_{mj} & \cdots & p_{mn} \end{matrix}$$

Observe that the *first* subscript indicates the *row*, and that the *second* subscript indicates the *column* in the matrix. For example, the second row contains the monthly profits made by each store during the second month, and the last column contains the monthly profits made by the nth or last store during the whole period under consideration.

EXAMPLE 2.8

Assume that m, n, and the values p_{ij}, $i = 1, 2, \ldots, m$, $j = 1, 2, \ldots, n$, have already been assigned values, as discussed in the preceding paragraph. Write a segment of a flow diagram that will compute and display values for:

1. the total profit P made by all stores over all the months, and
2. the greatest profit *pmax* made by any store in any month, and the corresponding month and store numbers, *imax* and *jmax*, respectively. Assume for simplicity that the greatest profit occurs for only one store in only one month.

The above is accomplished by first setting P to zero, and also pretending temporarily, as a basis for subsequent comparison, that the greatest profit was made by the first store in the first month—that is, $pmax = p_{11}$, $imax = jmax = 1$. Furthermore, it is necessary to establish a systematic way of examining each element p_{ij} in turn. As shown in Fig. 2.13, *nested iteration* is used. The first, or outer box, which terminates at the junction α, controls the row subscript i; the second, or inner box, which terminates at the junction β, controls the column subscript j. Then, to achieve (1), p_{ij} is simply added to the current value of P. And, to achieve (2), if p_{ij} exceeds the greatest profit *pmax* found thus far, p_{ij} is stored in *pmax*, and the corresponding values of i and j are put in *imax* and *jmax*, respectively; if not, the current values of *pmax*, *imax*, and *jmax* remain unchanged.

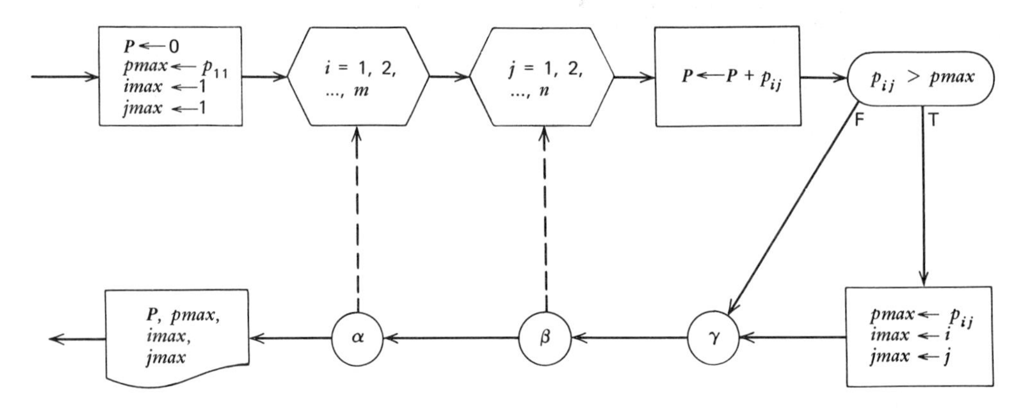

Figure 2.13 Determination of total and greatest monthly profits.

The sequence implied in Fig. 2.13 can be understood by noting that the inner (or innermost, if more than two) iteration loop must always be completed first. Thus, with $i = 1$ in the outer loop, and $j = 1$ in the inner loop, the necessary calculations are performed using p_{11}. Still with $i = 1$, j is incremented by one and so p_{12} is the next element under consideration. The column subscript j is further incremented repeatedly until its last value, $j = n$, is reached. Control now reverts to the outer loop, via junction α. Then, with $i = 2$, the inner loop is repeated in its entirety, and so on. The net effect of this nested iteration is to scan the elements of the matrix *row by row*. In this particular example, the two iteration boxes could have been interchanged, in which case the process would have been *column by column*. It is also conventional and more concise, but somewhat less clear, to combine the three

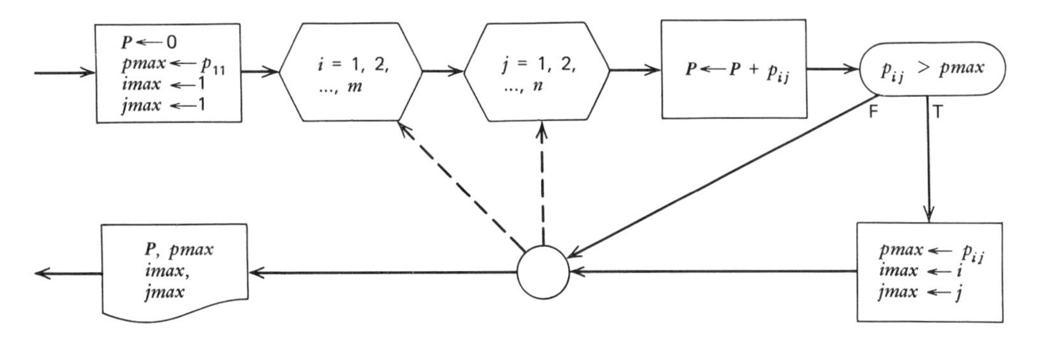

Figure 2.14 Alternative version of Fig. 2.13.

junctions α, β, and γ, into a single box, shown in Fig. 2.14.

2.9 Procedures

It frequently happens in numerical work that we wish to repeat a particular sequence of operations, but applied to different sets of variables, arrays, constants, and so on. There are certain operations, such as taking the square root or logarithm of a number, that are so standard as to have their own symbols. For example, $\sqrt{x}$, $\sqrt{7.95}$, and $\sqrt{b^2-4ac}$, all imply the same *procedure*, namely, that the square root of a quantity is being taken, but that the quantity itself, whether it is the value of the variable x, the constant 7.95, or the value of the expression b^2-4ac, may vary from one reference to the procedure to the next. Example 5.2 presents a definite algorithm for evaluating the square root of any positive real quantity. It is clearly desirable to preserve the generality of such an algorithm, for we can imagine the confusion that would arise if a whole set of such algorithms were to be developed, the first of which could only evaluate $\sqrt{x}$, the second could only evaluate $\sqrt{7.95}$, the third could only evaluate $\sqrt{b^2-4ac}$, and so on.

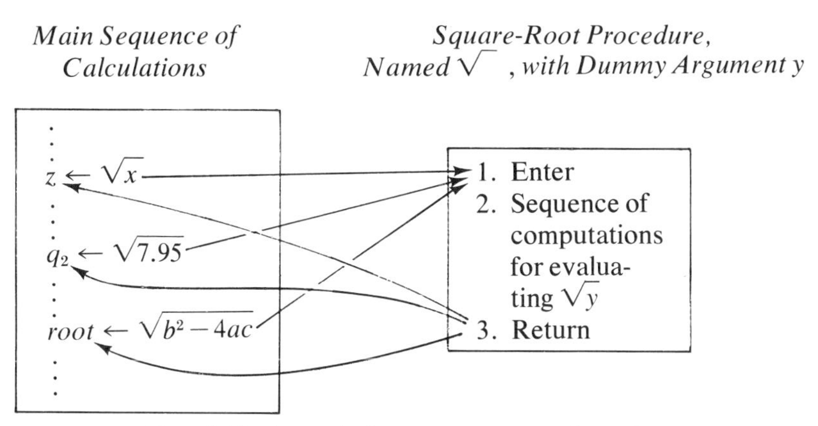

Figure 2.15 Repeated references to a square-root procedure.

Fig. 2.15 shows schematically how a procedure for evaluating square roots might stand in relation to a main sequence of calculations that makes several references to the procedure. Suppose that the square-root procedure, named $\sqrt{\ }$ for short, is formulated algebraically for finding the square root of a hypothetical quantity or *dummy argument* y. The understanding is that whenever a reference is made to $\sqrt{\ }$, y is automatically replaced wherever it appears in the sequence of calculations by the *value* of the actual quantity or *calling argument* whose square root is required. Thus the intention in the first reference to the procedure, which appears in the operation $z \leftarrow \sqrt{x}$, is:

(a) Leave the main sequence of calculations, enter the procedure $\sqrt{\ }$, and substitute the value of x (the calling argument) wherever y (the dummy argument) appears in the sequence of calculations.

(b) Perform these calculations, which now effectively find the square root of x.

(c) Return to the main sequence of calculations with the square root thus found, and store it in z.

The second and third references occur likewise, except that in the third, it is understood that the whole expression b^2-4ac is first evaluated, and that this single value is then transmitted to $\sqrt{\ }$ and used as the value of y.

In formulating a flow diagram for conveniently representing a procedure, which will be distinct from any flow diagram or diagrams in which reference is made to the procedure, the following basic items must be considered:

(a) A *name* must be selected for identifying the procedure. There are no hard-and-fast rules, although for clarity it is convenient to use a short mnemonic consisting of capital letters. For example, if the sign $\sqrt{\ }$ had not been invented, the name SQRT might be used for identifying the square-root procedure; however, the name XYZ could equally well be used, as long as it is quite clear that XYZ is indeed the name of the square-root procedure.

(b) An appropriate dummy argument or arguments must be selected; any convenient names can be used for the dummy arguments, which are usually names of variables or arrays. Note incidentally that a dummy argument can *never* be a *constant*, since this would preclude its replacement by a calling argument.

(c) It must be clear where the procedure starts and finishes. Circles inscribed with the words *Enter* and *Return* will be used here. The return box will also contain the value associated with the procedure that is to be returned to the main or referencing sequence.

EXAMPLE 2.9

Draw a flow diagram for a procedure that will compute the arithmetic mean or average of the numbers stored in a vector.

We shall arbitrarily decide to name the procedure AVG; there will be *two* dummy arguments in this case, one for the name of the vector (choose v, for example), and one for the number of elements it contains (choose m, for example). The flow diagram is given in Fig. 2.16.

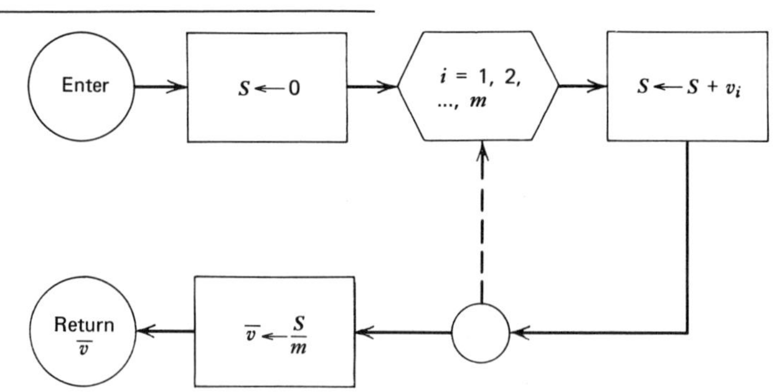

***Figure 2.16 Flow diagram for procedure* AVG(v, m).**

The particular way in which AVG is referenced from another flow diagram (probably one containing a main sequence of calculations, encompassed by circles inscribed with *Begin* and *End*) will depend on the problem at hand. Referring back to the problem of Example 2.7, if the individual scores for a class of n students are contained in $x_1, x_2, \ldots, x_n$, the average score $\bar{x}$ is found simply as

$$\bar{x} \leftarrow \text{AVG}(x, n).$$

Here, the calling arguments are x, the name of the vector, and n, the number of elements. At the time of referencing AVG, x, and n are automatically assumed to replace v and m, respectively, wherever they appear in the procedure. Note that the dummy and calling arguments must correspond in number, order, and type. In the present example, there are two arguments: the first is an array name and the second is an integer. The following reference, in which x and n are accidentally interchanged, would be improper (unless AVG were rewritten so that the order of its arguments were m, v, instead of v, m):

$$\bar{x} \leftarrow \text{AVG}(n, x).$$

A calling argument, in contrast to a dummy argument, does not necessarily have to be a single variable or array name. If it makes sense, it could be a constant or expression, for example. Thus, if we wish to find the average score of the first ten students in the class, the following reference could be used:

$$\bar{x} \leftarrow \text{AVG}(x, 10).$$

EXAMPLE 2.10

Two classes A and B contain na and nb students, respectively. Their scores on a particular examination are to be stored in the vectors a and b, respectively. Draw a main flow diagram that will accept this information and display a message as to which class has the higher average score; assume for simplicity that there is no chance of a tie.

The required flow diagram is given in Fig. 2.17. It refers to the procedure AVG developed in Example 2.9, and illustrates that the procedure reference need not necessarily appear in a substitution operation.

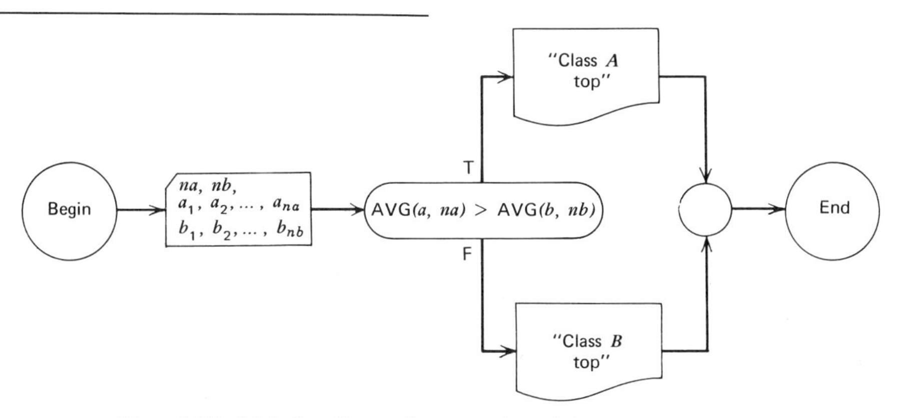

Figure 2.17 Main flow diagram for comparison of class average scores.

EXAMPLE 2.11

This final example illustrates how a procedure can be modified to return *two or more* values. Draw the flow diagram for a procedure, named PROFIT, that will start from dummy arguments m, n, and p, as defined in Example 2.8, and that will proceed to compute and return values for P, $pmax$, $imax$, and $jmax$, again as defined in Example 2.8.

Since four values must now be returned, and only one value (at most) can be associated directly with the name of the procedure, PROFIT, we simply include at least three of the four items in the argument list. There are two basic possibilities:

1. Associate one value, P, for example, with the name PROFIT, and include the

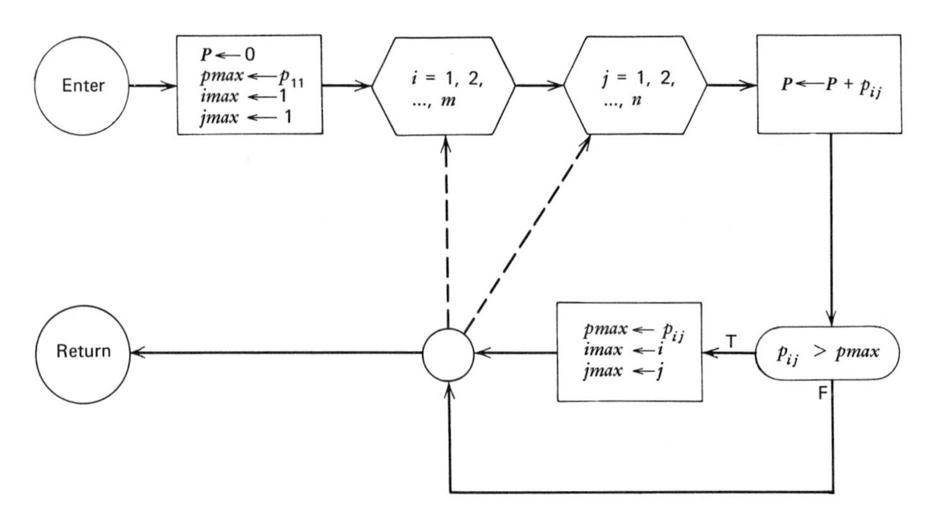

Figure 2.18 Flow diagram for procedure PROFIT *(m, n, p, P, pmax, imax, jmax).*

names of the other three in the list of arguments. A typical reference to PROFIT would then be

$$P \leftarrow \text{PROFIT}(m, n, p, pmax, imax, jmax),$$

it being understood that upon return from the procedure, the variables *pmax*, *imax*, and *jmax* will have been assigned their appropriate values. (Here, for simplicity, the calling and dummy arguments have been given the same names, but this is not essential, of course.)

2. Associate *no* value with the name PROFIT, and include all four names in the argument list. A typical reference would then be

$$\text{PROFIT}(m, n, p, P, pmax, imax, jmax).$$

Either (1) or (2) is equally acceptable. However, the flow diagram of Fig. 2.18 conforms to choice (2), to emphasize that no single value need be associated with the name of the procedure. Note that the arguments serve two purposes: to transmit information *to* the procedure, and also to retrieve information *from* it. Those arguments used for retrieving information will appear somewhere in the flow diagram to the left of a substitution arrow. It should be mentioned that upon return from PROFIT, the variables m and n, and the matrix **P** will still have their original values left intact. A reference of a form such as PROFIT$(P, pmax, imax, jmax, m, n, p)$, or PROFIT$(m, n, pmax, imax, jmax, p, P)$ could also be used as long as care is taken to preserve a one-to-one correspondence between the calling and dummy arguments.

Problems

2.1 A projectile is fired with velocity V at an angle θ above the horizontal. If the resistance of the air can be neglected, the horizontal and vertical distances traveled after a time t are:

$$x = ut,$$

$$y = vt - \frac{1}{2}gt^2,$$

where $u = V\cos\theta$ and $v = V\sin\theta$. Draw a flow diagram that will accept values for V and R, the desired horizontal range on flat terrain. Then determine whether the specified range can be achieved. If it can, display an appropriate value for θ and the time t taken to reach the target. If it cannot, display a warning message and the value of the maximum range R_{max} that *can* be achieved.

2.2 If the value accepted for x is 7.5, what will happen in the following sequence?

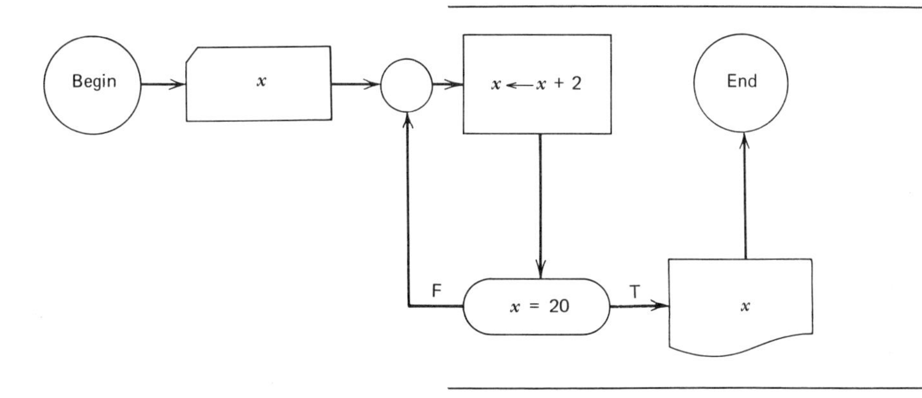

2.3 What operation does the following represent?

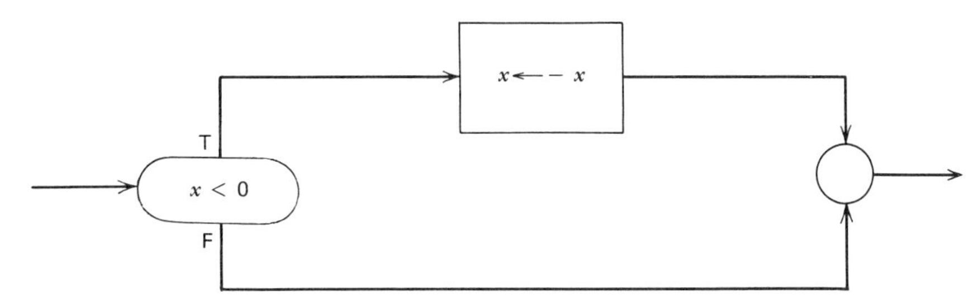

2.4 At the end of a month, when he is paid an income of \$$I$, an employee already has a current balance of \$$b$ in his checking account. If his new balance exceeds \$$d$, he wishes to transfer all this excess to his savings account, which currently has a balance \$$s$. If the new balance does *not* exceed \$$d$ (he anticipates that most of this will be needed for living expenses), he does not save for that month. Construct a flow diagram that will accept values for b, s, I, and d, carry out the desired manipulations, and display values for d, I, and the latest values for b and s. Assume that the checking account balance, after the current income has been deposited, always suffices to cover the monthly living expenses.

2.5 Modify Problem 2.4 for the case of an indigent employee, by incorporating the following additional features: (a) if the augmented or new balance \$$(b+I)$ falls below \$$d$, take enough *from* savings to bring this balance up to \$$d$, and (b) if there is not enough money left in savings to achieve (a), write a message to this effect.

2.6 What will be the value of S at the end of the following sequence if (a) $n = 9$, and (b) $n = 10$?

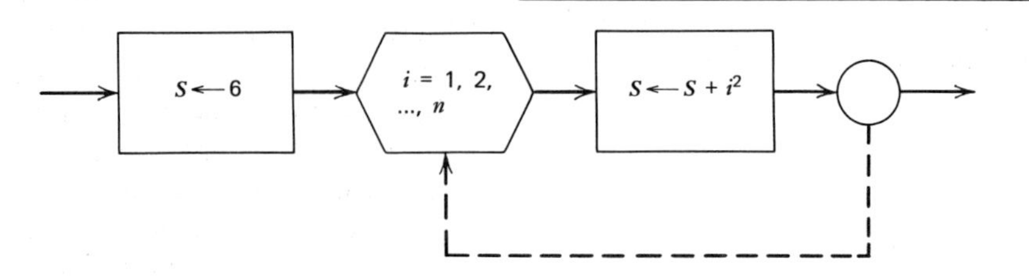

2.7 What would be the answer to Problem 2.6 if the iteration box were inscribed with $i = 3, 5, \ldots, n$ instead?

2.8 Consider the following flow diagram.

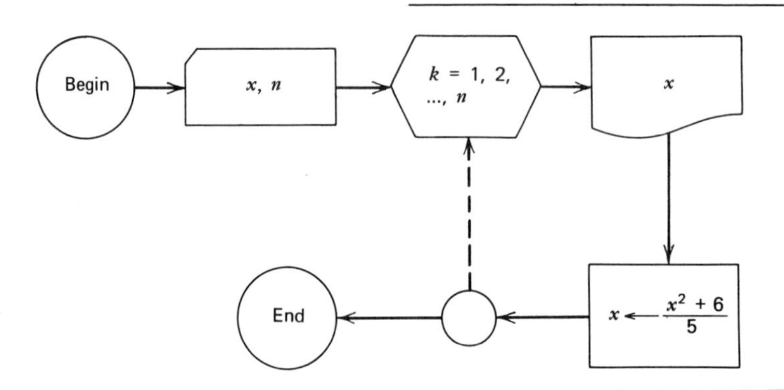

If the value accepted for x is 1.0 and the value accepted for n is 100, what values will be displayed for x during the first four iterations? Attempt to *estimate* (but do not compute) the value displayed for x during the last iteration (when $k = 100$).

2.9 Draw a flow diagram that accepts values for an integer n, a number x, and a small tolerance ε, and that proceeds to compute the sum

$$S = 1 + x + \frac{x^2}{2!} + \frac{x^3}{3!} + \cdots$$

The procedure should stop—and display the current value for S—either when the last term added is less than ε *or* when n such terms have been added together. What does the above series represent?

2.10 Consider the following matrix of values; the first subscript denotes the row (upper limit m) of a particular element and the second subscript denotes its column (upper limit n).

$$\begin{matrix} a_{11} & a_{12} & a_{13} & \cdots & a_{1n} \\ a_{21} & a_{22} & a_{23} & \cdots & a_{2n} \\ \vdots & \vdots & \vdots & & \vdots \\ a_{m1} & a_{m2} & a_{m3} & \cdots & a_{mn} \end{matrix}$$

Draw a flow diagram that will accept values for m and n, followed by all the elements in the matrix, row by row and one at a time; then, without performing any computations, display all these values, one at a time.

2.11 Refer to the matrix in Problem 2.10 and assume that all values are already in storage. Draw a flow diagram that will test to see if the matrix is square (same number of rows as columns) and, if it is, compute and display the sum of the squares of the diagonal elements, $S = a_{11}^2 + a_{22}^2 + \cdots$. If it is not square, compute and display the arithmetic average of all those elements in the matrix that are negative.

2.12 An instructor has m students in his class and gives a total of n examinations during the semester. Each examination carries a maximum of 100 points. The score is s_{ij} points for the ith student in the jth examination. Write an algorithm, in flow diagram form, that will accept values for m, n, and all the s_{ij}. Then compute the total score t_i and the corresponding average score a_i for each student ($i = 1, 2, \ldots, m$), *excluding* the examination on which the student obtained the *lowest* score. Finally, display all input data and paired values i and a_i for the whole class.

2.13 In Problem 2.12, discuss the feasibility of displaying the scores, i and a_i, in descending order of magnitude (by reordering or sorting the a_i), such that $a_1 \geqslant a_2 \geqslant \cdots \geqslant a_m$.

2.14 Consider the flow diagram of Fig. 2.12. If $n = 4$, $x_1 = 6$, $x_2 = 4$, $x_3 = 7$, and $x_4 = 3$, show in a table how the values of i and S vary at each stage of iteration.

2.15 Suppose that you have just been assigned a computer problem to work in a course. Draw a flow diagram giving the algorithm that shows the basic steps in the completion of your assignment, starting with the problem statement and concluding, hopefully, with the correct computer output.

2.16 Discuss whether or not in Example 2.8 the first box in Fig. 2.13 could have contained *only* the substitution operations $P \leftarrow 0$ and $pmax \leftarrow 0$, with the rest of the flow diagram unaltered.

2.17 A matrix with typical element a_{ij} contains m rows and n columns, as in Problem 2.10. Draw a flow diagram for a procedure, named MAXMIN, that will find the maximum and minimum values, *amax* and *amin*, respectively, in the matrix. Anticipate that a typical reference to the procedure will be

$$\text{MAXMIN}(m, n, a, amax, amin)$$

How would the procedure be referenced for the case of a matrix containing five columns and three rows, with typical element p_{ij}?

2.18 If you have studied Example 5.9, write the flow diagram for a procedure, named CUBEROOT, that will find the cube root of an argument y. That is, the procedure heading will be CUBEROOT(y). Give three valid references to the procedure.

2.19 In Ruritania, the individual income tax laws are simple. Consider a person with a gross annual income of $\$I$ who has n dependents; $\$w$ have already been withheld during the year by his employer, and the person wishes to know the amount of tax $\$t$ that is still to be paid.

The only deductions allowed amount to \$1000 for each dependent, up to a maximum of three. The resulting net income, $\$d$, is then taxed at a percentage rate of $d/1000$ unless d exceeds \$50,000, after which a fixed rate of 50% is used.

Draw a flow diagram that accepts values for I, n, and w, and that proceeds to compute and display the value of t; if t is negative, the message "Refund due" should be displayed.

2.20 Draw a flow diagram for a procedure, named TAX, that performs the Ruritanian income tax calculations described in Problem 2.19. The dummy arguments should be I, n, and w, and the procedure should: (a) return the value of t, and (b) display the message "Refund due" if t is negative. Thus, typical references to the procedure might be:

$$t \leftarrow \text{TAX}(I, n, w)$$
$$t \leftarrow \text{TAX}(10000, 0, 1200)$$
$$x \leftarrow \text{TAX}(75000, 4, 12000)$$

In the last two cases, what values would be stored in t and x?

2.21 Suppose that a person was born on the ith day of the jth month in the year k A.D. ($k \geqslant 1900$). Draw a flow diagram that represents the algorithm for computing and displaying the corresponding value for d, the day of the week on which his birth occurred (let Sunday = 1, Monday = 2, etc.).

2.22 Draw a flow diagram for representing the algorithm for some procedure with which you are familiar, such as: (a) deciding which courses to take next semester, or (b) the steps taken in driving to work, etc. Allow for all reasonable contingencies, such as courses being closed, car breaking down, and so on.

CHAPTER 3

A Synopsis of the FORTRAN-IV, WATFOR, and WATFIV Languages

3.1 Introduction

FORTRAN-IV is a procedure-oriented language (see Section 1.11) having a well-defined *syntax* or grammar that was developed to allow the unambiguous description and implementation of algorithms, particularly those of a numerical or scientific nature. Without question, FORTRAN-IV is the most popular programming language in use today. This popularity stems from three historical facts: (1) FORTRAN-IV is a direct descendent of FORTRAN-I, the first procedure-oriented language, (2) the FORTRAN family of languages was developed by the International Business Machines Corporation, whose sales have regularly accounted for more than half of all computer sales in the United States and probably around the world, and (3) in order to remain competitive in both hardware and software, other manufacturers have developed FORTRAN-IV compilers for their computing machines. As a result, virtually every manufacturer supplies a FORTRAN-IV compiler for each machine in his product line. Almost by default, FORTRAN-IV has become *the* universal computing language, while many "universal" computing languages have come and gone. This is not to say that FORTRAN-IV will never be displaced by better and more powerful procedure-oriented languages such as PL/I (also developed by IBM), but that its continued use for years to come seems assured.

We shall begin with a brief description of the development of the FORTRAN family of programming languages. The first mention of a "FORmula TRANslation System" seems to have occurred in an internal IBM report, dated late in 1954 [44]. The first translator for the original FORTRAN language, now called FORTRAN-I, was released in 1957 for use on the first of IBM's large-scale core-memory computers, the IBM 704. An improved translator that allowed the user to define his own functions and subroutines (see Sections 3.30 and 3.33) was released in 1958, and this expanded FORTRAN-I language was called FORTRAN-II. A FORTRAN-III language appears to have been developed and employed internally at various IBM installations, but was never released for customer use. In 1962, the FORTRAN-IV language specifications, essentially the ones currently in use, were published; by 1965 most manufacturers were committed to producing compilers for some version of FORTRAN for all of their machines.

Unfortunately, there has never been complete agreement among the various professional groups and computer manufacturers on exactly what the FORTRAN language rules are. Some attempts at standardization have been made. For example, documents describing a standard FORTRAN and BASIC FORTRAN, corresponding roughly to FORTRAN-IV and FORTRAN-II, respectively, have been prepared by the American Standards Association. In this case, BASIC FORTRAN is a *subset* of (that is, is completely contained in) ASA-FORTRAN [45]. However, each manufacturer seems to introduce a few nuances into his version of FORTRAN; to be fair, these are usually extensions of the standard language rather than violations of the stated standards. These variations lead to many "dialects" of FORTRAN. Often, the same manufacturer will provide compilers for different models of his computers that require the programmer to write in different FORTRAN dialects. It is, in fact, not at all unusual for the manufacturer to provide more than one compiler for the *same* machine, which are inconsistent in their interpretation of the syntactical rules for FORTRAN. And, of course, many compilers are not written by the manufacturer at all, but rather by a variety of software houses, industrial and governmental computing installations, and university computing groups.

Thus, there are many FORTRAN languages in use at the present time. This leads us to the conclusion that FORTRAN is not the machine-independent language that we would hope it to be. However, the variations from one dialect of FORTRAN to another tend to be rather small. With minimal effort it is usually possible to convert a FORTRAN program in one dialect to an equivalent one in another dialect. In some cases, translating programs have been developed to do this automatically. In this chapter, we shall use the term *FORTRAN* to refer to a particular FORTRAN-IV dialect used on the larger machines of the IBM 360/370 family of computers, and denoted as level G or level H FORTRAN-IV by IBM.

Most large computing installations will have more than one FORTRAN compiler to serve the differing needs of its customers. For example, at the University of Michigan three such compilers are available in public files (see Section 4.7) in the computing system library. Two of these translators are supplied by IBM and are called the G- and H-level compilers, respectively. Programs that adhere to the syntactical rules for FORTRAN outlined in this chapter can be translated by either compiler, and *results* produced by the object programs generated by these two compilers will be identical. The *object programs* will *not* be identical, however. The H-level translator is an "optimizing" compiler: that is, special effort is made to produce a very efficient object or machine-language program, in the sense of the execution time required; the price to be paid is that compilation with this compiler is much more expensive than with the G-level compiler. If the object program is expected to be run many times, then it may be worth the additional compilation expense to reduce running costs for the object program later on. For debugging runs, that is, runs for checking out programs, and for programs with short execution times such as student programs, its use could not be justified economically.

In response to the need for extremely fast translation of large numbers of university student programs, the staff of the Computing Center at the University of Waterloo, Ontario, has developed compilers for dialects of FORTRAN called WATFOR (WATerloo FORtran) and WATFIV [46]. These compilers are of the load-and-go type, that is, translation of the source program, and loading and execution of the object program are all handled by the compiler, which remains in the machine's memory during execution of the object program. The advantage of this approach is that the object program can be monitored continuously by the compiler, which can then produce excellent *run-time diagnostics* or error messages. The object program *cannot* be saved and is not particularly efficient, so that execution times will be longer than for object programs generated by the G-level IBM compiler and considerably longer than for those produced by the optimizing H-level compiler. The source program must be retranslated every time it is to be used. This may seem inefficient, but compilation times are very short, system overhead for the loading of the object program is virtually eliminated, and the total cost of producing an error-free program using the WATFOR or WATFIV compilers may be substantially lower than for the IBM compilers described earlier.

Because of their excellent diagnostic messages and their low computing charges for compiling, debugging, and executing student programs, the WATFOR and WATFIV compilers have achieved wide acceptance at major American and Canadian universities, particularly those with large IBM Series 360/370 computers. The WATFOR language is a dialect of FORTRAN-IV that includes virtually all the characteristics of IBM FORTRAN-IV, and some additional features, particularly in the area of input and output. The WATFIV language is an extension of the WATFOR language that includes WATFOR as a subset; that is, virtually all WATFOR programs are also WATFIV programs, but not *vice versa*. The major differences between WATFOR and WATFIV are the latter's character-manipulation features, a weak area in the WATFOR and IBM FORTRAN-IV languages. At the University of Michigan, a translator for WATFIV programs is available.

This information about the computing situation at the University of Michigan *vis-à-vis* FORTRAN compilers is included only to illustrate that engineers using large computing systems are likely to have a variety of FORTRAN dialects and compilers to choose from.

Because IBM FORTRAN-IV, WATFOR, and WATFIV are so widely used, we have elected to describe all three languages in a single chapter, and will refer to them as *FORTRAN*, *WATFOR*, and *WATFIV*, respectively. Unless otherwise stated, any syntactical rules ascribed to *FORTRAN* should be assumed to apply to *all three languages*. Features peculiar to one or two of the three languages will be so identified. The languages will be described rather tersely in synoptic rather than primerlike form. We would expect an instructor using our text to supplement this material with additional short examples. For those learning the material by self-study, we suggest reference [47] as an auxiliary text; that primer contains a large number of short examples of graded difficulty and uses the flow-diagramming conventions introduced in Chapter 2.

Before embarking on a description of FORTRAN, WATFOR, and WATFIV, it seems appropriate to discuss the nature of procedure-oriented languages and their similarities, if any, to natural languages. It is essential to understand that the syntax for a computer programming language is completely arbitrary. The syntactical rules that rigorously characterize what character strings constitute a grammatically legal program are usually drawn up by a committee. Thus the language rules depend on a variety of influences, not the least being personal opinions or prejudices of those formulating them. Language design must still be considered more an art than a science; consequently, some rules will seem very artificial or even silly, particularly to the beginning programmer. Nevertheless, *the rules are inviolable*, and must be followed to the letter. In this respect, programming languages are similar to artificial international languages, such as Esperanto and Interlingua, defined by committees of language specialists. As an illustration, Esperanto contains a rule on the formation of plural nouns: "*All* plural nouns end with the letter j; *only* plural nouns may end with the letter *j*." The rule is obviously quite arbitrary. Nevertheless, because it is inviolable, there is never any question about the identity of plurals; compare this with the irregularity of plural formation in English (books, sheep, men). The rigid syntax of such languages makes them much easier to learn than natural languages, which are replete with irregularities and ambiguities; of course the *value* of learning such artificial languages is open to question.

There are some analogies between a program written in a procedure-oriented language and a paragraph written in a natural language. For example, each is composed from a finite set of symbols that includes letters, digits, punctuation marks, and parentheses; in FORTRAN, this collection of symbols is called the *character set.* In a natural language, the basic element with a distinct meaning is called a word; words are assigned an identity as to type, such as noun, verb, or preposition. In FORTRAN we refer to the basic elements as *operands, operators, character strings, delimiters,* etc. For example, the FORTRAN statement X = 3.4 + Y contains the three operands X, 3.4, and Y with associated numerical values, and the two operators, = and +, which act on the numerical values associated with the operands; note the (rather tenuous) analogy between verbs and nouns in a natural language and operators and operands in FORTRAN. In a natural language, the next level of complexity beyond the individual word is usually called a phrase, whereas in FORTRAN the proper combination of operators and operands (for example, 3.4 + X) below the level of a complete thought is called an *expression.* A complete thought in a natural language is called a sentence, whereas the analog in FORTRAN is called a *statement.* A collection of FORTRAN statements that describes a complete algorithm is called a FORTRAN *program,* whereas the comparable entity in a natural language would be called a paragraph, or perhaps a chapter.

Two rather important differences between natural languages and well-designed programming languages should also be noted:

1. In a natural language there is often an unambiguous meaning associated with forms that violate the syntax or grammar. For example, the collection of English words "We ain't got none of them posters left" clearly violates elementary English grammar. While some might claim that since the grammar is violated, the word collection is not "English," the fact remains that anyone fluent in English would understand that the posters are no longer available. A comparable syntactical error in FORTRAN will render the FORTRAN statement *absolutely meaningless*, that is, only *perfect* FORTRAN is acceptable to the FORTRAN compiler. The compiler can translate *all* grammatically correct symbol strings; it cannot translate *any* grammatically incorrect symbol string.
2. In a natural language, grammatically correct sentences or a poorly written but grammatically correct paragraph can still contain ambiguity for the reader; this may occur because of a poor choice of phrasing or word order, or because the same symbol string may have several meanings associated with it, as in the sentence "I have a bill." Here, "bill" might refer to an invoice, a draft for a proposed statute, a piece of paper money, a public notice, and several less likely possibilities such as a bird's beak and a medieval weapon. The *meanings* of legal symbol strings are usually called the *semantics* of a language; clearly, natural languages have a very complex semantics. Unfortunately, the semantics of most computing languages has not been formalized adequately, but it is much simpler for a language such as FORTRAN than for a natural language. In general, the programmer should view any syntactically correct FORTRAN program as having only *one* meaning.

No doubt, the comments of the preceding paragraphs distort, rather carelessly, definitions sacred to linguistics and computing language designers. Attempts to carry these analogies between programming languages and natural languages much further would lead to a hopeless quagmire of contradictions, special cases, and endless footnotes. Nevertheless, there are obvious similarities between the programmer and the author. Each attempts to create a complete and unambiguous representation of an idea from a basic set of symbols using a formal set of syntactical rules and an associated semantics.

In order to gain a better understanding of the appearance of a FORTRAN program, we suggest that the reader glance rather quickly through Example 1 of Section 3.38, before continuing with Section 3.2. The example, involving the solution of quadratic equations, will be familiar to anyone with training in algebra. If the material in Chapter 2 has been mastered, the flow diagram should cause little difficulty. Next, the reader should scan through the printed listing of the punched-card deck, particularly the part labeled "FORTRAN program" and "data," and examine, briefly, the computer output on the following page. There is no intent that the reader should *understand* the material in Section 3.38 (that will come after studying the remainder of this chapter), but only that he gain some *feeling* for the nature of a FORTRAN program, and its relationship to the flow diagram for the algorithm.

Not every feature of the FORTRAN, WATFOR, and WATFIV languages is described in this text. The major topics not covered here are:

1. The COMPLEX mode and the manipulation of complex (imaginary) numbers.
2. The handling of magnetic tapes.
3. The direct-access input and output statements that allow the user to define and access sequential (serially organized) files on secondary storage devices such as magnetic drums, disk files, and data cells.

Details of these language features are covered in references [47] and [46] for FORTRAN, and WATFOR and WATFIV, respectively. Thus, the languages described here should be viewed as rather comprehensive subsets of the FORTRAN, WATFOR, and WATFIV languages. To help the reader to differentiate among features that are peculiar to one or two of the languages, a pale gray band appears at the edge of the page beside textual material and examples that do *not* apply equally to all three languages.

Certain features of the languages are clearly more important than others, particularly for beginning programmers. It is difficult to choose the "important" topics arbitrarily, however, since the choice is largely a function of the language used, the peculiarities of the local computing installation, the kinds of problems that are to be solved, and the emphasis that is to be given to particular areas such as input and output, the manipulation of character information, and the writing of subprograms. We shall leave this choice to the individual instructor or student.

3.2 Character Set

Only the characters listed in Table 1.10 may appear in FORTRAN, WATFOR, and WATFIV programs. The principal character subsets are:

Alphanumeric Characters. *Alphabetic:* the capital letters A through Z and $
Numeric: the decimal digits 0 through 9

Special characters.
(a) + − * / ( ) = . , '
(b) ¢ < | ! ¬ % _ > ? " # @ & ; :

Blanks may also appear, but in most contexts they have no significance. The characters in set (a) play an important role in the syntax of FORTRAN while those in set (b) normally appear only in character strings to be read or printed. In FORTRAN programs, *all* characters must be placed *on* the line; for example, A_1 and X^2 are improper.

3.3 Modes or Types

There are certain quantities, such as the number of a particular month in the year, or the number of points in a bridge hand, that can assume only integral or whole-number values. Other quantities, such as the pressure in a reactor or the mass of an electron (9.107×10^{-28} gram) may require fractional parts for their accurate representation, or may require a scale factor (for example, 10^{-28}) because they would otherwise be inconvenient to write in full. We differentiate between *integer* or *fixed-point* numbers on one hand, and *real* or *floating-point* numbers on the other. This distinction—that of different *modes* or *types*—is preserved in the computer, where integer numbers and real numbers are both stored and manipulated differently. In most real machines, arithmetical operations involving integer operands are executed more rapidly than are the corresponding operations on real operands, since no attention need be paid to scale factors and decimal point location. In any event, the programmer must know, at every stage in his program, the modes of the numbers he is using. The internal forms for integer and real numbers in the IBM 360/370 computers are described in Section 1.17.

Information of two other modes or types may also be stored and manipulated in FORTRAN programs. Quantities having truth values (*true* or *false*) are said to be of *logical* mode; these will be discussed later, in Sections 3.9 and 3.10. A *complex* mode is available for handling complex arithmetic, but will not be discussed in this text.

WATFIV, but not WATFOR or FORTRAN, has features to allow the manipulation of strings of symbols from the character set of the preceding section. These strings are assigned a *character* mode and are discussed in several later sections, in particular Sections 3.27 and 3.28.

3.4 Constants

A constant is a known quantity whose value is invariant.

Integer or *fixed-point* constants are numbers that consist of a string of digits. A decimal point is assumed to lie after the rightmost digit, *but must be omitted.* A plus sign is optional, but if the integer is negative it must be preceded by a minus sign. An integer constant of *standard length* (see Section 3.27) must lie in the range −2147483648 through +2147483647. This range is determined by the characteristics of the IBM 360 storage word, as discussed in Section 1.17, and would be different for the FORTRAN languages used on other manufacturers' equipment. Leading zeros, although permissible, are ignored. No commas to separate hundreds, thousands, and millions are permitted.

Examples. `365   +0365   −36174   0   700`

Real or *floating-point* constants of the *F-type* and of standard length are numbers that consist of a string of one to seven significant digits *with* a decimal point, again possibly preceded by a sign. Leading zeros and trailing zeros used to position the decimal point are not considered to be significant.

Examples. `3.14159   453.0   453.   −0.00006321702   0.0   21367250000.`

If necessary, an F-type mantissa may have appended to it a *scale factor* in the form of the letter E followed by an *integer* constant, producing an *E-type* real constant. The integer constant indicates the power of 10 by which the mantissa is to be multiplied. The absolute value of a real constant must lie between approximately 10^{-78} and 10^{75}, or be zero (more precisely, between $16^{-64} \doteq 0.5397605 \times$

10^{-78} and $16^{63} \doteq 0.7237005 \times 10^{76}$). As was the case for integer constants, this number range is established by the characteristics of the IBM 360 storage word (see Section 1.17) and may be different for FORTRAN languages used on other computers.

Examples.	*E-Type Constant*	*Interpretation*
	9.107E–28	9.107×10^{-28}
	9107.E–31	9.107×10^{-28}
	–1.9E4	$-19{,}000.0$

There are only two constants of *logical* mode

.TRUE. and .FALSE.

which will be mentioned further in Sections 3.9 and 3.10.

A sequence of from 1 to 255 of the characters listed in Section 3.2, enclosed in primes ('), is called a *character string* or *literal constant* (in WATFIV, the name *character constant* seems to be favored). If a prime is to appear as part of the *content* of the string, then *two* primes must be shown. Use of literal constants will be described in Sections 3.20, 3.21, and 3.28.

Examples. 'DIAL 763–0300', 'LADY LOVELACE', 'DON''T DO THAT'

3.5 Variables

A variable is a quantity that may assume different values at various times during the execution of a program. The *name* of a variable, which ultimately refers to a particular memory location where the *value* of that variable is stored, consists of one through six alphanumeric characters, of which the first must be alphabetic (a letter or $).

If the variable is of *integer* mode, that is, if the memory location associated with the name is to have an integer value, the first letter of the name must normally be either I, J, K, L, M, or N. Names beginning with any other alphabetic character normally denote variables of *real* mode. However, either of these rules can be overridden, if desired, by using an INTEGER or REAL *mode declaration*, described in Section 3.27. The use of suitably descriptive names whenever possible has obvious advantages.

Examples. (integer) I, MONTH, JBOND7,
(real) X, Y, PRESUR, TIME

It is also possible to have variables of *logical* mode, in which case the variables assume one of the values .TRUE. or .FALSE., as shown in Sections 3.9 and 3.10. The naming convention is the same as for the integer and real variables, discussed above; however, as there is no reserved first letter for the names of logical variables, all such variables *must* appear in a LOGICAL mode declaration (see Section 3.27).

In WATFIV, but *not* in WATFOR or FORTRAN, variables may be assigned a *character* mode. Such variables contain the hexadecimal codes for characters shown in Table 1.10, and are said to have *character value*. The naming conventions for these variables conforms to the rules given for variables of other modes. Since no first letter is reserved for the names of such variables, they must appear in a CHARACTER mode declaration, as described in Section 3.27.

Frequently, we wish to use a single *array name* for a whole series of variables, each of which is denoted by a *subscript* as being a particular member of that array.

In FORTRAN, the subscripts are enclosed in parentheses. For example, the pressures p_1, p_2, p_3, and p_4 at each point in a four-nodal piping network might be addressed as P(1), P(2), P(3), and P(4). Here, each element of the array shares the name P, but is assigned a unique memory location. The *name* of the array follows the usual rules for variable names with regard to length and mode. All of the elements of an array have the same mode.

Examples. VOLUME(NUMBER), X(5), B17QS(JACK)

In the first of these examples, assume that the integer variable NUMBER has the value 17. Then VOLUME(NUMBER) is the seventeenth element of the real array named VOLUME. A singly subscripted array is sometimes called a *vector*, in line with common mathematical notation.

Arrays may have more than one dimension, that is, involve more than one subscript, up to a maximum of seven. The individual subscripts are separated by commas.

Examples. A(I,J), TABLE(3,9), $SALES(ITEM,NSTORE,NWARE)

Here, A(I,J) is the element in the Ith row and the Jth column of the two-dimensional table named A. In line with the usual mathematical conventions, a two-dimensional array is sometimes called a *matrix*. The last of these examples is an element of the three-dimensional array $SALES. In some accounting program, $SALES(ITEM,NSTORE,NWARE) might be used to represent the total dollar sales of merchandise with catalog number ITEM to the NSTOREth customer from the NWAREth company warehouse.

The minimum value allowed for a subscript is one; the upper limit must be specified by the programmer in a DIMENSION declaration, as described in Section 3.25. The elements of FORTRAN arrays are stored in contiguous storage locations in the memory of the computer. Since the internal addressing structure of the memory is one-dimensional, arrays having more than one subscript must be reorganized into an equivalent one-dimensional sequence for internal bookkeeping purposes. The convention adopted by the FORTRAN, WATFOR, and WATFIV compilers is to let the element with all subscripts equal to unity be the first element stored. Thereafter, each subscript in turn is varied from its minimum to maximum value, starting with the first subscript, then the second, and so on. The $2 \times 3 \times 2$ array B would be saved in the order B(1,1,1), B(2,1,1), B(1,2,1), B(2,2,1), B(1,3,1), B(2,3,1), B(1,1,2), B(2,1,2), B(1,2,2), B(2,2,2), B(1,3,2), B(2,3,2). Note that this scheme leads to the storage of matrices in column-by-column rather than the more conventional row-by-row order.

3.6 Function References

Efficient programs for evaluating most of the common mathematical functions, such as the exponential function and the trigonometric functions, are already available in FORTRAN. These programs are usually stored in the computing system library, and have the characteristics of certain kinds of *procedures* already described in Section 2.9; that is, a FORTRAN function has a *name*, an *argument* or arguments, returns a *value* associated with its name, and can be brought into effect or *executed* at appropriate points in a FORTRAN program simply by *making reference* to it. The argument or arguments are enclosed in parentheses immediately following the name; commas are used to separate arguments, if more than one is required.

The *name* of a function is established according to the rules for naming variables (see Section 3.5), and the mode of the *value* of the function is determined by the first letter of the name in the usual way. Thus, of the functions named in the table below, only IABS returns an integer value; the others return real values. The *argument* or *arguments* for these functions are arithmetic *expressions* (see Section 3.7). The *mode* of each argument is specified *a priori* and cannot be changed by the programmer. In the table below, all arguments are of real mode with the exception of that for the function IABS, which must be of integer mode.

Some of the common functions* available in FORTRAN are:

Function Name	Mathematical Significance	Function Reference	Function Value
ABS	absolute value (magnitude)	ABS ($\mathscr{E}$)	$\|\mathscr{E}\|$
ALOG	natural logarithm	ALOG ($\mathscr{E}$)	$\log_e \mathscr{E}$
ALOG10	common logarithm	ALOG10 ($\mathscr{E}$)	$\log_{10} \mathscr{E}$
ARCOS	arccosine	ARCOS ($\mathscr{E}$)	$0 \leqslant \arccos \mathscr{E} \leqslant \pi$
ARSIN	arcsine	ARSIN ($\mathscr{E}$)	$-\frac{\pi}{2} \leqslant \arcsin \mathscr{E} \leqslant \frac{\pi}{2}$
ATAN	arctangent	ATAN ($\mathscr{E}$)	$-\frac{\pi}{2} \leqslant \arctan \mathscr{E} \leqslant \frac{\pi}{2}$
ATAN2	arctangent	ATAN2 ($\mathscr{E}_1, \mathscr{E}_2$)	$-\pi \leqslant \arctan(\mathscr{E}_1/\mathscr{E}_2) < \pi$
COS	cosine	COS ($\mathscr{E}$)	$\cos \mathscr{E}$
COSH	hyperbolic cosine	COSH ($\mathscr{E}$)	$\cosh \mathscr{E}$
COTAN	cotangent	COTAN ($\mathscr{E}$)	$\cot \mathscr{E}$
ERF	error function	ERF ($\mathscr{E}$)	$\mathrm{erf}\ \mathscr{E}$
ERFC	complementary error function	ERFC ($\mathscr{E}$)	$\mathrm{erfc}\ \mathscr{E}$
EXP	exponential function	EXP ($\mathscr{E}$)	$\exp \mathscr{E}$ or $e^{\mathscr{E}}$
GAMMA	gamma (factorial) function	GAMMA ($\mathscr{E}$)	$\Gamma(\mathscr{E})$
IABS	absolute value (magnitude)	IABS ($\mathscr{E}_{\mathscr{I}}$)†	$\|\mathscr{E}_{\mathscr{I}}\|$
SIN	sine	SIN ($\mathscr{E}$)	$\sin \mathscr{E}$
SINH	hyperbolic sine	SINH ($\mathscr{E}$)	$\sinh \mathscr{E}$
SQRT	square root	SQRT ($\mathscr{E}$)	$\sqrt{\mathscr{E}}$
TAN	tangent	TAN ($\mathscr{E}$)	$\tan \mathscr{E}$
TANH	hyperbolic tangent	TANH ($\mathscr{E}$)	$\tanh \mathscr{E}$

*See Section 3.27 for a list of double-precision functions.
†The argument for the function IABS must be an *integer* expression $\mathscr{E}_{\mathscr{I}}$; otherwise all arguments must be expressions of real mode.

Taking into account the mode assignments for the arguments, function references such as SQRT(4), SQRT(I*MEAN), and IABS(ALPHA) are *invalid*, while SQRT(4.), SQRT(XI*AVAG), and IABS(JACK) are syntactically *correct*. Arguments for the trigonometric functions must be given in *radians*, rather than *degrees*. Thus, COS(30.7) will have the value of the cosine of 30.7 radians; if the cosine of 30.7 degrees were required, then the proper function reference would be COS(30.7/57.29578).

The functions listed above are probably the most important ones for the average FORTRAN programmer. However, each computing installation will have many other special-purpose functions available in its program library (saved on secondary storage devices such as magnetic tapes, disk files, etc.). These are usually described in a catalog prepared by the computing center staff. Later, in Section 3.30, we shall see that provision is made for the programmer who wishes to write his own functions using the FORTRAN language.

Examples. SIN(X + Y*Z), EXP(VOL), ATAN2(X,Y), ALOG10(122.7553)

3.7 Arithmetic Expressions

So far, we have referred to arithmetic (and logical) constants, variables, and function references. Speaking loosely, these operands are the "nouns" from which we can begin to construct "sentences" or *statements* in FORTRAN. To qualify them, relate them to one another, and assemble them into phrases or *expressions*, we need *operators*—the "verbs," "prepositions," etc. of FORTRAN.

In decreasing order of precedence (decreasing order of processing by the computer), the FORTRAN *arithmetic operators* are:

Operator	Purpose	Example	Conventional Meaning
**	exponentiation	A**B	a^b
* and /	multiplication	3.17*A	$3.17a$
	division	X/BETA	x/β
+ and −	addition	X + Y	$x+y$
	subtraction	Z − 7.5	$z-7.5$
	unary +	+ Q	$+q$
	unary − (negation)	− Q	$-q$

Any *arithmetic* (real or integer) *constant* (c), *variable* ($\mathscr{V}$), or *function reference* ($\mathscr{F}(\mathscr{E})$) is, by definition, an *arithmetic expression* ($\mathscr{E}$). Here, $\mathscr{V}$ is a simple (unsubscripted) variable, or an element of an array. These simple arithmetic expressions can then be linked into more complicated *arithmetic expressions* by introducing the arithmetic operators and *parentheses.*

The rules for forming all valid FORTRAN arithmetic expressions, $\mathscr{E}$, are shown below; here, $\mathscr{E}$, $\mathscr{E}_1$, and $\mathscr{E}_2$ denote any arithmetic expressions already formulated using the rules. Note that the rules have a high degree of recursiveness or circularity. With the exception of arithmetic constants, variables, and function references, all expressions are described in terms of other expressions.

Rules for Forming Arithmetic Expressions

1. c	5. $+\ \mathscr{E}$	9. $\mathscr{E}_1 - \mathscr{E}_2$
2. $\mathscr{V}$	6. $-\ \mathscr{E}$	10. $\mathscr{E}_1 * \mathscr{E}_2$
3. $\mathscr{F}(\mathscr{E})$	7. $\mathscr{E}_1 ** \mathscr{E}_2$	11. $\mathscr{E}_1 / \mathscr{E}_2$
4. $(\mathscr{E})$	8. $\mathscr{E}_1 + \mathscr{E}_2$	

There is one additional rule: two or more arithmetic operators must *not* appear in juxtaposition. For example, X*–Y is illegal, but could be corrected by introducing parentheses: X*(–Y). The following are examples of legal FORTRAN expressions:

FORTRAN Expression	*Conventional Meaning*
X + Y	$x+y$
A + B*X + C*X**2	$a+bx+cx^2$
EXP(PHI*(EXP(THETA) – 1.0))	$e^{\phi(e^{\theta}-1)}$
SQRT(B*B – 4.0*A*C)	$\sqrt{b^2-4ac}$
ALOG(ABS(TAN(ALPHA)))	$\log_e \lvert\tan\alpha\rvert$
A/B*C	ac/b
A/(B*C)	$a/(bc)$

Note how an expression such as SQRT(B*B – 4.0*A*C) conforms to the basic rules in the earlier table. A, B, and C are expressions according to rule 2 and 4.0 is an expression according to rule 1. Hence B*B, and successively 4.0*A and 4.0*A*C are expressions according to rule 10. B*B = 4.0*A*C is then an expression conformimg to rule 9, so that the complete sequence SQRT(B*B – 4.0*A*C) is an expression according to rule 3.

In the evaluation of a legal arithmetic expression, individual operations are performed in an order determined by the *precedence* of the operators. Certain implied operations, such as the evaluation of function references and subscripts, have the highest precedence. These are followed, in decreasing order of precedence by: (1) exponentiation, (2) multiplication and division, and (3) addition, subtraction, unary plus and minus (negation). (A *unary* operator has just one operand.) For example, in the expression X**3 – 4.5* SQRT(Y), the operations will be performed in the order: function evaluation, exponentiation, multiplication, subtraction.

Among operators of equal precedence, the processing sequence is from *right* to *left* for exponentiation and from *left* to *right* for all others. Therefore, the expression 2.3*X + Y/Z – X**Y**3 will be evaluated in the following order: Y**3, X**(Y**3), 2.3*X, Y/Z, (2.3*X) + (Y/Z), ((2.3*X) + (Y/Z)) – (X**(Y**3)). All operators must appear explicitly; for example X × Y must be written as X*Y, not as XY, since the latter would be interpreted as a single variable named XY.

Note that the operator precedence order used in FORTRAN is identical with that normally used in algebraic notation. For example, in the algebraic expressions $x^3-4.5\sqrt{y}$ and $2.3x+y/z-x^{y^3}$, the operations would be implemented in exactly the order shown in the preceding paragraphs for the equivalent FORTRAN arithmetic expressions. Analogously, *parentheses* may be used to override the usual precedence rules. For example, in the algebraic expression $a+b(c^2-d/e)$, the parentheses insure that all operations *inside* the factor (c^2-d/e) will be performed before the multiplication operation *outside* the factor is implemented. Thus the sequence of operator processing is: exponentiation, division, subtraction, multiplication, addition. Without the parentheses, the order would be: exponentiation, multiplication, division, addition, subtraction.

Observe that *inside* the parentheses the usual precedence rules apply, that is, the order is: exponentiation, division, subtraction. The FORTRAN equivalent of the parenthesized expression is A + B*(C**2 – D/E); note that multiplication and exponentiation are not implied by position—each operator must appear explicitly. The third and last of the FORTRAN expressions in the preceding table illustrate the use of parentheses to override the usual operator precedence order.

The FORTRAN programmer will soon discover that parentheses are needed in FORTRAN arithmetic expressions exactly where they are required in the equivalent algebraic expressions. One minor exception occurs in the use of negative exponents for the exponentiation operator; these must be enclosed in parentheses to meet the rule that no two arithmetic operators may be juxtaposed. For example, one might write the algebraic expression $x^{-0.72}$, but the equivalent form X** −0.72 is not valid FORTRAN; the proper form is X**(−0.72). The general rule for parenthesis use is: any expression within a parenthesis pair will be completely evaluated before an operator outside the parenthesis pair will be applied to the parenthesized expression.

When the operands for a binary operation (here, *binary* implies that the operator requires *two* operands, rather than that the calculations are being performed in binary arithmetic) are both of real mode, the machine-language program generated by the compiler will employ the floating-point hardware to carry out the operation. For example, the expressions 3.5*X and ALPHA + BETA would be computed using the floating-point (real) multiplication and addition instructions, respectively; in both cases, the result of the computation will be of real mode.

In comparable fashion, when both operands for an operation are of integer mode, integer instructions will be used to calculate the value of the expression; the result will be of integer mode, so that any fractional part of the result will be lost (see Section 1.16 for a discussion of integer operations). The result is truncated to the next lower integer in absolute value. The programmer must exercise care in two cases, *division of one integer by another*, and *exponentiation of an integer by a negative integer*. For example 7/4 is evaluated as 1, −25/9 as −2, J**(−2) as zero, etc.

Digital computers do not normally have machine-language instructions for handling operands of different modes. Thus, while there will be an instruction for adding two integer operands, and another instruction for adding two floating-point or real operands, there will *not* be an instruction for addition of an integer operand to a floating-point operand. Nevertheless, FORTRAN *does* allow the mixing of integer and real operands in arithmetic expressions, so that expressions such as 3.4 + 4, I/ALPHA, and X**2 are syntactically correct. Whenever an operator is applied to two operands, *one* of *integer* mode and the *other* of *real* mode, the operand of *integer* mode is converted to *real* mode, the calculation is performed using *floating point arithmetic*, and the result is of *real* mode.

Thus, if even *one* operand in a complicated arithmetic expression is real, then the final value of the expression will be real. The programmer must be careful when he mixes modes, however, as parts of the expression may still be processed using integer operations. For example, 5/4 + 2.5 will be evaluated as follows: (1) since division has precedence over addition, the expression 5/4 will be calculated first, using *integer division* (both operands are integer constants) yielding the *integer* value 1; (2) before the addition operation can be performed, the integer 1 will be converted to real form automatically by instructions in the object program produced by the FORTRAN translator; (3) the floating-point addition operation will produce the *real* sum $1.+2.5=3.5$. If the expression were written as 5./4 + 2.5 or 5/4. + 2.5, or 5./4. + 2.5, the resulting value would be 3.75, since floating-point rather than integer division would have been used. To illustrate these points further, suppose that two integer variables, I and J, have values 4 and 10, respectively, and that a real variable, BETA, has a value 2.5.

Then:

Expression	Value	Mode
J/I	2	Integer
BETA*J/I	6.25	Real
J*BETA/I	6.25	Real
J/I*BETA	5.	Real

Note that the last expression appears to be identical with the second and third expressions; however, the order of processing produces different results.

The conversion of an integer operand to the equivalent real form requires several individual machine operations. Therefore, the programmer should not mix modes unnecessarily. An exception to this rule of thumb occurs in the raising of a real operand to an integral power. For example, X**3 would be preferred to the apparently identical expression, X**3.0. For small integer exponents the result is computed by repeated multiplication of the base. When a real exponent is used, the result is computed by calling on the function ALOG to evaluate the logarithm of the base, next multiplying the logarithm by the exponent, and then calling on the function EXP to generate the antilogarithm. The latter procedure requires more computing time than the former, and is invalid when the base is negative.

Mode conversions can be effected directly by referencing one of the functions IFIX and FLOAT. If $\mathscr{E}_{\mathscr{R}}$ and $\mathscr{E}_{\mathscr{I}}$ are, respectively, expressions of real and integer mode, then the function reference IFIX($\mathscr{E}_{\mathscr{R}}$) has as its value the integer-mode equivalent of the real-mode argument $\mathscr{E}_{\mathscr{R}}$. The function reference FLOAT($\mathscr{E}_{\mathscr{I}}$) has as its value the real-mode equivalent of the integer-mode argument $\mathscr{E}_{\mathscr{I}}$. Thus, SQRT(FLOAT(I)) is a valid FORTRAN function reference, while SQRT(I) is not.

3.8 Subscripts

Subscripted variables or arrays have been introduced in Section 3.5. Each subscript must ultimately be interpreted as an integer. For example, in a two-dimensional table, there is a first row and a second row, but no row with a fractional index between them. In most cases, the programmer will use simple integer constants or variables for subscripts.

Examples. TABLE(2,3), A(I,J,K), TEMP(LAST)

However, FORTRAN allows *any arithmetic expression* formulated according to the rules outlined in Section 3.7 to be used as a subscript, provided only that it have a value in the valid range, described below. Therefore, subscripts may involve other subscripted variables or quite arbitrary integer expressions.

Examples. BETA(J(K)), X(N), TABLE(3*I+4,(J+K)/2), MARY(J**2–MARY(K*MOE+J))

The individual subscript expressions are evaluated using integer arithmetic.

Even real or mixed-mode expressions are allowed.

Examples. X(3.5*J), Y(P*ZETA–2.5), A(SQRT(R**2 + S**2) , JOE)

The subscript expressions are evaluated in the usual way, and the real value is then truncated to the next smaller integer in magnitude.

The smallest subscript for any array is 1; the largest value for each subscript is specified in a DIMENSION declaration (see Section 3.25). Therefore, to be considered valid the integer value of each subscript expression must be larger than zero and less than or equal to the maximum value specified in the dimensioning information for the array.

Examples (illegal). A(–4,J), X(0)

3.9 Simple Logical Expressions

We have already mentioned the two (and only two) logical constants, .TRUE. and .FALSE., and the fact that a logical variable $\mathscr{V}_{\mathscr{B}}$ may assume either of these values. The programmer will soon discover, however, that he makes more frequent use of *logical expressions* than of logical constants or variables in the making of decisions.

A simple logical expression, also known as a simple *Boolean* expression, is of the form

$$\mathscr{E}_1 \;.\mathscr{R}.\; \mathscr{E}_2$$

Here, $\mathscr{E}_1$ and $\mathscr{E}_2$ are any arithmetic expressions and $.\mathscr{R}.$ is a *relational operator*, available in six forms:

Relational Operator	*Algebraic Equivalent*
.LT.	$<$
.GT.	$>$
.EQ.	$=$
.LE.	$\leqslant$
.GE.	$\geqslant$
.NE.	$\neq$

For example, if we have variables X = 1., Y = 3.5, I = 2, and INT = 7, then:

Simple Logical Expression	*Value*
I .EQ. INT	.FALSE.
INT/X .GE. Y	.TRUE.
X*Y .NE. INT + 1.	.TRUE.

An additional form of simple logical expression, involving character variables and/or literal constants and the relational operators, is available in WATFIV (but not in FORTRAN or WATFOR), and is discussed in Section 3.28.

3.10 Compound Logical Expressions

Compound logical expressions, also known simply as *logical* or *Boolean expressions*, can be formed from suitable combinations of logical constants, variables, and function references, simple logical expressions, and logical operators.

Using $\mathscr{B}$, $\mathscr{B}_1$, and $\mathscr{B}_2$ to denote logical expressions, either simple or compound, the FORTRAN *logical operators* .NOT., .AND., and .OR. are defined as follows, in decreasing order of precedence:

Logical Operator	*Use*
.NOT.	Logical negation (a unary operator)
.AND.	The expression $\mathscr{B}_1$ .AND. $\mathscr{B}_2$ has the value .TRUE. if and only if both $\mathscr{B}_1$ *and* $\mathscr{B}_2$ are *true*. Otherwise, the expression has the value .FALSE.
.OR.	The expression $\mathscr{B}_1$ .OR. $\mathscr{B}_2$ has the value .TRUE. if either $\mathscr{B}_1$ or $\mathscr{B}_2$ (or both) is *true*. Otherwise, the expression has the value .FALSE.

The .AND. and .OR. operators and their truth tables are described in detail in Section 2.6.

A single logical constant or variable represents the simplest form of logical expression. If the script symbols have the meanings already established, and $\mathscr{F}_\mathscr{B}$ is the name of a function having a logical value, the following represent all possible FORTRAN logical expressions:

$\mathscr{F}_\mathscr{B}$( )	$\mathscr{E}_1$.$\mathscr{R}$. $\mathscr{E}_2$	($\mathscr{B}$)	$\mathscr{V}_\mathscr{B}$	
.NOT.$\mathscr{B}$	$\mathscr{B}_1$ .AND. $\mathscr{B}_2$	$\mathscr{B}_1$ .OR. $\mathscr{B}_2$	.TRUE.	.FALSE.

These rules and the precedence-ordering of the logical operators are illustrated by evaluating the following logical expression, shown to have the value .TRUE. if, for example, X = 9.6, Y = 67.4, and I = 3:

```
X.GT.Y   .AND.   I.NE.2   .OR.   .NOT.   SQRT(Y).LE.4.7
.FALSE.          .TRUE.                  .FALSE.
                                  .TRUE.
       .FALSE.
                    .TRUE.
```

3.11 Complete Precedence Table for Operators

The FORTRAN operators may be listed in a single table, shown below, in decreasing order of precedence. Parentheses appear at the top of the list to indicate their role in overriding the usual precedence order. The evaluation of functions and subscripts has been included to indicate their high priorities.

	( )
	Evaluation of function references and subscripts
Arithmetic	**
	*, /
	+, − (binary and unary)
Relational	.LT., .GT., .EQ., .LE., .GE., .NE.
Logical	.NOT.
	.AND.
	.OR.
Assignment	= (see Section 3.14)

3.12 Executable and Nonexecutable Statements

To write valid algorithms in FORTRAN, the programmer must adhere to standard statement types of rigid basic structure that may involve arithmetic and logical expressions, some words from the English vocabulary, and punctuation and grouping marks. The basic statement types are few in number (see Appendix C for a complete list), and once the idea of proper formulation of arithmetic and logical expressions has been mastered, the language fits into a simple pattern.

There are two fundamentally different types of sentences or *statements* in the language, executable and nonexecutable. The *executable* statements are those that cause the FORTRAN translating program to generate machine-language instructions that will be executed when the object or machine-language version of the FORTRAN source program is loaded into the memory of the computer. *Nonexecutable* statements, also known as *declarations*, are special statements that do not cause machine code to be generated, but which give information to the translating program concerning the modes of variables, the amount of memory to be assigned to subscripted variables, and so on.

In the sections that follow, a basic set of FORTRAN statement types will be described, using the following general script references for convenience:

Script Reference	*Meaning*
$\mathcal{A}$	Dummy argument for a subprogram.
$\mathcal{A}\dagger$	Calling argument for a subprogram.
$\mathcal{B}$	Logical (Boolean) expression.
c	Arithmetic or logical constant.
$'c'$	Literal (character) constant.
$\mathcal{C}$	Character expression.
$\mathcal{E}$	Arithmetic (real or integer) expression.
$\mathcal{E}_{\mathcal{I}}$	Integer expression.
$\mathcal{F}$	Subprogram (function or subroutine) name.
$\mathcal{I}$	Integer constant or unscripted integer variable.
$\mathcal{L}$	A list, usually of variable names.
m	Input/output device number.
$\mathcal{N}$	NAMELIST name or labeled COMMON block name.
$\mathcal{S}$	Executable statement.
$\mathcal{T}$	Type or mode.
$\mathcal{V}$	Arithmetic (real or integer) variable.
$\mathcal{V}_{\mathcal{B}}$	Logical (Boolean) variable.
$\mathcal{V}_{\mathcal{C}}$	Character variable.
$\mathcal{V}_{\mathcal{I}}$	Integer variable.
η	Statement number or label; an array name in some circumstances.

Script symbols not already introduced in preceding sections will be described when they are first used. Any printed or typed characters, such as words and punctuation marks, shown in the general formulation of each statement type are part of the *basic statement structure* and must *not* be altered in any way.

3.13 Format for FORTRAN Statements

The FORTRAN statements to be described in Section 3.14 *et seq.* will normally be punched on a series of cards, with one statement to a card. In the width of a card there are 80 columns, each of which will ultimately be left blank or will be punched with holes corresponding to a *single* character. The card format is shown in Fig. 3.1 (not to scale):

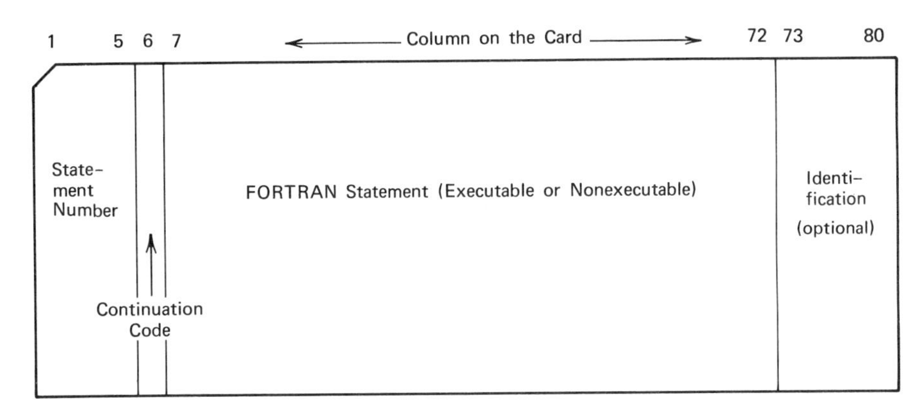

Figure 3.1 Punched-card format for FORTRAN statements.

Column	*Contents*
1	May contain the letter C. If so, the content of the card will be treated as a *comment*, and will be ignored by the compiler (see Section 3.24).
1–5	May contain a *statement number* or *label*. The statement number may be positioned anywhere in the five-column field. It is often necessary to make specific reference to particular statements in a program. A statement may be uniquely identified by prefixing it with an integer from 1 through 99999. The statement numbers need not be in any particular sequence, but a given statement number can correspond to only *one* statement. If no reference is made to a statement elsewhere in the program, then that statement is usually (and preferably) *not* assigned a number; any *executable* statement *may* be given a statement number, however. Labels should not appear on any *nonexecutable* statement, *except* for a FORMAT declaration, described in Section 3.20.
7–72	Contains a FORTRAN declaration or executable statement. If the statement is very long, it may be continued in columns 7–72 of one or more subsequent cards, up to a maximum of 19; in this case, each subsequent card must contain a *continuation code* in the form of *any* character other than 0 (or a blank) in column 6. We suggest that the digits 1–9 be used. FORTRAN allows up to 19 continuation cards per statement; the number permitted in WATFOR and WATFIV is assigned by the local computing installation.
73–80	Ignored by the translating program. May contain an identifying number for ordering the card deck, for example; frequently left blank.

WATFIV, but *not* FORTRAN or WATFOR, allows the programmer to punch *more than one* statement on a single card. This feature is particularly useful for reducing the size of the deck required for a program that is *already* checked out and is expected to be used rather often. We recommend that this feature *not* be used by beginning programmers, as it complicates rather than simplifies the debugging of programs, and tends to obscure the sequential information flow of the program.

The WATFIV program must be punched as follows:

1. All comments (see Section 3.24) and FORMAT declarations (see Section 3.20) must be punched in the usual way.
2. The usual rules for card continuation apply.
3. All statements and declarations must be punched in columns 7–72.
4. More than one statement can be punched in columns 7–72, with the exception of the FORMAT declaration. Each statement is separated from the one following it by a semicolon (;).
5. Statement numbers may appear in columns 1–5 as usual, or may follow a semicolon. In the latter case, the statement number must be followed by a colon (:). A statement number may *not* be continued from one card to the next.

Example. If all the comment cards, with the exception of the first, are ignored, the program of page 155 could be punched in the condensed card format as follows:

```
         1         2         3         4         5         6         7         8
12345678901234567890123456789012345678901234567890123456789012345678901234567890


C     SAMPLE PROGRAM NO. 1.    ROOTS OF QUADRATIC EQUATION.
    1 READ(5,100,END=999)A,B,C;DISCR=B*B-4.*A*C;WRITE(6,200)A,B,C,DISCR
      IF(ABS(A).GT.1.E-10)GOTO2;X=-C/B;WRITE(6,203)X;GOTO1;2:IF(DISCR.
     1GE.0.)GOTO3;WRITE(6,201);GOTO1;3:X1=(-B+SQRT(DISCR))/(2.*A);X2=(
     2-B-SQRT(DISCR))/(2.*A);WRITE(6,202)X1,X2;GOTO1;999:CALL EXIT
  100 FORMAT (3F10.3)
  200 FORMAT ('1     ROOTS OF QUADRATIC EQUATION, WITH'/
     1        '0     A      = ', F10.3/ 5X, 'B      = ', F10.3/
     2        '      C      = ', F10.3/ 5X, 'DISCR  = ', F10.3)
  201 FORMAT ('0     DISCRIMINANT NEGATIVE - ROOTS IMAGINARY')
  202 FORMAT ('0     X1     = ', F10.5/ 5X, 'X2     = ', F10.5)
  203 FORMAT ('0     COEFFICIENT A IS NEAR ZERO - ONE ROOT ONLY'/
     1        '0     X      = ', F10.5)
      END
```

In this example, we have used the maximum possible compaction of the executable statements, reducing the number of cards in that part of the program from 15 to 4.

3.14 Assignment Statements

The *arithmetic substitution* or *assignment* statement is of the general form

$$\mathscr{V} = \mathscr{E}$$

in which the variable $\mathscr{V}$ (possibly subscripted) and the arithmetic expression $\mathscr{E}$ are of either integer or real mode. The value of $\mathscr{E}$ is first computed and then assigned to $\mathscr{V}$; that is, the variable assumes the new computed value. If $\mathscr{V}$ and $\mathscr{E}$ have the same mode, the substitution is straightforward. However, if $\mathscr{E}$ is real and $\mathscr{V}$ is integer, then the value of $\mathscr{E}$ is truncated to the next lower integer in absolute value before substitution occurs. Conversely, if $\mathscr{V}$ is real and $\mathscr{E}$ is integer, then the value of $\mathscr{E}$ will be converted to the equivalent real representation with no change in value. The equals sign, =, is known as the *substitution* or *assignment operator*.

WATFOR and WATFIV, but *not* FORTRAN, permit an extended arithmetic assignment statement of the form

$$\mathscr{V}_1 = \mathscr{V}_2 = \ldots = \mathscr{V}_n = \mathscr{E}$$

where the $\mathscr{V}_i, i = 1, 2, \ldots, n$, are integer or real variables. The assignment operators are implemented from right to left, and mode conversions are carried out in the usual way for each one.

Valid Assignment Statement	*Invalid Assignment Statement*	
X = Y	X = Y = Z = 5.3*B	(valid in WATFOR and WATFIV but not FORTRAN)
INT = 4.9	4.9 = INT	
DISCR = SQRT(B*B − 4.*A*C)		
OHMS = VOLTS/AMPS	VOLTS/AMPS = OHMS	
A(I,J+2) = EXP(T(4))	EXP(T(4)) = A(I,J+2)	(EXP = exponential function – p. 65)
I = I + 1	I + 1 = I	

The last valid example demonstrates that the assignment statement does not mean equality in the usual sense. Rather, the statement I = I + 1 will cause the value of I to be incremented by unity. The equals sign is in a sense a poor choice for the substitution operator. The left-pointing arrow ← would be preferable, as it indicates the dynamic nature of the operation. Unfortunately, this character is not available in the FORTRAN character set of Section 3.2. However, when representing algorithms by flow diagrams, we shall continue to use ← to denote substitution (see Section 2.3).

The *logical substitution* or *assignment* statement has the form

$$\mathscr{V}_{\mathscr{B}} = \mathscr{B}$$

in which $\mathscr{V}_{\mathscr{B}}$ and $\mathscr{B}$ are now of logical mode. In the following, SWITCH and ANSWER must both be logical variables:

Example.
```
ANSWER = .TRUE.
SWITCH = ANSWER .AND. T.GT.TMAX
```

WATFOR and WATFIV, but *not* FORTRAN, permit an extension of the logical substitution statement of the form

$$\mathscr{V}_{\mathscr{B}_1} = \mathscr{V}_{\mathscr{B}_2} = \cdots = \mathscr{V}_{\mathscr{B}_n} = \mathscr{B}$$

where the $\mathscr{V}_{\mathscr{B}_i}$ are logical variables.

Example.
```
ANSWER = SWITCH = T.LE.SQRT(TMAX) .OR. ALPHA*X.GE.7.5
```

WATFIV, but *not* FORTRAN or WATFOR, has a *character assignment* statement with one of the forms:

$$\mathscr{V}_{\mathscr{C}} = \mathscr{C}$$

where $\mathscr{C}$ is a *character expression* consisting of one of the following:

1. A character variable, $\mathscr{V}_{\mathscr{C}}$.
2. A literal or character constant, '*c*'.
3. A parenthesized character expression, $(\mathscr{C})$.

Character assignment statements having more than one assignment operator are not allowed.

Examples.

```
LETTER = 'A'
MESSAG = 'BREAK IN CASE OF FIRE'
A      = LETTER
```

Here, the variables LETTER, MESSAG, and A are assumed to be of character mode and of sufficient length to hold the character strings (see Section 3.27).

3.15 Transfer Statements

Statements are normally executed in sequence, but the *unconditional transfer* statement

GO TO η

at any point in the program will cause the statement numbered η to be executed next. The programmer is likely to find two principal uses for such a statement: (1) after processing a set of data, to return to the beginning of a program to read another set of data (for example, GO TO 10 could be used to return to a READ statement numbered 10); and (2) to bypass a segment of a program under certain circumstances, often in conjunction with the logical IF statement of the next section.

A multiple branch can be achieved at any point by the *computed* GO TO statement

GO TO $(\eta_1, \eta_2, \ldots, \eta_n)$, $\mathscr{V}$

in which $\mathscr{V}_{\mathscr{I}}$ is an *unsubscripted integer variable* that normally assumes a value from 1 through n, the number of statement labels. For example, if the current value of INT is 2, then

```
GO TO (1, 7, 4, 10), INT
```

will transfer control to the statement numbered 7. Should $\mathscr{V}_{\mathscr{I}} < 1$ or $\mathscr{V}_{\mathscr{I}} > n$, control passes automatically to the next statement. WATFOR, but *not* WATFIV or FORTRAN, requires that $\mathscr{V}_{\mathscr{I}}$ be *greater than* zero.

3.16 Conditional Statements

It is frequently necessary to take a certain course of action if a prescribed condition is met—and to avoid such action if the condition is not met. In such cases, we normally use the *conditional* or *logical* IF statement, of general form

IF $(\mathscr{B})$ $\mathscr{S}$

Here, $\mathscr{B}$ is a logical expression, and $\mathscr{S}$ is any *one executable* statement, except another logical IF statement or a DO statement (see Section 3.17). The action of the logical IF statement, using the flow diagramming convention introduced in Section 2.4, is illustrated in Fig. 3.2.

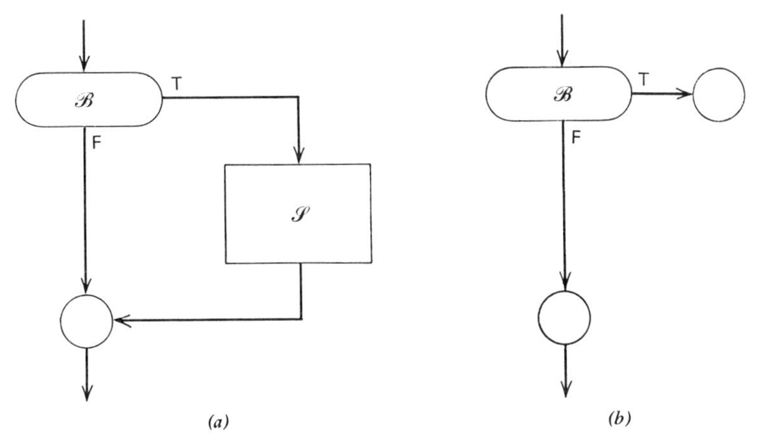

Figure 3.2 Conditional branching.

$\mathscr{B}$ is evaluated and if found to be *true*, $\mathscr{S}$ is performed; however, if $\mathscr{B}$ is *false*, no action is taken. The two paths will usually recombine as shown in Fig. 3.2*a* (the small circle represents a junction point), and control passes to the next executable statement following the logical IF statement. Sometimes, as in example (b) below, $\mathscr{S}$ will be a transfer statement to another point in the program; in such cases, illustrated in Fig. 3.2*b*, the "true" branch will not recombine with the "false" branch.

Examples.

(a) Suppose that the function ABS is unavailable; replace X with its absolute value:

```
IF (X .LT. 0.)  X = -X
```

(b) If T exceeds TMAX or the logical variable PRINT is false, transfer to the statement numbered 50:

```
IF (T.GT.TMAX .OR. .NOT.PRINT)  GO TO 50
```

(c) If K does not equal KMAX, interchange X and Y; otherwise, do nothing. Note that since *three* statements are required for the interchange, a transfer statement is needed in the construction:

FORTRAN Statements

```
      IF (K .EQ. KMAX) GO TO 5
         HOLDX = X
         X = Y
         Y = HOLDX
5     (next statement)
```

Flow Diagram

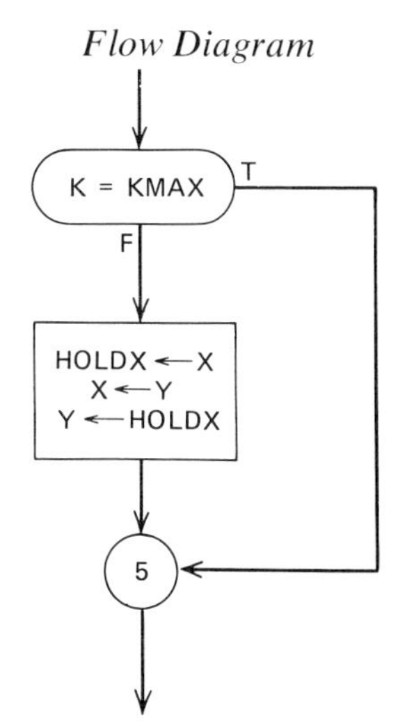

Here, some statements have been indented for ease of reading. Except in special cases to be mentioned later (see Sections 3.20 and 3.28, in particular), blanks are ignored in FORTRAN statements and may be inserted at will to make statements more readable.

(d) If alternative action is to be taken on *either* outcome, *two* transfer statements will be needed in the construction. For example, if K does not equal KMAX, interchange X and Y; otherwise, store the square root of B in A:

FORTRAN Statements

```
      IF  (K .EQ. KMAX)  GO TO 5
            HOLDX = X
            X = Y
            Y = HOLDX
      GO  TO 10
5         A = SQRT(B)
10        (next statement)
```

Flow Diagram

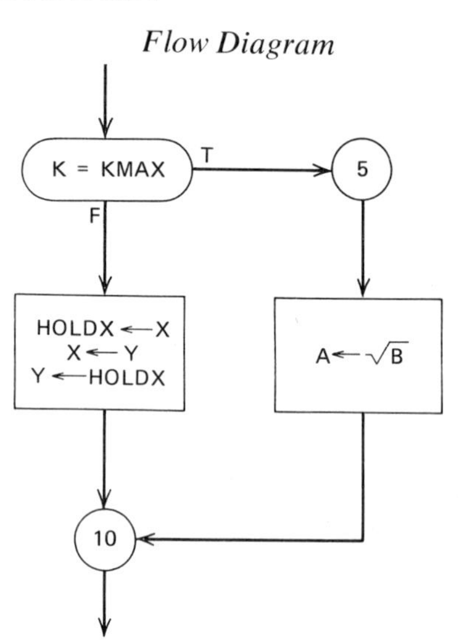

(e) It is frequently desirable to nest conditional sequences inside one another, particularly when three or more different courses of action are contemplated. For example, the following program segment will achieve the steps indicated by the flow diagram:

FORTRAN Statements

```
      IF (I .LT. 0) GO TO 20
         IF  (K .EQ. KMAX)  GO TO 5
               HOLDX = X
               X = Y
               Y = HOLDX
         GO  TO 10
5              A = SQRT(B)
10       GO  TO 25
20    I = 0
25    (next statement)
```

Flow Diagram

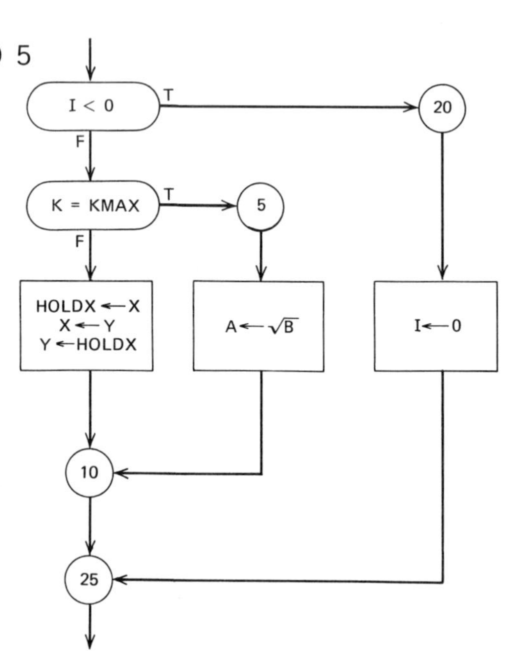

Note that instead of asking, "Is I less than zero?" we can also ask the complementary question, "Is I greater than or equal to zero?" Working along these lines, the reader should soon discover that there are several alternative constructions for achieving the same result. Note that the logic above would be unchanged if the statement GO TO 10 were replaced by GO TO 25; the latter would be preferred as it is computationally slightly more efficient.

A transfer to one of *three* possible statements can also be achieved by the *conditional transfer* or *arithmetic* IF statement:

$$\text{IF } (\mathscr{E})\ \eta_1,\ \eta_2,\ \eta_3$$

According to whether the value of the arithmetic expression $\mathscr{E}$ is negative, zero, or positive, control will be transferred to the statement numbered η_1, η_2, or η_3, respectively. If a two-way branch were desired, then two of the three statement numbers would be identical. With $\eta_1 = \eta_2 = \eta_3 = \eta$, the effect would be simply: GO TO η. The arithmetic IF statement was the only conditional statement in the early FORTRAN languages. Although it finds occasional use, it has largely been superseded by the logical IF statement, which is generally more flexible and readable.

3.17 Iteration Statement

Algorithms frequently contain one or more segments that are to be processed repeatedly a finite number of times. This procedure is called *iteration* and has already been described in Section 2.7. The sequence of FORTRAN statements used to implement the repeated processing is called a *loop*. There is normally some variable, called the *iteration variable* or *index*, that assumes a different value on each *pass* through the loop (that is, each time the sequence is processed or executed). Iteration loops can be handled by using only the substitution, conditional, and transfer statements.

For example, the mean $\bar{x}$ of n numbers $x_1, x_2, \ldots, x_n$ could be computed as follows, using the formula

$$\bar{x} = \left[\sum_{i=1}^{n} x_i\right] \Big/ n.$$

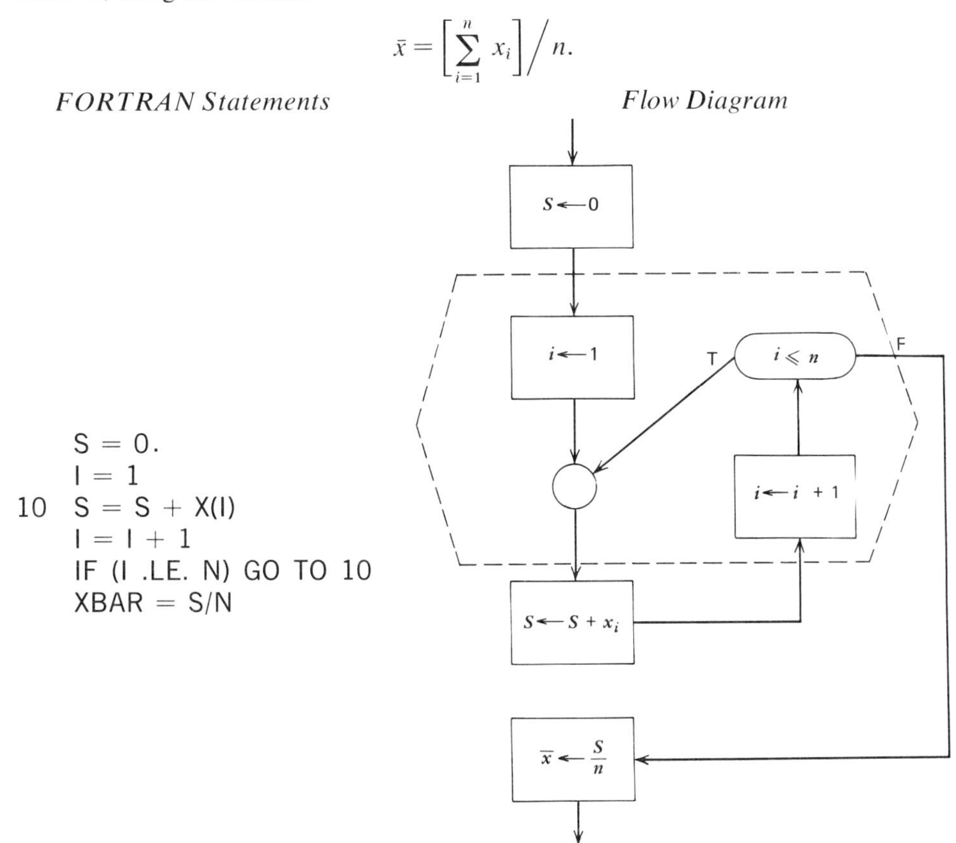

Note that the second, fourth, and fifth FORTRAN statements are concerned with *initializing* and *incrementing* the index or counter I, *testing* to see if I is less than or equal to the upper limit N and *transferring back* to an earlier statement in the loop if it is. This type of situation occurs frequently, and these four operations may be replaced by a single *iteration statement* whose general form is

$$\text{DO}\ \eta\ \mathscr{V}_{\mathscr{I}} = \mathscr{I}_1,\ \mathscr{I}_2,\ \mathscr{I}_3$$

$\updownarrow$ (One or more statements, forming

η the *range* or *scope* of the iteration)

Here, $\mathscr{I}_1$, $\mathscr{I}_2$, and $\mathscr{I}_3$ are unsigned *positive* (≥ 1) *integer constants* or *unsubscripted integer variables* with positive values. The *unsubscripted integer variable index* $\mathscr{V}_{\mathscr{I}}$ is first set equal to $\mathscr{I}_1$ and the subsequent statements, up to and *including* the statement numbered η (the *range* of the iteration), are executed. $\mathscr{V}_{\mathscr{I}}$ is then incremented by $\mathscr{I}_3$ and a test is made to see if it *exceeds* the upper limit $\mathscr{I}_2$. If so, control is transferred to the statement immediately *following* the one numbered η. If $\mathscr{V}_{\mathscr{I}}$ does not exceed $\mathscr{I}_2$, the range of the iteration is performed again. This procedure of incrementing $\mathscr{V}_{\mathscr{I}}$ by $\mathscr{I}_3$, testing against $\mathscr{I}_2$, and performing the range of iteration is continually repeated until $\mathscr{V}_{\mathscr{I}}$ eventually exceeds $\mathscr{I}_2$, whereupon control passes to the statement immediately following the one numbered η.

If the increment $\mathscr{I}_3$ is simply unity, then the following special form is also available:

$$\text{DO}\ \eta\ \mathscr{V}_{\mathscr{I}} = \mathscr{I}_1,\ \mathscr{I}_2$$

Thus, the mean of n numbers x_1 through x_n could alternatively be computed as follows:

FORTRAN Statements *Flow Diagram*

```
      S = 0.
      DO 10  I = 1, N
 10   S = S + X(I)
      XBAR = S/N
```

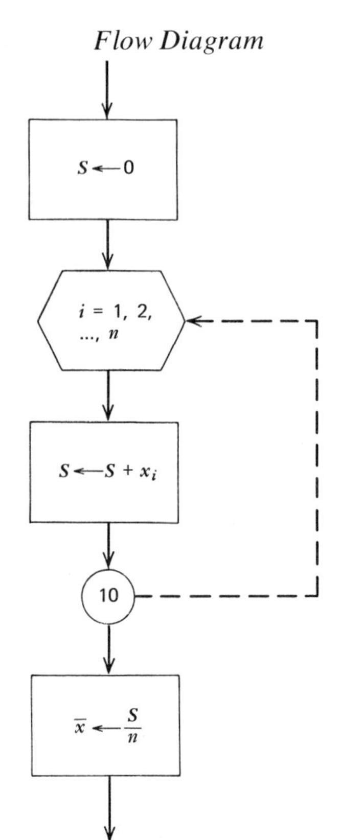

Observe that we use the hexagonal box introduced in Section 2.7 to denote iteration. It is essentially a simplified notation for the *initialization*, *incrementing*, and *testing* boxes that were contained inside the dotted hexagon of the previous flow diagram.

There are some additional operating peculiarities and programming restrictions that must be observed when using the DO statement:

(a) Because the index $\mathscr{V}_{\mathscr{I}}$ is incremented by $\mathscr{I}_3$ and tested against the upper limit $\mathscr{I}_2$ at the *end* of each pass, the iteration loop will be executed at least *once*, with $\mathscr{V}_{\mathscr{I}} = \mathscr{I}_1$, even when $\mathscr{I}_1 > \mathscr{I}_2$.

(b) A one-way transfer can be made completely out of a DO loop, but this must not lead to the DO statement itself.

(c) If a one-way transfer is made out of the DO loop, the index $\mathscr{V}_{\mathscr{I}}$ will have the value it had at the time of the transfer, and may be used as an ordinary integer variable. If the natural exit (when the index is sequenced through *all* specified values) is taken, the value of the index $\mathscr{V}_{\mathscr{I}}$ is *not* necessarily the value that it had on the last pass through the loop; the programmer should assume that the value of $\mathscr{V}_{\mathscr{I}}$ is *unknown*.

(d) The index or any of the associated indexing parameters must not be modified during the execution of the range; they may be changed *outside* the range, however.

(e) The *last* statement in the range (the statement with the label η) must be *executable* and must *not* be a GO TO, STOP, RETURN, DO, or arithmetic IF statement, or a logical IF statement containing any of these statement types. (In WATFIV, the last statement may be a logical IF statement containing a GO TO, STOP, RETURN, or an arithmetic IF, but *not* a DO statement.)

(f) DO statements may appear within the ranges of other DO statements. However, the ranges of the DO statements must not overlap; that is, all of the statements in the range of an inner DO statement must be within the range of an outer DO statement. The DO loops are said to be *nested*.

(g) A one-way transfer must not be made *into* the range of a DO loop from *outside* that loop. In the only exception to this rule, FORTRAN, but *not* WATFOR or WATFIV, allows a transfer *out* of the range of the *innermost* DO loop, and then a subsequent transfer *back into* the range of *that* DO, provided that none of the iteration parameters $\mathscr{V}_{\mathscr{I}}$, $\mathscr{I}_1$, $\mathscr{I}_2$, or $\mathscr{I}_3$ is modified outside the range.

Examples.

(a) Compute the product P of all elements above the main diagonal of the N × N square matrix A (see page 85 and Section 8.2 for examples of typical matrices).

```
      P = 1.
      NM1 = N - 1
      DO 5  I = 1, NM1
      IP1 = I + 1
      DO 5  J = IP1, N
    5 P = P*A(I,J)
```

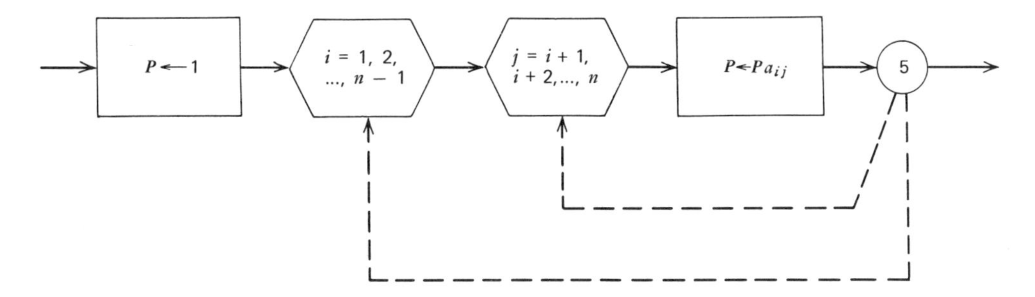

Note that two loops may terminate with the *same* statement; this does not violate restriction (f) above, or interfere with normal sequencing of the iteration variables. An iteration statement such as

```
      DO 5  I = 1, N - 1
```

is unacceptable because the upper limit (as well as the lower limit and increment, if specified) must not be expressions containing arithmetic operators. However, the following is acceptable:

```
      NM1 = N - 1
      DO 5  I = 1, NM1
```

(b) Compute NUMBER, the number of elements in the vector K(1), K(2), . . . , K(N) that have even subscripts and are also even in value:

```
      NUMBER = 0
      DO 3  INT = 2, N, 2
    3 IF (K(INT)/2*2 .EQ. K(INT))  NUMBER = NUMBER + 1
```

3.18 CONTINUE Statement

FORTRAN has an executable statement

```
      CONTINUE
```

that causes *no* calculation to be done. However, it may be, and nearly always is, prefixed by a statement number; hence, it may serve as a junction point in a program. Its only significant role is to serve as the last statement in the range of a DO statement: (1) to avoid violation of restriction (e) listed in Section 3.17, and (2) to allow the conditional execution of statements at the end of an iteration loop; this case is illustrated in the following example.

Example. Compute SUM2 and SUM3, the sums of the squares and cubes, respectively, of the positive elements of the vector V(1), V(2), . . . , V(M):

```
      SUM2 = 0.
      SUM3 = 0.
      DO 20  I = 1, M
      IF (V(I) .LE. 0.)  GO TO 20
         SUM2 = SUM2 + V(I)**2
         SUM3 = SUM3 + V(I)**3
   20 CONTINUE
```

3.19 General Input and Output Statements

Most computer programs that solve engineering problems are written for the processing of numerical values of data or *input* in order to yield numerical values of results or *output*. Of course, many problems require the manipulation of essentially nonnumerical information, such as logical quantities or strings of characters, as well. We are concerned in this and in the following two sections with the FORTRAN, WATFOR, and WATFIV statements that will cause a program to accept data and to display the results. The most significant variations among these three FORTRAN-IV dialects occur in the area of input and output, so the reader should take care to note which of the features described are available in the language of interest to him.

As discussed in Chapter 1, there are many ways of communicating with a digital computer, especially if the machine is large; consequently, input and output may involve a variety of devices, such as a teletypewriter, a magnetic-tape unit, or a television screen. At present, however, the majority of computer users are likely to submit their input data on punched cards to a *card reader*, and to receive their output on successive pages of paper from a *line printer*. Such units will be assumed as standard in the following discussion of I/O (input and output) statements. In most cases, however, these statements would need no modification in order to relate them to other I/O devices.

The most general input and output statements, available with only minor differences in FORTRAN, WATFOR, and WATFIV, for reasons that will be obvious later, are called the *formatted* input and output statements. The input statement has one of the forms:

READ (m, η_1) $\mathscr{L}$
READ (m, η_1, END $= \eta_2$) $\mathscr{L}$
READ (m, η_1, ERR $= \eta_3$) $\mathscr{L}$
READ (m, η_1, END $= \eta_2$, ERR $= \eta_3$) $\mathscr{L}$

Here, $\mathscr{L}$ is a list of *variables* whose values are to be read from the input device with the *device number* m. The device number must be either an *unsigned integer constant* or an *integer variable* of standard length (see Section 3.27) whose value is nonnegative; ordinarily the device number is in the range 0 through 9. Although device number assignments are rather arbitrary, most computing installations have standardized on a value of m equal to 5 for denoting a *card reader*.

Information as to how the data values are *arranged* on the data cards is supplied in the form of a special code, called the *format*. Usually, the format for a READ statement is specified in a separate statement called a *FORMAT declaration*, as described in the following section. The statement number η_1 is attached to the FORMAT declaration to associate the format with the READ statement. In some cases, however, the format may be saved in character form in successive elements of an *array* whose *name* is η_1; the use of such variable formats is discussed in Section 3.20.

The entries END $= \eta_2$ and ERR $= \eta_3$ are *optional* and may appear alone or together in either order. Their functions will be described in the latter part of this section.

Example. READ (5, 101) X, Y, I, A(I+1), TEMP(4)

As a result of the appearance of this statement in a FORTRAN program, the object program produced by the FORTRAN compiler would read five values from a punched card or cards in the card reader. The five values would replace, in order, the values currently present in the memory locations associated with the variables X, Y, I, A(I+1), and TEMP(4). The layout of these data items on the punched card or cards would be described in a FORMAT declaration having the statement number 101. Note that it would be essential to read a value for I before reading a value for A(I+1), unless I had itself previously been read or assigned a value earlier in the execution of the program. Only *variables* may appear in an input list. Clearly, the attempted reading of values for more general expressions, such as X + Y, would lead to difficulty, since no memory locations are assigned for storing expression values.

The general formatted output statement

WRITE (m, η) $\mathscr{L}$

causes the values of all the elements of the output list $\mathscr{L}$ to be transcribed from storage onto the output device whose number is m according to a format usually specified in a FORMAT declaration with the statement number η. As was the case for the READ statement above, m must be an unsigned integer constant or an integer variable of standard length. A value of m equal to 6 is conventionally used to denote a line printer, the most common output device. As described in Section 3.20, variable formats are allowed for output as well as input, in which case η is the *name* of an array containing a format specification in coded character form.

The elements of all *input* lists* are separated by *commas* and may be any of the following:

(a) Simple variable names: X, NANCY, SWITCH.
(b) Names of arrays: TABLE, MATRIX.
(c) Single elements of arrays: TABLE(5), MATRIX(I+3,J).
(d) Array segments (described below).

When the name of an array appears, *all* of the elements of the array, as specified in a DIMENSION declaration (see Section 3.25), are implied. For example, if TABLE is a singly subscripted array having ten elements and MATRIX is a doubly subscripted array having five rows and seven columns, then the statement

```
READ (5,177)  TABLE, MATRIX
```

will cause ten (real) numbers to be read into successive elements of TABLE and 35 (integer) numbers to be read *column-by-column* into successive elements of MATRIX.

Output lists* in *FORTRAN* are identical in form with input lists. However, *WATFOR* and *WATFIV* allow *expressions* as well as variables to appear in *output* lists. Any arithmetic or logical expression conforming to the rules of Sections 3.7 or 3.10 may appear, subject only to the restriction that a left parenthesis *not* be the *first* character in an element of the list. WATFIV, but not WATFOR, allows character variables and character (literal) constants to appear in output lists.

Examples.

Legal in FORTRAN, WATFOR, and WATFIV:

```
WRITE (6,205)  X, MEAN, ALPHA(7), MATRIX
```

Illegal in FORTRAN, WATFOR, and WATFIV:

```
WRITE (6,207)  X, (X+Y)/7.8, 1.E-5, ALPHA(7), A.LE.SQRT(P), SIN(X)
```

*It is legal, though very unusual, for a READ statement to have *no* input list, in which case cards in the input deck may be skipped under format control as described on page 92. The WRITE statement, however, will frequently appear without a list; this occurs when only literal information (characters) is to be printed (headings for a table, for example), as described in Section 3.20.

Legal in WATFOR and WATFIV, but illegal in FORTRAN:

```
WRITE (6,207)  X, +(X+Y)/7.8, 1.E–5, ALPHA(7), A.LE.SQRT(P), SIN(X)
```

Legal in WATFIV, but not in WATFOR and FORTRAN:

```
WRITE (6,275)  'X = ', X, 'A.LE.SQRT(P) = ', A.LE.SQRT(P)
```

An *array segment* may appear as an element of either an input list or an output list and is defined formally as follows:

$$\text{Array Segment} \equiv (\overline{\mathscr{L}}, \mathscr{V}_{\mathscr{I}} = \mathscr{I}_1, \mathscr{I}_2, \mathscr{I}_3)$$

Here, $\overline{\mathscr{L}}$ is an input or output list as described on the previous page. Thus the list in an array segment may contain other, imbedded, array segments. The appearance of an array segment in a list $\mathscr{L}$ is equivalent to the hypothetical sequence:

$$\begin{array}{ll} & \text{DO } \eta \ \mathscr{V}_{\mathscr{I}} = \mathscr{I}_1, \mathscr{I}_2, \mathscr{I}_3 \\ \eta & \text{(add list } \overline{\mathscr{L}} \text{ to the input or output list } \mathscr{L}) \end{array}$$

That is, the list of variables $\overline{\mathscr{L}}$ will be accumulated on the input or output list $\mathscr{L}$ as many times as the corresponding DO loop would be performed. This is not mere repetition, since the list of variables will almost always contain the index $\mathscr{V}_{\mathscr{I}}$ as a subscript. This type of construction is, therefore, particularly useful when reading or printing arrays. As with the DO statement, omission of $\mathscr{I}_3$ implies that the incremental value is unity.

As an example, we consider array segments derived from the following two-dimensional array or matrix A, which has M rows and N columns:

```
                        Column J
                           ↓
         A(1,1)  A(1,2)..........A(1,J)..........A(1,N)
         A(2,1)  A(2,2)..........A(2,J)..........A(2,N)
           .       .               .               .
           .       .               .               .
           .       .               .               .
Row I →  A(I,1)  A(I,2)..........A(I,J)..........A(I,N)
           .       .               .               .
           .       .               .               .
           .       .               .               .
         A(M,1)  A(M,2)..........A(M,J)..........A(M,N)
```

Examples of Array Segments	*Meaning*
(A(1,J), J = 1, N)	All elements in the first row.
(A(1,J), J = 1, N, 2)	Every element with an odd column subscript in the first row.
(A(I,J), I = 1, M)	All elements in the Jth column.
((A(I,J), J = 1, N), I = 1, M)	Entire matrix, row-by-row.
((A(I,J), I = 1, M), J = 1, N)	Entire matrix, column-by-column.

Still referring to the same matrix, the output statement

```
WRITE (6,201) M, N, (I, (A(I,J), J = 1, N), I = 2, 4)
```

would cause the following sequence to be printed:

(a) The maximum row and column subscripts, M and N.
(b) The number 2, followed by all the elements in row 2.
(c) The number 3, followed by all the elements in row 3.
(d) The number 4, followed by all the elements in row 4.

Similarly, the entire matrix might be read row by row using the statement

```
READ (5,127) M, N, ((A(I,J), J = 1,N), I = 1,M)
```

If an output list contains an array *name* without subscripts, then, as described above, the entire array will be printed. For arrays with more than one subscript, this will usually produce an undesirable ordering of the output values.

Example.
```
DIMENSION  A(2,3)
WRITE (6,202)  A
```

Here, the six elements of the matrix A will appear in the order A(1,1), A(2,1), A(1,2), A(2,2), A(1,3), A(2,3), *column by column*, corresponding to the normal FORTRAN ordering of arrays in storage, in which the leftmost subscript is modified first, the next leftmost is modified second, etc. Since output occurs in lines *across* the page, this is a poor method for printing matrices. A better approach would be to print the elements row by row as

```
WRITE (6,202) ((A(I,J), J = 1, 3), I = 1, 2)
```

or even by

```
    DO 5  I = 1, 2
5   WRITE (6,205) (A(I,J), J = 1, 3)
```

When the READ statement contains the *optional* entries

END = η_2
ERR = η_3

as shown in the first part of this section, then the reading of data proceeds in the usual way until one of two conditions prevails: (1) an error in the transmission of the data from the input device is encountered, or (2) no more data cards are available.

In the first instance, control will pass directly to the statement with the number η_3 for possible corrective action by the program. If the ERR = η_3 entry is not present, then execution of the program will be terminated immediately.

In the second instance, control will pass to the statement with the number η_2 and program execution continues. If the END = η_2 entry is not present, then execution of the program will cease immediately. For additional information on ways of terminating a program, see Section 3.22.

Example. READ (5,101, END = 999, ERR = 752) $\mathscr{L}$

Here, a transmission error encountered while reading values for elements of list $\mathscr{L}$ from device 5 according to the FORMAT declaration with statement number 101 will cause control to be transferred to the statement with number 752; an attempt to read a data card after the last card has already been read will cause control to be transferred to the statement with number 999.

Input and Output Statements with Default Device Assignments. In the general input and output statements, the programmer must specify a device number, a FORMAT declaration statement number (or, alternatively, the name of an array containing the format specification), and an input or output list. Most computing installations have some convention for assigning a FORTRAN device number to the standard input device (usually a card reader) and to the standard output device (usually a line printer). If this is the case, then the programmer who wishes to use the standard devices may use the alternative input and output statements

READ η, $\mathscr{L}$
PRINT η, $\mathscr{L}$

Here, η is the statement number of the associated FORMAT declaration, and $\mathscr{L}$ is an input or output list. No device number can be specified.

As mentioned before, the usual device number assignment is 5 for a card reader and 6 for a line printer, so that the following would normally be equivalent:

```
{READ (5,100)  N, A, BETA(5), (X(JACK), JACK = 1, N)
{READ  100, N, A, BETA(5), (X(JACK), JACK = 1, N)

{WRITE (6,200) ALPHA, TEMP(N)
{PRINT  200, ALPHA, TEMP(N)
```

The list $\mathscr{L}$ may be omitted, in which case the comma should also be omitted, and the statements have the appearance

READ η
PRINT η

Punching Cards. Card output may be produced using the general formatted output statement

WRITE (m, η_1) $\mathscr{L}$

where the device number m corresponds to that of the card punch, η_1 is the statement number of an associated FORMAT declaration or the name of an array containing format information, and $\mathscr{L}$ is the output list. There does not appear to be a standard device number for the card punch (corresponding to 5 for the card reader and 6 for the printer) that is used by most computing installations; the programmer should check with his computing center staff for the proper device number.

Corresponding to the input and output statements for reading cards from the standard card reader or printing lines on the standard printer, described above, there is an output statement for writing on the standard card punch

PUNCH η, $\mathscr{L}$

Here, η is the statement number of a FORMAT declaration, and $\mathscr{L}$ is the output list. Care must be taken to assure that the number of columns in each output record, established by the format specification of the next section, does not exceed 80, since each card is limited to 80 columns of characters.

3.20 FORMAT Declaration

(This section may be skipped, if format-free input and output statements are to be used exclusively.)

The last section was concerned with specifying the *names* of the particular variables whose values are to be read from or printed on a specified input or output device. The accompanying statement

η FORMAT (Format Specification)

governs *how* these values will appear on the punched card input or printed page output. Here, η is a statement number corresponding to that specified in the READ, WRITE, PRINT, or PUNCH statement.

Since most of the following remarks apply equally well to input and output, we shall call the information on a single card, or a single line of output, a *unit record.* The maximum size of a unit input record would normally be the 80 columns of a punched card. In most cases, the maximum size of a unit output record will correspond to the number of columns on the line printer, usually 120 or 132. If punched output is being produced, then the maximum size of the unit output record will be 80 columns.

A record is subdivided into one or more *fields*, each of which is in turn a group of one or more *columns.* A format specification consists of one or more field specifications, separated by commas, showing how each record is thus subdivided, and also indicating what type of information is to be found in each field. There must be a one-to-one correspondence between the format specification and the order and modes of the variables in the relevant input or output list. Thus, unless special provision is made, variable *names* will *not* appear on the record.

The field specifications available for *numerical* information include the following:

Field Specification	*Interpretation*
nIm	n *integer* fields, each m columns wide, each containing an integer constant placed to the extreme right of its field.
nF$m.d$	n *real* fields, each m columns wide, each containing an F-type constant, with the decimal point assumed to be located d columns to the left of the extreme right of its field.
nE$m.d$	n *real* fields, each m columns wide, each containing an E-type constant. The integral exponent must be placed to the extreme right of the field. The decimal point in the mantissa is assumed to be located d columns to the left of the letter E.
nX	A *skip field*, consisting of n blank columns. (If used for input, these columns are ignored no matter what they contain.)

Note: if n is omitted from an I, F, or E specification, a single field is assumed (that is, $n = 1$); n may not be omitted from an X specification.

Examples (the symbol □ is inserted to emphasize a blank space; the vertical bars are shown to help delineate the field):

<table>
<tr><th>Field Specification</th><th>Possible Contents</th><th>Value Assigned</th></tr>
<tr><td>I5</td><td>|□5768|</td><td>5768</td></tr>
<tr><td>I5</td><td>|□□−6□|</td><td>−60</td></tr>
<tr><td>F9.5</td><td>|□□3.14159|</td><td>3.14159</td></tr>
<tr><td>F9.5</td><td>|□□3.142□□|</td><td>3.142</td></tr>
<tr><td>E12.3</td><td>|□□□□−6.731E5|</td><td>-6.731×10^5</td></tr>
<tr><td>E12.4</td><td>|□□−0.6731E+6|</td><td>-6.731×10^5</td></tr>
<tr><td>E12.3</td><td>|□□−67.310E04|</td><td>-6.731×10^5</td></tr>
</table>

For *input*, the decimal point may be *omitted* from a number read from an E or F field, in which case it will be presumed to be located according to the field specification. If the point is in a position not agreeing with the field specification, the latter will be overridden. For example, the following would be acceptable alternatives for the third and fifth specifications above:

```
|3.14159□□|          |□□□□□−6731E5|
|□□□314159|          |□−6.731□□E□5|
```

In the E-type field, there is considerable flexibility available on input, as shown by the equivalence of the last three numbers in the above table. The output for such a number, however, normally appears with the first significant digit of the mantissa immediately to the right of the decimal point and with the exponent adjusted accordingly, that is, in normalized scientific form. Optional plus signs will not appear on output. If fewer decimal places are requested than are actually available in the machine's memory, then the figures of least significance will be lost. Obviously a correct field specification must specify enough columns to be able to contain all the characters requested, including digits, decimal points, and minus signs. Special care is required when specifying the width of E fields on output. E fields are printed in the form *s*0.*xxx*E*see* where *s* is a blank or a minus sign, *ee* is the integral exponent, and *xxx* are the *d* significant digits in the mantissa. Thus, m must not be smaller than $d+7$. The width of a field can often be overspecified to advantage, resulting in several blank spaces separating the numbers, which are then easier to interpret. When an output field is underspecified (for example, attempting to print a value 9872 using an I3 field) a row of asterisks is printed in the field to indicate the error.

Example.

The statements

```
      READ (5,100) I, J, Y, EMASS
100   FORMAT (2I5, F10.4, 5X, E15.3)
```

would be appropriate for reading values from a card punched as follows:

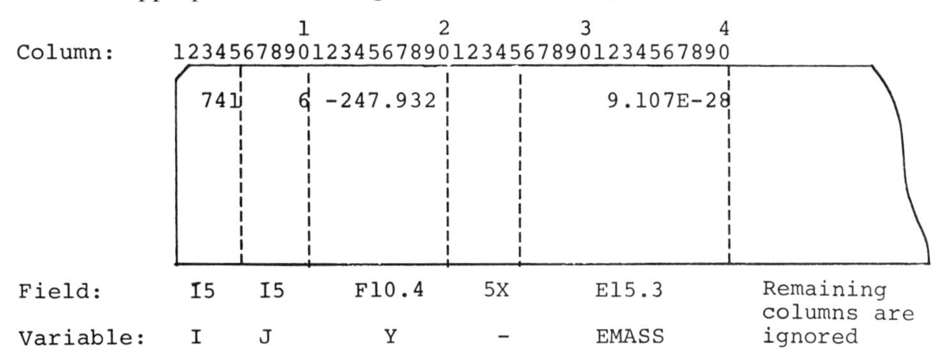

After the READ statement has been processed, I, J, Y, and EMASS would have the values 741, 6, −247.932, and 9.107×10^{-28}, respectively. If columns 21 through 25 have no punches, then an equivalent format specification would be

```
100    FORMAT (215, F10.4, E20.3)
```

That is, blank spaces can usually be incorporated as part of an overspecified field width.

If the format specification of a unit record contains a repeated field pattern, then an abbreviated specification is allowed; the part of the specification that is to be repeated is enclosed in parentheses and is preceded by the number of repetitions. For example, the following are equivalent:

```
101    FORMAT (I5, F10.4, 5X, E15.4, F10.4, 5X, E15.4, I10)
101    FORMAT (I5, 2(F10.4, 5X, E15.4), I10)
```

If more than one unit record is to be described in a single input or output statement, three courses of action are available:

(a) If the unit records are to be identical, the format specification need be given once only.

(b) If the unit records are different, slashes (/) must be incorporated into the specification to delineate the end of each unit record.

(c) If the first one or more unit records are different from the last unit records, which are identical, then extra parentheses must be inserted around the *last* part of the specification. This parenthesized part of the format is used repeatedly for remaining elements of the input/output list.

Examples.

(a) Print the N × N matrix A(1,1) through A(N,N) with ten F10.5 numbers per line:

```
       WRITE (6,51) ((A(I,J), J = 1, N), I = 1, N)
 51    FORMAT (1X, 10F10.5)
```

In the event that N equals 10, this would result in one row of the array being printed per line of output. With N equal to 6, for example, each line would still contain 10 numbers, but the output would be rather inconvenient to interpret, consisting of a mixture of rows and partial rows in sequence. The following statements, however, would ensure that each line of the array begins in the second column of a new output line. The reason for insertion of the one-column skip field will become apparent in the next part of this section.

```
       DO 10  I = 1, N
 10    WRITE (6,66) (A(I,J), J = 1, N)
 66    FORMAT (1X, 10F10.5)
```

Note that if N equals 6, for example, the remaining four items in the format specification will be ignored, and that each new execution of the WRITE statement uses the format from its beginning.

(b) Read values for B, C, D, IE, F, and G, punched on three cards as follows:

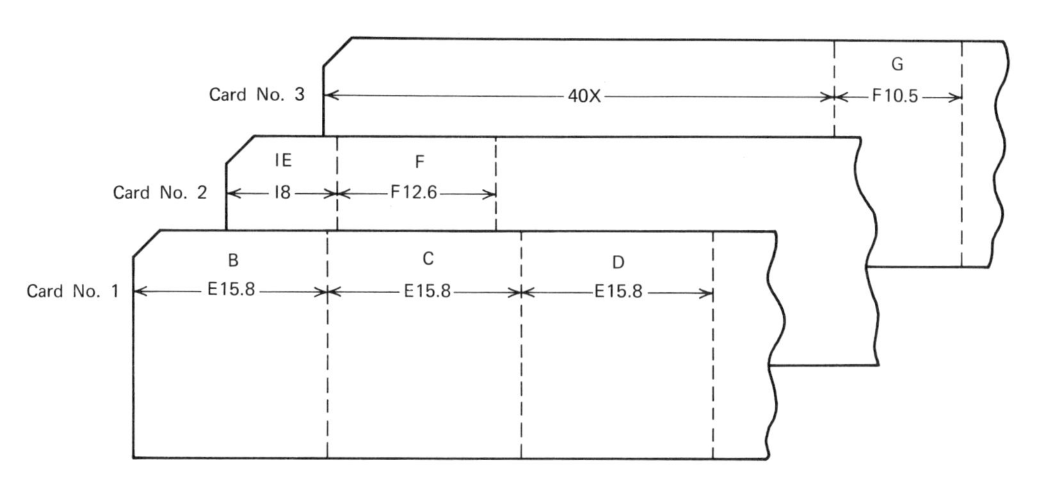

Card Number	*Contents of the Card*
1	B, C, and D as three E15.8 numbers in columns 1 to 15, 16 to 30, and 31 to 45.
2	IE as an I8 number in columns 1 to 8, followed by F as an F12.6 number in columns 9 to 20.
3	G as an F10.5 number in columns 41 to 50.

The appropriate statements would be

```
      READ (5,105) B, C, D, IE, F, G
  105 FORMAT (3E15.8/ I8, F12.6/ 40X, F10.5)
```

(c) Read B, C, D, IE, F, and G, punched on the first three cards as in (b), followed by cards containing IE pairs of values X(I), Y(I), I = 1, 2, . . . , IE. The values are arranged in alternating fashion, four pairs per card, with each card having the same record format, F6.3, E14.4, F6.3, E14.4, F6.3, E14.4, F6.3, E14.4 as shown below:

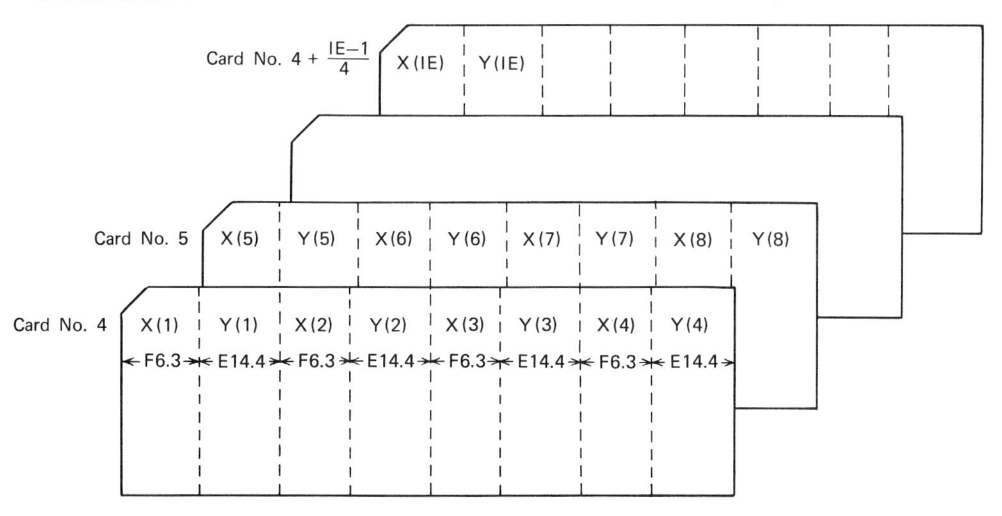


```
      READ (5,105)  B, C, D, IE, F, G, (X(I), Y(I), I = 1, IE)
  105 FORMAT (3F15.8/ I8, F12.6/ 40X, F15.5/
     1  (4(F6.3, E14.4)))
```

Each data set will contain a total of 4 + (IE − 1)/4 cards (this expression should be truncated to the next lower integer). All the cards containing the paired values will be read using the record format 4(F6.3, E14.4), since it is enclosed by parentheses and appears at the end of the complete format specification. Note that the digit 1 appearing in column 6 of the last line is a continuation code, as described in Section 3.13, so that the characters on the line are part of the FORMAT statement started on the preceding line. Since IE will not necessarily be an integral multiple of four, the last data card in the set may not be full; this causes no difficulty, as reading ceases as soon as the input list is *satisfied*, that is, as soon as values for each variable have been read.

The rules for the scanning of input formats may be summarized as follows: When a READ statement is executed, the associated format is scanned from the beginning, and from left to right. A data card is read, and partitioned into fields according to the format for the first record; contents of the fields are converted to the specified internal form (real, floating point, etc.) and assigned, in order, to variables in the input list. This procedure is followed for additional records, their associated formats, and appropriate variables in the list until (a) the list is satisfied or (b) the end of the format specification is reached, but the list is not satisfied. If situation (a) applies, then both reading and format scanning ceases. If case (b) applies, the format is scanned from *right* to *left* until the *rightmost level-one left parenthesis* is encountered. Here, the appropriate left parenthesis is the one that matches the rightmost right parenthesis *inside* the format specification; for example, in the declaration FORMAT (( ) ( )), the appropriate parenthesis is
↑
marked with an arrow. Reading resumes with the *next* data card using the format specification starting with this level-one left parenthesis (any repetition digits *preceding* the parenthesis are also included). If the end of the format is again encountered before the list is satisfied, this is just another instance of case (b). If *no* level-one left parenthesis appears, then the next data card is read, and the format scan proceeds from the *beginning* of the format specification.

A word of caution is in order concerning the modes of the variables and their associated format specifications. Information is taken from the cards, converted as indicated in the specification, and saved in the memory location assigned to the variable. Usually, *no check* is made to determine that the information stored is of the proper mode for the variable. For example, if the user inadvertently uses an I10 specification for the information stored in a real variable named X, then normally no indication of the error will be made at the time. Almost invariably, however, such a mismatch of modes will cause problems later in the calculations. The user must take special care to insure that the data, format specification, and input list are compatible in their order of appearance and modes.

An empty input record (a card to be skipped over) can be indicated with two slashes in sequence in the format specification. For example, suppose that values are to be read for two variables, X and Y, and that there are three data cards in each set of data. Each variable is assigned to an F10.5 field, X starting in column 16 of the first card and Y starting in column 31 of the third card. Then an appropriate input statement and FORMAT declaration would be:

```
      READ (5,120)  X, Y
120   FORMAT (15X, F10.5//30X, F10.5)
```

The Printing of Literal Fields (Character Information). It is frequently desirable on output to print messages and to identify numerical values with appropriate labels.

This can be effected by incorporating *literal constants* directly into the format specification. As described in Section 3.4, a literal constant is a sequence of characters, sometimes called a *character string*, enclosed in primes (apostrophes) as follows:

$$'m_1m_2m_3 \ldots m_n'$$

Here, the n characters, *including blank spaces*, constitute the content of the string, which in output formats we shall call a *literal field*; the characters will be printed *automatically*.

Alternatively, the literal field may be specified in the form of a *Hollerith field* or *string* having the form

$$n\text{H}m_1m_2m_3 \ldots m_n$$

This form for literal constants was introduced into the early FORTRAN languages, and is retained in FORTRAN-IV, but has largely been replaced by the more convenient prime-delimited character string. [Incidentally, Hollerith was the inventor of the punched card as we know it, and designed many of the early business machines, particularly those used in tabulating the United States censuses in the late nineteenth century.]

Example. Suppose that at some stage in a program that computes pressures at various nodes in a piping network, N has the value 7 and PPSIG(7) has the value 2.793. Then the statements

```
      WRITE (6,29) N, PPSIG(N)
   29 FORMAT (' AT NODE NUMBER ', I3, 5X, ' PRESSURE = ', F7.3)
```

would cause the following to be printed:

|□AT□NODE□NUMBER□□□7□□□□□□PRESSURE□=□□□2.793

The symbol "□" is inserted to emphasize a blank space. The vertical line at the left represents the left-hand margin, that is, the blank space before AT would be in column 1. If the Hollerith field equivalent of the literal field were used, then the FORMAT statement for this example would appear as

```
   29 FORMAT (16H AT NODE NUMBER , I3, 5X, 12H PRESSURE = ,
     1 F7.3)
```

We shall not refer to the Hollerith field again.

Example. A format specification may consist entirely of a literal field. The corresponding WRITE statement will then have a blank list:

```
      WRITE (6,52)
   52 FORMAT (' THE INPUT DATA NOW FOLLOW')
```

In this case, the message is simply printed without any accompanying numerical values.

Carriage Control. Single line spacing on the printer is achieved by ensuring that there is a *blank space in column 1* of each line. This can be achieved in several ways, for example:

(a) By making sure that the width of the first field is overspecified.
(b) By starting the format specification with a skip field.
(c) By starting the format specification with a blank in a literal field.

Thus, if we wish to print J (known to be no more than five digits in length, possibly also with a minus sign), the following will ensure a blank space in column 1:

```
      WRITE (6,200) J
200   FORMAT (I7)  or  (1X, I6)  or  (' ', I6)
```

Note that in examples given under *literal fields* above and in those appearing on page 90 the output lines also started with a blank in column 1.

Carriage control other than single line spacing can be achieved by placing an appropriate character in column 1 by means of a literal field. The following codes are used at most installations.

Character Code	*Interpretation*
□ (Blank)	Single space the carriage before printing the line.
0	Double space the carriage before printing the line.
–	Triple space the carriage before printing the line.
+	Do not advance the carriage before printing the line.
1	Skip to the top of the next page before printing the line.
2	Skip to the next half-page before printing the line (not allowed by many WATFOR and WATFIV compilers).

The carriage-control character is *not* printed with the remainder of the formatted line. The *second* character in the line appears in the second column of the printed output.

Examples. The statements

```
      WRITE (6,210) JOBNO
210   FORMAT ('1THESE ARE THE RESULTS FOR JOB NUMBER', I5)
```

would insure the printing of the line

```
      THESE ARE THE RESULTS FOR JOB NUMBER      7
```

at the top of a fresh page starting in column 2 of the line (assuming that the current value of the variable JOBNO is 7), whereas the statements

```
      WRITE (6,211)
211   FORMAT ('1')
```

would merely advance the carriage to the top of the next page, without any printing.

It is important to note that *each* new line must have its own carriage-control character in the first column. Thus, to write out a labeled value of X at the top of a new page and unlabeled values of Y and Z two lines below, the following statements would accomplish the desired effect:

```
      WRITE (6,215) X, Y, Z
215   FORMAT ('1X = ', F10.3/ '0', 2E15.4)
```

As a final illustration, the first example in this section might alternatively be written as

```
      WRITE (6,29) N, PPSIG(N)
29    FORMAT (' AT NODE NUMBER ', I3/ '0PRESSURE = ', F7.3)
```

which would cause the following to be printed

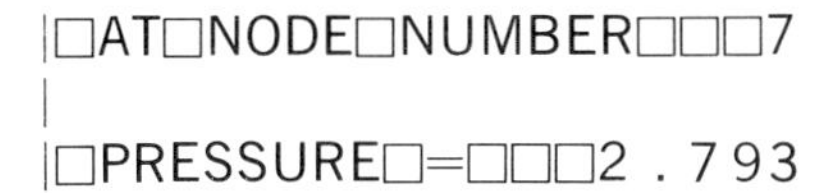
|□AT□NODE□NUMBER□□□7
|
|□PRESSURE□=□□□2 . 793

If the first character in a line to be printed is *not* one of the valid carriage-control characters, then at most installations the carriage will be single spaced, and the entire line, *including* the first character, will be printed starting in column 2 of the output record. Note that in this case the maximum number of columns specified in the format for the unit record must be one smaller than the width of the printer carriage. When the output record is to be *punched*, the carriage control character has no significance; *all* characters in literal fields, including any character in column one of the record, will be punched.

The rules for scanning output formats are identical with those for scanning input formats, except that records are *assembled* according to the format specification and sent to the line printer for display. Literal fields are processed automatically, even when they follow the field associated with the last variable in the output list.

Example. Suppose that an integer variable N and a set of N paired values X(I), Y(I), I = 1, 2, . . . ,N, are to be printed with labels so that the output page appears as follows:

```
                                    Column of printed output

Line on                     1         2         3         4         5
the page           12345678901234567890123456789012345678901234567890
                  ________________________________________________
    1             | THE VALUES OF THE nn DATA SETS ARE:
    2             |
    3             | X( 1) = xxxxx.xxxx      Y( 1) = yyyyy.yyyy
    4             | X( 2) = xxxxx.xxxx      Y( 2) = yyyyy.yyyy
    5             |         .                       .
    6             |         .                       .
    .             |         .                       .
    .             |         .                       .
  N+2             | X(nn) = xxxxx.xxxx      Y(nn) = yyyyy.yyyy
```

Here, *nn* is the value of N and *xxxxx.xxxx* and *yyyyy.yyyy* are the values of the variables X(I), Y(I), I = 1, 2, 3, . . . , N. An appropriate output statement and FORMAT declaration would be:

```
      WRITE (6,200) N, (I, X(I), I, Y(I), I = 1, N)
200   FORMAT ('1THE VALUES OF THE ', I2, ' DATA SETS ARE:'//
     1  (' X(', I2, ') = ', F10.4, 5X, 'Y(', I2, ') = ', F10.4))
```

Note that two slashes in sequence indicate that an empty record or blank line is to be printed.

Reading and Printing Variable Alphabetic (Character) Information. It is often desirable to read and write alphabetic information of a variable nature. For example, one important piece of information required for a payroll calculation and check-writing program would be the employee's name. FORTRAN allows symbolic information to be stored internally in character format, using the hexadecimal codes shown in Table 1.10. Since the coding scheme uses two hexadecimal characters to represent each external symbol, one byte of internal storage is required for each such symbol. Thus words of length one, two, four, and eight bytes can store the internal representations of at most one, two, four, and eight characters, respectively. *Any FORTRAN variable* of proper length (see Section 3.27), regardless of mode (integer, real, logical), may be used to store variable alphabetic information; unless the length specification parameter of Section 3.27 is employed, the FORTRAN translator assigns four bytes of storage to every variable, regardless of mode.

The format specification for input and output of character (alphabetic) information is

nAm

where n is the number of fields containing characters from the character set of Section 3.2, and m is the number of columns in each field; if n is equal to one, it may be omitted. Let b be the number of bytes in a variable storage word (the standard word length for integer, real, and logical variables is four bytes). Then, if $m \leq b$, the m characters are stored left-adjusted in the storage word on input and blanks are added to fill out the word on the right; on output, the leftmost m stored characters are retrieved from the storage word and printed. If $m > b$, the leftmost $m-b$ characters in the field are discarded, and the rightmost b characters are stored on input; on output, the b characters are printed right-justified in the field with leading blanks.

Example. Suppose that a data card contained the following characters in the first 50 columns

```
         1         2         3         4         5
12345678901234567890123456789012345678901234567890

RINGO     STARR       214282104 BEATLES    GB         31
```

Then a READ statement and an accompanying FORMAT statement might appear as follows:

```
      READ (5,100) FIRST, LAST1, LAST2, (SS(I), I = 1, 3),
     1  GROUP(1), GROUP(2), NATNLT, NAGE
  100 FORMAT ( A4, 5X, 2A4, 2X, A3, A2, A4, 1X, 2A4, 2X, A2,
     1  7X, I2 )
```

Then the variables would contain the following character strings (b = blank) after the above card is read:

FIRST	'RING'	SS(3)	'2104'
LAST1	'STAR'	GROUP(1)	'BEAT'
LAST2	'R*bbb*'	GROUP(2)	'LES*b*'
SS(1)	'214*b*'	NATNLT	'GB*bb*'
SS(2)	'28*bb*'	NAGE	31

Note that the content of NAGE is the *integer* 31. All other information is stored in the encoded character form of Table 1.10, even the digits 214. An output statement and format to print the information might appear as follows:

```
      WRITE (6,200) FIRST, LAST1, LAST2, GROUP(1), GROUP(2),
     1 (SS(I), I = 1, 3), NAGE, NATNLT
 200  FORMAT ( '1NAME: ', A1, '. ', 2A4 / ' MUSICAL GROUP: ',
     1 2A4 / '0SOCIAL SECURITY NUMBER: ', A3, '–', A2, '–', A4 /
     2 ' AGE: ', I2 / ' NATIONALITY: ', A2 )
```

The printed output for the information above would appear as follows:

```
NAME: R. STARR
MUSICAL GROUP: BEATLES

SOCIAL SECURITY NUMBER: 214–28–2104
AGE: 31
NATIONALITY: GB
```

In WATFIV, which allows variables to be assigned *character* mode, the preceding example could be simplified somewhat, since individual character variables may be assigned any length from 1 to 255 characters. The same data card could be read using the following CHARACTER declaration (see Section 3.27) and READ statement:

```
      CHARACTER FIRST*4, LAST*8, SS*4, GROUP*8, NATNLT*2
      READ (5,101) FIRST, LAST, (SS(I), I = 1, 3), GROUP, NATNLT,
     1 NAGE
 101  FORMAT ( A4, 5X, A8, 2X, A3, A2, A4, 1X, A8, 2X, A2, 7X, I2 )
```

Note that the A format specification is used for input of alphabetic information for variables of character mode. The content of the variables for the same data would be:

FIRST	'RING'	SS(3)	'2104'
LAST	'STARR*bbb*'	GROUP	'BEATLES*b*'
SS(1)	'214*b*'	NATNLT	'GB'
SS(2)	'28*bb*'	NAGE	31

A WATFIV output statement and an associated FORMAT declaration that would produce printed results identical with that shown above is:

```
      WRITE (6,201)'1NAME: ',FIRST,' . ',LAST,' MUSICAL GROUP: ',
     1 GROUP, '0SOCIAL SECURITY NUMBER: ',
     2 (SS(I), I = 1, 3), ' AGE: ', NAGE, ' NATIONALITY: ', NATNLT
 201  FORMAT ( A7, A1, A2, A8 / A16, A8/ A25, A3, '–', A2, '–', A4/
     1 A6, I2 / A14, A2 )
```

Note that since WATFIV allows literal constants to appear in the output list, literal fields (including any carriage control characters) can be removed from the format specification; the FORMAT declaration must then contain A field specifications to correspond to any literal constants in the list. This is not a particularly useful variation on the preceding output FORMAT, since the number of characters in each literal constant must be counted accurately.

Additional Format Specifications. Three additional format specifications that are used less frequently than those already described are:

*n*L*m*
*n*G*m*.*d*
T*m*

The first specification, *n*L*m*, is used for reading and printing values of *logical* variables. Here, *n* is the number of logical fields, each *n* columns wide. In an *input* field, the first appearance of a T or F in the field causes the value .TRUE. or .FALSE., respectively, to be assigned to the pertinent variable. On output, the letter T or F will be written right-adjusted in the output field of width *m*.

The G or *general* specification can be used for reading or writing variable values of *any* mode; *n* is the number of such fields, *m* is the number of columns in each field, and *d* is the number of decimal places (for real variables) for input. The .*d* part of the specification may be deleted if the variables are logical or integer in mode. On the data card, the data may be punched using either the I, E, F, or L forms for constants, provided that the data form used has a mode appropriate for the pertinent variable in the input list. For example, if the input statement and the associated format are as follows:

```
      LOGICAL  SWITCH
      READ (6,200)  ALPHA, BETA, MAX, SWITCH
200   FORMAT ( 4G10.3 )
```

and the data card has the appearance

```
         1         2         3         4         5
12345678901234567890123456789012345678901234567890

          27344   1.459E-8          100       T
```

then the variables would be assigned the values:

ALPHA	27.344
BETA	1.459×10^{-8}
MAX	100
SWITCH	.TRUE.

When used as an *output* specification, the .*d* portion of the specification has no significance for integer and logical data. For real data, *d* is the *number of significant digits* that will be printed. The information is automatically scaled to either E or F form, depending on the magnitudes of the numbers involved.

The T or *transfer* specification, determines the column position on an input card or output line where the next field in the format specification is to begin. It is analogous to the tab facility on a typewriter. For example, the output statement and associated format

```
      WRITE (6,200)
200   FORMAT (' ', T20, 'INVENTORY', T60, '1968', T40, 'STORE 10' )
```

would produce an output line with the following appearance:

```
         1         2         3         4         5         6         7
1234567890123456789012345678901234567890123456789012345678901234567890

                   INVENTORY           STORE 10            1968
```

Note that the *m* in the specification T*m* causes a tab-stop to column *m*, and that successive transfer specifications need not proceed from left to right across the line.

Variable Formats. As indicated in Section 3.19, the FORMAT statement number in the general formatted input and output statements may be replaced by the *name* of an array or a singly subscripted variable that contains the complete format in coded-character form. Since character information can be read from data cards using the A format specification, it is possible to have *variable formats* that are read during execution of the translated FORTRAN program. For example, the statement sequence

```
      DIMENSION  ALPHA(5)
      READ (5,100)  ALPHA
100   FORMAT (5A4)
      READ (5,ALPHA) X, DELTA, GAM, MAX, EPS
```

will cause the format for the second READ statement to be read as data by the first READ statement. If the first data card has the following appearance

```
         1         2         3
123456789012345678901234567890

(3F10.5, I6, E14.5)
```

then the resulting input will cause the literal strings '(3F1', '0.5,', ' I6,', ' E14', '.5) ' to be assigned, in order, to the five elements of the array ALPHA. Note that the format specification includes a left parenthesis as the first character and a right parenthesis as the last character.

The use of variable formats adds considerable flexibility to a program, as the programmer need not commit himself to a specific input format at compilation time. WATFIV, but not FORTRAN or WATFOR, allows the programmer to create variable formats in character mode variables without reading them from data cards; this feature is described in Section 3.28.

3.21 Format-Free Input and Output

For the beginning programmer, the preparation of FORMAT statements (see Section 3.20) to accompany the input and output statements of Section 3.19 is probably the most formidable task in the writing of a working FORTRAN program. Even experienced programmers dislike the tedium of preparing formats, particularly when the program is only going to be used a few times or when the appearance of the output is not particularly crucial. Unfortunately, if the data needed by a program is already punched in some arbitrary arrangement on data cards, or if attractively labeled reportlike output is essential, then there is no alternative to the use of formatted input and output described in the preceding two sections.

When formatted input and/or output is not essential, FORTRAN, WATFOR, and WATFIV allow the programmer the option of reading and/or writing without specifying an accompanying format(s). Unfortunately, the format-free input and output statements for FORTRAN and WATFOR are completely incompatible. Therefore, programs containing the WATFOR format-free statements will not compile correctly if the FORTRAN translator is called, and programs containing the FORTRAN format-free statements will be unacceptable to the WATFOR compiler. The WATFIV language, however, includes both the FORTRAN and WATFOR format-free input and output statements, so that programs written in either WATFOR or FORTRAN will often be acceptable to the WATFIV compiler. Formatted and format-free input and output statements may be used in the same program.

Format-Free Input and Output Using WATFOR and WATFIV. The format-free input statement, sometimes called the *simple* input statement, available in both WATFOR and WATFIV but *not* in FORTRAN, is

```
READ, ℒ
```

Here, the comma following the word READ is mandatory, and the list $\mathscr{L}$ is an ordinary FORTRAN input list, already described in Section 3.19.

Examples.

```
READ, N, ALPHA, P(7), Q(N), (X(I), I = 1, N)
READ, X, Y, Z
READ, N, M, ((A(I,J), J = 1, M), I = 1, N)
```

The READ, $\mathscr{L}$ statement has *no* accompanying FORMAT declaration. The data card or cards containing values for the input list variables are format-free in the sense that field widths have no particular significance; the input device is assumed to be the card reader and is defaulted to device number 5 at most computing installations. The data items are simply punched in the order specified by the list $\mathscr{L}$, separated by either a *comma* or any number of *blanks* or both. Thus, if a data card had the following appearance

```
123 . . .
7,1.745E–5, 16.7, 6., 1.2, 4.,5.,6. 7.4,1.2E6 16.
```

and the input statement were the first of the above examples, then the values read for the variables in the list would be:

N	= 7	X(3)	= 5.0
ALPHA	= 1.745×10^{-5}	X(4)	= 6.0
P(7)	= 16.7	X(5)	= 7.4
Q(7)	= 6.0	X(6)	= 1.2×10^{6}
X(1)	= 1.2	X(7)	= 16.0
X(2)	= 4.0		

Note that both blanks and commas have been used as delimiters, and that the constants are punched in the mode corresponding to the variable type. When the comma is used as a delimiter, it should *immediately* follow the final character in the preceding entry; *no* comma should appear after the last number in the input list.

Each time an unformatted READ statement is executed, data values are read, starting with a *new* data card; reading continues over more than one card, if necessary, until values have been read for all variables in the list. Any information on the last data card that is punched to the right of the value assigned to the last variable in the input list will be lost. Each constant must be contained on one card, so that the *splitting* of a constant between two data cards is *not allowed*. Thus, the data shown above could have been punched on three cards having the following content.

```
Card 1:  7,         1.745E-5,  16.7
Card 2:       6.,             1.2,
Card 3:  4. 5. 6. 7.4 1.2E6,    16.
```

In WATFOR, only integer constants, F-type and E-type real constants, the logical constants .TRUE. and FALSE., and the letters T and F (alternative forms for the logical constants) may appear on the data cards. Literal constants are *not* allowed. If the same constant, c, appears m times in sequence on a card, an abbreviated construction $m*c$ is allowed. For example, if we wished to read the value 7 for the integer variable NUMBER, the value .TRUE. for the logical variable SWITCH, the value 4.5 for the first five elements of the array A, and the value -1.7×10^{-15} for the next five elements of the array A, the input statement would be

```
READ, NUMBER, SWITCH, (A(I), I = 1, 10)
```

and the content of the data card could be

```
7 T 5*4.5 5*-1.7E-15
```

WATFIV, which allows variables to be assigned a *character* mode, as described in Sections 3.5, 3.27, and 3.28, allows such variables to appear in the input list; the corresponding data values are punched in the form of *literal constants*. Suppose that the character variables and their lengths equivalent to the number of characters or bytes are as follows:

Name	*Length*
BEAST	8
SEX	1
HAIR	5
EYES	5
KIND	10

Then if the CHARACTER declaration (see Section 3.27) and the unformatted input statement

```
CHARACTER  BEAST*8, SEX, HAIR*5, EYES*5, KIND*10
READ, BEAST, SEX, HAIR, EYES, KIND, AGE
```

appear in a program, three sets of data might be punched on cards as follows:

```
'DOG' 'M' 'BROWN' 'BROWN' 'BEAGLE' 1.5
'CAT', 'F', 'BEIGE', 'BLUE', 'SIAMESE', 3.
'RABBIT', 'M', 'WHITE', 'PINK' 'UNKNOWN' 2.25
```

If the length of the character variable exceeds the length of the assigned literal constant, blanks are automatically padded on the right when the constant is stored in the memory. If the number of characters in the literal constant exceeds the length of the character variable, any extra characters are trimmed from the right end of the constant.

WATFIV, but not WATFOR, allows the format-free input of data from *any* device. The general formatted input statement of Section 3.19 is modified by replacing η_1 (the statement number of the associated FORMAT declaration or the name of an array containing the format in character form) with an asterisk. Thus the most general form, including the optional END and ERR parameters, is

READ (m, *, END=η_2, ERR=η_3) $\mathscr{L}$

Assuming that device number 5 is the standard input device, the following WATFIV input statements are equivalent:

READ, $\mathscr{L}$
READ (5,*) $\mathscr{L}$

The format-free output statement, sometimes called the *simple* output statement, available in both WATFOR and WATFIV but *not* in FORTRAN, is

PRINT, $\mathscr{L}$

Here, the comma following the word PRINT is mandatory, and the list may be any WATFOR or WATFIV output list, already described in Section 3.19.

In WATFOR, variables and expressions of integer, real, and logical mode are allowed to appear in the list. The only restriction is that when expressions are used, the *first* character must *not* be a left parenthesis. Thus (A+B)/2. would be illegal while A/2. + B/2. or + (A+B)/2. would be valid.

Examples.

```
PRINT, N, ALPHA, P(7), Q(N), (X(I), I = 1, N)
PRINT, X, SQRT(X), 17.23, A*SIN(B) + Q
PRINT, 5, A+B.GE.6.2*BETA, 72.4E-7
```

The values of the expressions are printed, in order, across the output page. Integer, real, and logical expressions are printed using I12, E16.7, and L8 specifications, respectively, with one blank between successive fields. *No labeling is produced or is permitted*, that is, literal constants such as 'X = ' are not legal list elements.

WATFIV, on the other hand, permits both *literal constants* and variables of *character* mode to appear in the output list. This allows the programmer to label his output variables easily, without the necessity of using the A format specification. For example, if the integer variable NUMBER has the value 6, and the real variables X and Y have the values 7.5 and −0.6, respectively, then the statement

```
PRINT, 'NUMBER =', NUMBER, ', X.LT.Y =', X.LT.Y
```

would produce printed output having the appearance

```
NUMBER =                7 , X.LT.Y =          F
```

As an additional example, illustrating the use of character variables, the output statement

```
PRINT, 'THE', HAIR, BEAST, 'HAS', EYES, 'EYES'
```

could be placed following the READ statement in the example shown on page 101. If the third data card had just been read, then the following would be printed:

```
THE WHITE RABBIT  HAS PINK  EYES
```

The literal constants and character variables are automatically separated by a single blank. The programmer has no control over the positioning of the line on the printed page when the simple output statement is used exclusively.

In WATFIV, but *not* in WATFOR, the programmer may specify the output device number on which format-free output is to be written. The general formatted output statement of Section 3.19 is modified by replacing the statement number of the FORMAT declaration, η_1, with an asterisk. Thus, if device number 6 is the standard output device, the following are equivalent:

PRINT, $\mathscr{L}$
WRITE (6,*) $\mathscr{L}$

Both WATFOR and WATFIV allow the format-free output of information on the standard card-punch unit. The statement has the form

PUNCH, $\mathscr{L}$

where $\mathscr{L}$ is any acceptable WATFOR or WATFIV output list.

Format-Free Input and Output Using FORTRAN and WATFIV. FORTRAN and WATFIV, but *not* WATFOR, allow input and output statements of the form

READ (m,$\mathscr{N}$, END=η_2, ERR=η_3)
WRITE (m,$\mathscr{N}$)

that require no corresponding FORMAT declaration. Here, as in Section 3.19, m is the number of the device to be used for input or output, and the *optional* END=η_2 and ERR=η_3 entries cause a transfer to statements η_2 or η_3 in the event the end of data is encountered or a data transmission error occurs, respectively. $\mathscr{N}$ is the *name* of a *list* $\mathscr{L}$ of simple *unsubscripted variables* and/or *array names* whose values are to be read or written. The list $\mathscr{L}$ is defined in the declaration

NAMELIST /$\mathscr{N}$/ $\mathscr{L}$

More than one list may be defined in a single NAMELIST declaration as in

NAMELIST $/\mathcal{N}_1/\mathcal{L}_1/\mathcal{N}_2/\mathcal{L}_2/\ldots/\mathcal{N}_n/\mathcal{L}_n$

All NAMELIST declarations must precede the first executable statement in a program. A given name $\mathcal{N}$ may be used for both input and output, and a particular variable may appear in more than one NAMELIST.

If the NAMELIST declaration and associated input statement are

NAMELIST $/\mathcal{N}/\mathcal{V}_1, \mathcal{V}_2, \ldots, \mathcal{V}_n$
READ $(m, \mathcal{N})$

then the data card(s) containing values for the variables have the following general appearance

12...

First Card: &$\mathcal{N}$ $\mathcal{V}_1 = c_1$, $\mathcal{V}_2 = c_2$,

(Here come any intermediate cards.)

Last Card: ..., $\mathcal{V}_n = c_n$ &END

Column 1 of each card in the data set must *not* be used. The first card in the set must contain & in column 2, followed immediately, with no intervening blanks, by the NAMELIST name $\mathcal{N}$. The last card in the data set must end with the characters &END (no intervening blanks). Between these delimiters, *any* or *all* of the variables in the list, $\mathcal{V}_i$, $i = 1, 2, \ldots, n$, may be assigned values, in *any* order, by punching the name of the variable, an equal sign, and a constant, in sequence (an *assignment segment*):

$$\mathcal{V}_i = c_i$$

Except for literal constants, which may be assigned to variables of any mode, there must be agreement between the mode of the constant c_i and the mode of the variable $\mathcal{V}_i$. Thus,

```
X=7
I=9.5
```

would normally be illegal, while

```
X=7.
I=9
BETA='XYZ'
KIND='BEAR'
S=.TRUE.
```

would be acceptable, provided that S were of logical mode. The letters T and F may be used as abbreviations for the logical constants .TRUE. and .FALSE..

When more than one assignment segment appears, each must be separated from the one following by a comma. At least one blank must appear after the NAMELIST name and before the terminating characters &END. A comma may appear after the final segment, but is optional. In general, it is best to avoid embedded blanks in a segment, and to place the comma immediately after the final character in a segment (FORTRAN allows blanks to appear in places where WATFIV does not).

Example. We wish to read values for the variables N, A, B, and GAMMA, and the elements of the 3 by 4 matrix T, dimensioned using the DIMENSION declaration described in Section 3.25. Let the NAMELIST name be DATA. Then the appropriate declarations and input statement would be

```
DIMENSION  T(3,4)
NAMELIST / DATA / N, A, B, GAMMA, T
READ (5,DATA)
```

Suppose that in the first data set N, A, B, and GAMMA are to be assigned the values 10, 457.62, 0.00567, and 6.543×10^{-8}, respectively, while the matrix T is to have the values

$$\begin{matrix} 1.1 & 4.1 & 6.6 & 0.0 \\ 2.0 & 5.7 & 6.6 & 0.0 \\ 0.0 & 6.6 & 0.0 & 1.0 \end{matrix}$$

Then the corresponding data cards might appear as follows:

First Card

```
12...
 &DATA N=10,     A=457.62, B=0.00567,          GAMMA=6.543E-8,
```

Second Card

```
12...
 T=1.1,2.0,0.0,4.1,5.7, 3*6.6, 3*0.0,1.0 &END
```

Note that spacing of the segments is not crucial, that arrays are read *columnwise* in the usual FORTRAN fashion, and that the construction $m*c$ indicates that the constant c is to be repeated m times.

Suppose that none of the values read in the first data set were altered before the second execution of the READ statement, and that in the second data set, the variables N, A, B, and GAMMA are to have the values 12, 457.62, 0.00567, 7.555, respectively, and that T remains unchanged, except for the elements T(2,2) and T(3,4) which are to have the new values 7.6 and −0.5, respectively. Then an appropriate card for the second data set would be

```
12...
 &DATA N=12, GAMMA=7.555,  T(2,2)=7.6,  T(3,4)=-0.5 &END
```

Note that only those variables whose values are to be changed need be read.

To illustrate the reading of logical and character information, consider the following statements and an associated data card; the variables VALVE, FIRST, and LAST are variables of logical, real, and integer mode, respectively.

```
NAMELIST / GO / VALVE, FIRST, LAST
READ (5,GO)
```

12...

```
 &GO  FIRST='SAM',   LAST='KERN',   VALVE=T. &END
```

The contents of FIRST and LAST will be the same as if the characters SAM and KERN had been read using the format specifications A3, A4. In WATFIV, which allows character mode variables, one would normally declare FIRST and LAST to be of character mode.

Format-free output, produced by a WRITE statement of the form

WRITE (m,$\mathcal{N}$)

has essentially the appearance of the data cards for the corresponding READ statement. In this case, the values of variables in the list with name $\mathcal{N}$ are printed, in the order of appearance in the NAMELIST declaration, enclosed by the two delimiters &$\mathcal{N}$ and &END. The name of each variable is printed, followed by an equal sign and the current value, scaled appropriately for readability. When an array name appears in the list, *all* elements of the array are printed. In FORTRAN, values of integer, real, and logical variables are printed in the corresponding mode, even if the contents of such variables are coded characters. Therefore, the characters read into the variables FIRST and LAST in the preceding example could *not* be printed out using the NAMELIST output statement; of course, a formatted output statement could be used, provided the associated format contained A field specifications. In WATFIV, however, characters *can* be written directly by the NAMELIST output statement, provided that the corresponding variables are of *character* mode.

The programmer has no control over the specific appearance of the output, except that the order of the printed values is the same as the order of the variables in the corresponding NAMELIST statement.

3.22 Terminating Program Execution—STOP Statement

Execution of the object program produced by a FORTRAN, WATFOR, or WATFIV compiler can be terminated in a variety of ways. Most programs are written to process more than one set of data values; in fact, the number of such data sets is usually arbitrary. In such cases, the natural way to terminate computation is with the attempted execution of an input statement, after the last data set has already been processed. As discussed in Section 3.19, the addition of the END option to the READ statement is one way to detect when the last of the data cards has been read. If the END option is used, the programmer retains control of the machine when the end of the data file is encountered by transferring to another statement in his FORTRAN program. Often, the programmer wishes to discontinue all computation when the end of data is detected, in which case he has the option of: (a) doing nothing special, or (b) using the END option in the READ statement and transferring to a statement that will terminate execution somewhat more gracefully. Either approach is considered to be perfectly acceptable programming practice.

If the END option is *not* used, then execution will be terminated automatically when a READ statement is encountered but there are no more data cards left. When the "end-of-data" condition is encountered, the WATFOR compiler will write the comment (see Section 3.39 and Appendix E)

```
UN–1  'END OF FILE ENCOUNTERED'
```

and the WATFIV compiler will write the comment (see Appendix F)

```
UN–1  'END OF FILE ENCOUNTERED (IBM CODE IHC217)'
```

In IBM systems using the standard G-level compiler, the following message will be written (see Appendix D) for FORTRAN programs:

```
IHC217  END OF DATA SENSED
```

Since such messages are somewhat system-dependent, variations on these messages may be encountered.

Control can be passed directly to the operating system at *any* time (see Chapter 4), resulting in immediate termination of execution of the object program, with any of the FORTRAN statements:

```
CALL EXIT
STOP
STOP c
```

Here *c* is an unsigned integer constant of five digits or less. In the first instance, no message will be written. In the latter two cases, one of the messages

```
      STOP
or    STOP c
```

will be written. If the last of the three forms is used, then the WATFOR compiler will print the additional comment

```
PS–0  'STOP WITH OPERATOR MESSAGE NOT ALLOWED.
       SIMPLE STOP ASSUMED'
```

while the WATFIV compiler will print the message

```
PS–0  'OPERATOR MESSAGES NOT ALLOWED:SIMPLE STOP
       ASSUMED FOR STOP, CONTINUE ASSUMED
       FOR PAUSE'
```

The STOP or STOP *c* statements can be used to terminate execution at any time, not just after encountering the end of the data. In fact, many programs have no data at all, and this is the natural way to stop computation. The computer is not actually stopped as a result of execution of the STOP statement; only the execution of the program containing the STOP statement is terminated. The END declaration, described in the next section is *nonexecutable*, and hence must *not* be used to terminate program execution.

3.23 END Declaration

The declaration

```
END
```

must appear once, as the very *last* statement, in *every* subprogram. It is a signal to the compiler that no more statements appear in that subprogram. In common with other nonexecutable statements, the END declaration should not be numbered, nor should it be the last statement in execution of the program. This means that a transfer statement, a STOP statement, or a call for return to the operating system will almost invariably be the final executable statement preceding the END declaration.

3.24 Comments

A comment prefixed by the letter C will cause no action other than to reproduce a message in the listing of the *source program* produced by the compiler. For example, the cards

```
C     *** COMPUTE RMS VALUE OF CURRENT ***
C     *** CURRENT IN AMPERES ***
```

might be used at some stage to assist the reader in interpreting a program. Each comment card must have a C in column 1; a continuation code is *not* allowed. Note that the comment is *not* written as part of the output from the *object program* generated by the compiler.

3.25 DIMENSION Declaration

Sufficient memory locations must be reserved for the individual elements of a subscripted array by using the declaration

$$\text{DIMENSION } \mathscr{V}(\mathscr{I}_1, \mathscr{I}_2, \ldots, \mathscr{I}_n)$$

where $\mathscr{V}$ is the array name, and $\mathscr{I}_1, \mathscr{I}_2, \ldots, \mathscr{I}_n$ are *unsigned integer constants* equal to the *largest* subscript values likely to be used for each of the n subscripts. For one-dimensional arrays, $\mathscr{I}_2, \ldots, \mathscr{I}_n$ would be deleted. The smallest subscript values are always assumed to be *unity*, but it is not essential to make complete use of all the storage space thus assigned. More than one array may be dimensioned in a single DIMENSION declaration. For example,

```
DIMENSION A(5,10), STRESS(15)
```

would reserve 50 memory locations for the elements A(1,1) through A(5,10) and 15 locations for the elements STRESS(1) through STRESS(15). The DIMENSION declaration for an array must precede the first mention of the array name in the program.

3.26 DATA Declaration

It is frequently useful to be able to *preset* or *initialize* values of variables (possibly subscripted, as long as the subscripts are constants) *before execution* of the program. The DATA declaration can be used for this purpose; at its simplest, it has the form

$$\text{DATA } \mathscr{L} / \; c_1, c_2, \ldots, c_n \; /$$

Here, $\mathscr{L}$ represents a list of variables or array names that will be preset with the n constants $c_1, c_2, \ldots, c_n$. For example,

```
      DATA X, Y(6), INT, FIRST, LAST, SWITCH/ –6.7, 1.93E–6, 49,
     1  'JOHN', 'DOE', .TRUE. /
```

would preset X with the value −6.7, Y(6) with 1.93×10^{-6}, INT with 49, FIRST with the characters "JOHN", LAST with the characters "DOE", and SWITCH with the logical value .TRUE.. The vector Y would appear in a DIMENSION declaration and the variable SWITCH in a LOGICAL declaration (see the next section). Variables of standard length and of any mode may be preset with character information in the form of literal constants having no more than four characters. Otherwise the mode of the variable and its assigned value must agree. In WATFIV, character information normally would only be assigned to variables of character mode (see Section 3.27).

To preset entire arrays, the array name only is given; also, the construction $m*c$ may be used to repeat the constant c a total of m times. For example, the sequence

```
      DIMENSION A(3,3),  B(5,10)
      DATA A, B/ 6*1., 1.5, 2.7, –6.2, 16.3, –14.6, 48*0. /
```

would preset the first two columns of matrix A with the value 1., A(1,3) with 1.5, A(2,3) with 2.7, A(3,3) with −6.2, B(1,1) with 16.3, B(2,1) with −14.6; the remaining 48 elements of the matrix B would be assigned zeros. Every element of a named array must be assigned a value.

The pattern of slashes can be repeated, in which case a comma must precede every list of variables except the first. For example,

```
      DIMENSION B(5,10) Y(10)
      DATA X, Y(6)/ –6.7, 1.93E–6/, B/ 16.3, –14.6, 48*0./,
     1 SWITCH/ .TRUE./
```

The DATA declaration must not be confused with the process of accepting input data values from cards by means of a READ statement. The DATA declaration is *nonexecutable*, but the READ statement is *executable*. Variables initialized with the DATA declaration may, however, be modified subsequently as any other variable might be. In WATFOR and WATFIV, the list $\mathscr{L}$ may contain array segments identical in form to that described in Section 3.19. For example, the statement

```
      DATA (B(I), C(I), I = 1, 9, 2) / 5*5.3, 5*–0.7 /
```

would assign the value 5.3 to B(1), C(1), B(3), C(3), and B(5), and the value −0.7 to C(5), B(7), C(7), B(9), and C(9). There are some restrictions, discussed in Sections 3.34 and 3.35, on the assignment of initial values to variables in blank or labeled COMMON blocks.

3.27 Mode Declarations, Length, Precision, and Range

As described in Section 3.5, the first letter in the name of a variable normally establishes its mode; if the first letter is I, J, K, L, M, or N, the variable is of integer mode; otherwise it is of real mode. There are no reserved letters for logical mode. If desired, we can override this convention by using one or more *mode* or *type* declarations of the form

```
REAL 𝓛
INTEGER 𝓛
LOGICAL 𝓛
```

Here, the simple variable or array names comprising the list $\mathscr{L}$ are understood by the compiler to be real, integer, or logical, as specified, *regardless* of their initial letters. A mode declaration must precede the first appearance of the variable or array name in the program. Occasionally (see Section 3.30), a function name, without arguments, may also be included as part of the list $\mathscr{L}$.

Examples.

```
REAL  LAMBDA,  LENGTH,  I
INTEGER  COUNT
LOGICAL  SWITCH,  ANSWER
```

In WATFIV, but *not* in WATFOR and FORTRAN, variables may be declared to be of *character* mode with the declaration

```
CHARACTER  𝓛
```

Character variables named in the list $\mathscr{L}$ are assumed to contain one character, unless another optional length is selected, as described below.

Precision, Range, and Word-Length. In Chapter 1, we described the organization of the IBM 360/370 computer store. The *byte*, the smallest addressable memory element, consists of eight binary bits, equivalent to two hexadecimal characters. Depending on the mode or type of information and the desires of the FORTRAN programmer, computer words assigned to hold the values of constants or variables may be of length 1, 2, 4, or 8 bytes. In 360 parlance, a word of length 1, 2, 4, or 8 bytes is called a *byte*, *half-word*, *full word*, or *double word*, respectively. The *standard* word size assigned by the FORTRAN, WATFOR, and WATFIV compilers for all integer, real, and logical constants and variables is the full word. In WATFIV, variables of character mode are assigned a standard length of one character or byte. The programmer may choose, as options, other word sizes. The selection of lengths and magnitudes, where applicable, is shown in the following table (these options may differ for compilers run on non-IBM 360 computers).

Mode (type)	*Number of Bytes*	*Approximate Magnitude Range*	*Precision (Decimal Digits)*
Standard integer	4	0 through 2147483647	9+
Optional integer	2	0 through 32767	4+
Standard real	4	$0, 16^{-65}$ through 16^{63} $0, 10^{-78}$ through 10^{75}	7+
Optional real	8	$0, 16^{-65}$ through 16^{63} $0, 10^{-78}$ through 10^{75}	16+
Standard logical	4	.TRUE. or .FALSE.	
Optional logical	1	.TRUE. or .FALSE.	
Standard character (WATFIV only)	1		
Optional character (WATFIV only)	2–255		

Constants of *integer* and *logical* mode are always assigned to words of standard length. *Real constants* of the F type are always assigned to standard words by FORTRAN compilers. Real constants of the F type with more than seven significant digits are stored in the double-word or *double-precision* form by the WATFOR and WATFIV compilers. If it is necessary to insure that a real constant be stored in double-precision form, the exponential type of real constant should be used, with the letter E replaced by the letter D. Up to 16 significant digits may appear. Examples of double-precision constants are:

FORTRAN, WATFOR, WATFIV: 1.7D10,
1.789654323889447D0, −6.655D−22, 1.D0
WATFOR, WATFIV: 1.789654323889447, −0.00123456789

If word lengths other than the standard are to be used for *variables*, several options are open to the programmer. The most straightforward approach is to modify the mode declarations as follows:

REAL*8 $\mathscr{L}$
INTEGER*2 $\mathscr{L}$
LOGICAL*1 $\mathscr{L}$
CHARACTER*b $\mathscr{L}$ (WATFIV only)

Here, the variables in the list $\mathscr{L}$ will be assigned the appropriate mode and storage of the indicated length, in bytes. In the character declaration, b must be an unsigned integer constant in the range 1 through 255.

Examples.

```
REAL*8  ALPHA, MEAN, BETA
INTEGER*2  G, N
LOGICAL*1  SWITCH, BOOL
CHARACTER*5 WORD, INDEX     (WATFIV only)
```

A statement completely equivalent to the declaration REAL*8 is the declaration

DOUBLE PRECISION $\mathscr{L}$

Example. DOUBLE PRECISION ALPHA, MEAN, BETA

Individual variables within a mode declaration may also be specified to be of particular length by appending **b* to the variable name, where *b* is the number of bytes to be assigned to that particular variable. If present, **b* overrides the assumed length for other variables in the list. For example, the declaration

```
REAL*8  ALPHA, MEAN*4, BETA
```

will cause ALPHA and BETA to be assigned to double words while MEAN would be designated as a full-word real variable.

On occasion, it is desirable to declare that *all*, or perhaps most, variables are to be of nonstandard mode or length. The IMPLICIT declaration is provided for this purpose. For example, if all variables whose names begin with the letters A, B, D, E, and F are to be double-precision real variables, all variables beginning with the letters M, N, and P are to be integers assigned to half-words, and all variables beginning with R are to be full-word integers, while the usual rules apply for variables beginning with other letters, the declaration

```
IMPLICIT  REAL*8(A, B, D–F), INTEGER*2(M, N, P), INTEGER(R)
```

would accomplish the desired mode and length assignments. In WATFIV, but not in FORTRAN or WATFOR, first letters may be reserved for variables of character mode; for example,

```
IMPLICIT  CHARACTER*27(C, S–U)
```

would assign character mode to variables whose first letters are C, S, T, and U; each variable would be of length 27.

The IMPLICIT statement is particularly useful in converting a program from single- (four-byte) to double- (eight-byte) precision arithmetic for all calculations. If the usual naming conventions have been used for all variables, this conversion from single- to double-precision can be accomplished with the single statement

```
IMPLICIT  REAL*8(A–H, O–Z, $)
```

One word of caution is in order here. The IMPLICIT statement also assigns the given mode to any *function references* starting with the indicated letters. Such an assignment will almost always result in computational problems unless care is also taken to insure that the functions are modified to accept double-precision arguments and to return double-precision values. This is a simple matter when the programmer is supplying his own FORTRAN functions, and is described in Section 3.30. The IMPLICIT declaration *cannot* change the single-precision nature of the *library* functions listed in Section 3.6, however. The double-precision equivalents of these functions, discussed in the next part of this section, must be referenced instead. Thus, if the preceding IMPLICIT declaration appeared in the program, the following would be illegal

```
A = SQRT(B)
C = 7.5*SIN(3.2*X + Y)
```

whereas

```
A = DSQRT(B)
C = 7.5*DSIN(3.2*X + Y)
```

would be acceptable.

When present, there must be only *one* IMPLICIT statement in a program. It must be the *first* statement in a main program and the *second* in a function or subroutine subprogram (see Sections 3.30 and 3.33). Any assignments made in a mode or DOUBLE PRECISION declaration override the implied assignments of an IMPLICIT declaration.

When operands of different lengths are mixed in arithmetic expressions, all operands are normally converted to the mode of the operand of greatest precision; the result is also of that precision. If X, Y, M, and N were of type real-full, real-double, integer-half, and integer-full, respectively, the modes and lengths of the results for some typical operations would be as shown below.

Expression	*Mode*	*Length*
M + M	Integer	Half
2*M	Integer	Full
2.*M	Real	Full
2.D0*M	Real	Double
M*N	Integer	Full
X/Y + M	Real	Double
M − X	Real	Full
M − Y	Real	Double
5*DSQRT(Y)	Real	Double
5.D0*SQRT(X)	Real	Double
M*SIN(X)	Real	Full

Double-Precision Function References. With the exception of the function IABS, each of the functions listed in Section 3.6 requires a real argument of standard length (four bytes) and returns a real value of standard length. They are called *single-precision* functions. A comparable set of functions, called *double-precision* functions, is also available. These require real arguments of double length and return real values of double length. The names, listed in the same order as those of the single-precision functions in Section 3.6, are:

```
DABS      DERF
DLOG      DERFC
DLOG10    DEXP
DARCOS    DGAMMA
DARSIN    DSIN
DATAN     DSINH
DATAN2    DSQRT
DCOS      DTAN
DCOSH     DTANH
DCOTAN
```

Function references are of the same form as before. For example,

```
DSQRT(3.D0)
```

has the value $\sqrt{3}$, accurate to approximately 16 significant decimal digits.

Assigning Dimensions and Initial Values in Mode Declarations. The mode declarations may be used to specify the dimensions and initial values as well as the modes and types of variables. The general form is

$$\mathscr{T}*b \quad \mathscr{V}_1*b_1(\bar{d}_1)/ \ \bar{x}_1 \ /, \ \ldots \ /, \ \mathscr{V}_n*b_n(\bar{d}_n)/ \ \bar{x}_n \ /$$

Here $\mathscr{T}$ is one of the *types* INTEGER, REAL, LOGICAL, or CHARACTER (WATFIV only); $\mathscr{V}_i$, $i = 1, 2, \ldots, n$ are *variable names* that are to be of type $\mathscr{T}$ and of length b_i bytes; b_i is optional, and, if absent, is assigned the value b. If $\mathscr{V}_i$ is a subscripted variable, then the *dimensioning information* is given by $(\bar{d}_i)$ where d_i is a string of integers, separated by commas, and equivalent to those that would appear in a DIMENSION declaration for the same variable (see Section 3.25). The optional entries / $\bar{x}_i$ / are the *initial values* to be assigned to $\mathscr{V}_i$, and are equivalent to those that would appear in a DATA declaration for the same variable. If b does not appear, then the standard length of page 111 is used.

Example.

```
REAL*8  ALPHA*4(10,5,7)/ 350* 1./, MEAN, DELTA(10)/
1     2*5.D0,8*0.D0/
```

is equivalent to the following:

```
REAL*8  MEAN, DELTA
DIMENSION ALPHA(10,5,7), DELTA(10)
DATA  DELTA, ALPHA / 2*5.D0, 8*0.D0, 350*1. /
```

We shall not use the compact form of the mode declaration in the examples in this text.

3.28 Manipulating Character Information

FORTRAN and WATFOR are rather weak in their character-handling capability. Character information can be assigned to variables of integer, real, or logical mode in one of three ways: (1) preset in a DATA declaration, (2) read using the general formatted input statement with an accompanying A format specification, or (3) (for FORTRAN only) read in the form of a literal constant using the unformatted or NAMELIST input statement. Variable character information can be written only by using the general formatted output statement with an accompanying A format specification.

An example assignment that uses the DATA declaration is

```
DIMENSION  N(3)
DATA  N / 'USA', 'USSR', 'GB' /
```

If N is a variable of standard length, the literal constants must not consist of more than four characters each. Examples illustrating the reading and printing of character information with the formatted input and output statements are shown on pages 96 through 99. The reading of literal constants using unformatted input is shown on pages 101–103 and 105.

Direct calculations involving literal constants are not allowed in either FORTRAN or WATFOR. For example, the following would be incorrect syntactically:

```
N(3) = 'GB'
IF (NATNLT .EQ. 'GB')  WRITE (6,200)
```

A limited amount of computation may be carried out among *variables* containing character strings, provided that the variables involved are of the *same* mode, either integer or real but *not* logical. Thus, if N(3) had been assigned the character string 'GB' in a DATA declaration, as shown above, then the following would be acceptable:

```
M = N(3)
IF ( NATNLT .EQ. N(3) )  WRITE (6,200)
```

After the first of these statements is executed, the variable M will contain the same character string as N(3). In the second example, the WRITE statement would be executed only if the variable NATNLT contained the same character string as N(3).

If comparison of character-valued variables other than for equality (sometimes called *string-matching*) is desired, then only *integer* mode variables should be used. Advantage can be taken of the fact that the hexadecimal equivalents of the alphabetic characters shown in Table 1.10 increase in magnitude with the order of the alphabet. For example, if the N integer variables NAME(1), NAME(2), . . . , NAME(N) contain four-letter words in no particular order, the statement sequence

```
      NM1 = N - 1
      DO  2  I = 1, NM1
      IP1 = I + 1
      DO  2  J = IP1, N
      IF ( NAME(I) .LE. NAME(J) )  GO TO 2
         NSAVE = NAME(I)
         NAME(I) = NAME(J)
         NAME(J) = NSAVE
    2 CONTINUE
```

would sort the N words into alphabetical order in the NAME array. The algorithm used here is quite simple. The word in NAME(1) is compared, in turn, with each of the other elements of the NAME array; whenever a word pair is found to be out of order, the two words are interchanged. After one complete sweep through the array, the first word in alphabetical order is in NAME(1). Next, NAME(2) is compared with the elements of the array with larger subscripts in similar fashion; after the second pass, the word second in alphabetical order will be in NAME(2). The procedure is repeated until a total of N − 1 such passes through the array have been completed.

Character Manipulation Using WATFIV. In contrast to the rather limited character-handling capability of both FORTRAN and WATFOR, the WATFIV language allows rather general manipulation of character strings. As already described in Sections 3.5, 3.20, and 3.27, WATFIV allows variables to be assigned a character mode; literal constants are also considered to be of character mode. The reading and writing of literal constants and character variables is covered in Sections 3.20 and 3.21.

Let $\mathscr{C}$ be a *character expression* consisting of one of the following forms:

$$\mathscr{V}_{\mathscr{C}}$$

$$'c'$$

$$(\mathscr{C})$$

Here, $\mathscr{V}_{\mathscr{C}}$ is a simple character variable or the element of a character array, and $'c'$ is a literal or character constant.

Character assignment statements of the form

$$\mathscr{V}_{\mathscr{C}} = \mathscr{C}$$

are allowed. For example, the following would be valid WATFIV statements, assuming that LETTER and ALPHA have been declared to be of character mode.

```
LETTER(2) = 'B'
LETTER(4) = ('D')
ALPHA = LETTER(2)
```

Two character expressions $\mathscr{C}_1$ and $\mathscr{C}_2$ may appear with one of the relational operators of Section 3.9 in simple logical expressions of the form

$$\mathscr{C}_1 \; . \mathscr{R} . \; \mathscr{C}_2$$

For purposes of comparison, the character expressions are treated as if they were integers, converted according to the hexadecimal codes of Table 1.10. For example, the logical expression

```
'A' .LT. 'B'
```

will always be true, whereas

```
'MARY' .LT. 'JACK'
```

will always be false. If one of the operands is shorter in length than the other, the shorter operand is padded on the right with blanks to produce two operands of equal length. Thus,

```
'CHURCHILL' .GT. 'HITLER'
```

is equivalent to

```
'CHURCHILL' .GT. 'HITLER
```

Because the hexadecimal equivalents of the characters increase in the order of the alphabetic characters, these simple logical expressions allow the programmer to alphabetize character strings containing letters and blanks rather easily. For example, the program segment from page 115 could be made into a complete program for alphabetizing as many as 100 words of up to 9 letters, as follows:

```
      CHARACTER*9  NSAVE, NAME
      DIMENSION  NAME(100)
    1 READ, N, (NAME(I), I = 1, N)
      NM1 = N - 1
      DO 2 I = 1, NM1
      IP1 = I + 1
      DO 2 J = IP1, N
      IF (NAME(I) .LE. NAME(J))  GO TO 2
         NSAVE = NAME(I)
         NAME(I) = NAME(J)
         NAME(J) = NSAVE
```

```
 2 CONTINUE
   PRINT, N, (NAME(I), I = 1, N)
   GO TO 1
   END
```

Note that to change the length of the words to be alphabetized, only the CHARACTER declaration need be modified. In the earlier example, however, the number of characters is limited by the word sizes of integer variables allowed, either two or four, depending on whether the optional or standard length is chosen.

Character expressions may appear in compound logical expressions as well as simple ones.

Example.
```
IF ( WORD(1).EQ.'BLACK' .AND. WORD(2).EQ.'CAT' )
1 GO TO 5
```

One of the most interesting features of the WATFIV language is the capability of performing *core-to-core* input and output operations. As the name implies, reading and writing operations are confined to different parts of the core memory. In effect, information *already* stored in memory can be manipulated under format control and written into other parts of the memory; individual characters can be examined and manipulated. It is important to remember that the reading and writing operations do *not* involve either card readers or line printers as have all the input and output statements discussed previously.

The *core-to-core output statement* has the form

WRITE $(\mathscr{V}_{\mathscr{C}}, \eta)$ $\mathscr{L}$

where $\mathscr{V}_{\mathscr{C}}$ is a simple character variable, the name of a character array, or an element of a character array, η is the statement number of an associated FORMAT declaration (or the name of a character array or an element of a character array where the format specification is stored in coded character form), and $\mathscr{L}$ is a legal WATFIV output list (see Section 3.19). This statement causes the values associated with the elements of the list $\mathscr{L}$ to be prepared as unit output records under control of the format specification in the usual way. Instead of being written as successive lines on a printer, however, the successive unit output records are written into successive elements of the character array $\mathscr{V}_{\mathscr{C}}$. If only one output record is involved, then $\mathscr{V}_{\mathscr{C}}$ may be a simple variable. If $\mathscr{V}_{\mathscr{C}}$ is an element of a character array, then the first output record is written into that element, the second output record is written into the next element, etc. The number of characters in any record should not exceed the length of the character variable; if fewer characters are generated for any record, then blanks are padded on the right to fill the corresponding character variable.

One of the major uses of the core-to-core WRITE statement is in the preparation of execution-time formats for use in other ordinary input and output statements. Suppose that we wish to prepare an output format of the form

(1X, *nn*F*mm.j*)

in the first three elements of the character array OUTPUT, where each element of OUTPUT is of length five characters, and *nn*, *mm*, and *j* are the values of three integer variables N, M, and J, respectively (to be computed in the program). Then the program would have the following general appearance:

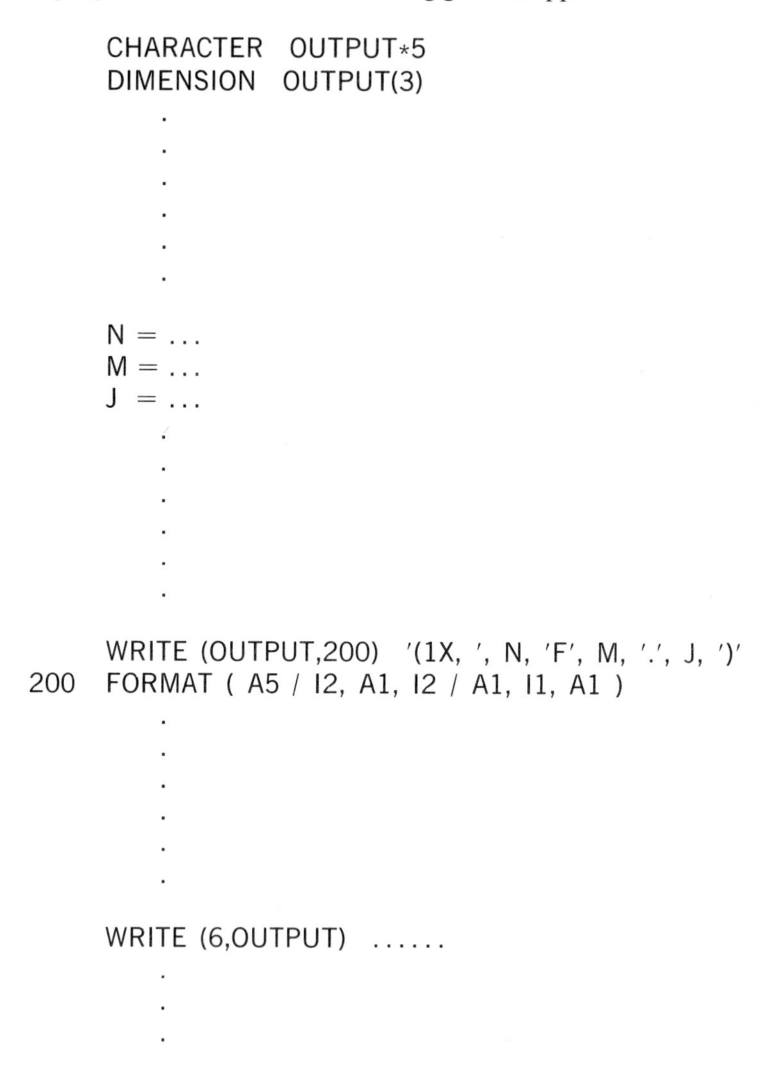

```
      CHARACTER  OUTPUT*5
      DIMENSION  OUTPUT(3)
          .
          .
          .
          .
          .
          .
      N = ...
      M = ...
      J = ...
          .
          .
          .
          .
          .
          .
      WRITE (OUTPUT,200)  '(1X, ', N, 'F', M, '.', J, ')'
  200 FORMAT ( A5 / I2, A1, I2 / A1, I1, A1 )
          .
          .
          .
          .
          .
          .
      WRITE (6,OUTPUT)  ......
          .
          .
          .
```

If the values computed for N, M, and J were 6, 15, and 4, respectively, then the contents of the elements of OUTPUT would be as follows:

```
OUTPUT(1)      '(1X,□ '
OUTPUT(2)      '□6F15'
OUTPUT(3)      '.4)□□ '
```

so that the format used in the last WRITE statement would be

```
(1X,□□6F15.4)
```

Example. As a more complicated example, suppose that we wish to prepare a plot of the sine function at intervals of 10° so that the printer output has the appearance shown on page 119.

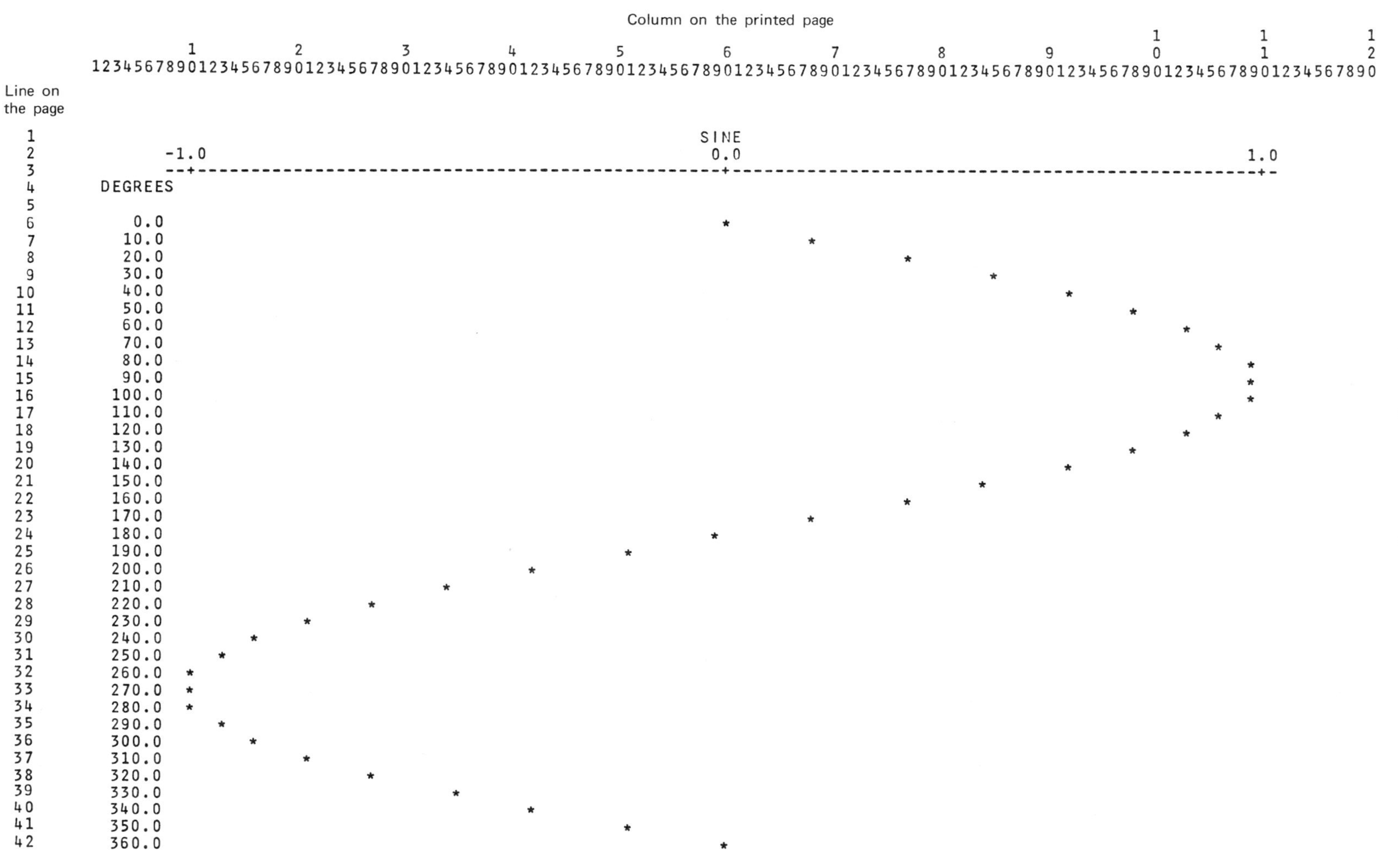
Column on the printed page
Line on the page
SINE
-1.0
0.0
1.0
DEGREES
0.0
10.0
20.0
30.0
40.0
50.0
60.0
70.0
80.0
90.0
100.0
110.0
120.0
130.0
140.0
150.0
160.0
170.0
180.0
190.0
200.0
210.0
220.0
230.0
240.0
250.0
260.0
270.0
280.0
290.0
300.0
310.0
320.0
330.0
340.0
350.0
360.0

The following is a complete WATFIV program that will produce the desired graphical output.

```
      CHARACTER  GRAPH*10
      DIMENSION  GRAPH(2)
      DATA  GRAPH / '(1X,F6.1,T', '  ' /
      DEGPR = 180./3.141593
      WRITE (6,200)
      DO 3  I = 1, 37
      DEGREE = 10*(I-1)
      NCOL = 60. + 50.*SIN(DEGREE/DEGPR)
      WRITE (GRAPH(2),201)  NCOL, ','''*''')'
    3 WRITE (6, GRAPH) DEGREE
      STOP
  200 FORMAT ('1', T58, 'SINE' / ' ', T8, '-1.0', T59, '0.0',
     1  T109, '1.0' / ' ', T8, '--+', 49('-'), '+', 49('-'),
     2  '+-' / ' DEGREES' / ' ')
  201 FORMAT (I3, A5)
      END
```

Here, DEGPR is the number of degrees per radian and DEGREE is the angle in degrees; note that the function SIN requires its argument to be in radians. I is an integer variable that allows us to calculate the successive values of DEGREE easily, and also serves to count the number of lines in the part of the graph containing the plotting character, an asterisk. NCOL is the column number on the paper where the plotting character should be printed, for each successive line. The core-to-core WRITE statement prepares the format for the ordinary WRITE statement following it.

The format for the ordinary WRITE statement is in a character array named GRAPH that has two elements of length 10, preset with the values.

```
GRAPH(1)        '(1X,F6.1,T'
GRAPH(2)        '          '
```

in the DATA statement. The first three declarations could be replaced by the more compact form, described in Section 3.27:

```
CHARACTER GRAPH*10(2)/'(1X,F6.1,T', '  '/
```

The core-to-core WRITE statement places the characters

nnn,'*')□□

into GRAPH(2), where *nnn* is the number of the column in which the asterisk is to be printed. The content of GRAPH(1) does not change from its initial value. For example, for the 9th line on the printed output, the angle is 60°, the sine function has the value 0.5, and NCOL has the value 85. Then the format that will be used by the final WRITE statement will be

```
(1X,F6.1,T□85,'*')
```

so that the asterisk will be printed in column 85.

The *core-to-core input statement* has the form

$$\text{READ } (\mathscr{V}_{\mathscr{C}}, \eta) \ \mathscr{L}$$

where $\mathscr{V}_{\mathscr{C}}$ is a simple character variable, the name of a character array, or an element of a character array; η is the statement number of a FORMAT declaration (or the name of a character array or an element of a character array containing a format specification); and $\mathscr{L}$ is a valid WATFIV input list (see Section 3.19).

This statement causes the character content of successive elements of $\mathscr{V}_{\mathscr{C}}$ to be read as successive unit input records under control of the associated format specification. The information is assigned to the elements of the list in order; effectively, the contents of successive elements of $\mathscr{V}_{\mathscr{C}}$ are treated exactly as cards would be, assuming that ordinary READ statements were being used.

Example. Suppose that we want to know if the third character of a word saved in the eighth element of the array NAME is the letter B. Assume that each element of NAME has up to ten characters. Then the following statements might appear in the WATFIV program:

```
      CHARACTER  NAME*10, LETTER
      DIMENSION  NAME(100)
           .
           .
           .
           .
      READ (NAME(8),100) LETTER
  100 FORMAT (2X, A1)
      IF (LETTER .EQ. 'B')  GO TO 10
           .
           .
           .
```

The third letter of NAME(8) is read into the character variable of length 1 named LETTER. Presumably, appropriate action will be taken at the statement numbered 10.

An important use of the core-to-core input statement is the rereading of card images, with possibly different formats. Recall that when a card is read with a formatted input statement, it is read only *once*, and that the card must be punched in strict accordance with the format specification and the input list.

Example. Suppose that for each employee, a company maintains personnel information on punched cards. We wish to read the cards and tabulate each employee's name, social security number, age, and sex on the printer. The employee cards containing this information have been punched in two different formats as follows:

	Format 1 card columns	*Format 2 card columns*
Name	5–25	2–20
S. S. No.	41–49	21–29
Age	30–31	30–31
Sex	72	50

The names are punched as alphabetic characters, the social security number and age as integers, and the sex as either the letter M (male) or F (female). The card deck also contains other employee information cards of no interest at the moment. The format for each card is indicated by a punch appearing in column 80. The letter A appears in column 80 of cards punched according to format 1, the digit 2 appears in column 80 of cards punched according to format 2, and all other cards have different characters in column 80. A complete WATFIV program to handle this problem follows:

```
        CHARACTER  CARD*79, FCODE, NAME*21, SSNO*9, AGE*2, SEX
        WRITE (6,200)
      1 READ (5,100)  CARD, FCODE
        IF (FCODE .EQ. '2')  GO TO 2
        IF (FCODE .NE. 'A')  GO TO 1
        READ (CARD,101)  NAME, AGE, SSNO, SEX
        GO TO 3
      2 READ (CARD,102)  NAME, SSNO, AGE, SEX
      3 WRITE (6,201)  NAME, SSNO, AGE, SEX
        GO TO 1
    100 FORMAT (A79, A1)
    101 FORMAT (4X, A21, 4X, A2, 9X, A9, 22X, A1)
    102 FORMAT (1X, A19, A9, A2, 18X, A1)
    200 FORMAT ('1', T5, 'NAME', T26, 'SOC.SEC.NO.',
       1 T40, 'AGE', T47, 'SEX' / ' ')
    201 FORMAT (' ', A21, 4X, A9, 4X, A3, 5X, A1)
        END
```

All variables have been assigned character mode. Here CARD contains the content of the first 79 columns of a card and FCODE contains the character punched in column 80. Other data items are assigned obvious names.

3.29 EQUIVALENCE Declaration

It is occasionally convenient to refer to the same storage location by two or more different names. This is achieved by using the declaration

$$\text{EQUIVALENCE}\ (\mathscr{V}_1, \mathscr{V}_2, \ldots, \mathscr{V}_n)$$

Here, the $\mathscr{V}_i$, $i = 1, 2, \ldots, n$ may be: (1) simple variable names, (2) array names, (3) an element of an array, or (4) an element of a multidimensional array with a single equivalent subscript. The listed variables may be of any mode; modes of individual entries need not agree.

Examples.

```
EQUIVALENCE (SPEED, VEL)
EQUIVALENCE (AVG, NANCY, SWITCH)
EQUIVALENCE (GO, Y(2), X(5))
```

In the first of these examples, the variables SPEED and VEL could be used interchangeably. In the second example, assuming that AVG is of real mode, that NANCY is of integer mode, and that SWITCH is of logical mode, it would be unreasonable to use the three variables interchangeably. In this case, the same location would serve to hold values of the variables of three different modes at different times during execution of the program. In the third example, the specification that Y(2) is to be coincident with X(5) automatically assigns relative locations to all other elements of the *X* and *Y* arrays; that is, this statement also equates Y(1) with X(4), Y(3) with X(6), etc. Care must be taken to avoid destroying the normal sequential arrangement of arrays.

Example (invalid). EQUIVALENCE (Y(2), X(5), Y(4))

The most important use of the EQUIVALENCE statement is to save storage in a program requiring large arrays that are used during only part of the execution of the program. For example, suppose that a large matrix A, dimensioned to have 200 rows and 10 columns, is needed only during the first part of a program's execution, while two other arrays, a matrix M with 30 rows and 40 columns and a singly subscripted variable DELTA with 300 elements, are not required until later in the execution of the program. Then the 2000 full words (8000 bytes) assigned to A is more than adequate to hold the 1200 words of M and the 300 words of DELTA. Appropriate EQUIVALENCE declarations for this situation are:

```
      EQUIVALENCE  (A,M), (A(1,7), DELTA)
or    EQUIVALENCE  (A(1,1), M(1,1)), (A(1,7), DELTA(1))
or    EQUIVALENCE  (A(1), M(1)), (A(1201), DELTA)
```

The first element of an array can be indicated by the array name alone. Care must be taken to assure that the usual FORTRAN storage conventions have been used in assigning equivalent locations. In this example, M will occupy the first 1200 locations assigned to A. Because two-dimensional arrays are stored column by column, this means that the 1201st element of A is A(1,7). Elements of multidimensional arrays may be specified with an equivalent single subscript that takes into account the storage scheme used, as is illustrated by the final example above. Note that more than one equivalence group can be assigned in a single EQUIVALENCE declaration.

3.30 FUNCTION Subprograms

Before embarking on the material in this and the following two sections, we suggest that the student reread Section 2.9 describing procedures. As indicated on page 49, a *procedure* is a self-contained algorithm having:

(1) A *name*.
(2) A set of *dummy arguments*.
(3) A beginning, or *entry point*.
(4) An end or *return point*.

A procedure is activated when *reference* to it is made in another algorithm that supplies *actual* or *calling arguments* to replace the corresponding dummy arguments.

As illustrated by Examples 2.11a and 2.11b some procedures have *values* associated with their *names* while others do not. In FORTRAN, procedures are called *subprograms*; those whose names have associated values are called *FUNCTION* subprograms, while those that do not are called *SUBROUTINE* programs. We shall discuss only FUNCTION subprograms in this section; SUBROUTINE subprograms will be described in Section 3.33.

Many of the common mathematical functions are needed so often that procedures for them are already available in function subprogram form in the

FORTRAN library. Indeed, all of the functions of Sections 3.6 and 3.27 have been supplied with the FORTRAN compiler as part of the software support for the computing installation. To activate one of these FORTRAN function subprograms, we need only refer to it in *another* FORTRAN program (up to this point only programs *without* names have been discussed – in FORTRAN parlance, such programs are called *main subprograms*, but most programmers simply call them *main programs*). For example, the SQRT, ABS, and SIN functions can be invoked simply by including a proper function reference in an arithmetic expression.

Examples.
```
X = SQRT(Y+Z)
Z(7) = 3.9*ABS(SIN(ANGLE)) - 9.2*AMP
```

Such function subprograms have many advantages, among the more important being: (1) the programmer need not prepare his own algorithms for evaluating the common mathematical functions; (2) the functions can be used with complete confidence that they have been thoroughly tested and will produce the specified results; (3) the same function may be referenced in different FORTRAN programs and, indeed, may be called several times with different arguments in the same program; and (4) by relegating some parts of the total algorithm to function subprograms, complex algorithms can be segmented into individual parts that are easier to debug and to document.

Some other advantages, less obvious to the FORTRAN programmer, are: (1) only *one* copy of the object program for a function subprogram need be stored with the object version of a calling program, even when several references are made to it; (2) the *same* copy of the object program for the function can be referenced by more than one subprogram (a main program or other function or subroutine subprograms); and (3) because the algorithms for many of the most common functions and computational tasks are already available as library functions, the overall efficiency of a computing installation is greatly increased (imagine the problems that would be encountered if every programmer had to write his own program to calculate square roots, or logarithms, or the trigonometric functions!).

It frequently happens that a programmer would like to segment his algorithm into parts that are not already available as procedures in function or subroutine subprogram form in the FORTRAN library. In such cases, the programmer may write his own procedures using FORTRAN. We begin by describing the appearance of a FORTRAN function subprogram.

Every user-defined FORTRAN function subprogram must contain some *function definition statements* to specify the name of the function, its dummy arguments, and its entry and return points, and a *function body*, consisting of FORTRAN executable statements and declarations that are required to evaluate the function. The general appearance of every FORTRAN function is:

$\mathscr{T}$ FUNCTION $\mathscr{F}*b$ $(\mathscr{A}_1, \mathscr{A}_2, \ldots, \mathscr{A}_n)$

```
.....
.....   FORTRAN executable statements and declarations needed to
.....       evaluate the function (the body of the function)
.....
```

$\mathscr{F} = \mathscr{E}$ or $\mathscr{B}$ or $\mathscr{C}$

```
RETURN
END
```

Here, $\mathscr{F}$ is the *name* of the function; the naming convention is the same as that for variables. $\mathscr{T}$ is the mode or *type* of value returned by the function; it is an *optional entry and, if present, must be either* INTEGER, REAL, LOGICAL, or CHARACTER (WATFIV only). If the type is not specified, then the naming conventions for variables apply; that is, if the name begins with one of the letters I, J, K, L, M, or N, the value is a four-byte integer, otherwise it is a four-byte real number.

If the type is omitted, then the entry $*d$ must *not* appear. If the type is specified, however, then $*d$ may appear as an option, and indicates the length, in bytes, of the function value. Only those combinations of standard and optional type and length listed in Section 3.27 may appear. Some examples should help to clarify the situation:

FUNCTION Statement	*Function Value* Mode or Type	*Function Value* Length (Bytes)	
FUNCTION SAM(. . . .)	Real	4	
INTEGER FUNCTION SAM(. . . .)	Integer	4	
INTEGER FUNCTION SAM*2(. . . .)	Integer	2	
LOGICAL FUNCTION SAM*1(. . . .)	Logical	1	
REAL FUNCTION SAM*8(. . . .)	Real	8	
CHARACTER FUNCTION SAM(. . . .)	Character	1	WATFIV only
CHARACTER FUNCTION SAM*25(. . . .)	Character	25	WATFIV only
LOGICAL FUNCTION SAM(. . . .)	Logical	4	
FUNCTION SAM*8(. . . .)	Illegal		
INTEGER FUNCTION SAM*8(. . . .)	Illegal		

The type of value returned by the function may also be specified by an IMPLICIT declaration (see Section 3.27); if present, the IMPLICIT declaration must be the *second* statement in the function definition, immediately following the FUNCTION statement. The function type can also be specified in an explicit mode declaration within the function definition.

Examples. (1)
```
FUNCTION MARY(. . . .)
IMPLICIT REAL*8(M)
```

(2) (WATFIV only)
```
FUNCTION WORD(. . . .)
IMPLICIT CHARACTER*5(S–Z)
```

(3)
```
FUNCTION MARY (. . . .)
REAL*8  MARY
```

If the function name has a type and length that do not conform to the usual naming conventions for variables, then the function name must appear in a mode declaration in the *calling* program where it is referenced. For example, if the double-precision function MARY were defined as shown in the first or third of the examples, then one of the declarations

```
REAL*8  MARY
DOUBLE PRECISION MARY
```

must appear in the calling program.

The entries $\mathscr{A}_1, \mathscr{A}_2, \ldots, \mathscr{A}_n$ are the *n dummy arguments* for the function and must be unsubscripted variable or array names or the dummy names of subroutine or other function subprograms. The dummy arguments that are variable, array, or function names may be of any standard or optional type and word-length (see Section 3.27). The modes and lengths of names that do not conform to the usual naming conventions may be specified implicitly with the IMPLICIT declaration or explicitly with an appropriate mode declaration in the body of the function. Every function must have at least one argument.

Example.
```
REAL FUNCTION PROF*8(DELTA, MAX, N, X, NUMBER, M, Y)
IMPLICIT  REAL*8(A–M)
REAL  N
INTEGER  M, X*2
```

In this case, the following mode and length assignments will be made:

Name	*Mode or Type*	*Length (Bytes)*
PROF	Real	8
DELTA	Real	8
MAX	Real	8
N	Real	4
NUMBER	Integer	4
M	Integer	4
X	Integer	2

All dummy arguments that are array names *must be dimensioned* in the body of the function. Detailed rules follow later in this section.

When a reference to the function appears in another program, each of the *dummy* arguments is replaced by the *actual* or *calling* argument that appears in the same position in the argument list in the calling program (the argument list in the calling program is sometimes termed the *calling sequence*). The *body* of the function is then executed.

Inside the body of the function, the name $\mathscr{F}$ may be considered to be the name of a simple variable having the mode and word-length assigned to $\mathscr{F}$. The name $\mathscr{F}$ *without attached arguments* may appear virtually anywhere in the *executable* statements of the body of the function that a simple variable of the same type and length might appear (one exception: $\mathscr{F}$ may not appear as a calling argument for another function). There must be no function reference to the function $\mathscr{F}$; that is, $\mathscr{F}$ must not call on itself. The function name must appear as the left member of an appropriate *assignment* statement, *at least once* in the body of the function. Thus one of the following statement forms must appear somewhere in the body of the function (such statements may appear more than once):

Mode or Type of $\mathscr{F}$	*Assignment Statement*
Integer or Real	$\mathscr{F} = \mathscr{E}$
Logical	$\mathscr{F} = \mathscr{B}$
Character (WATFIV only)	$\mathscr{F} = \mathscr{C}$

When the execution of statements in the body is complete, and a value has been assigned to $\mathscr{F}$ as indicated above, the return to the calling program is effected by the statement

```
RETURN
```

The value present in the memory location assigned to $\mathscr{F}$ in the body of the function will be the value that is assigned to the function reference in the calling program. There must be at least one RETURN statement in every function; there may be more than one. Once the function is called, or *entered*, calculation continues in the body of the function until a RETURN statement is encountered. The function then transfers control back to the calling program.

The last statement in every function subprogram (and indeed in every subprogram) is the declaration

```
END
```

The END statement must not be numbered, or treated as an executable statement or as a substitute for a RETURN statement.

The names selected for the dummy arguments and for all the variables in the body of the function have no relationship whatsoever with variables of the same names in the main program or other function or subroutine subprograms *unless* they are associated with one another by occupying the same relative positions in the lists of calling and dummy arguments. With this one exception, a variable named X in one subprogram is unrelated to a variable named X in another subprogram, since each subprogram is compiled separately. (The programmer has the option of deliberately equating variables in different subprograms with the COMMON declaration, to be discussed in Section 3.34). In fact, function subprograms are sometimes called *external functions* to emphasize that they are programs external to the programs that refer to them. This feature of FORTRAN subprograms allows the programmer to choose quite arbitrary names for his dummy arguments and other *actual* variables in the function without fear that a particular name has already been given some other meaning in another program. The term *actual* indicates that variables not listed as dummy arguments are assigned memory space in the usual way for FORTRAN variables.

The algorithm for computing the value of the function reference appears as the body of the function and may contain any of the FORTRAN statements described in the preceding sections. (It is unusual, however, for input statements to appear in functions.) A typical approach to the writing of a FORTRAN function is shown by the following simple example:

Example. Suppose that we wish to compute the mean, $\bar{x}$, of n numbers, $x_1, x_2, \ldots, x_n$ according to the formula

$$\bar{x} = \left(\sum_{i=1}^{n} x_i\right) \Big/ n.$$

Using obvious FORTRAN names for the variables, the following FORTRAN statements would accomplish the task, assuming that the values of N and X(1), . . . , X(N) are available in storage:

```
      SUM = 0.
      DO 1  I = 1, N
    1 SUM = SUM + X(I)
      XBAR = SUM/N
```

If in a program we are likely to need the means of the elements of several arrays, possibly with different numbers of elements, then the above procedure could be conveniently incorporated into a function. Let us name the function AVG, so that the value returned will be real and of standard word length (four bytes). The dummy arguments required are the name of the array, X, and the number of elements to be averaged, N. The complete function definition would be:

```
      FUNCTION AVG(X, N)
      DIMENSION  X(50)
      SUM = 0.
      DO 1  I = 1, N
    1 SUM = SUM + X(I)
      AVG = SUM/N
      RETURN
      END
```

Note that the dummy argument X has been dimensioned, as required, and that the name of the function, AVG, appears in the left member of an assignment statement preceding the return to the calling program.

In the *calling* program, which will be either a main (sub)program, a subroutine subprogram, or another function subprogram, the function reference must appear as part of or the whole of an arithmetic, logical, or character expression, depending on the mode of the function value. The function reference will have the general form

$$\mathscr{F}\ (\mathscr{A}_1\dagger,\ \mathscr{A}_2\dagger,\ \ldots,\ \mathscr{A}_n\dagger)$$

where the $\mathscr{A}_i\dagger$, $i = 1, 2, \ldots, n$ are the *actual* or *calling* arguments that are associated with the dummy arguments $\mathscr{A}_i$, $i = 1, 2, \ldots, n$ of the function. The actual arguments $\mathscr{A}_i\dagger$ in the calling sequence must correspond in *number*, *order*, and *mode* with the dummy arguments $\mathscr{A}_i$ in the FUNCTION statement for the function. If a dummy argument is the *name* of an *array*, then the corresponding calling argument must also be the *name* of an *array* (or possibly an element of an array, as described below). If the dummy argument is the name of a function, then the corresponding calling argument must be the name of a function (this possibility is discussed in Section 3.31). If the dummy argument is a *simple variable* of integer, real, logical, or character (WATFIV only) mode, then the corresponding calling argument may be any *expression* of that mode.

Examples of references to the function AVG.

```
(1) XMEAN  = AVG(X, N)
(2) AVGPOT = AVG(POT, M)
(3) BSUM   = 5.*AVG(B, 5)
(4) Y(7)   = 3./AVG(BETA, N+2)
(5) SVAR   = (SUMSQ - N*AVG(Z, INT)**2)/(N-1.)
(6) IF (AVG(GAM, 2 + J*K) .LE. X/P + 7.5)  P = SQRT(DEL)
```

In example (1), we compute the average of the first N elements of an array named X and assign the value to the variable XMEAN. Here, it is merely coincidental that the names of the calling arguments are identical with the names of the dummy arguments used in AVG. In (2), AVG calculates the average value of the first M elements of an array named POT. Examples (3) and (4) show that when the dummy argument is a simple variable (N), then the corresponding calling argument may be an expression of the same mode such as 5 or N+2. In (4), the value of N used in the function is two greater than the value of N in the calling program. In example (5), the average of the first INT elements of an array Z is computed and used in the evaluation of a rather complicated expression. The last example illustrates that a reference to AVG may occur anywhere that an arithmetic expression is appropriate, in this instance as part of a simple logical expression.

Example. Two classes A and B contain NA and NB students, respectively. Their scores on a particular examination are stored in the vectors A and B, respectively. Write a program that will accept this information and display a message as to which class has the higher average score; assume for simplicity that there is no chance of a tie.

A complete main program that calls on the function AVG twice, and that solves this problem using the algorithm of Fig. 2.17 is:

```
      DIMENSION  A(50), B(50)
      READ (5,100)  NA, NB, (A(I), I = 1, NA), (B(I), I = 1, NB)
      IF (AVG(A,NA) .GT. AVG(B,NB))  GO TO 2
      WRITE (6,200)
      STOP
    2 WRITE (6,201)
      STOP
  100 FORMAT (2I5/(10F8.1))
  200 FORMAT ('0CLASS B TOP')
  201 FORMAT ('0CLASS A TOP')
      END
```

As is shown in Example 2.11 on page 51 procedures may compute more than one result, in which case some of the dummy arguments may be used to return *results* to the calling algorithm. Similarly, values assigned to dummy arguments in a FORTRAN function are in fact assigned to the actual arguments in the calling sequence corresponding to those dummy arguments. For example, any time that a statement of the form

$$\mathscr{A}_i = \mathscr{E}$$
$$\mathscr{A}_i = \mathscr{B}$$
$$\mathscr{A}_i = \mathscr{C} \text{ (WATFIV only)}$$

appears in the body of the function, the value of the expression on the right-hand side of the assignment statement is stored in the memory location assigned to the *calling argument* that corresponds to the dummy argument $\mathscr{A}_i$. In such cases, the calling argument $\mathscr{A}_i$† must be a simple variable, an array name, or an element of an array. It must *not* be a more general expression or a function name, since there would be no associated memory location in which to store the value. The assignment of some results to dummy arguments is illustrated by the following example.

Example. A company runs N stores, and records the monthly profit for each store over a period of M months. Denote the profit made during the Ith month by the Jth store as P(I,J), and suppose that the P(I,J) have been assigned values for $1 \leqslant I \leqslant M$, $1 \leqslant J \leqslant N$. Write a function, named PROFIT, that will return:

(a) as its value, the total profit made by all the stores over all the months, and
(b) by assignment to dummy arguments, the greatest profit PMAX made by any

store in any month, and the corresponding month and store numbers, IMAX and JMAX, respectively. Assume for simplicity that PMAX occurs for only one store in only one month.

A complete FORTRAN function that solves this problem using the algorithm of Fig. 2.13 is shown below. The FORTRAN function PROFIT corresponds to the symbol P in the flow diagram while the array named P is equivalent to the symbol p; otherwise, there is an obvious correspondence between the symbols in the diagram and the FORTRAN variable names in the problem statement.

```
      FUNCTION PROFIT(M, N, P, PMAX, IMAX, JMAX)
      DIMENSION  P(24,50)
      PROFIT = 0.
      PMAX = P(1,1)
      IMAX = 1
      JMAX = 1
      DO 1  I = 1, M
      DO 1  J = 1, N
      PROFIT = PROFIT + P(I,J)
      IF (P(I,J) .LE. PMAX)  GO TO 1
         PMAX = P(I,J)
         IMAX = I
         JMAX = J
    1 CONTINUE
      RETURN
      END
```

The only significant difference between this function and the flow diagram of Fig. 2.13 is that the complementary relational operator has been used in the test, so that the true and false paths in the diagram have been interchanged. This avoids the necessity of introducing an additional statement, and also illustrates the use of the CONTINUE statement at the end of a DO loop and of a common range for the nested iteration statements.

Note that the function name PROFIT has been used as an ordinary real variable in the body of the function, and that values will be assigned by the function to the calling arguments corresponding to the dummy arguments PMAX, IMAX, and JMAX. A valid reference to the function PROFIT in a calling program, provided that $PROF were appropriately dimensioned (see below) would be:

```
      $ = PROFIT(NMONTH, NSTORE, $PROF, PMAX, NMMAX, NSMAX)
```

The Dimensioning of Arrays in Subprograms. *Every* subscripted variable that appears in a function must be dimensioned in the function. The dimensioning of arrays that are *dummy* arguments for the function is handled somewhat differently from the dimensioning of arrays that are not.

Subscripted variables that are *not* dummy arguments must be dimensioned in the usual way (see Section 3.25). Storage for such arrays appears inside the object program that is generated by the FORTRAN compiler when the function is translated. Consequently, the precise arrangement of the array must be known before the function is compiled; the maximum value of each subscript must be specified as an integer constant.

Dummy arrays, that is, array names that appear in the dummy argument list, require *no* storage. All dummy arguments are replaced by calling arguments when reference is made to the function. Hence, it would appear that the actual arrays should be dimensioned in the calling program and that the dummy arrays should not be dimensioned at all. In fact, the calling array must be dimensioned in the calling program *and* the dummy array must be dimensioned in the function. There are two reasons for this apparent inconsistency.

First, since the compiler translates each subprogram independently, it must be able to distinguish between subscripted variables and function references by examining *only* the FORTRAN statements in the function definition. For example, the translator must be able to identify the symbol string ALPHA(2,JACK,M) as either: (1) the element of the ALPHA array with subscripts 2, JACK, and M, or (2) a reference to a function named ALPHA with three arguments, 2, JACK, and M. If ALPHA appears in a DIMENSION statement then the potential ambiguity is eliminated.

The second reason is that multidimensional arrays are stored in sequence, as if they were singly subscripted variables, because of the serial addressing structure used in computer memories (see the discussion in Section 3.29). For singly subscripted variables this causes no difficulty. If the address of the first element is known, then the address for any other element can be computed with no additional information. For multiply subscripted variables, however, this is not the case. Knowing the address of the first element and even the total space assigned to the variable is insufficient to establish the address corresponding to a particular element; the shape of the array must be known. For example, if ALPHA is the dummy name for an array in the calling program with a total of 1200 memory locations, it is essential to know if the calling array has dimensions 1×1200, 2×600, 3×400, $4 \times 300, \ldots, 300 \times 4$, 400×3, 600×2 or 1200×1. The DIMENSION statement supplies this information.

Either fixed or variable dimensions are allowed for dummy arrays. If *fixed dimensions* are used, then the DIMENSION declaration appears with an integer constant as the upper limit for each dimension. Such constants should, in general, imply a *total space* requirement for the array *no longer than* that of any *actual array* that might be used as a calling argument. If the compiler checks to insure that all referenced array elements lie inside the storage region for the array (the WATFOR and WATFIV compilers normally do, while the FORTRAN compiler does not), then this precaution insures that no locations outside the region will be used as array elements. For example, in the function PROFIT shown on page 130, the dummy array P is assigned the fixed dimensions of 24 rows and 50 columns. Even if the reference to PROFIT in the calling program were

```
$ = PROFIT( 12, 40, $PROF, $MAX, IMAX, JMAX )
```

the array $PROF should still be dimensioned in the calling program to contain at least $24 \times 50 = 1200$ elements. To avoid the possibility of any difficulty, the dimensions could be made to agree *exactly*. That is, if the DIMENSION declaration in the calling program is

```
DIMENSION $PROF(24,50)
```

then there will be a one-to-one correspondence between every element of the array $PROF and the dummy array P. Note that if the first two arguments were 12 and 40 as indicated above, only 480 of the 1200 elements in $PROF would be used.

Should $PROF be dimensioned differently, for example,

```
      DIMENSION  $PROF(12,100)
```

then the programmer must take into account the column-by-column storage of arrays. In this case, for example, $PROF(1,2) will be assigned to the 13th element of $PROF, which would correspond to P(13,1), the 13th element of P in the function; the element P(1,2) corresponds to $PROF(1,3), etc. On the other hand, if $PROF were dimensioned with the declaration

```
      DIMENSION  $PROF(24,100)
```

the correspondence of the elements would remain intact. Clearly, if the dimensioning is not identical in the function and the calling program, then the programmer must exercise considerable caution. This problem does not arise for linear arrays. Thus in the function AVG of page 128, the following declaration would be acceptable:

```
      DIMENSION  X(1)
```

There must be essential agreement between the dimensions of a multidimensional array in a calling program and its corresponding dummy subprogram argument as described above. If fixed dimensions are used for multidimensional dummy arrays, then the generality of the subprogram is severely limited. Whenever the dimensions of an array in a calling program are changed, the DIMENSION declaration in the subprogram must be changed to retain the equivalence of subscripts in the calling and dummy arrays. Fortunately, this problem can be overcome by using *variable dimensions* in the *subprogram*. Only integer variables that appear as dummy arguments (or that are assigned to COMMON storage, as described in Section 3.34) may appear in DIMENSION declarations in subprograms. For example, if two additional dummy arguments are added to the argument list for the function PROFIT of page 130, so that the FUNCTION and DIMENSION statements are

```
      FUNCTION PROFIT (M, N, P, PMAX, IMAX, JMAX, NROWS,
     1 NCOLS)
      DIMENSION  P(NROWS,NCOLS)
```

then the calling program might contain the statements

```
      DIMENSION  $PROF(24,50)
            .
            .
            .
      $ = PROFIT (NMONTH, NSTORE, $PROF, PMAX, NMMAX,
     1 NSMAX, 24, 50)
            .
            .
```

or perhaps

```
      DIMENSION  $PROF(12,75)
           .
           .
           .
      $ = PROFIT (10, 50, $PROF, PMAX, NMMAX, NSMAX, 12, 75)
```

and no change would need be made in the DIMENSION declaration in PROFIT. The combination

```
      DIMENSION  $PROF(12,75)
           .
           .
           .
      $ = PROFIT (10, 50, $PROF, PMAX, NMMAX, NSMAX, 10, 50)
```

would lead to inconsistent subscription, however. The variable dimensions should correspond to the actual dimensions, and not to the number of elements in the array that are being used. Variable dimensions may be used *only* in function and subroutine subprograms and *never* in main programs.

In FORTRAN and WATFIV, but *not* WATFOR, any array *element* may be used as a calling argument corresponding to a *dummy array name* in a function. The calling array element is treated as if it were the *first* element of the dummy array; the correspondence between elements of the calling and dummy arrays can be established by accounting for the sequential storage scheme used by FORTRAN. For example, suppose that we wished to find the average value $\bar{a}$ of elements in the matrix A that lie in the jth column between the kth and mth rows inclusive; that is

$$\bar{a} = \frac{\sum_{i=k}^{m} a_{ij}}{m-k+1}.$$

Then, using obvious FORTRAN names, the function AVG of page 128 could be called to compute $\bar{a}$ in a calling program as follows:

```
      ABAR = AVG(A(K,J), M-K+1)
```

This is feasible because matrices are stored column by column, so that the correspondences between the calling array A and the dummy array X are: (A(K,J), X(1)), (A(K+1,J), X(2)), . . . , (A(M,J), X(M-K+1)). This feature of FORTRAN and WATFIV should be used only if the storage allocation scheme for arrays is thoroughly understood by the programmer.

Multiple Entry Points—the ENTRY Statement. Subprograms that have several statements in common or that require large local (that is, not dummy) arrays may often be combined into one subprogram with several different entry points, one for each subprogram name. This merging of routines leads to more compact FORTRAN programs, and may save considerable storage. For example, the algorithm for computing the sine function is clearly very similar to that for computing the cosine function, since $\cos(x) = \sin(x+\pi/2)$. And, in fact, the library functions SIN and COS of Section 3.6 are not independent functions, but rather are two different entry points to a single subprogram.

The entry point for a function with a single name is the FUNCTION statement itself. When a function has more than one entry point, the ENTRY statement is used; such functions have the following general appearance:

$\mathscr{T}$ FUNCTION $\mathscr{F}_1{*}b$ $(\mathscr{A}_{11}, \mathscr{A}_{12}, \ldots, \mathscr{A}_{1n_1})$

.
.
.
.

ENTRY $\mathscr{F}_2$ $(\mathscr{A}_{21}, \mathscr{A}_{22}, \ldots, \mathscr{A}_{2n_2})$

.
.
.
.

ENTRY $\mathscr{F}_3$ $(\mathscr{A}_{31}, \mathscr{A}_{32}, \ldots, \mathscr{A}_{3n_3})$

.
.
.
.

ENTRY $\mathscr{F}_m$ $(\mathscr{A}_{m1}, \mathscr{A}_{m2}, \ldots, \mathscr{A}_{mn_m})$

.
.
.
.

END

Here, the $\mathscr{F}_j$, $j = 1, 2, \ldots, m$, are the m different function names, and the $\mathscr{A}_{ji}$, $i = 1, 2, \ldots, n_j$, are the n_j dummy arguments associated with the jth function name $\mathscr{F}_j$. The FUNCTION statement has already been described in detail in earlier parts of this section. The modes and word lengths of the values associated with the function names need not be identical. If the names selected do not conform to the usual naming conventions, the mode may be assigned explicitly in a mode declaration or implicitly with an IMPLICIT declaration immediately following the FUNCTION statement.

The dummy argument lists for the various entries need not be identical although the same dummy argument may appear in more than one list. The first appearance of a dummy argument in the subprogram must not precede its first appearance in a dummy argument list, however. For example, the following would be illegal:

```
FUNCTION ALPHA (X, Y, Z)
      .
      .
      .
Q = X + Y
      .
      .
      .
```

```
      .
ENTRY BETA (Q, N, X, Y)
      .
      .
      .
```

When a reference to one of the names is made in a calling program, the calling argument list must agree in number, order, and type with the dummy argument list appearing in the ENTRY statement for that name. In FORTRAN and WATFOR, each entry name is treated as a simple variable in the function. The value returned to the calling program is specified in the usual way; the function name must appear on the left-hand side of an assignment statement before the return to the calling program. In WATFIV, a slightly different approach is allowed. A single memory word is used to save the value to be returned by the function. Consequently, when a statement of the form

$$\mathscr{F}_j = \mathscr{E}$$

or

$$\mathscr{F}_j = \mathscr{B}$$

or

$$\mathscr{F}_j = \mathscr{C}$$

is executed for *any j*, the same location is used, regardless of which entry name was originally referenced in the calling program.

Example. Write a FORTRAN function with two entry names, AVGA and AVGD, that have common dummy argument lists (M,N,X). When AVGA is referenced, the elements of the array X from X(M) to X(N) inclusive are to be sorted into *ascending* algebraic order; when AVGD is referenced, the elements X(M) to X(N) inclusive are to be sorted into *descending* algebraic order. The value returned by the function for either entry should be the average value of the Mth through the Nth element of X. The sorting algorithm used will be that already described on page 115 for alphabetizing a list of words.

```
      FUNCTION AVGA (M, N, X)
      DIMENSION X(N)
      LOGICAL ASCEND
      ASCEND = .TRUE.
      GO TO 1
      ENTRY AVGD (M, N, X)
      ASCEND = .FALSE.
    1 NM1 = N - 1
      AVGA = 0.
      DO 3  I = M, NM1
      IP1 = I + 1
      DO 2  J = IP1, N
      IF (X(I) .LE.X(J) .AND. ASCEND .OR. X(I).GE.X(J) .AND.
     1 .NOT.ASCEND)  GO TO 2
         XISAVE = X(I)
         X(I) = X(J)
         X(J) = XISAVE
    2 CONTINUE
    3 AVGA = AVGA + X(I)
      AVGA = (AVGA + X(N))/(N-M+1)
      AVGD = AVGA
      RETURN
      END
```

In this example, the logical variable ASCEND is set to *true* when the function AGVA is called and to *false* when AVGD is called. The name AVGA is used in the function as a simple variable to accumulate the sum of the array elements and is then set equal to the arithmetic mean of the array elements. AVGD is then assigned the same value, so that the average value will be returned to the calling program, whichever entry name was used in the original function reference. In WATFIV, the statement AVGD = AVGA would be unnecessary, since the *same* memory location is used to store the value of AVGA and AVGD.

3.31 EXTERNAL Declaration

Dummy function names may appear in the dummy argument list of a FORTRAN subprogram, in which case the corresponding entry in the argument list in the calling program must be a function name. For example, we might write a function named SIMPS that employs Simpson's rule on N equal subintervals of the interval of integration $\mathsf{A} \leqslant x \leqslant \mathsf{B}$ (see page 325), to estimate the value of the integral

$$\int_{\mathsf{B}}^{\mathsf{A}} f(x)\,dx.$$

Let the FUNCTION statement in the FORTRAN subprogram SIMPS be of the form

```
FUNCTION SIMPS ( A, B, N, F )
```

where A, B, and N are dummy variables and F is a *dummy function name*. If we wished to have SIMPS evaluate the integral

$$\int_0^{\pi} \cos x\,dx$$

then the appropriate FORTRAN statement in the calling program would be

```
AREA = SIMPS ( 0., 3.141593, N, COS )
```

FORTRAN, WATFOR, and WATFIV require that any subprogram name that appears as the argument for *another* subprogram must be listed in an EXTERNAL declaration in the *calling* program. The EXTERNAL statement has the form

$$\text{EXTERNAL } \mathscr{F}_1, \mathscr{F}_2, \ldots, \mathscr{F}_n$$

where the $\mathscr{F}_i$, $i = 1, 2, \ldots, n$, are function names.

Examples.
```
EXTERNAL COS
EXTERNAL MYFUN, SQRT
```

3.32 Statement Functions

If a function named $\mathscr{F}$ with dummy arguments $\mathscr{A}_1, \mathscr{A}_2, \ldots, \mathscr{A}_n$ can be described by a single arithmetic expression $\mathscr{E}$, then FORTRAN allows an alternative to the function subprogram definition form of Section 3.30. Called an *arithmetic statement function*, $\mathcal{F}$ may be defined with a declaration of the form

$$\mathcal{F}\ (\mathcal{A}_1, \mathcal{A}_2, \ldots, \mathcal{A}_n) = \mathcal{E}$$

The defining statement must precede the first executable statement in the program in which the function is referenced.

Example. An arithmetic statement function named RAD is defined as follows:

```
RAD(X,Y,Z) = Y**2 - 4.0*X*Z
```

Then a reference to the function in the substitution statement

```
RADIC = SQRT(RAD(A,B,C))
```

would cause the variable RADIC to be assigned the value $\sqrt{B^2 - 4 \times A \times C}$.

The arithmetic expression $\mathcal{E}$ may involve other functions, even other statement functions, provided they have been defined in preceding declarations. If the functions COT and F have been defined with the statements

```
COT(X) = COS(X)/SIN(X)
F(X) = 4.0*X**2 + COT(X+0.5)
```

then a reference to F of the form

```
Z = F(D + E**2 - 5.37)
```

would cause Z to be assigned the value

$$4(D + E^2 - 5.37)^2 + \cos(D + E^2 - 4.87)/\sin(D + E^2 - 4.87).$$

The arithmetic statement function is an *internal function*, in that it may be referenced *only* by the main program, function subprogram, or subroutine subprogram in which it is defined. The arithmetic statement function name may not be used as an argument for an external function. In a sense, this restricts the generality of the statement function. However, because the statement function is *internal* to another program, it has access to *all* variables in that program. Thus, the expression $\mathcal{E}$ may contain both *dummy* variables and *actual* variables in the program. For example, when the function defined by

```
ALPHA(X,Y) = X - 3.*Y + B/C
```

is referenced in the statement

```
Q = 5.2*ALPHA(A,P)
```

Q will be assigned the value $5.2(A - 3P + B/C)$, where the current values of actual variables B and C are used.

Logical statement functions of the form

$$\mathcal{F}_{\mathcal{B}}\ (\mathcal{A}_1, \mathcal{A}_2, \ldots, \mathcal{A}_n) = \mathcal{B}$$

are also allowed, while in WATFIV, but not in WATFOR or FORTRAN, statement functions of character mode

$$\mathcal{F}_{\mathcal{C}}\ (\mathcal{A}_1,\ \mathcal{A}_2,\ \ldots,\ \mathcal{A}_n) = \mathcal{C}$$

may be defined.

Example.
```
LOGICAL ALTB
ALTB(A,B) = A .LT. B
```

In FORTRAN and WATFOR, the statement function definition must not contain subscripted variables; in WATFIV, however, subscripted variables may appear to the right-hand side of the equals sign.

Example.
```
FUN(A,B,I) = A*B*X(I)/Y(I)
```

The same dummy arguments may be used in any number of statement function definitions; all statement function definitions must precede the first executable statement.

3.33 SUBROUTINE Subprograms

A FORTRAN *subroutine subprogram* is a procedure with a name, a set of dummy arguments, an entry point, a return point, and a body consisting of FORTRAN executable statements and declarations. The most significant difference between a FORTRAN subroutine and a FORTRAN function is that a function returns a *value* that is associated with its name; a subroutine does not. All results of calculations in the subroutine must be returned to the calling program through the argument list. The key statements in the definition of a subroutine are

SUBROUTINE $\mathcal{F}\ (\mathcal{A}_1,\ \mathcal{A}_2,\ \ldots,\ \mathcal{A}_n)$

......
......
......
...... } FORTRAN executable statements and declarations needed to perform the subroutine calculations (the *body* of the subroutine)

RETURN
END

The subroutine name $\mathcal{F}$ follows the same rules as for naming variables, except that the first letter has no special significance, since there is no mode associated with the subroutine name. No assignment statement of the form $\mathcal{F} = \mathcal{E}$ precedes the RETURN statement.

Since a subroutine has no value, it may *not* be referenced implicitly in an arithmetic, logical, or character (WATFIV) expression. Instead, it must be invoked explicitly in a calling program with the statement

CALL $\mathcal{F}\ (\mathcal{A}_1\dagger,\ \mathcal{A}_2\dagger,\ \ldots,\ \mathcal{A}_n\dagger)$

where the $\mathcal{A}_i\dagger$, $i = 1, 2, \ldots, n$, are the n *actual* or *calling arguments* that will replace the *dummy arguments* $\mathcal{A}_i$, $i = 1, 2, \ldots, n$, during execution of the subroutine. Most of the material in Section 3.30 describing the behavior of function subprograms carries over directly to subroutine subprograms. The number, order, and modes of calling and dummy arguments must agree. Variable dimensioning of dummy arrays is permitted. There must be at least one RETURN statement, but there may be more than one. There may be any number of ENTRY statements with different subroutine names.

Example. Write a subroutine, named CMPLEX, that will compute the magnitude r and the angle θ in the polar representation of a complex number having real and imaginary parts x and y. Assume that r and θ are given by:

$$r = (x^2 + y^2)^{1/2}$$
$$\theta = \tan^{-1}(y/x)$$

Use the names X, Y, R, and THETA, with obvious interpretations.

The FORTRAN statements defining the subroutine CMPLEX are:

```
SUBROUTINE CMPLEX (X, Y, R, THETA)
R = SQRT(X**2 + Y**2)
THETA = ATAN2( Y, X )
RETURN
END
```

Some possible statements appearing in calling programs are:

```
CALL CMPLEX (X, Y, R, THETA)
CALL CMPLEX (XREAL, YIMAG, RADIUS, PHASE)
CALL CMPLEX (2.9, B - 7.6, X(4), Y(4))
CALL CMPLEX (A - B*SIN(ALPHA), Z(29), R, PHI)
```

In each case, the first two arguments serve to supply values *to* the subroutine. In general, such arguments may be expressions, in which case values will be computed for them before entering the subroutine. The last two arguments indicate where the computed values returned *from* the subroutine are to be stored. Such arguments *must* be variable names, possibly with subscripts. For example, the following call would be meaningless, because there is no indication as to where the computed values are to be stored:

```
CALL CMPLEX (X, Y, 2.0, A + B)
```

The following calls would also be improper:

```
CALL CMPLEX (X, Y, R, THETA, EPS)–too many arguments
CALL CMPLEX (I, J, R, THETA)        –incorrect mode for the first
                                     two arguments
```

Example. Write a subroutine named MAXMIN that will find the maximum and minimum values AMAX and AMIN in the matrix A having M rows and N columns and elements A(1,1) through A(M,N). The names M, N, A, AMAX, and AMIN are the dummy names to be used in the subroutine; the actual arguments in the calling program may, of course, have different names. Assume that the actual array will be dimensioned to have M rows and N columns.

The defining FORTRAN statements are:

```
      SUBROUTINE  MAXMIN (M, N, A, AMAX, AMIN)
      DIMENSION  A(M,N)
      AMAX = A(1,1)
      AMIN = A(1,1)
      DO 2  I = 1, M
      DO 2  J = 1, N
      IF (A(I,J) .LE. AMAX)  GO TO 1
      AMAX = A(I,J)
      GO TO 2
    1 IF (A(I,J) .LT. AMIN)  AMIN = A(I,J)
    2 CONTINUE
      RETURN
      END
```

The algorithm used here assigns the first element A(1,1) to both AMAX and AMIN. Thereafter, each element of the array is checked against AMAX and AMIN. Anytime an element larger than AMAX or smaller than AMIN is found, AMAX or AMIN is assigned the new value. Some possible calling statements are:

```
      CALL MAXMIN (M, N, VELOC, HIGH, FLOW)
      CALL MAXMIN (5, 6, AA, AAMAX, AAMIN)
```

Variable dimensions have been used in the subroutine. Therefore, in the first case, the values of M and N should be the values of the row and column dimensions for the array VELOC. Similarly, the array AA in the second example should be dimensioned to have 5 rows and 6 columns.

A subroutine may be defined to have *no* arguments. For example, the subroutine EXIT, available in the FORTRAN library, is executed when the statement

```
      CALL EXIT
```

appears in the calling program (see Section 3.22). If there are no arguments, then the SUBROUTINE statement has the form

SUBROUTINE $\mathscr{F}$

This situation usually arises only when all of the information needed by the subroutine is available in COMMON storage, described in the next section.

Normally, when a RETURN statement is encountered in a subroutine, control passes back to the calling program with execution resuming at the statement immediately following the CALL statement. In subroutine but *not* function subprograms, the RETURN statement may be modified to have the form

RETURN $\mathscr{I}$

where $\mathscr{I}$ is either an unsigned integer constant or an integer variable of standard length. The parameter $\mathscr{I}$ allows the subroutine to return control to particular statement numbers in the calling program. This *alternate return* feature is implemented rather awkwardly, as follows. Asterisks may appear as elements of the dummy argument list:

$$\text{SUBROUTINE } \mathscr{F} \; (\mathscr{A}_1, \; \mathscr{A}_2, \; *, \; \mathscr{A}_4, \; *, \; \ldots, \; \mathscr{A}_n)$$

Whenever asterisks appear in the dummy argument list, statement numbers in the calling program must appear in the actual argument list in the CALL statement:

$$\text{CALL } \mathscr{F} \; (\mathscr{A}_1\dagger, \; \mathscr{A}_2\dagger, \; \&\eta_1, \; \mathscr{A}_4\dagger, \; \&\eta_2, \; \ldots, \; \mathscr{A}_n\dagger)$$

In this case there are just two statement numbers, η_1 and η_2, shown, but there could be any number in any argument position; the character "&" must precede each statement number.

If the modified RETURN statement

$$\text{RETURN } \mathscr{I}$$

is encountered in the subroutine and $\mathscr{I}$ has the value 1 then control returns to the statement in the calling program with the statement numbered η_1; if $\mathscr{I}$ has the value 2 then control is returned to the statement numbered η_2 in the calling program, and so forth.

Example. Write a WATFIV subroutine named TESTC that examines a dummy variable of character mode and length 1 named CHAR and determines if it contains a letter, a blank, a digit, or a special character. If the character is a letter, then the normal return should be used; if the character is a blank, a digit, or a special character, then the alternate return feature of FORTRAN subroutines should be used to return to three different numbered statements in the calling program.

Knowledge of the hexadecimal equivalents of the characters shown in Table 1.10 is useful in solving this problem. All of the symbols of interest appear in groups within ranges of the hexadecimal equivalents as follows:

	Range			
	Lower Limit		*Upper Limit*	
Symbol Subset	*Character*	*Hexadecimal Equivalent*	*Character*	*Hexadecimal Equivalent*
letters	A	C1	Z	E9
digits	0	F0	9	F9
blank		40	—	—
special characters	¢	4A	"	7F

In this case, we are considering the dollar sign to be a special character rather than a letter.

A WATFIV subroutine to identify the character type follows:

```
SUBROUTINE TESTC (CHAR,*,*,*)
CHARACTER  CHAR
IF (CHAR.GE.'A' .AND. CHAR.LE.'Z')  RETURN
IF (CHAR.EQ.' ')  RETURN 1
IF (CHAR.GE.'0' .AND. CHAR.LE.'9') RETURN 2
RETURN 3
END
```

Suppose that the following statements appear in a calling program.

```
        CHARACTER  SYMB
             .
             .
             .
             .
        CALL TESTC (SYMB,&10,&15,&27)
    5   ........
             .
             .
             .
             .
   10   ........
             .
             .
             .
             .
   15   ........
             .
             .
             .
             .
   27   ........
             .
             .
             .
             .
```

Then, after TESTC is called, it will return control to the statements numbered 5, 10, 15, or 27, if CHAR contains a letter, blank, digit, or special character, respectively. Here, we have tacitly assumed that only the hexadecimal equivalents of legitimate characters could appear in the variable SYMB.

3.34 COMMON Declaration

Ordinarily, variables in subprograms are local to their subprograms. Thus, a variable named ALPHA in a function has no direct relationship to a variable named ALPHA in a calling program. Of course, the argument list for the function allows a direct relationship to be established if ALPHA is a dummy variable in the function. Sometimes it is desirable to establish a direct relationship without passing the information through the argument list. The COMMON declaration allows the user to do this. In its simplest form, the statement has the appearance

$$\text{COMMON } \mathscr{V}_1, \mathscr{V}_2, \ldots, \mathscr{V}_n$$

where the $\mathscr{V}_i$, $i = 1, 2, \ldots, n$, are the names of simple variables or arrays. If the same COMMON statement appears in a main program and one or more associated subprograms, then the listed variables will be identical in all programs. Thus, the COMMON statement has the effect of transmitting data implicitly (indirectly) from one subprogram to another.

The COMMON statement allows a more general approach than indicated above, in that it is not necessary to use the *same* names for the common variables in the various subprograms. Only the *position* of the name in the COMMON statement list is crucial. For example, if the statements

```
COMMON  X, Y, Z
COMMON  ALPHA, BETA, Z
```

occurred in a main program and a function, respectively, then the values of ALPHA, BETA, and Z in the function would be identical with the values of X, Y, and Z, respectively, in the main program. If array names appear in the COMMON list, then such variables must be dimensioned in both programs; usually the dimensions will be compatible. In essence, the FORTRAN compiler sets aside a region of memory called *blank* common for storage of values associated with the variables in the COMMON lists. The variables are ordered linearly in the common region, with arrays occupying the appropriate number of locations in the string of memory words.

An extension of the notion of the blank common block, called *labeled* common, is also available. The COMMON statement then has the form

COMMON $/\mathcal{N}/\ \mathcal{V}_1,\ \mathcal{V}_2,\ \ldots,\ \mathcal{V}_n$

where $\mathcal{N}$ is the *name* of the common block; the naming convention is the same as for variables (see Section 3.5). The major use of labeled common is to allow certain subprograms to have "private" common regions, not shared by all subprograms in a job. If desired, blank common and more than one named common block can be defined in a single COMMON declaration. For example, the statement

```
COMMON  A, B / DELTA / BAT, G / Q / BETA, AB
```

assigns A and B to blank common, BAT and G to the common region with the name DELTA, and BETA and AB to the common region with the name Q.

Special care is required when COMMON variables appear in EQUIVALENCE statements. Since the variables within a COMMON block are assigned to storage in the order of their appearance within the block, and since different COMMON blocks cannot overlap, variables within one block or variables in two different blocks cannot be made equivalent in an EQUIVALENCE statement. Other variables may be made equivalent to COMMON variables, however, provided that there is no implied shifting of the first element in a COMMON block. For example, the statements

```
DIMENSION  ALPHA(5)
COMMON  BETA, A, B
EQUIVALENCE  (A, ALPHA(3))
```

would not be permitted, since the element ALPHA(1) could not be brought into the COMMON region without changing the address assigned to the first COMMON variable, BETA.

For reasons too complex to explain here, program efficiency may be considerably enhanced if variables within COMMON blocks are listed in order of decreasing word size; that is, all double-word (eight-byte) variables should appear first, all full-word (four-byte) variables should appear next, etc.

3.35 BLOCK DATA Subprograms

The WATFOR and WATFIV compilers allow the user to initialize the values of variables assigned to either labeled or unlabeled COMMON storage (see Section 3.34) using the DATA declaration of Section 3.26 in *any* program segment. FORTRAN, however, does not allow such assignments for any variable appearing in *unlabeled* COMMON. Variables in *labeled* COMMON blocks may be assigned initial values, but only in a separate subprogram, called the BLOCK DATA subprogram, which has the form

```
BLOCK DATA
    .
    .
    .
    .
END
```

These delimiting statements must enclose any essential COMMON, DIMENSION, mode (type) and DATA declarations. For example, to initialize the variables A and B in the labeled block ALPHA, where A and B are each arrays with two elements, with the values A(1) = 2.3, A(2) = −5.7, B(1) = 1.3×10^{-7}, and B(2) = 0.345, the BLOCK DATA subprogram would have the form

```
BLOCK DATA
DIMENSION A(2), B(2)
COMMON / ALPHA / A, B
DATA  A, B / 2.3, -5.7, 1.3E-7, 0.345 /
END
```

The BLOCK DATA subprogram has no name, cannot be called by other subprograms, and must not contain any executable statements. Its sole purpose is the initialization of variables in labeled COMMON blocks. This subprogram is not needed in WATFOR or WATFIV programs. It is available, however, and does have the virtue of concentrating most of the information about COMMON variables in a single program segment.

3.36 Order of Statements in a FORTRAN Program

The order of executable statements in a FORTRAN program depends on the algorithm being implemented in program form. The order of nonexecutable statements or declarations, however, depends to some extent at least on the construction of the compiler being used. The WATFOR and WATFIV compilers are "one-pass" compilers, that is, the FORTRAN source program is scanned only once from the first statement to the last; object code for each statement is produced as it is encountered. In contrast, the more common multipass compilers may examine the statements in a subprogram more than once before generating the final object code. Obviously the one-pass compiler may require that certain restrictions on the ordering of declarations be adhered to by the programmer.

To simplify the problem, we suggest the following order for statements within a FORTRAN *main subprogram*; most compilers allow considerably more flexibility, but this order should satisfy even the most restrictive ordering rules:

```
IMPLICIT
          ⎧CHARACTER
          ⎪DOUBLE PRECISION
Explicit  ⎨INTEGER
  Type    ⎪LOGICAL
          ⎩REAL
DIMENSION
COMMON
EQUIVALENCE
DATA
NAMELIST
EXTERNAL
Statement Functions
     .
     .
     .
Executable
Statements
     .
     .
     .
FORMAT
END
```

FORMAT declarations may appear *anywhere* in the program *after* an IMPLICIT declaration and *before* the END declaration.

The statement order for FUNCTION, SUBROUTINE, or BLOCK DATA subprograms is identical, except that the FUNCTION, SUBROUTINE, or BLOCK DATA declaration must be the *first* statement in the respective subprogram.

3.37 Deck Arrangement for a Batch Job

As described in detail in Chapter 4, a complete card deck ready to be processed by the computer is called a *batch job*. Since the supervision of the processing of a job is handled by a set of computer programs called the *operating system*, rather than by a human operator, each job deck must contain certain command lines for the operating system programs in addition to source program statements and data cards.

Typically, the command lines, which must be written in the *command language*, begin with a special character, such as a dollar sign or a slash. Each command contains a keyword indicating the nature of the operation desired (RUN, COPY, COMPILE, etc.) and some parameters associated with the operation (such as device numbers or the name of a program to be executed). The operating systems, and hence the command languages, in current use vary considerably from machine to machine and from installation to installation. For the processing of a simple job involving the translation of a FORTRAN program and execution of the object program with supplied data, the computing installation will normally have available a standard set of command cards; the novice programmer can begin to use the computing system without learning the command language in detail.

A complete FORTRAN source program consists of one main subprogram (there must be one and only one main subprogram), any number of FUNCTION and SUBROUTINE subprograms, and possibly one BLOCK DATA subprogram. Although many FORTRAN compilers will accept a source program with the subprogram segments in any order, virtually any translator will accept the subprogram segments in the order: (1) BLOCK DATA subprogram, (2) main subprogram, and (3) FUNCTION and SUBROUTINE subprograms, in any order; each subprogram segment must be terminated with its own END declaration.

In almost any operating system, the job deck involving translation by the standard FORTRAN compiler will have roughly the following appearance:

Deck Arrangement for a Batch Job

[Command line(s) with the user's name and account number, and some indication of the user's estimates of computing time, number of pages to be printed, etc.]
[A command line to load and execute the FORTRAN translator.]

```
⎡BLOCK DATA
⎢       BLOCK DATA subprogram
⎣END
⎡
⎢       FORTRAN main subprogram
⎣END
⎡FUNCTION or SUBROUTINE
⎢       First FORTRAN function or subroutine subprogram
⎣END
⎡FUNCTION or SUBROUTINE
⎢       Second FORTRAN function or subroutine subprogram
⎣END
⎡ .
⎢ .
⎢ .
⎣END
⎡FUNCTION or SUBROUTINE
⎢       Last FORTRAN function or subroutine subprogram
⎣END
```

[A command line to indicate the end of the FORTRAN source program.]
[A command line to load and execute the object program.]

```
⎡
⎢       Data cards for the object program
⎣
```

[A command line to indicate the end of data for the object program.]
[A command line to indicate the end of the job.]

A load-and-go compiler such as the WATFOR or WATFIV compiler will usually have a somewhat different set of command lines, since the compiler stays in the memory with the object program; but the deck will have the same general arrangement, including the ordering of the subprograms in the FORTRAN source program.

The operating system used on the University of Michigan IBM 360/67 computing system, called MTS (Michigan Terminal System), is described in Chapter 4. Several MTS batch jobs that involve translation of FORTRAN and WATFIV source programs are included in the next section.

3.38 Example FORTRAN, WATFOR, and WATFIV Programs

This section includes five complete example programs, ready for processing as batch jobs by the Michigan Terminal System, an operating system for the University's IBM 360/67 computing system, described in detail in the next chapter. In the listings of the jobs, each line represents the content of one punched card in the job deck. Lines beginning with a dollar sign are either MTS commands or (for the WATFIV programs) command lines for the load-and-go WATFIV compiler; in each case, the line type is indicated at the right edge of the listing.

Three problems are solved; each includes a problem statement, an outline of the solution method, a flow diagram for the solution algorithm using the flow-diagramming conventions shown in Appendix A, a listing of the complete batch job deck, computed results, and a brief discussion. Two of the problems are solved using both FORTRAN and WATFIV; the third is solved using a WATFIV program only.

Example Program No. 3.1 illustrates the use of a FORTRAN program to find the roots of quadratic equations whose coefficients are read as data. The general formatted input and output statements are used; the FORTRAN G-level compiler, named *FTN, is used to translate the source program.

Example Program No. 3.2 is identical to Example Program No. 3.1, except that the WATFIV compiler, named *WATFIV, is used to translate the source program.

Example Program No. 3.3 employs a main (sub)program and a subroutine subprogram to find the largest and smallest elements of an array, read as data. The NAMELIST format-free input and output statements are used, and the program is translated using the IBM G-level compiler, *FTN.

Example Program No. 3.4 solves the same problem as Example Program No. 3.3. However, all input and output statements are of the simple WATFOR-WATFIV format-free type. The WATFIV compiler, *WATFIV, is used to translate the source program.

Example Program No. 3.5 analyzes a set of test scores for a class of students whose names are read as part of the data. Some statistics for the set of scores are calculated and the class list is ordered in descending sequence by test score and alphabetically by student name. The program illustrates the use of some character manipulation features of the WATFIV language, including the core-to-core input and output statements.

EXAMPLE 3.1

ROOTS OF A QUADRATIC EQUATION (FORTRAN)

Problem Statement

Write a FORTRAN program that reads values for a, b, and c, and then finds the roots of the quadratic equation

$$ax^2 + bx + c = 0.$$

Method of Solution

There are three possible alternative solutions:

1. One-root case. When $a = 0$, the equation is linear and the solution is

$$x = -\frac{c}{b}.$$

   In the program, any a smaller in magnitude than 10^{-10} is assumed to be effectively zero.
2. Imaginary-roots case. When $b^2 - 4ac$ is negative, the quadratic equation has two imaginary roots (complex conjugate roots). The program indicates that the roots are imaginary, but does not calculate them.
3. Two real roots case. When $b^2 - 4ac$ is not negative, the quadratic equation has two real roots given by

$$x_1 = \frac{-b + \sqrt{b^2 - 4ac}}{2a},$$

$$x_2 = \frac{-b - \sqrt{b^2 - 4ac}}{2a}.$$

Flow Diagram

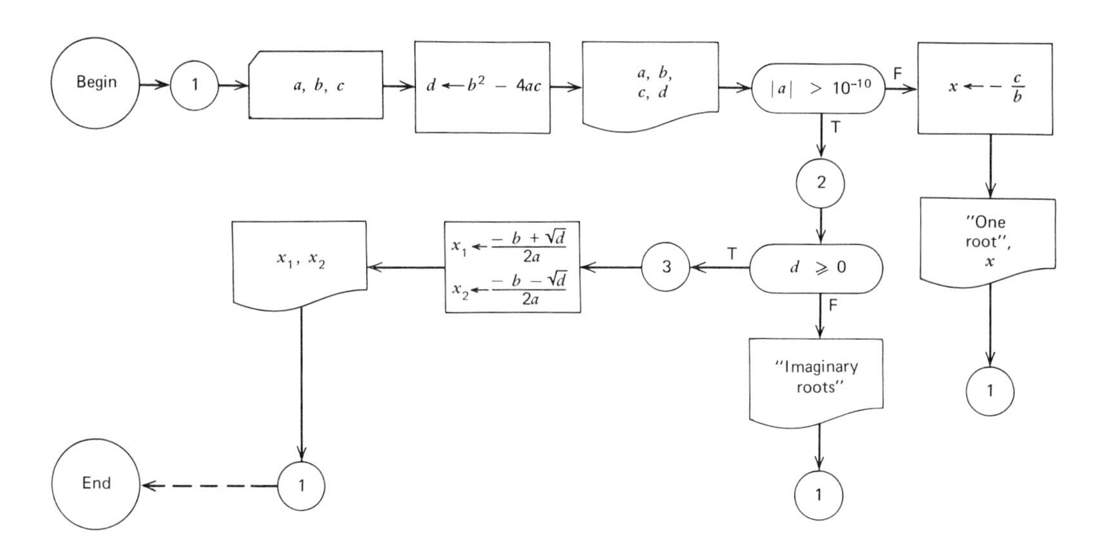

FORTRAN Implementation

List of Principal Variables

Program Symbol	Definition
A	Coefficient, a.
B	Coefficient, b.
C	Coefficient, c.
DISCR	$d = b^2 - 4ac$.
X	Root for the one-root case, x.
X1, X2	Roots for the two-root case, x_1, x_2.

Batch Job Deck

The listing below is a complete batch job deck for solution of the problem on the University of Michigan Computing System. Commands for the operating system (MTS) begin with a dollar sign. The program is written in FORTRAN using the standard formatted input and output statements. The IBM G-level compiler, *FTN, is used to translate the source program.

```
$SIGNON CCNO T=10 P=20 C=0 'YOUR NAME'
PASSWD
$RUN *FTN
C
C         SAMPLE PROGRAM NO. 1.    ROOTS OF QUADRATIC EQUATION.
C
C      ..... READ AND CHECK INPUT DATA .....
    1  READ (5,100,END=999)  A, B, C
       DISCR = B*B - 4.*A*C
       WRITE (6,200)  A, B, C, DISCR
C
C      ..... INVESTIGATE CASE OF A ESSENTIALLY ZERO .....
       IF (ABS(A) .GT. 1.E-10)  GO TO 2
          X = -C/B
          WRITE (6,203)  X
          GO TO 1
C
C      ..... REJECT CASE OF IMAGINARY ROOTS .....
    2  IF (DISCR .GE.0.)  GO TO 3
          WRITE (6,201)
          GO TO 1
C
C      ..... COMPUTE AND PRINT ROOTS X1 AND X2 .....
    3  X1 = (-B + SQRT(DISCR))/(2.*A)
       X2 = (-B - SQRT(DISCR))/(2.*A)
       WRITE (6,202)  X1, X2
       GO TO 1
C
  999  CALL EXIT
C
C      ..... FORMATS FOR INPUT AND OUTPUT STATEMENTS .....
  100  FORMAT (3F10.3)
  200  FORMAT ('2     ROOTS OF QUADRATIC EQUATION, WITH'/
     1        '0     A        = ', F10.3/ 5X, 'B        = ', F10.3/
     2        '      C        = ', F10.3/ 5X, 'DISCR    = ', F10.3)
  201  FORMAT ('0     DISCRIMINANT NEGATIVE - ROOTS IMAGINARY')
  202  FORMAT ('0     X1       = ', F10.5/ 5X, 'X2       = ', F10.5)
  203  FORMAT ('0     COEFFICIENT A IS NEAR ZERO - ONE ROOT ONLY'/
     1        '0     X        = ', F10.5)
C
       END
$ENDFILE
$RUN -LOAD 5=*SOURCE* 6=*SINK*
      1.0         1.0        -6.0
      1.0        -5.0         6.25
      4.46        2.957       0.901
      0.0         2.957       0.901
$ENDFILE
$SIGNOFF
```

MTS | FORTRAN program | MTS | data | MTS

Computer Output

```
ROOTS OF QUADRATIC EQUATION, WITH

A       =       1.000
B       =       1.000
C       =      -6.000
DISCR   =      25.000

X1      =     2.00000
X2      =    -3.00000



ROOTS OF QUADRATIC EQUATION, WITH

A       =       1.000
B       =      -5.000
C       =       6.250
DISCR   =       0.000

X1      =     2.50000
X2      =     2.50000



ROOTS OF QUADRATIC EQUATION, WITH

A       =       4.460
B       =       2.957
C       =       0.901
DISCR   =      -7.330

DISCRIMINANT NEGATIVE - ROOTS IMAGINARY



ROOTS OF QUADRATIC EQUATION, WITH

A       =       0.000
B       =       2.957
C       =       0.901
DISCR   =       8.744

COEFFICIENT A IS NEAR ZERO - ONE ROOT ONLY

X       =    -0.30470
```

Discussion of Results

The four data sets yield, respectively, two different real roots, two identical real roots, imaginary roots, and one real root. All possible paths through the program have been tested. With the exception of a data set for which both a and b have zero values (an inconsistent case for which we could have tested), the program should be able to handle any arbitrary combination of real values for a, b, and c.

The first MTS command line requests access to the computer with the user numer CCNO and a specified time, page, and card limit of 10 seconds, 20 pages, and no cards, respectively. The second card contains the user password, which serves as a secret code for computer access. The third command instructs the operating system to load the FORTRAN translating program, saved under the name *FTN on disk storage, into the memory, and to begin its execution by the processor.

In MTS, the files or devices from which the compiler *FTN expects to read its FORTRAN source program, to write the source program listings and diagnostics, and to save the object program may be specified as parameters on the $RUN *FTN command. If no parameters are specified, then by default, in a batch job, the FORTRAN source program is read from the batch card reader (called *SOURCE* in MTS parlance), the program listings and diagnostics are written on the batch printer (called *SINK*), and the object program is saved in a temporary file (see Section 4.7) named –LOAD on disk storage. The first $ENDFILE command informs the translator that the complete FORTRAN program has been read.

The next command

```
$RUN –LOAD 5=*SOURCE* 6=*SINK*
```

tells the operating system to load the object program (written by the compiler into the file –LOAD) into memory and to begin its execution. The entries 5=*SOURCE* and 6=*SINK* indicate that any references to input device 5 [for example, in the statement READ (5,100,END=999) A,B,C] should be directed to the card reader (*SOURCE*) and that any references to output device 6 [for example, in the statement WRITE (6,200) A, B, C, DISCR] should be directed to the line printer (*SINK*). As we shall see in the next chapter, devices 5 and 6 might be equated with files on disk storage, or other devices (such as a teletypewriter) as well. The final $ENDFILE command signals the object program that all data cards have been processed. Finally, the command $SIGNOFF causes the operating system to terminate processing of the batch job. Notice that the overall structure of the deck follows quite closely that outlined in Section 3.37.

EXAMPLE 3.2

ROOTS OF A QUADRATIC EQUATION (WATFOR OR WATFIV)

Problem Statement

Solve the problem of Example 3.1 using a WATFOR or WATFIV program.

Method of Solution

Since the general input and output statements have been used in the FORTRAN program of Example 3.1, no changes need be made in the program to convert it to a WATFOR or WATFIV program; only the command cards to the operating system need be changed. In the following batch job, the WATFIV compiler, named *WATFIV, is used to translate the source program.

Batch Job Deck

```
$SIGNON CCNO T=10 P=20 C=0 'YOUR NAME'                                    MTS
PASSWD
$RUN *WATFIV SCARDS=*SOURCE* SPRINT=*SINK* 5=*SOURCE* 6=*SINK*
$COMPILE                                                                  *
C
C         SAMPLE PROGRAM NO. 2.   ROOTS OF QUADRATIC EQUATION.
C
C      ..... READ AND CHECK INPUT DATA .....
    1  READ (5,100,END=999)  A, B, C
       DISCR = B*B - 4.*A*C
       WRITE (6,200)  A, B, C, DISCR
C
C      ..... INVESTIGATE CASE OF A ESSENTIALLY ZERO .....
       IF (ABS(A) .GT. 1.E-10)  GO TO 2
          X = -C/B
          WRITE (6,203)  X
          GO TO 1
C
C      ..... REJECT CASE OF IMAGINARY ROOTS .....
    2  IF (DISCR .GE.0.)  GO TO 3
          WRITE (6,201)
          GO TO 1
C                                                                         WATFIV
C      ..... COMPUTE AND PRINT ROOTS X1 AND X2 .....                      program
    3  X1 = (-B + SQRT(DISCR))/(2.*A)
       X2 = (-B - SQRT(DISCR))/(2.*A)
       WRITE (6,202)  X1, X2
       GO TO 1
C
  999  CALL EXIT
C
C      ..... FORMATS FOR INPUT AND OUTPUT STATEMENTS .....
  100  FORMAT (3F10.3)
  200  FORMAT ('2      ROOTS OF QUADRATIC EQUATION, WITH'/
      1         '0      A        = ', F10.3/ 5X, 'B        = ', F10.3/
      2         '       C        = ', F10.3/ 5X, 'DISCR    = ', F10.3)
  201  FORMAT ('0      DISCRIMINANT NEGATIVE - ROOTS IMAGINARY')
  202  FORMAT ('0      X1       = ', F10.5/ 5X, 'X2       = ', F10.5)
  203  FORMAT ('0      COEFFICIENT A IS NEAR ZERO - ONE ROOT ONLY'/
      1         '0      X        = ', F10.5)
C
       END
$DATA                                                                     *
     1.0       1.0       -6.0
     1.0      -5.0        6.25                                            data
     4.46      2.957      0.901
     0.0       2.957      0.901
$STOP                                                                     *
$ENDFILE                                                                  MTS
$SIGNOFF
```

*The lines $COMPILE, $DATA, and $STOP are commands for the WATFIV compiler, *WATFIV, and are *not* MTS commands.

Computer Output

The results for this program are identical to those shown on page 152.

Discussion of Results

Because the WATFIV compiler is a load-and-go compiler, only one $RUN command is needed; once the compiler is loaded and in execution, it manages the writing and loading of the object program without further MTS instructions (the object program is written directly into the memory). In the University of Michigan system, the WATFIV compiler reads the source program and any *compiler commands* from a device called SCARDS and produces the source program listings and any diagnostics on the SPRINT device. Thus the command

```
$RUN *WATFIV SCARDS=*SOURCE* SPRINT=*SINK* 5=*SOURCE* 6=*SINK*
```

causes the operating system to load and initiate execution of the WATFIV compiler stored in disk memory in a file named *WATFIV. Once it is in execution, the compiler will read the WATFIV source program from the card reader (SCARDS=*SOURCE*), and write the program listing and any diagnostics on the line printer (SPRINT=*SINK*). Once the compiler initiates execution of the object program, any reference to device 5 (in the READ statement) will cause the card reader to be accessed (5=*SOURCE*); any references to device 6 (in the WRITE statements) will be directed to the line printer (6=*SINK*).

The three commands denoted with asterisks (*) in the program listing are not commands to the operating system, MTS, but are commands for the WATFIV compiler. The command

```
$COMPILE
```

causes the compiler to enter its translation phase. The object program is saved directly in memory. When the command

```
$DATA
```

is encountered, the WATFIV compiler initiates execution of the object program. The command

```
$STOP
```

terminates execution of the object program. The MTS command $ENDFILE indicates that the WATFIV segment of the job is ended. At other computing installations, both the operating system commands and the compiler commands for either the WATFOR or WATFIV translators may differ from those in the example batch job of the previous page. Nevertheless, the program and data portions of the job will be unchanged, and the structure of the job deck will be quite similar.

EXAMPLE 3.3

FINDING THE LARGEST AND SMALLEST ELEMENTS OF AN ARRAY
A FORTRAN SUBROUTINE—NAMELIST INPUT AND OUTPUT

Problem Statement

Write a FORTRAN subroutine subprogram, named MAXMIN, with dummy argument list

(TOTAL, A, BIGA, SMALLA)

that finds the largest element, BIGA, and the smallest element, SMALLA, of the first TOTAL (integer) elements of the one-dimensional array A. Then write a main (sub)program to test MAXMIN that reads values for N and the first N elements of the array X, that is, X(1), . . . , X(N), and then calls on MAXMIN to find the largest and smallest of X(1), . . . , X(N), and to assign the values to XHIGH and XLOW, respectively. Use the format-free NAMELIST input and output statements in the main program.

Method of Solution

A simple way to find the largest and smallest elements is to assign the value of A(1) to both BIGA and SMALLA, and thereafter to examine each element of A in turn, A(2), . . . , A(N). Any time that a value of A(I), I = 2, 3, . . . , N, is smaller than SMALLA or larger than BIGA, replace SMALLA or BIGA, respectively, by A(I).

Flow Diagram

Main Program

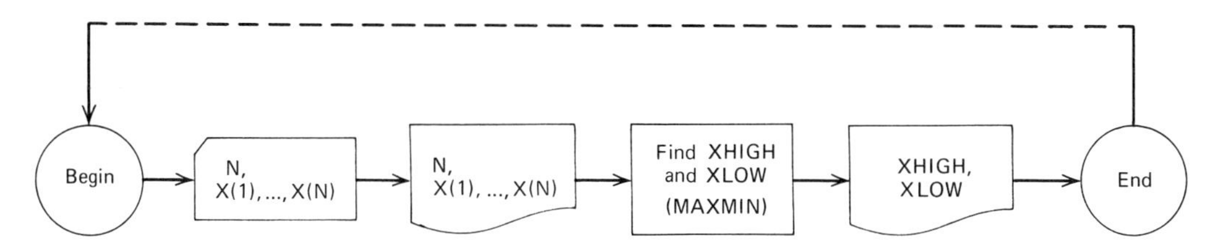

Subroutine MAXMIN (TOTAL, A, BIGA, SMALLA)

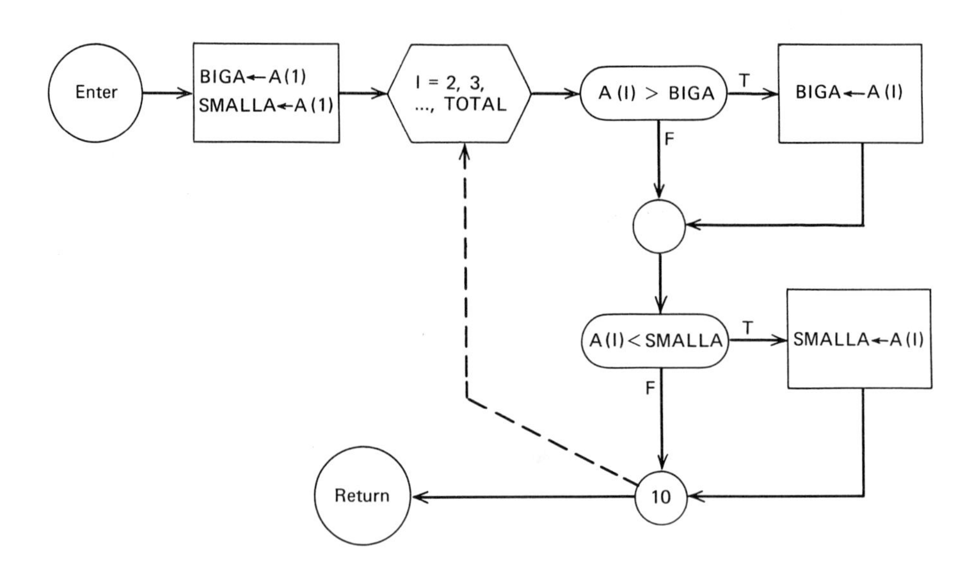

Batch Job Deck

```
$SIGNON CCNO T=10 P=20 C=0 'YOUR NAME'
PASSWD
$RUN *FTN
C
C        SAMPLE PROGRAM NO 3.   THIS MAIN PROGRAM CALLS UPON THE
C        SUBROUTINE MAXMIN TO FIND THE SMALLEST (XLOW) AND LARGEST
C        (XHIGH) ELEMENTS IN THE ARRAY X(1), X(2), ..., X(N).
C
      DIMENSION X(20)
      NAMELIST / DATA / N, X / HILO / XHIGH, XLOW
C
C     ..... READ AND PRINT THE X ARRAY .....
    1 READ (5,DATA,END=10)
      WRITE (6,DATA)
C
C     ..... CALL UPON SUBROUTINE MAXMIN AND PRINT RESULTS .....
      CALL MAXMIN (N, X, XHIGH, XLOW)
      WRITE (6,HILO)
      GO TO 1
C
   10 CALL EXIT
C
      END
      SUBROUTINE MAXMIN (TOTAL, A, BIGA, SMALLA)
C
C        THIS SUBROUTINE FINDS THE SMALLEST (SMALLA) AND THE LARGEST
C        (BIGA) OF THE FIRST TOTAL ELEMENTS OF THE ARRAY A.
C
      INTEGER  TOTAL
      DIMENSION  A(TOTAL)
C
C     ..... ASSUME A(1) IS THE LARGEST AND THE SMALLEST ELEMENT .....
      BIGA = A(1)
      SMALLA = A(1)
C
C     ..... CHECK EACH ELEMENT AGAINST CURRENT HIGH AND LOW VALUES .....
      DO 10  I = 2, TOTAL
      IF (A(I) .GT. BIGA)   BIGA = A(I)
   10 IF (A(I) .LT. SMALLA)   SMALLA = A(I)
      RETURN
C
      END
$ENDFILE
$RUN -LOAD 5=*SOURCE* 6=*SINK*
 &DATA N=5, X=-67.4, 76.9, 0., 46.2, 1.735, 15*0. &END
 &DATA N=10, X= 896.8, 57., 64.7, 436., 99.7, -473.666,
 2*1.1, 44.4, 44.2 &END
$ENDFILE
$SIGNOFF
```

MTS | FORTRAN main program | FORTRAN subroutine MAXMIN | MTS | data | MTS

Computer Output

The NAMELIST output statement uses the full carriage width of the line printer. The results are shown below.

```
&DATA
N=            5,X= -67.399994      ,  76.899994    , 0.0          ,  46.199997   ,
 0.0              , 0.0            , 0.0           , 0.0          , 0.0          ,
 0.0              , 0.0            , 0.0           , 0.0          , 0.0
&END
&HILO
XHIGH=   76.899994     ,XLOW= -67.399994
&END
&DATA
N=           10,X=  896.79980      ,  57.000000    ,  64.699997   ,  436.00000   ,
   1.0999994      ,  44.399994     ,  44.199997    , 0.0          , 0.0          ,
 0.0              , 0.0            , 0.0           , 0.0          , 0.0
&END
&HILO
XHIGH=   896.79980     ,XLOW= -473.66577
&END
```

Discussion of Results

Note that the NAMELIST output statements print the NAMELIST name before each list of output values and that the user has no control over the appearance of the output. When an array name appears in an output NAMELIST, the *entire* array is printed; this is the reason for filling out the 20 elements of the X array with zeros in the two data sets. This program is also a **WATFIV** program, and could be translated by the WATFIV compiler without changing the source program or data cards. The MTS command lines would have to be changed, however. The program would not be acceptable to a WATFOR compiler, since the WATFOR language does not allow NAMELIST input and output statements.

The operating system commands for this job are identical to those used in the batch job for Example Program 3.1.

```
 1.7349997   , 0.0          , 0.0
0.0          , 0.0          , 0.0
```

```
 99.699997   , -473.66577   ,  1.0999994
0.0          , 0.0          , 0.0
```

EXAMPLE 3.4

FINDING THE LARGEST AND SMALLEST ELEMENTS OF AN ARRAY
A WATFIV SUBROUTINE–WATFOR/WATFIV SIMPLE INPUT AND OUTPUT

Problem Statement

Solve the problem of Example 3.3 using a WATFIV program; use the WATFOR/WATFIV simple (format-free) input and output statements in the main program.

Batch Job Deck

```
$SIGNON CCNO T=10 P=20 C=0 'YOUR NAME'
PASSWD
$RUN *WATFIV
$COMPILE
C
C        SAMPLE PROGRAM NO 4.   THIS MAIN PROGRAM CALLS UPON THE
C        SUBROUTINE MAXMIN TO FIND THE SMALLEST (XLOW) AND LARGEST
C        (XHIGH) ELEMENTS IN THE ARRAY X(1), X(2), ..., X(N).
C
      DIMENSION X(20)
C
C     ..... READ AND PRINT THE X ARRAY .....
    1 READ, N, (X(I), I=1,N)
      PRINT, N, (X(I), I=1,N)
C
C     ..... CALL UPON SUBROUTINE MAXMIN AND PRINT RESULTS .....
      CALL MAXMIN (N, X, XHIGH, XLOW)
      PRINT, XHIGH, XLOW
      GO TO 1
C
   10 CALL EXIT
C
      END
      SUBROUTINE MAXMIN (TOTAL, A, BIGA, SMALLA)
C
C        THIS SUBROUTINE FINDS THE SMALLEST (SMALLA) AND THE LARGEST
C        (BIGA) OF THE FIRST TOTAL ELEMENTS OF THE ARRAY A.
C
      INTEGER  TOTAL
      DIMENSION  A(TOTAL)
C
C     ..... ASSUME A(1) IS THE LARGEST AND THE SMALLEST ELEMENT .....
      BIGA = A(1)
      SMALLA = A(1)
C
C     ..... CHECK EACH ELEMENT AGAINST CURRENT HIGH AND LOW VALUES .....
      DO 10  I = 2, TOTAL
      IF (A(I) .GT. BIGA)   BIGA = A(I)
   10 IF (A(I) .LT. SMALLA)   SMALLA = A(I)
      RETURN
C
      END
$DATA
5, -67.4,76.9,0.,46.2,1.735
10 896.8 57. 64.7 436. 99.7
-437.666 1.1 1.1 44.4 44.2
$STOP
$ENDFILE
$SIGNOFF
```


*The lines $COMPILE, $DATA, and $STOP are commands for the WATFIV compiler, *WATFIV, and are *not* MTS commands.

Computer Output

The WATFOR/WATFIV simple (format-free) output statement uses the full carriage width of the line printer. The results are shown below.

```
             5  -0.6739999E 02    0.7689999E 02    0.0000000E 00   0.4620000E 02
0.7689999E 02   -0.6739999E 02
            10   0.8968000E 03    0.5700000E 02    0.6470000E 02   0.4360000E 03
0.1100000E 01    0.4439999E 02    0.4420000E 02
0.8968000E 03   -0.4376660E 03
```

Discussion of Results

In order to convert the FORTRAN program of Example 3.3 into an equivalent WATFOR or WATFIV program using the simple input and output statements, it is only necessary to remove the NAMELIST statement, to replace the NAMELIST input and output statements with simple input and output statements, and to re-punch the data cards to conform to the WATFOR/WATFIV specifications. As before, the asterisks at the right edge of the listing indicate that the associated line contains a command for the WATFIV compiler. In the $RUN *WATFIV command, no device assignments appear; the default assignments for SCARDS and SPRINT are *SINK* and *SOURCE*, respectively (see page 157), so this command is equivalent to

```
$RUN *WATFIV SCARDS=*SOURCE* SPRINT=*SINK*
```

Note that unlike the NAMELIST output of the preceding example, no labels are printed with the numerical results; only the array elements of interest are printed (compare this with the results for Example 3.3). Since this WATFIV program is also a WATFOR program, it could be translated and executed by a WATFOR compiler without change; the operating system commands and possibly the compiler commands would need to be altered, however.

```
0.1735000E 01

0.9970000E 02   -0.4376660E 03   0.1100000E 01
```

EXAMPLE 3.5

TEST SCORE ANALYSIS AND CHARACTER MANIPULATION USING A WATFIV PROGRAM

Problem Statement

Write a WATFIV program to analyze test scores for a class containing an undetermined number of students. Each data card will contain a student's first name, starting in column 1, his last name, following the first name with exactly one intervening blank, and his test score, an integer, right adjusted in a field of width 5 in columns 26 to 30. The total number of characters in the first and last names must not exceed 23, so that the last name must not extend beyond column 24; long last names should be truncated at column 24 on the card.

First, order the scores in descending sequence. Print the number of students, the high score and the name of the student with the high score, the low score and the name of the student with the low score, the median score (if there is an even number of students, average the two middle scores), the average score for the class, and the list of student names with their test scores, ordered in descending sequence by test score.

Next, reorder the student names so that the last name appears first, followed by the first name with an intervening comma and blank. Now alphabetize the names according to last names and then first names. Finally, print the list of students and their associated scores in alphabetic order by student names.

Method of Solution

Let the full name and test score for the ith student be designated $name_i$ and t_i, respectively. Then a WATFIV character array of length 25 can be used to hold the student names, read directly from the first 25 columns of the data cards; the restriction that the last name not be punched beyond column 24 assures that there will be a blank in column 25 of each card. Hence, $name_i$ will always contain, from left to right, the first name, exactly one blank, the last name, and at least one blank where the names are those of the ith student. This will prove useful when it is necessary to establish which columns contain the first name and which the last name, since the first and second occurrences of a blank can be used as delimiters. An integer array can be used to store the test scores.

Since the number of students is not specified ahead of time (that is, does not appear as part of the data), some other approach must be used to determine when the last of the student records has been read. Assuming that only one data set is to be processed per execution of the program, the END option in the formatted READ statement can be used to good advantage. A counter can be initialized to 1, and incremented by 1 each time that a READ operation takes place. When the end-of-data is sensed (see Section 3.19), control can be retained through the END option, and the counter will have a value one greater than the number of students in the class, n.

To gather the requested statistics and to print the class list in descending order by test score, it is a simple matter to order the scores using the sorting algorithm described on page 115 if the names are interchanged in the $name$ array each time that the test scores are interchanged in the t array, then the name in $name_i$ will always correspond to the score in t_i. Once the scores are ordered, the high score and the name of the student with the high score, h, will be stored in t_1 and $name_1$, respectively; the low score, l, and the name of the student with the low score will be stored in t_n and $name_n$, respectively. For n odd, the median score, m, will be $t_{[(n+1)/2]}$, while for n even, the median score (taken as the average of the two middle scores) will be given by $[t_{n/2}+t_{(n/2+1)}]/2$. By taking advantage of the truncation in integer division in FORTRAN, the median score for n either even or odd can be computed with the single expression

$$m = [t_{(n/2+1)} + t_{[(n+1)/2]}]/2.$$

The average test score, $\bar{t}$, can be computed as s/n, where s is the sum of the n test scores.

The ordering of the class alphabetically by student name can be accomplished by first isolating those columns containing the first name and those containing the last name. The following program first reads the 25 characters of $name_i$ into the 25 elements of a character array, *temp*, each containing a single character, such that the jth character of $name_i$ is stored in $temp_j$, using the core-to-core input statement. The *temp* array is then scanned element by element to find the first and second occurrences of a blank, allowing the indices of the last character in the first name, *lf*, the first character in the last name, *fl*, and the last character of the last name, *ll*, to be established. The original content of $name_i$ is then blanked out and the new content becomes $temp_{fl}, \ldots, temp_{ll}$, a comma, a blank, and $temp_1, \ldots, temp_{lf}$, in order, so that the information in the ith element of name consists of the last name followed by the first name with an intervening comma and blank.

Once the order of the given and surnames has been reversed, the entire list can be alphabetized using the algorithm described on page 116.

WATFIV Implementation

List of Principal Variables

Program Symbol	**Definition**
FL	Column index of the first character in the last name on the original data card, *fl*.
H	The high test score, *h*.
L	The low test score, *l*.
LF	Column index of the last character in the first name on the original data card, *lf*.
LL	Column index of the last character in the last name on the original data card, *ll*.
N	The number of students in the class, *n*.
NAME	A character array of student names, $name_i$, $i = 1, 2, \ldots, n$. Each element of the array has a length of 25 characters.
M	The median test score, *m*.
S	The sum of the test scores, $s = \sum_{i=1}^{n} t_i$.
SAVNAM	A character variable of length 25, used to save a name temporarily, *savnam*.
SAVSCR	A variable used to save a test score temporarily, *savscr*.
T	An array of test scores, t_i, $i = 1, 2, \ldots, n$.
TBAR	The average test score, $\bar{t} = s/n$.
TEMP	A character array of length one with 25 elements; the *j*th character of one element of the *name* array is saved in $temp_j$.

Flow Diagram

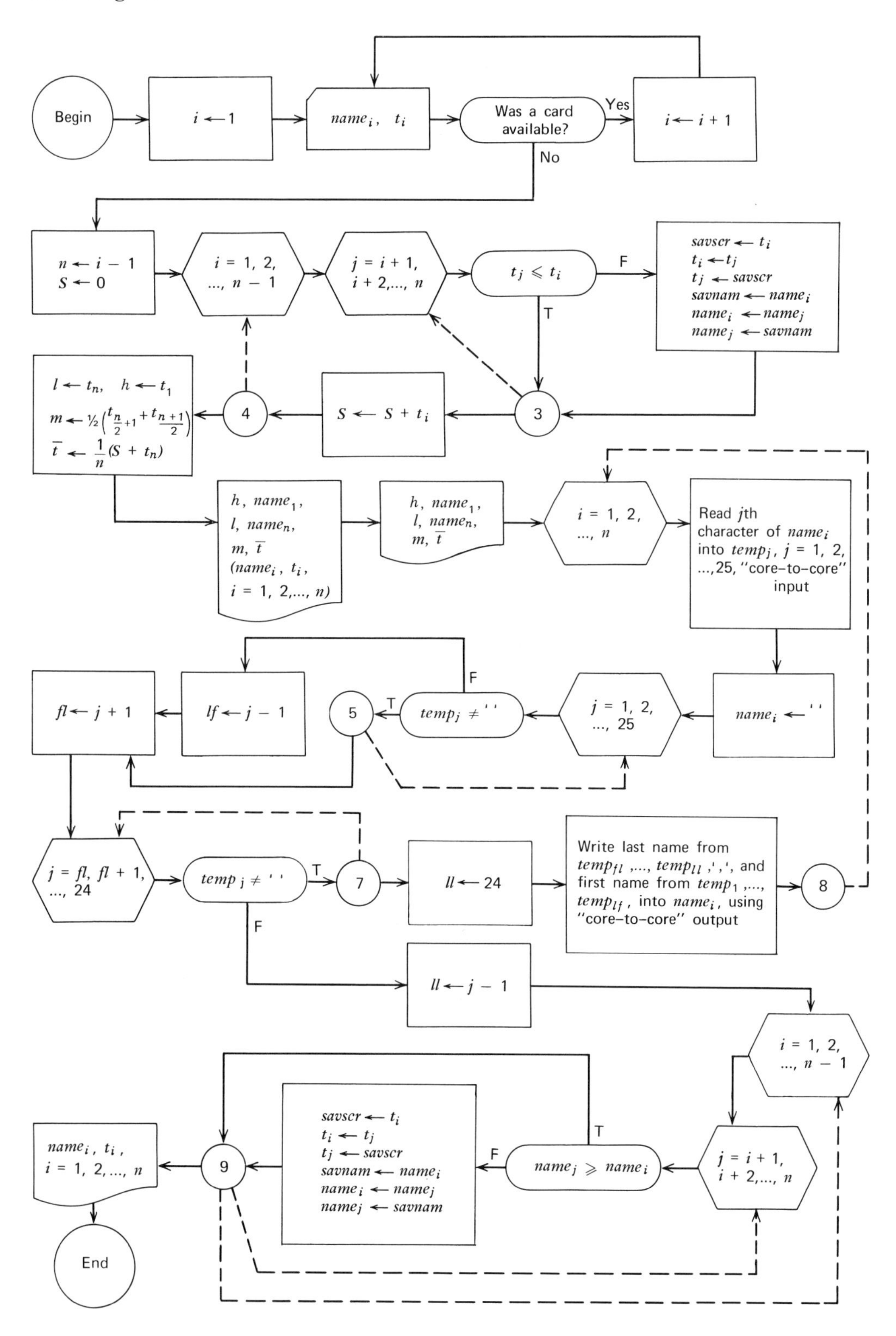

Batch Job Deck

```
$SIGNON CCNO T=10 P=20 C=0 'YOUR NAME'                                MTS
PASSWD
$RUN *WATFIV SCARDS=*SOURCE* SPRINT=*SINK* 5=*SOURCE* 6=*SINK*
$COMPILE
C                                                                       *
C          EXAMPLE PROGRAM NO. 5.  ORDERING A SET OF STUDENT TEST SCORES
C          IN DESCENDING ORDER BY SCORE AND IN ALPHABETICAL ORDER BY LAST
C          NAME, THEN FIRST NAME.
C
      INTEGER  FL, H, S, SAVSCR, T(100)
      REAL  M
      CHARACTER  NAME*25(100), SAVNAM*25, TEMP*1(25)
C
C     ..... READ STUDENT NAMES ( FIRST - LAST) AND SCORES UNTIL ALL
C           HAVE BEEN READ .....
      I = 1
    1 READ (5,100,END=2)  NAME(I), T(I)
      I = I + 1
      GO TO 1
C
C     ..... DETERMINE THE NUMBER OF STUDENT RECORDS READ .....
    2 N = I - 1
C
C     ..... ORDER THE RECORDS IN DESCENDING SEQUENCE BY TEST SCORE.
C           ACCUMULATE TEST SCORE SUM .....
      NM1 = N - 1
      S = 0
      DO 4  I=1,NM1
      IP1 = I + 1
      DO 3  J=IP1,N
      IF (T(J).LE.T(I))   GO TO 3
         SAVSCR = T(I)
         T(I) = T(J)
         T(J) = SAVSCR                                              WATFIV
         SAVNAM = NAME(I)                                           program
         NAME(I) = NAME(J)
         NAME(J) = SAVNAM
    3 CONTINUE
    4 S = S + T(I)
C
C     ..... DETERMINE THE LOW, HIGH, MEDIAN, AND AVERAGE SCORE .....
      L = T(N)
      H = T(1)
      M = (T(N/2 + 1) + T((N+1)/2))/2.
      TBAR = 1.*(S + T(N))/N
C
C     ..... PRINT CLASS ORDERED BY TEST SCORE .....
      PRINT 200, N, 'IN DESCENDING SEQUENCE BY TEST SCORE
     1 H, NAME(1), L, NAME(N), M, TBAR, 'NAME', 'SCORE'
      PRINT 201,   (NAME(I), T(I), I=1,N)
C
C     ..... PRINT HEADINGS FOR ALPHABETICALLY ORDERED CLASS LIST .....
      PRINT 200, N, 'ALPHABETICALLY BY STUDENT NAME (LAST - FIRST)',
     1 H, NAME(1), L, NAME(N), M, TBAR, 'NAME', 'SCORE'
C
C     ..... COPY INDIVIDUAL LETTERS OF NAME(I) INTO ELEMENTS OF TEMP.
C           SCAN TEMP FOR INDICES CONTAINING THE FIRST NAME (1 - LF)
C           AND THE LAST NAME (FL - LL).  THEN WRITE THE LAST
C           NAME, A COMMA, AND THE FIRST NAME BACK INTO NAME(I) .....
      DO 8  I=1,N
      READ (NAME(I),101)  TEMP
      NAME(I) = ' '
         DO 5  J=1,25
            IF (TEMP(J).NE.' ')  GO TO 5
               LF = J-1
               GO TO 6
```

Batch Job Deck (continued)

```
   5      CONTINUE
   6   FL = J + 1
          DO 7  J=FL,24
             IF (TEMP(J).NE.' ')  GO TO 7
                LL = J - 1
                GO TO 8
   7      CONTINUE
       LL = 24
   8   WRITE (NAME(I),101)  (TEMP(J), J=FL,LL), ',', ' ',
     1    (TEMP(J), J=1,LF)
C
C      ..... ALPHABETIZE THE LIST OF NAMES AND REORDER THE SCORES .....
       DO 9  I=1,NM1
       IP1 = I + 1
       DO 9  J=IP1,N
       IF (NAME(J).GE.NAME(I))  GO TO 9
          SAVSCR = T(I)
          T(I) = T(J)
          T(J) = SAVSCR
          SAVNAM = NAME(I)
          NAME(I) = NAME(J)
          NAME(J) = SAVNAM
   9   CONTINUE
C
C      ..... PRINT THE CLASS LIST IN ALPHABETICAL ORDER .....
       PRINT 201,  (NAME(I), T(I), I=1,N)
C
       CALL EXIT
C
C      ..... FORMATS FOR INPUT AND OUTPUT STATEMENTS .....
 100   FORMAT ( A25, I5 )
 101   FORMAT ( 25A1 )
 200   FORMAT ( '1SUMMARY OF ', I3, ' TEST SCORES ORDERED' / ' ', A45 /
     1    '0HIGH SCORE    = ', I3, '  ( ', A25, ')' /
     2    ' LOW SCORE     = ', I3, '  ( ', A25, ')' /
     3    ' MEDIAN SCORE  = ', F6.2, / ' AVERAGE SCORE = ', F6.2 /
     4    '0' 5X, A4, 18X, A5 / ' ' )
 201   FORMAT ( ' ', A25, 3X, I3 )
C
       END
$DATA
MARY JOHNSON                22
HARVEY JOHNSON              61
HARRY JOHNSON               42
HARRY JOHNS                 92
HARRY JOHNSTON              15
ROBERT KADLEC               21
EVERETT SONDREAL            46
HARVEY JOHNS                92
JAMES WILKES                 5
BRICE CARNAHAN             100
ALFRED JONES                77
FANNY HILL                  97
OLIVER TWIST                84
GEORGE WASHINGTON           69
ALEXANDER WILLIAMS          35
PATRICIA BERNSTEIN          60
$STOP
$ENDFILE
$SIGNOFF
```

WATFIV program

*

data

*

MTS

*The lines $COMPILE, $DATA, and $STOP are commands for the WATFIV compiler, *WATFIV, and are *not* MTS commands.

Computer Output

```
SUMMARY OF  16 TEST SCORES ORDERED
IN DESCENDING SEQUENCE BY TEST SCORE

HIGH SCORE    = 100  ( BRICE CARNAHAN              )
LOW SCORE     =   5  ( JAMES WILKES                )
MEDIAN SCORE  =  60.50
AVERAGE SCORE =  57.38

     NAME                      SCORE

BRICE CARNAHAN                  100
FANNY HILL                       97
HARRY JOHNS                      92
HARVEY JOHNS                     92
OLIVER TWIST                     84
ALFRED JONES                     77
GEORGE WASHINGTON                69
HARVEY JOHNSON                   61
PATRICIA BERNSTEIN               60
EVERETT SONDREAL                 46
HARRY JOHNSON                    42
ALEXANDER WILLIAMS               35
MARY JOHNSON                     22
ROBERT KADLEC                    21
HARRY JOHNSTON                   15
JAMES WILKES                      5
```

```
SUMMARY OF  16 TEST SCORES ORDERED
ALPHABETICALLY BY STUDENT NAME (LAST - FIRST)

HIGH SCORE    = 100  ( BRICE CARNAHAN              )
LOW SCORE     =   5  ( JAMES WILKES                )
MEDIAN SCORE  =  60.50
AVERAGE SCORE =  57.38

     NAME                      SCORE

BERNSTEIN, PATRICIA              60
CARNAHAN, BRICE                 100
HILL, FANNY                      97
JOHNS, HARRY                     92
JOHNS, HARVEY                    92
JOHNSON, HARRY                   42
JOHNSON, HARVEY                  61
JOHNSON, MARY                    22
JOHNSTON, HARRY                  15
JONES, ALFRED                    77
KADLEC, ROBERT                   21
SONDREAL, EVERETT                46
TWIST, OLIVER                    84
WASHINGTON, GEORGE               69
WILKES, JAMES                     5
WILLIAMS, ALEXANDER              35
```

Discussion of Results

This problem could be solved using FORTRAN or WATFOR, but only with considerable clumsiness. The WATFIV features allowing character arrays of varying length, and the core-to-core input and output statements make the solution of the problem rather simple. Note that both the comma and the blank have smaller hexadecimal codes than the letters (see Table 1.10). Therefore, it is not necessary to alphabetize separately the last names and then the first names; only one alphabetizing algorithm is required.

3.39 Diagnostic Error Messages

To assist the programmer in his development of grammatically and logically correct FORTRAN, WATFOR, or WATFIV programs, error messages, called *diagnostics*, may be issued by programs operating at several different levels of the software support for the hardware. For the FORTRAN programmer, messages concerning language errors are printed by the FORTRAN compiler. Messages related to computational errors may be produced by library subprograms or by operating system programs (see Chapter 4).

The FORTRAN compilers that generate recoverable object code, that is, compilers that produce an object program that can be saved or punched on cards for execution at any future time, usually print a listing of the FORTRAN source program, and flag any offending statements with terse messages regarding the likely source of the error; for example,

"VARIABLE NAME TOO LONG"
"MORE LEFT PARENS THAN RIGHT PARENS"

In some cases, an error cannot be detected until the entire program has been examined. The diagnostic message then appears at the end of the listing of the source statements; for example:

"LAST STATEMENT IN DO-LOOP NOT FOUND"

Since the FORTRAN compilers used on different computers have been written by different groups at different times, the diagnostics produced by a particular compiler will usually be unique. The message format and the list of all messages likely to be produced by the IBM level-G FORTRAN compiler (used on most of the larger IBM 360 computers) are shown in Appendix D.

If no language errors are present in the source program, an object program can be generated by the compiler and subsequently loaded into the computer's memory for execution. (In some computing systems, the compiler will generate a partial object program even when some language errors are present. It may be possible to load and to execute such a program up to the point of the first fault. This arrangement is not common, however.) Once an apparently sound object program is in the execution phase, additional messages may appear. One major source of such problems is the calling of subprograms with improper arguments. For example, a reference to SQRT(X) for X negative is not permitted. Normally, the subprogram detecting the error will print an appropriate message such as

"SQRT ARGUMENT NEGATIVE"

and then call directly on the operating system to discontinue execution of the program. Error messages of this type for programs produced by the level-G IBM FORTRAN compiler are listed in Appendix D.

Certain abnormal conditions encountered during the execution of an object program can cause a temporary or permanent suspension of computation, called a *program interrupt*. A message indicating the nature of the problem is normally printed to assist the programmer in finding the source of the trouble. For example, an attempted division by zero might cause a message such as

"ZERO DIVISOR AT LOCATION 14530"

to be printed by the operating system. The address of the offending instruction is often printed as part of the message. A list of program-interrupt messages for programs compiled by the level-G IBM FORTRAN compiler appears in Appendix D.

Load-and-go compilers, such as WATFOR and WATFIV, handle the printing of all diagnostics during both the compilation and execution phases of program processing. The error-testing facilities of the WATFOR and WATFIV compilers are very comprehensive, and are a principal reason for the popularity of the two languages. Complete lists of diagnostics for WATFOR and WATFIV are shown in Appendices E and F, respectively. At compile time there are three levels of messages: *extension*, *warning*, and *error*.

An *extension* message is printed whenever a statement contains valid WATFOR or WATFIV constructions that are invalid in IBM FORTRAN. These messages are issued so that the programmer can recognize easily statements that will not be acceptable to other FORTRAN compilers.

A *warning* message is printed whenever the compiler has made some assumption about the program. For example, if a variable name contains more than six characters, the compiler will truncate the name to the first six characters and continue translation and execution. Such "minor" language errors should always be corrected before the program is used again.

An *error* message is printed when a language error, serious enough to prevent execution, is encountered.

During execution, the printing of undefined variables, that is, variables that have not been assigned values previously, is permitted. In WATFOR, an undefined full-word integer value appears as −2139062144 while an undefined real value appears as −0.4335017E–77. In WATFIV, any undefined value is printed as a string of U's.

In some implementations of these compilers, only the *error codes* shown in Appendices E and F are printed; the programmer must then refer to the alphabetically arranged list for a more complete description of the problem. For example, the error code SV–0 stands for the messages

"WRONG NUMBER OF SUBSCRIPTS"

and

"WRONG NUMBER OF SUBSCRIPTS SPECIFIED FOR VARIABLE ________"

in WATFOR and WATFIV, respectively.

The WATFOR and WATFIV compilers assign an identifying number to each executable statement. This number is called an internal statement number (ISN) and is unrelated to the FORTRAN statement number or label and is used only for reference purposes in a diagnostic *subprogram trace*, printed whenever execution of a program is terminated. For example, the following might be printed:

```
***ERROR***  SS–3  SUBSCRIPT IS OUT OF RANGE
PROGRAM WAS EXECUTING LINE 15 IN ROUTINE JACK
  WHEN TERMINATION OCCURRED
PROGRAM WAS EXECUTING LINE  7 IN ROUTINE M/PROG
  WHEN TERMINATION OCCURRED
```

These messages indicate that the error leading to termination occurred in line 15 of the subprogram named JACK, which was called in line 7 of the main program.

The programmer should be aware that diagnostic comments, particularly those generated during compilation, may not pinpoint the problem precisely. Clearly, the compiler writers cannot prepare an error message for every error that might conceivably be committed by all programmers. In addition, it often happens that an error in one statement causes the translator to detect apparent (but not real) errors in other statements. This is particularly true when the first error occurs in a nonexecutable statement. The reason for this is that as a statement is scanned from left to right, the appearance of one error may cause the translator to abandon further examination of the statement. Thus an error in a DIMENSION statement

```
DIMENSION  A(10), BETA(5,3(, DEL(7,3,2)
```

may cause other statements such as

```
A(7) = 10.5
X = DEL(1,1,3)/SQRT(Y)
```

to be flagged with error messages. Once the DIMENSION statement is corrected, all of the related error messages will disappear. The point here is that the programmer must analyze error messages to determine if the error is real or simply related to some other already detected error.

Problems

3.1 Indicate which of the following are integer, F-type real, E-type real, literal, logical, or invalid constants:

(a) 10.
(b) CONST
(c) – 39
(d) 6.3e–6
(e) .693E09
(f) 0.693E+09
(g) – 2147843646
(h) 2.7E(56)
(i) PI
(j) –723.4E–80
(k) 37,560
(l) 6.023*10**23
(m) '1234'
(n) '1'.56
(o) 'IT''S'
(p) 'TRUE'

3.2 Examine the following list of alleged variable names. State the normal mode of those that are valid; give the individual reason or reasons why the remainder are invalid.

(a) XXXXXX
(b) V[6]
(c) I,J&K
(d) STRESS(999)
(e) XXXXXXX
(f) V(–4, 2, J)
(g) $VALUE
(h) 37FEET
(i) X1
(j) LINE
(k) .TRUE.
(l) βVALUE
(m) XPLUS2
(n) 'VAR'
(o) FALSE
(p) REAL
(q) INTEGR(4.7)
(r) X+2

3.3 Choosing obvious variable names, write FORTRAN arithmetic expressions that are equivalent to the following:

(a) $\dfrac{a+b}{c+d}$

(b) $\dfrac{de}{e+fg}$

(c) $\dfrac{y}{2\pi r}$

(d) $1+x+\dfrac{x^2}{2!}+\dfrac{x^3}{3!}+\dfrac{x^4}{4!}$

(e) $\dfrac{a^{z^3}-exp\,(y^{-0.7})}{\sin^{-1}(x/\pi)}$

(f) $n(n+1)/2$

(g) $a\times 10^{0.9}\times 0.9^{10}$

(h) $\dfrac{\sin\,(\alpha+|\beta|)^2}{\log_e \gamma+\delta}$

(i) $\dfrac{1}{x}\left[\dfrac{2}{x}\left(\dfrac{3}{x}-1\right)\right]$

(j) $\dfrac{|\sigma+\sin^2(x+y)\cos(x+\sqrt{z})|}{(\beta+xz)^{0.217a}}$

3.4 If $I=6$, $J=7$, $K=-2$, $X=2.25$, and $Y=4.75$, state the value and mode of the following arithmetic expressions:

(a) (X + Y)/J
(b) I**K
(c) I – J/K
(d) (I – J)/K
(e) SQRT(X) – IABS(K)
(f) J/3*K
(g) J/3.*K
(h) X**(–K)
(i) ABS(K*Y)

3.5 Find the errors, if any, in the following arithmetic expressions:

	FORTRAN Expression	*Intended Meaning*
(a)	LOG(X(I) + SINE(ALPHA))	$\log_e(x_i+\sin\alpha)$
(b)	(A/D*B/E/(F/C)	abc/def
(c)	(TAN.BETA – 1.)**16GAMMA	$(\tan\beta-1)^{16\gamma}$
(d)	0.79E(2Y+1)	7.9×10^{2y}

(e)	`A*B**C*D**E`	$ab^c d^e$
(f)	`A**B**2`	a^{2b}
(g)	`SIN**(-1)(COS**2(THETA))`	$\sin^{-1}(\cos^2(\theta))$
(h)	`A(I)**-3.1*B(J)*C**2(K)`	$a_i^{-3.1} b_j c_k^{\ 2}$
(i)	`3*SQRT(J*K) - X**3.0`	$3\sqrt{jk} - x^3$
(j)	`-7.4E    -5E**E*EXP(E*E)`	$-7.4 \times 10^{-5} e^e \exp(e^3)$

3.6 Choosing obvious variable names, write FORTRAN logical expressions that are true when:

(a) x is greater than $5z + \sin p$
(b) z lies between $x+2$ and $x+7$, inclusive
(c) x and y and z are equal
(d) x and y and z are of equal magnitude
(e) $x < y < z$
(f) $|x| > \epsilon$
(g) $\alpha^2 \neq \beta$ and $\delta = 0$
(h) either $a = 0$ or b is negative, or both
(i) either $a = 0$ or b is negative, but not both
(j) j is an integral multiple of 3
(k) $j + k$ is an odd integer
(l) two of the variables, x, y, and z, but not all three, have equal magnitudes

3.7 In conventional notation, show the meaning that the FORTRAN translator will ascribe to the expression

```
E*1.3E - 4*EXP(E).EQ.E**E*4./EXP(2.*E)**E
```

3.8 If I = 2, J = −3, X = 0.81, and TOBE (a logical variable) is false, evaluate the following logical expressions:

(a) `TOBE .OR. .NOT. TOBE`
(b) `I.LE.3 .AND. SQRT(X)*6.EQ.4.5 .OR. TOBE`

3.9 If $\mathscr{B}_1$, $\mathscr{B}_2$, and $\mathscr{B}_3$ are logical expressions, can there possibly be any difference between the values of the following expressions?

(a) $\mathscr{B}_1$ .AND. $\mathscr{B}_2$ .OR. $\mathscr{B}_3$
(b) $\mathscr{B}_3$ .OR. $\mathscr{B}_2$ .AND. $\mathscr{B}_1$

3.10 Consider the crosshatched regions of the xy plane shown below. If X and Y are FORTRAN variables whose values are the x and y coordinates of a point in the plane, write a logical expression for each figure that is true if and only if the point (X,Y) lies *inside* or *on the boundary* of the crosshatched region. *Hint*: the equation of a circle of radius r is $x^2 + y^2 = r^2$.

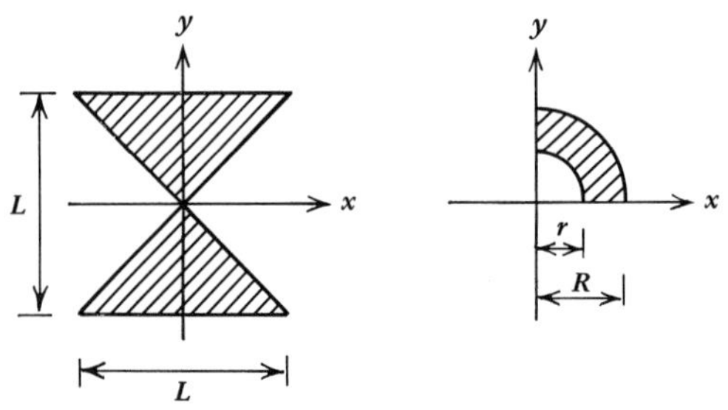

Figure P3.10.

3.11 Let the figure below represent an electrical network with switches numbered 1 through 10. Let the off/on status of each switch be represented by the value of a logical variable, such that logical true indicates that the switch is closed (on) and logical false indicates that the switch is open (off). Write a FORTRAN logical expression that will be true if and only if there is at least one closed path from point *A* to point *B*. Check your expression for several different switch settings.

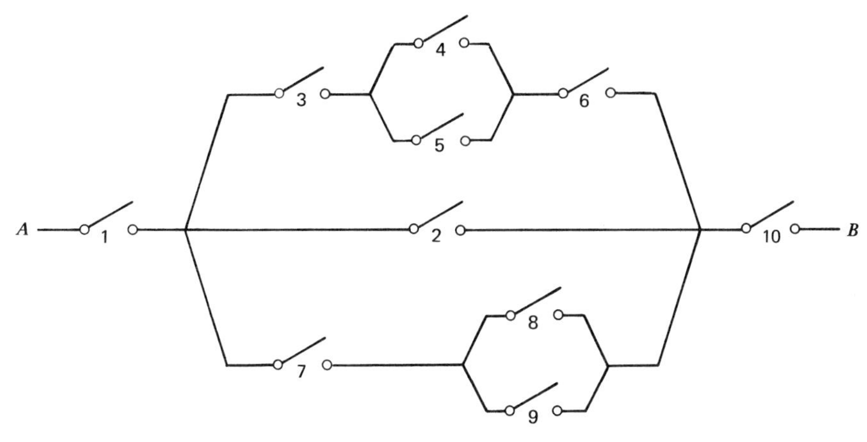

Figure P3.11.

Suppose that the network is viewed as a piping system with valves numbered 1 through 10. Let the closed/open status of each valve be represented by the value of a FORTRAN logical variable as above, but this time let logical true indicate that the valve is closed and logical false indicate that the valve is open. Write a FORTRAN logical expression that will be true if and only if there is at least one open path for flow from point *A* to point *B*.

3.12 Explain the difference, if any, between the following FORTRAN expressions:

(a) X = Y + Z and X .EQ. Y + Z
(b) A .NE. B and .NOT. A .EQ. B

3.13 Which, if any, of the following are valid FORTRAN, WATFOR, or WATFIV assignment statements?

(a) X → Y
(b) X ← Y
(c) I = 4.78 − I + I − 4.78
(d) COLOR = 'BLACK'
(e) ALPHA = BETA = DELTA − Q = JACK
(f) A + B = SQRT(D − E)
(g) BOOL = D710*P.LE.DELTA
(h) SWITCH = .TRUE.
(i) SQRT(4.0) = 2.0

3.14 If A = 5.3, B = 3.2, JACK = 4 and GO is a logical variable whose value is false, what values will be assigned to the variables I, K, X, Y, and Z (a logical variable) in the following FORTRAN, WATFOR, or WATFIV assignment statements?

(a) I = A
(b) X = JACK + A
(c) I = JACK + A

(d) X = I = Y = A
(e) Z = A.LT.B.OR..NOT.GO
(f) X = B = Y = A

3.15 Consider the variables W, X, Y, and Z. Write FORTRAN statements that will store the value of X in W, Y in X, Z in Y, and W in Z.
3.16 Write a FORTRAN statement that is equivalent to:

$$y = y_i + (x - x_i)\,\frac{y_{i+1} - y_i}{x_{i+1} - x_i}.$$

3.17 Write a FORTRAN statement that will increase the value of the variable NVAL to the next higher even integer.
3.18 Write FORTRAN statements that will add one to the value of ISN, ISZ, or ISP, according to whether the value of VOLT(16) is negative, zero, or positive, respectively.
3.19 Write a single FORTRAN assignment statement that will assign the value true to the logical variable BOOL if I is an integral multiple of J and the value false otherwise.
3.20 Draw a flow diagram that is equivalent to the one on page 78, except that the first conditional statement to be tested is $I \geqslant 0$ instead of $I < 0$. Write FORTRAN statements that correspond to this new flow diagram.
3.21 Write FORTRAN statements that will store X in XR if FX and FXR have the same sign, but that will store X in XL if they do not have the same sign.
3.22 Indicate the errors, if any, in the following iteration statements that would be unacceptable to the FORTRAN translator:

(a) DO 67.5 I = 1, 2, 3
(b) DO 4, INT = 3, 2, 1
(c) DO 11, J(I) = −4, 17
(d) DO 11 COUNT = 1, N+1

3.23 Write FORTRAN statements that will store N!, the factorial of a positive integer N, in the variable NFACT:

$$\text{NFACT} = 1 \times 2 \times \cdots \times \text{N} = \prod_{i=1}^{\text{N}} i.$$

3.24 Write the FORTRAN statements that will evaluate any nth-degree polynomial with real coefficients $a_1, a_2, \ldots, a_{n+1}$ for any argument x,

$$p_n(x) = a_1 + a_2x + a_3x^2 + \cdots + a_nx^{n-1} + a_{n+1}x^n,$$

using the nested evaluation scheme discussed on page 291,

$$p_n(x) = (\cdots((a_{n+1}x + a_n)x + a_{n-2})x + \cdots + a_2)x + a_1.$$

3.25 Using obvious variable names, write FORTRAN statements that will compute the mean, $\bar{x}$, and the standard deviation, s, using the following formulas:

$$\bar{x} = \frac{\sum_{i=1}^{n} x_i}{n}, \qquad s = \sqrt{\frac{\sum_{i=1}^{n} (x_i - \bar{x})^2}{n-1}}.$$

3.26 Write FORTRAN statements that will reverse the order of elements in a vector A(1), . . . , A(N), that is, that will interchange A(1) and A(N), A(2) and A(N–1), etc.

3.27 Write FORTRAN statements that will sort, into ascending sequence, the elements of the vector B(N), . . . , B(M), where N < M. The sorting should be done *in place*, that is, no additional vectors should be used.

3.28 For the matrix A shown on page 85, let M = N, and write FORTRAN statements that will:

(a) Store the sum of the main diagonal elements in the variable SPUR.

(b) Store the individual number of negative, zero, and positive elements below the main diagonal in the variables NNEG, NZER, and NPOS, respectively.

3.29 The number of items of type I (= 1, 2, . . . , L) sold by store J (= 1, 2, . . . , M) for month K (= 1, 2, . . . , 12) is designated as NSALES(I,J,K). The L × M × 12 three-dimensional array NSALES contains all possible such information. The profit on the sale of each item I is $PROF(I). Write FORTRAN statements that will compute over the twelve months:

(a) ITEM(I), the number of items of each type sold by all the stores combined.

(b) $TOTAL, the total profit made by all the stores combined.

3.30 Given a matrix with elements A(1,1), . . . , A(M,N), write FORTRAN statements to rearrange the elements in descending order, by rows, that is, place the largest element in A(1,1), the next largest element in A(1,2), and so forth, with the smallest element in A(M,N).

3.31 V, MAT, and TRI are the names of real arrays of maximum dimensions 10, 10 × 10, and 10 × 10 × 10. Write the declarations necessary to reserve space, assign proper mode, and preset all these array elements to zero.

3.32 Write the appropriate FORTRAN declarations to indicate that all variables beginning with the letters A to J, P, and X are of type real and length double, all variables beginning with the letters R and Z are integers of length two, all variables beginning with the letter T are of type logical and length one, and that all other variables follow the usual naming conventions.

3.33 The variables A, I, K, and ST are to be integers of standard length. Write the declarations to assign the modes of these variables and to preset them with the values 1, 2, 2, and 5, respectively.

3.34 Write the FORTRAN statements to reserve storage for the subscripted variables in the following program sequences, where N is 100 or less.

```
      DO 100  I=1,N                   DO 50    JACK=2,N,2
  100 C(I) = A(I)*B(I)**2          50 C(2*JACK + 2) = JACK
```

```
      N2 = 2*N + 1
      DO 5  M=1,N2,3
    5 ALPH(M + 1) = BETA(N2 - M + 1)
```

3.35 If the following declarations appeared in a FORTRAN program, what values would be assigned to D(4,1), D(1,2), D(4,5), BETA(1) and BETA(5)?

```
      DIMENSION  BETA(5), D(4,6)
      DATA  D, BETA / 2*5., 3*7.3, 15*-5.7, 0., -5.1, 3*-7.3E4,
     1  3*2., 1.6 /
```

3.36 Suppose that the following declarations appeared in a FORTRAN program:

```
IMPLICIT REAL*8(A, F–H), INTEGER (P)
INTEGER SAM, X, GO*2, MONTH*2
REAL ALPHA, GO2
DOUBLE PRECISION PAUL, MANDY
LOGICAL G, SW*1
DIMENSION SW(10), GO(2)
```

What would be the modes and lengths of the following FORTRAN expressions?

(a) GO2 + GO(2)
(b) SAM*MONTH + AL
(c) IABS(PS – NANCY) – DSIN(MANDY)
(d) X.LT.GONE.AND.SW(7)
(e) PENNY*GO(1)
(f) GO(1)/INT
(g) FUN(PAUL) – SAM/P
(h) 5.37D–6*GO(2) – N*SQRT(ALPHA – BETA)

3.37 Explain in detail what each of the following declarations accomplishes:

```
 INTEGER*2 D*4(4,4,2)/4*2,8*–3,20*0/,MAN(20),JOE,FIX*4(3,2)/
1 6*5/
 REAL MEAN(3,4,7)/42*1.E–35,42*–1.E–35/,AL*8(10)/5*5.2D0,
1 5*0.D0/
```

3.38 Write a FORTRAN input statement and an accompanying FORMAT declaration to read from one card values for three variables N, X, and Y where the range and accuracy (the number of digits) are –999 to 9999, 2.675 to 19.646, and 0.001×10^{-17} to 0.999×10^{-17}, respectively. Using your format, show the precise appearance of the data card you would punch if the values –469, 6.417, and 0.987×10^{-17} are to be assigned to N, X, and Y, respectively.

3.39 The following information is to be punched on data cards:

Variable	*Field Specification*	*Columns on the Card*
N	I5	1–5
M	I5	6–10
A	F7.3	11–17
B	F7.3	18–24
C	F7.3	25–31
D	F7.3	32–38
X	E11.4	39–49
Y	E11.4	50–60
Z	E11.4	61–71

Write the input and FORMAT statements required to:
(a) Read in the entire contents of the card.
(b) Read in N, A, B, C, X, and Z.
(c) Read in M, D, and X.

3.40 Two vectors A and D, each with N elements, are to be read from data cards. The first card contains a value for N in the first three columns. The vectors are punched on subsequent cards in F6.4 fields, ten fields per card, with the elements of A and D in alternating sequence, that is, in the order A(1), D(1), A(2), D(2), etc. Write the input and format statements required to:

(a) Read N and all the elements of A and D.

(b) Read N and all the elements of D.

(c) Read N and all the elements of A and D having odd subscripts.

(d) Read N and all the elements of A having odd subscripts and all the elements of D having even subscripts.

3.41 Print the elements of a 20 × 12 matrix BETA in matrix form (by rows) with double spacing between rows. The minimum field specification for each element is F7.2.

3.42 Print the elements of the N by M matrix BOOM in matrix form (by rows) with field specification for each element, F10.4. The elements should be printed across the page with no more than 12 per line. If M is larger than 12, remaining elements in the row should be printed on subsequent lines, single spaced. Each new row of elements should be printed starting at the left edge of the page, triple spaced below the elements of the preceding row. Write the FORTRAN statements required to do this.

3.43 Write FORTRAN statements that will cause the printer to skip a double space and print the message "Reactor Explodes" starting in column 23 if T exceeds TMAX, but that will do nothing otherwise.

3.44 Write FORTRAN statements that will determine whether the variable N contains a prime number or not, and that will print a suitable message in either event.

3.45 Suppose that variables N, X, Y, and Z are to be read as data from cards and then printed immediately in I5, F10.3, E15.4, and L2 fields starting in columns 5, 16, 43, and 72, respectively, on line number 3 of the next page in the printer. Subsequently, these data are used to compute new variables P, M, and BOOL which are to be printed on the *same* line in F7.4, I3, and L1 fields starting in columns 30, 39, and 65, respectively. Write the FORTRAN output statements and FORMAT declarations to accomplish this.

3.46 What would be the appearance of the FORMAT declaration and of the printed output if the G field specification had been used for all variables in Problem 3.45?

3.47 Write the appropriate input/output statement and accompanying FORMAT declarations to inplement the following:

(a) You wish to read variables N, EPS, and A(1), . . . , A(N) from cards arranged as follows:

	Variable	*Card Columns*	*Decimal Places*
Card 1	N	1–5	–
	EPS	6–20	3 with exponent
	P(5)	31–40	5 without exponent
	A(1)	41–50	4 without exponent
	A(2)	51–60	4 without exponent
Card 2	A(3)	1–10	4 without exponent
	A(4)	11–20	4 without exponent
	A(5)	21–30	4 without exponent
	A(6)	31–40	4 without exponent
	A(7)	41–50	4 without exponent

Card 3	A(8)	1–10	4 without exponent
	.	.	.
	.	.	.
	.	.	.
	A(12)	41–50	4 without exponent

Card 4 etc. 5 elements of the vector A with the format of Cards 2 and 3.

(b) You wish to write some column headings and then write appropriate variable values under the headings as shown below:

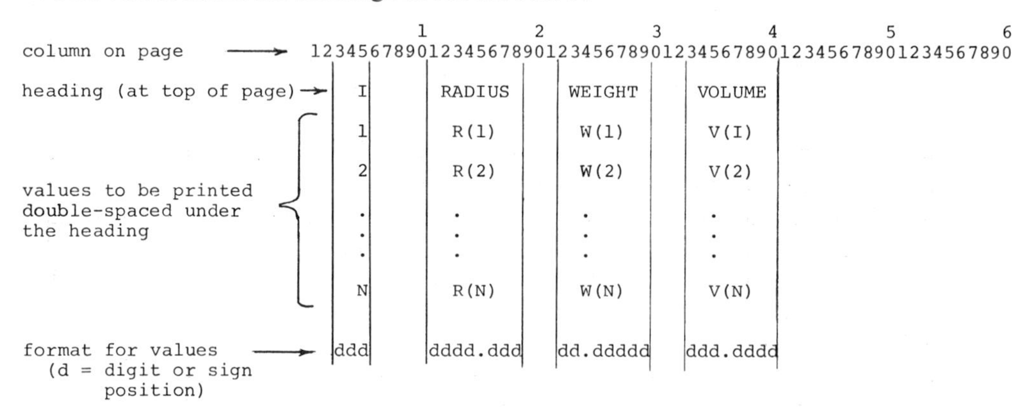

3.48 Using the variable format features of FORTRAN, write the FORTRAN statements that would print the variables N, M, X(1), . . . , X(N), Y(1), . . . , Y(M) according to a format having the form (2I*w*, *n*F*w*.3, T*c*, *m*E15.*d*), where *w*, *n*, *c*, *m*, and *d* have the values of the *integer* variables W, N, C, and D, respectively. Include all statements required to prepare the format in the format variable array FORM.

3.49 If you wished to solve the test score analysis problem of Example 3.5 using either FORTRAN or WATFOR, what modifications would you make in the WATFIV program shown in the text?

3.50 Using FORTRAN variable names of your choice, write a complete FORTRAN program for the algorithm developed in Problem 2.1.

3.51 Refer to Problems 2.4 and 2.5. Prepare a card format for the employee data containing values for *b*, *s*, *I*, *d*, and the employee's name. Write a program that reads the data cards, one at a time, implements the algorithm developed in Problem 2.5, and writes out all pertinent information (including the name). The final printed results should be a tabular, reportlike listing of the information for all employees, with suitable labels for column headings.

3.52 Medical records for all students in an elementary school are punched on cards as follows:

Columns on the Card	*Content*
1–12	Last name, starting in column 1.
15–27	First name, starting in column 15.
29	Middle initial.
32	Sex (M or F).
35	Grade (1, 2, 3, 4, 5, or 6).
39–40	Age in years (I2 field specification).
42–43	Height in inches (I2 field specification).
45–47	Weight in pounds (I3 field specification).

50–52	School days absent caused by illness during the current year (I3 field specification).
53	Vision correction? T indicates student wears glasses; F indicates no correction.
54	Vision impairment index: 0, perfect vision; 1, some impairment; . . . 5, very serious impairment.
60–69	Childhood illness indicators, each column represents one common illness. For example, column 60 might correspond to mumps (T if student has contracted mumps, F otherwise).

Write a FORTRAN program (or programs) that reads the arbitrarily arranged data cards for the entire school population (the number of students is unknown) and produces as printed output one or more of the following:

(a) The number of students in each class and the total number in the six classes.

(b) An alphabetical listing of all students and of any or all classes in the order: last name, first name, middle initial.

(c) Separate alphabetical listings as in (b) for boys and girls.

(d) The minimum, maximum, average, and median weight and height for each class level and each age level according to sex.

(e) The number and names of any students with a vision impairment index of 2 or greater whose vision has not been corrected.

(f) The average number of days absent among boys in grade 5 who are 11 years old.

(g) The names of all girls who are more than ten percent below the average weight for the corresponding age and who have a history of the childhood disease encoded in column 66 of the medical record.

3.53 In its simplest form, a concordance is an alphabetical listing of all the words appearing in a written text (for example, *the Bible*), usually with the frequency of occurrence for each word. Assume that a text has been typed onto punched cards such that words are always separated from one another by one or more blanks, a comma, a left or right parenthesis, a colon, a semicolon, an exclamation point, a question mark, quotation marks, or a period. Any hyphens should be considered part of a hyphenated word; that is, no word will be broken so that it appears on more than one card.

Write a WATFIV program that will read a text of arbitrary length from data cards, and prepare a complete concordance for it. The total number of words should be printed. To test the program, punch the first two paragraphs of Chapter 1 onto cards. *Note*: This problem can be solved using FORTRAN or WATFOR, but only with considerable difficulty.

3.54 Choosing appropriate names for the problem variables, transcribe the algorithm prepared as a solution to Problem 2.12 into a complete FORTRAN program. Prepare a set of test data to check out the program.

3.55 Consider the problem of accelerating an automobile of mass m lb_m from a standing start to a speed of v ft/sec along a straight horizontal road in t seconds flat. Assuming that the accelerating force is constant, use Newton's second law to determine:

a, the acceleration, in ft/sec^2.
f, the accelerating force, in lb_f.
x, the distance traveled, in ft.

Write a FORTRAN program that:

(a) reads values for m, v, and t in the indicated units.
(b) prints out the data values read in (a).
(c) calculates s, the velocity v in miles per hour.
(d) computes a, f, and x in the indicated units.
(e) finds w, the total work expended by the engine, in ft-lb_f, ignoring all inefficiencies and frictional losses.
(f) computes p, the average power (w/t), in horsepower.
(g) prints s, a, f, x, w, and p.
(h) returns to (a) to read another data set.

Insofar as is possible, your program should print the data and results in tabular form. The following conversion factors will prove useful:
1 horsepower = 550 ft lb_f/sec.
1 lb_f = 32.2 ft lb_m/sec^2.

Suggested test data

m, lb_m:	2000,	3000,	4000,	3000,	3000,	3000,	1916.4,	3920
v, ft/sec:	88,	88,	88,	88,	88,	88,	54.6,	100
t, sec:	10,	10,	10,	20,	5,	1,	12.9,	15.5

3.56 When an incompressible fluid is pumped at a steady rate through a pipe from point 1 to point 2, as shown in Fig. P3.56, the pressure *drop* is given by:

$$\Delta p = p_1 - p_2 = \frac{\rho}{g_c}(g\Delta h + lw).$$

in which lw in ft^2/sec^2, is the lost work per unit mass flowing due to frictional effects:

$$lw = \frac{4fu_m{}^2L}{D}.$$

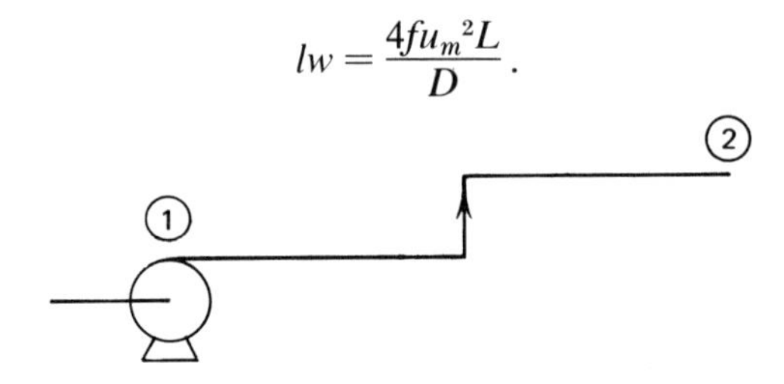

Figure P3.56.

Here:

ρ = fluid density, $lb_m/ft.^3$
Δp = pressure drop, $lb_f/ft.^2$
p_1, p_2 = pressure at points 1 and 2, $lb_f/ft.^2$
g = gravitational acceleration, 32.2 ft/sec.^{2}
Δh = elevation *increase*, $h_2 - h_1$, ft.
L = pipe length, ft.
D = pipe diameter, ft.
u_m = mean velocity of fluid, ft/sec.
g_c = conversion factor, 32.2 lb_m ft/lb_f sec.^{2}

For smooth circular pipes, the *friction factor f* depends only on the *Reynolds number*

$$Re = \frac{\rho u_m D}{\mu},$$

where μ is the fluid viscosity in lb_m/ft sec. The correlations are:

Laminar flow ($Re \leqslant 2000$); $f = 8/Re$.

Turbulent flow ($Re > 2000$): $f = 0.0395\, Re^{-1/4}$.

Write a program that reads values for L, ρ, D, Δh, (in the given units), μ (in centipoise), and Q, the volumetric flow rate in gallons per minute. The program should then print the data values, compute Re, f, lw, u_m, and Δp (in both lb_f/ft^2 and lb_f/in^2), print the results, and then read another data set. Useful conversion factors are:

$$1 \text{ ft}^3 = 7.48 \text{ gal.}$$
$$1 \text{ lb}_m/\text{ft sec} = 1488 \text{ centipoise.}$$

Suggested test data

L (ft)	ρ (lb_m/ft^3)	D (in)	Δh (ft)	μ (centipoise)	Q (gpm)
100	80.2	0.622	0	74.6	4
100	80.2	3.068	0	74.6	4
100	80.2	0.622	10	74.6	4
100	80.2	3.068	10	74.6	4
10000	62.4	0.622	0	1	100
10000	62.4	3.068	0	1	100
10000	62.4	0.622	50	1	100
10000	62.4	3.068	50	1	100

3.57 Examine the following FORTRAN program with its accompanying data cards:

Program:

```
    1 READ (5,100)   I, J
      N = 1
      IF ( I.LE.J .OR. J.EQ.0 )   GO TO 3
      DO 2    K = 1, J
    2 N = N*(I - K + 1)/K
    3 WRITE (6,200)   I, J, N
      GO TO 1
      END
```

```
Column on the card:          1         2         3         4         5         6
                    123456789012345678901234567890123456789012345678901234567890
Data:                   0    0
                        1    0
                        1    1
                        2    0
                        2    1
                        2    2
                        3    0
                        3    1
                        3    2
                        3    3
```

(a) There are two essential statements missing from the program. Write the statements and show where you would put them in the program.
(b) What value will be printed for N for each of the ten data sets?
(c) Write the value of the function computed by the program in symbolic form.
(d) What is the common name and notation for the function of part (c)?
(e) Change the program so that only format-free input and output statements are used. How would you change the data cards?

3.58 The following FORTRAN program is intended to compute the horizontal and vertical displacements, s_h and s_v, respectively, of a projectile with horizontal and vertical velocities, u and v, respectively, at time $t=0$, according to the formulas

$$s_h = ut,$$
$$s_v = vt - gt^2/2,$$

where g is the gravitational acceleration.

```
                                                      Card
                                                       No.
     SMAPLE PROBLEM FOR PROJECTILE                       1
     NAMELIST  INPUT/U, V, T/ DISPLAY/ SH, SV            2
   1 READ (6,INPUT)                                      3
     DO 2,  I = 1, 10                                    4
     T = I*0.1                                           5
     SH = UT                                             6
     SV = T*[V - ½*GT]                                   7
   2 PRINT (5,DISPLAY)                                   8
     GO TO 1                                             9
     STOP                                               10
```

Assuming that the data are properly arranged on cards, point out all FORTRAN errors that would prevent the above program from (a) compiling, (b) executing to produce *correct* results. Use the line identification numbers at the extreme right when referring to statements in your answers.

3.59 Write a complete FORTRAN function, named NFACT, with a single argument N, that computes the value N! (see Problem 3.22).

3.60 Consider the following FORTRAN program:

```
      DIMENSION  A(10)
    1 READ (5,100)  N, X
      NP1 = N + 1
      READ (5,101)  (A(I), I=1,NP1)
      P = A(NP1)
      IF ( N.LE.0 )  GO TO 3
      DO 2  I=1,N
    2 P = P*X + A(NP1-I)
    3 WRITE (6,200)  N, X, P, (A(I), I=1,NP1)
      GO TO 1
  100 FORMAT ( I5, F10.4 )
  101 FORMAT ( 8F10.4 )
  200 FORMAT ('1', I10, 2F10.4 / (' ', 8F12.4) )
      END
```

(a) In one sentence, what does this program accomplish?

(b) What is the common mathematical notation for the variable P in the program?

(c) If N, X, A(1), A(2), and A(3) have the values 2, 1.5, 1., 2.5, and −0.5, respectively, what would be the final value of P produced by the program?

3.61 Examine the following FORTRAN program carefully:

```
Program:                    DIMENSION  A(100)
                          1 READ (5,100)  N, X
                            NP1 = N + 1
                            READ (5,110)   (A(I), I=1,NP1)
                            P = EVAL (N, A, X)
                            WRITE (6,200)  N, X, P, (A(I), I=1,NP1)
                            GO TO 1
                        100 FORMAT ( I5, 15X, F10.3 )
                        110 FORMAT ( 5F10.4 )
                        200 FORMAT ('2', 'N = ', I5, 5X, 'X = ', F10.3, 5X,
                           1'P = ', F15.4/ ('0', 5F15.3 ) )
                            END
                            FUNCTION EVAL (N, A, X)
                            DIMENSION A(100)
                            EVAL = 0.
                            NP1 = N + 1
                            DO 2   I = 1, NP1
                          2 EVAL = EVAL + A(I)*X**(I-1)
                            RETURN
                            END

                                  1         2         3         4         5         6
Column on the card:      123456789012345678901234567890123456789012345678901234567890

Data cards:                  2                     2000
                           3.000     -1.000     2.000
```

(a) In a few words, what does this program accomplish?

(b) What value does EVAL return to the calling program? The answer should be written in symbolic form.

(c) What value will be printed for P for the given set of data?

(d) Draw a diagram, clearly indicating exact column and line positions, that shows the appearance of the computer output for the given set of data.

(e) What changes would you make in the program to allow format-free input and output? Show the input and output statements and associated declarations that would be required. How would you change the data cards?

(f) Would the program produce the correct result for X equal to zero? If not, how would you modify the program to allow for this possibility?

3.62 Suppose that we have N tabulated pairs of values of A (in ascending order) and B:

```
          A(1)      B(1)
          A(2)      B(2)
           .         .
  AVAL→    .         .    →BVAL
           .         .
          A(N)      B(N)
```

Given a value AVAL, the following program segment will interpolate linearly (see Section 6.3) in the table to find a corresponding value BVAL:

```
      DIMENSION A(25), B(25)
      DO 1  I=2,N
      1F ( A(I) .GT. AVAL .OR. I .EQ. N )  GO TO 2
1     CONTINUE
2     I = I - 1
      BVAL = B(I) + (AVAL - A(I))*(B(I+1) - B(I))/(A(I+1) - A(I))
```

(a) Add the necessary FORTRAN statements to convert this program segment into a function named DRAW that will perform the linear interpolation and return the interpolated result as the value of a reference to DRAW in another program.

(b) If, in a main program, we have ten tabulated values of specific heat CP(1), . . . , CP(10) at the corresponding temperatures T(1), . . . , T(10), and we wish to obtain the value of specific heat, CPVAL, at a specified temperature, TGIVEN, what would be the appropriate call on the function DRAW? Assume that the temperatures are in ascending order.

3.63 Suppose a function $f(x)$ has the appearance shown in the diagram below. Write a FORTRAN function named F, with the single argument X, that calculates the value of $f(X)$ and returns it as the value of the function reference $F(\mathscr{E})$, where $\mathscr{E}$ is any real expression.

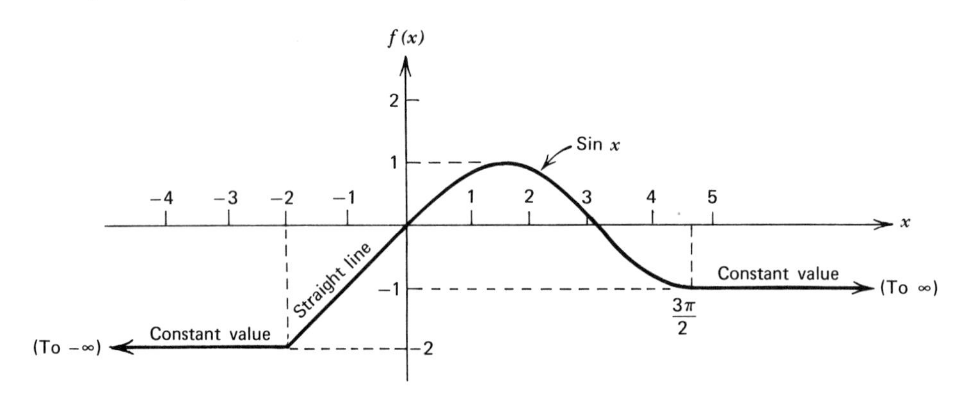

Figure P3.63.

3.64

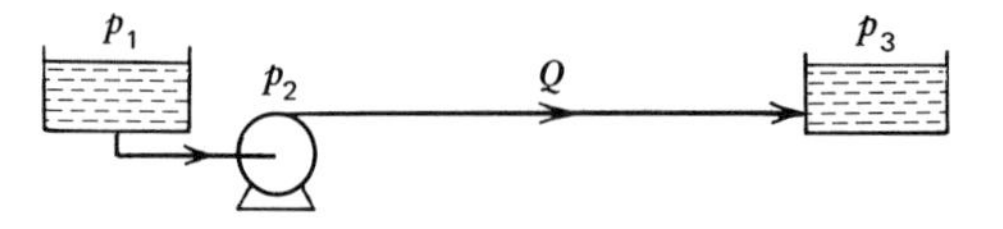

Figure P3.64.

The centrifugal pump shown in Fig. P3.64 is used for transferring Q gpm (gallons/min) of a liquid of density ρ (lb_m/cu ft) and viscosity μ (lb_m/ft sec) from one tank to another, with both tanks at the same level. The pump raises the pressure of the liquid from p_1 (atmospheric pressure) to p_2, but this pressure is gradually lost because of friction inside the long pipe of length L ft, and, at the exit, p_3 is back down to atmospheric pressure.

The pressure drop in the pipe, denoted by Δp, also the increase in pressure across the pump (since $p_1 = p_3$), is given by (see page 269, for example):

$$\Delta p = p_2 - p_1 = p_2 - p_3 = 2.16 \times 10^{-4} \frac{f_M \rho L Q^2}{D^5} \text{ lb}_f/\text{sq in.}$$

where D is the inside pipe diameter in inches. To a first approximation (for a smooth pipe), the Moody friction factor f_M depends on the Reynolds number Re in the pipe, given by

$$Re = 0.03404 \frac{\rho Q}{\mu D} \qquad \text{(dimensionless).}$$

For $Re \leqslant 2000$, $f_M = 64/Re$ (laminar flow); for $Re > 2000$, the Blasius equation holds: $f_M = 0.316 Re^{-0.25}$ (turbulent flow).

Assuming a combined pump and motor efficiency of 70%, the required rate of energy input to the pump can be shown to be

$$P = 6.214 \times 10^{-4} Q \Delta p \quad \text{kW.}$$

Clearly, the selection of a small pipe diameter will result in high pumping costs and low piping costs; on the other hand, a large pipe diameter will result in low pumping costs and high piping costs.

The initial installed cost of pipe, $\$_{pipe}$, is a function of the diameter and length of the pipe:

$$\$_{pipe} = 0.34\, L D^{1.43} \quad \text{dollars.}$$

The initial installed cost of the pumping equipment, $\$_{pump}$, is a function of the capacity Q and the pressure drop Δp (usually called the *head*):

$$\$_{pump} = 43.3 Q^{0.435} (\Delta p)^{0.3} \quad \text{dollars.}$$

The cost of pumping the fluid for one hour is given by

$$\$_{elect} = 0.0215 P \quad \text{dollars,}$$

that is, electricity costs 2.15¢ per kwh.

Pipes are available commercially with the following inside diameters in inches:

0.364, 0.622, 1.049, 2.067, 3.068, 4.026, 6.065, 7.981, 10.02, 12.09

Write a FORTRAN program that will read and print values for Q, L, ρ and μ (the variable names Q, L, RHO, and MU are suggested). For *each* of the pipe diameters listed above, the program should compute values for Re, f_M, Δp, P, $\$_{pipe}$, $\$_{pump}$, $\$_{power}$, and $\$_{total}$, where $\$_{power}$ is the pumping cost for a *five*-year period (24 hours per day, 365 days per year), and $\$_{total}$ is the sum of the installed equipment costs and the cost of operating the pump for five years (we will write off the capital costs in five years):

$$\$_{total} = \$_{pipe} + \$_{pump} + \$_{power}.$$

Suggested variable names are RE, FM, DELTAP, PKW, $PIPE, $PUMP, $POWER, and $TOTAL. All of these variables should be printed for each diameter. Since some of these values may vary over several orders of magnitude for different pipe diameters, all should be printed using the E format specification. The ten given pipe diameters should be preset in the vector DIAM (in DIAM(1)... DIAM(10)) using the DATA declaration (see Section 3.26).

Suggested test data

Liquid	Q (gpm)	L (ft)	RHO (lb_m/cu ft)	MU (lb_m/ft sec)
Water	480	10,560	62.4	0.000677
Kerosene	50	200	51.5	0.001656
Molasses	160	1,000	80.2	0.105

In order to minimize the total cost, $\$_{total}$, what size pipe do you recommend for each of the three cases?

3.65 A rocket of total initial mass M_0 lb_m, including fuel, is fired vertically upwards (in the positive z direction) from the surface of a heavenly body. The subsequent motion of the rocket is primarily determined by two major competing factors: (a) the upwards thrust T lb_f imparted by the exhaust gases escaping from the motor nozzle as long as the motor is operating, and (b) the gravitational acceleration a_g ft/sec^2 in the negative z direction, always present and assumed constant, while the rocket is in the vicinity of the body.

Let m lb_m/sec be the rate of fuel consumption while the motor is firing, v ft/sec be the velocity of the rocket at any time t sec after the initial firing, and u ft/sec be the velocity of the escaping gases relative to the rocket.

The rocket motor is fired for a total time t_b sec after launching, until all the fuel is exhausted. At this particular moment of burnout, let the velocity and vertical distance traveled be v_b ft/sec and z_b ft, respectively.

Using Newton's laws of motion, the following equations can be shown to apply before and after burnout time t_b:

Before Burnout

$$T = \frac{um}{32.2}, \qquad a = \frac{um}{M_0 - mt} - a_g,$$

$$v = u \ln\left(\frac{M_0}{M_0 - mt}\right) - a_g t,$$

$$z = ut\left[1 + \left(\frac{M_0}{mt} - 1\right)\ln\left(1 - \frac{mt}{M_0}\right)\right] - \frac{a_g t^2}{2}.$$

After Burnout

$$T = 0, \qquad a = -a_g,$$

$$v = v_b - a_g(t - t_b),$$

$$z = z_b + v_b(t - t_b) - \frac{a_g(t - t_b)^2}{2}.$$

Here, a is the acceleration vertically upwards of the rocket.

(a) Write a FORTRAN program that reads values for the following variables as data:

Program Symbol	*Meaning*
U	u, ft/sec.
M	m, lb_m/sec.
AG	a_g, ft/sec^2.
MEMPTY	Mass of the rocket *without* fuel, lb_m.
MFUEL	Mass of fuel initially loaded in the rocket, lb_m.
TMAX	Maximum time, t_{max}, sec, for which the motion of the rocket is to be studied (see below).
DELTAT	Time increment, Δt, sec (see below).

(b) Print, with labels, the variables read in step (a) above.
(c) Compute and print values for M_0 and t_b.
(d) Compute and print values for t, T, a, v, z, and M_r (the instantaneous mass of the rocket) for all times t from zero to t_{max} in increments of Δt. *Do not use any subscripted variables.* As soon as these values are computed at each particular value of t, print them in tabular form under the headings TIME, THRUST, ACCLN, VELOCITY, HEIGHT, and MASS. See page 387 for an illustration of a similar table.
(e) Return to step (a) and read another set of data.

Suggested test data.

Use the following sets of test data (supplemented by additional sets of your own choosing, if you wish):

U	M	AG	MEMPTY	MFUEL	TMAX	DELTAT
8350	29000	32.2	1.9×10^6	4.2×10^6	200	5
6000	50	5.4	3000	12000	400	10
6000	50	5.4	3000	12000	100	10
8350	29000	0	1.9×10^6	4.2×10^6	200	5
6000	50	32.2	3000	12000	400	10

What vehicles do you think are represented by the first two data sets? Discuss the significance of the results for the last two data sets.

Notes. Start the results for each data set at the top of a new page. All the variables mentioned above are of type REAL.

3.66 This problem concerns finding the economically optimum thickness of insulation to be installed on steam supply lines. With no insulation, the heat losses to the outside air might be prohibitively high; on the other hand, there is no point in paying for excessive amounts of unnecessary insulation.

Consideration of heat conduction from the hot pipe wall across the insulation, and subsequent heat convection from the outer surface of the insulation to the surrounding air, shows that Q, the rate of heat loss in BTU/hr per foot length of pipe, is given by:

$$Q = \frac{2\pi k(T_s - T_i)}{\ln (d_2/d_1)} = \pi d_2 h(T_i - T_a).$$

Here,

k = thermal conductivity of the insulation, BTU/hr ft °F,
h = heat-transfer coefficient for convection from the outer surface of the insulation to the surrounding air, BTU/hr sq ft °F,
d_1 = outside diameter of the pipe, also the inside diameter of the insulation, ft,
d_2 = outside diameter of the insulation, ft,
T_s, T_i, T_a = temperatures of the steam, the outer surface of the insulation, and the air, respectively, °F.

The particular insulation to be used is available in thicknesses ranging from 0 (no insulation) to 5 in., in half-inch increments. It costs $13.50 per cubic foot of insulation delivered at the site, and the installation costs amount to $42.00 per 100 ft of pipe (independent of the thickness used). A figure of $3.72 is put on every million BTUs of heat lost from the pipe. The installed cost of the insulation is to be written off over a five-year period.

Eliminate T_i from the equations, to give a new expression for Q that does not involve T_i. Assuming that $k = 0.043$ and $h = 5.7$, write a program that will accept values for d_1, T_s, and T_a as data (the variable names D1, TS, and TA are suggested), and that will proceed to compute and print values for the following items for each thickness of insulation in turn:

(a) T, the thickness of the insulation, in.,
(b) VOLI, the volume of insulation used, cu ft,
(c) QBTU, the heat loss, BTU,
(d) $LOSS, the dollar value placed on the heat loss,
(e) $INSUL, the cost of the insulation, including installation,
(f) $TOTAL, the sum of (d) and (e).

Each of the quantities (b) through (f) should be based on 1000 ft of pipe for the five-year write-off period.

Suggested test data.

D1 (in.)	TS (°F)	TA(°F)
3.500	300.0	0.0
3.500	450.0	0.0
5.563	300.0	0.0
5.563	450.0	−50.0
5.563	450.0	0.0
5.563	450.0	50.0

What thickness of insulation do you recommend in each case?

3.67 Write a FORTRAN program that reads values for the variables XLOW, DELTAX, and XHIGH, and then prepares a table of functional values, $\sin(x)$, $\cos(x)$, $\ln(x)$, $\log_{10}(x)$, e^x, and $\sqrt{x}$ for all values of x = XLOW, XLOW + DELTAX, XLOW + 2 × DELTAX, . . . , XHIGH. Columns for x and the various functional values should be appropriately labeled.

3.68 Write and test one or more of the following FORTRAN subprograms for use in your private program library. Use the variable dimensioning feature where possible.

(a) Subroutine ASORT(A, M, N): This routine sorts into algebraically ascending order all elements of the array A between A(M) and A(N), inclusive. A second entry, DSORT(A, M, N) sorts the same elements into descending order.

(b) Subroutine MASORT(A, MR, MC, NR, NC, NROWS, NCOLS): This routine sorts all of the elements in the matrix A between A(MR,MC) and A(NR,NC) into ascending order, row by row; NROWS and NCOLS are the row and column dimensions of A in the calling program. A second entry, MDSORT(A, MR, MC, NR, NC, NROWS, NCOLS) sorts the same elements into descending order.

(c) Subroutine REVERS(A, M, N): This routine reverses the order of elements of an array A between A(M) and A(N), inclusive.

(d) Subroutine AMINMX(A, M, N, MIN, MAX): This routine finds the algebraically smallest and largest elements of an array A between A(M) and A(N), inclusive, and assigns the corresponding *subscript* values to MIN and MAX, respectively.

(e) Subroutine MMINMX(A, MR, MC, NR, NC, NROWS, NCOLS, MRMIN, MCMIN, MRMAX, MCMAX): This routine examines all elements of a matrix A with row and column dimensions of NROWS and NCOLS, respectively, row by row between elements A(MR,MC) and A(NR,NC), inclusive. The row and column subscripts of the algebraically smallest element are assigned to MRMIN and MCMIN, respectively; the row and column subscripts for the algebraically largest element are assigned to MRMAX and MCMAX, respectively.

(f) Subroutine INSERT(A, N, X, J): This routine increments the subscripts of all elements of an array A having N elements, whose subscript on entry is greater than or equal to J; X is assigned as the new element with subscript J, and N is incremented by 1 upon return to the calling program.

A second entry, REMOVE(A, N, X, J), removes the Jth element of the array A having N elements, and assigns its value to X. Subscripts associated with elements from A(J+1) through A(N) are reduced by 1 and N is decremented by 1 before returning to the calling program.

(g) Function FINDIT(A, M, N, X, ISUB): This routine searches the array A between elements A(M) and A(N), inclusive, for all occurrences of elements having the value X. The number of repetitions of X is returned as the *value* of the function, and subscripts of the duplicated elements are assigned to successive elements of the array ISUB, starting with ISUB(1).

(h) Function INSORT(A, N, X): This routine searches through the array A, having N elements and assumed to be sorted into either ascending or descending order, and inserts the value X into the location in the array that maintains element order. N is incremented by 1 and elements are "pushed down" (their subscripts are increased by 1) as necessary to make room for the inserted element. The subscript of the newly inserted element is returned as the value of the function.

(i) Function INTERP(A, N, X): This routine searches through an array A having N elements, assumed to be sorted into ascending or descending sequence, until the value X is found to lie inclusively between two adjacent elements in the array. The value of the subscript of the first of these two elements is returned as the value of the function. If the value of X is smaller than the smallest element or larger than the largest element, the value returned is 0 or N, respectively.

3.69 The distance between two points in three-dimensional space, **x** and **y**, with coordinates x_1, x_2, x_3, and y_1, y_2, y_3, respectively, is given by

$$\|\mathbf{x}-\mathbf{y}\| = \sqrt{(x_1-y_1)^2+(x_2-y_2)^2+(x_3-y_3)^2},$$

and the direction cosines of the line from **x** to **y** are

$$d_1 = (x_1 - y_1)/\|\mathbf{x}-\mathbf{y}\|,$$
$$d_2 = (x_2 - y_2)/\|\mathbf{x}-\mathbf{y}\|,$$
$$d_3 = (x_3 - y_3)/\|\mathbf{x}-\mathbf{y}\|,$$

Write a FORTRAN function named VECTOR that has arguments (X, Y, D) where X, Y, and D are each arrays corresponding to **x**, **y**, and **d** (the three elements d_1, d_2, d_3). The length $\|\mathbf{x}-\mathbf{y}\|$ should be returned as the value of VECTOR.

3.70 In the parametric representation of a plane curve, the x and y coordinates may be represented as functions of a parameter u, for example,

$$x = x(u),$$
$$y = y(u),$$

where $0 \leqslant u \leqslant 1$.

In the special case of the parabola, there are several control points that allow a rather simple parametric representation. Let **p**, **q**, **r**, **s**, and **t** be points with x coordinates p_1, q_1, r_1, s_1, and t_1, respectively, and y coordinates p_2, q_2, r_2, s_2, and t_2 and located as shown in Fig. P3.70. Here **p** and **q** are end points of a parabolic arc, **t** is the point of intersection of lines tangent to the parabola at the points **p** and **q**, and **r** is the bisection point of the chord between points **p** and **q**. It can be shown that **s** is the bisector of the parabola between points **p** and **q**. It can further be shown that placement of the points **p**, **q**, and **t** is sufficient to completely determine the parabola's x and y coordinates as follows:

$$x(u) = (q_1 - 2t_1 + p_1)u^2 + 2(t_1 - p_1)u + p_1$$
$$y(u) = (q_2 - 2t_2 + p_2)u^2 + 2(t_2 - p_2)u + p_2$$

For $0 \leqslant u \leqslant 1$, $x(u)$ and $y(u)$ trace out the parabola between the end points **p** and **q**.

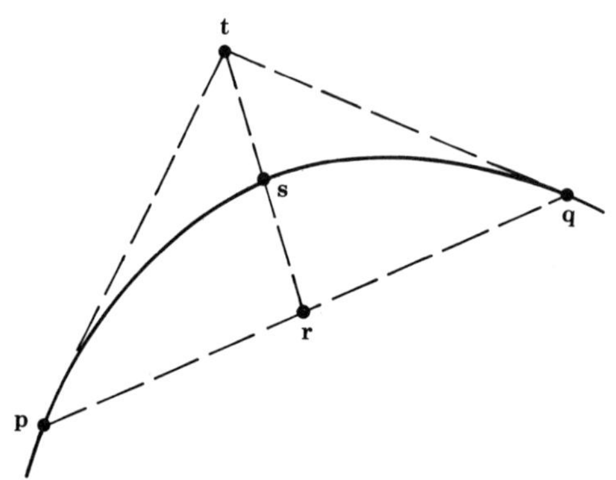

Figure P3.70.

Write a FORTRAN subroutine named PARAB with arguments (N, P, Q, T, X, Y) that will calculate the x and y coordinates of a parabola for N + 1 equally spaced values of the parameter u on the interval $0 \leqslant u \leqslant 1$, given the coordinates of the three points **p**, **q**, and **t** in the arrays P, Q, and T, each having two elements (P(1) corresponds to p_1, P(2) to p_2, etc.). The coordinates x_i and y_i, $i = 1, 2, \ldots,$ N + 1, should be assigned to the ith elements of the X and Y arrays, respectively.

3.71 Consider the possibility of a billion cards, each inscribed with one of the billion possible decimal fractions ranging from 0 to 0.999999999 in increments of 0.000000001. If the cards are placed in a large container that is then well shaken, and a card is withdrawn at random, then the number z_1 appearing on that card will be a *random variable* that has an equal probability of lying anywhere (in increments of 10^{-9}) between 0 and 0.999999999 (or, for all practical purposes, between 0 and 1). If the card is replaced, and the container again shaken, the withdrawal of a second card will lead to a second random variable, z_2. The procedure may be repeated indefinitely, generating a whole sequence of random numbers.

Random numbers are important because, as discussed in Problem 3.72, they can be used for the simulation of operations in which there is an element of chance. It is clearly desirable to be able to generate random numbers on the computer, and we now discuss just one of the many available methods.

The *additive method* for generating random numbers, uniformly distributed between 0 and 1, uses the algorithm

$$x_i = x_{i-1} + x_{i-n}\ (mod\ m), \qquad i = n+1, n+2, \ldots$$

in which all quantities are integers, and the magnitude of the modulus m is restricted by the available word length of the machine being used. (The operation $y\ mod\ m$ means that m is divided as many times as possible into y, and only the remainder is kept; for example, $9\ mod\ 5 = 4$, and $85\ mod\ 23 = 16$.) The required random number between 0 and 1 is then the fraction $z_i = x_i/m$.

The method requires that $x_1, x_2, \ldots, x_n$ be preset with uniformly distributed random integers whose values lie between 0 and m. Green, Smith, and Klem [29] indicate that n should be at least 16 to ensure completely random properties of the sequence, although a value as low as $n = 6$ can suffice for most purposes. Note that as soon as x_i has been computed, x_{i-n} is no longer needed. Thus, by suitable ordering and replacement, it is unnecessary to extend the x array beyond its original n storage locations.

Write a subroutine named RANADD that will implement the above procedure. If possible, the call should be of the form

```
CALL RANADD (Z)
```

where Z is returned as the random number.

If $n = 16$ and $m = 10^9$, Table P3.71, taken from a table of random digits [30], will probably give suitable starting values for x_1, $x_2, \ldots, x_{16}$; these values should be preset using a DATA declaration.

Table P3.71

356,120,997	988,919,262	437,131,844	459,225,517
939,989,737	666,850,663	149,716,880	491,221,612
789,764,838	252,429,765	378,688,294	737,326,323
016,664,811	951,830,262	566,387,005	194,272,481

These values can be truncated for use with machines whose word length does not permit integers as large as 10^9.

Test the subroutine by calling it many times from a short main program; for each call, print the value of Z. The subroutine RANADD may also be used in Problem 3.72.

3.72 Use the random number generator subroutine RANADD developed in

Problem 3.71 to simulate either: (a) the queueing problem described below, or (b) any operation of your choice in which there is an element of chance.

The queueing problem concerns the simulation of an n-bay gasoline station. Based on past experience, one car can be expected to arrive at the station every m minutes *on the average*. However, the time interval between cars is random but can be simulated as follows. Assuming that a car has just arrived, a call on RANADD will yield a random fraction z between 0 and 1; the next car is then taken to arrive $2mz$ minutes later. Further calls on RANADD will similarly indicate the times of arrival of subsequent cars. (Note that since z is uniformly distributed between 0 and 1, its average value is 1/2, so the average time interval between arrivals is $2 \times m \times 1/2 = m$.)

When a car arrives, it proceeds to the bay with the least number of cars waiting; if no bay has less than q cars waiting (including the one being serviced), the new arrival gives up in disgust and drives away. The time to service each car is s minutes, assumed constant.

Write a program that will read values for n, m, q, s, and $tmax$, and then simulate the operation of the gasoline station for a period of $tmax$ minutes. The output should include the following information:

(a) The total number of cars serviced at each bay.
(b) The average queue length at each bay.
(c) The total idle time for each bay.
(d) The total number of cars that drive away in disgust.

Suggested Test Data

$s = 5$ min, $tmax = 180$ min, with all 18 combinations of $n = 1, 3, 5$, $m = 1, 3, 5$, and $q = 2, 3$.

3.73 Solve Problem 3.72 with one additional complication. The service time s is also to be a random variable, uniformly distributed over the range $s - d$ to $s + d$. Suggested values for s and d are 5 min and 2 min, respectively.

3.74 Write a FORTRAN subroutine named PLOT with dummy arguments (XMIN, DELTAX, XMAX, FMIN, FMAX, CHAR, YLABEL, XLABEL, F) that prepares a plot similar to that shown on page 119 for any function $f(x)$ on the interval XMIN $\leqslant x \leqslant$ XMAX using increments of DELTAX for x. The functional value, $f(x)$, will be computed by another subprogram, a FORTRAN function whose dummy name is F. The variables FMIN and FMAX are the minimum and maximum ordinate values (corresponding to -1.0 and $+1.0$ for the sine function in the graph of page 119. CHAR is an integer variable containing the character code for the plotting character (see Table 1.10) in the first character position. YLABEL and XLABEL are arrays of 10 elements containing characters to be used as labels for ordinate and abscissa, respectively; the abscissa label should be written, one character per line, along the left edge of the printer output page.

Write a main program that calls on the subroutine, using the following data to check out the program:

XMIN	DELTAX	XMAX	FMIN	FMAX	CHAR	YLABEL	XLABEL	$f(x)$
0.0	0.1 (rad)	3.2	−1.0	1.0	*	'SINE'	'ANGLE'	sin (x)
1.0	0.2	7.0	−1.5	2.0	$	'2*E**X*SIN(X)'	'X'	$2e^x \sin(x)$

Note that the second of these data sets will require that a FORTRAN function be written to compute the functional values, $2e^x \sin(x)$.

3.75 Modify the subroutine PLOT of Problem 3.74 to print a grid consisting of dashes for the ordinate printed at every tenth abscissa value and plus signs for three grid lines in the direction of the abscissa, one line at the minimum ordinate value, one at the maximum ordinate value, and one halfway between the two extremes. If a plotting character coincides with a grid line character, print only the plotting character.

Would it be possible to overprint a grid character with a coincident plotting character?

3.76 Assume that the x and y coordinates for N points are available in the arrays X and Y such that the coordinates of the first point are in X(1) and Y(1), those for the second point are in X(2) and Y(2), etc., and that the x coordinates are arranged in ascending algebraic order (but not necessarily uniformly spaced). Write a plotting routine named PTPLOT with arguments (XMIN, DELTAX, XMAX, FMIN, FMAX, CHAR, YLABEL, XLABEL, N, X, Y) that plots the points in the manner of the plot of page 119. Here, the first eight arguments are the same as in Problem 3.74. The plotting character should be positioned as near as possible to the proper ordinate and abscissa locations.

Write a main program to test the program that reads the data for the tar and nicotine content of several brands of cigarettes shown on page 311, and calls upon PTPLOT to plot the points.

3.77 The following are some suggestions for improving the flexibility of the plotting programs discussed in the preceding four problems:

(a) Indicate when any point falls outside the minimum and maximum values for the ordinate or abscissa.

(b) Allow more than one function to be plotted on the same graph, using different plotting characters for each function.

(c) Allow the user to specify the number and spacing of the grid lines on the graph.

(d) Allow the user to specify the formats for printed values of ordinate and abscissa.

(e) When the coordinates of the data points are specified as arguments, have the routine scan the values and determine limits for the coordinates automatically. Presumably it would be desirable to have the values chosen so that grid lines correspond to "nice" values for the variables.

(f) Allow the user to specify the plotting of multivalued functions, for example, the unit circle.

(g) Allow the user to specify that non-Cartesian coordinate systems be used, for example, polar coordinates, multicycle log-log or semi-log coordinates.

3.78 Discuss the possibility of using a large FORTRAN array as if it were a piece of paper upon which the grid lines, ordinate and abscissa labels and numerical values, and plotting characters are "drawn" by a FORTRAN plotting subprogram. What would be some of the advantages and disadvantages of such a scheme? Comment on the possibility of having more than one entry to the subprogram.

CHAPTER 4
An Introduction to Computer Operating Systems

4.1 Introduction

The *software* support of a computing installation is the collection of all programs that are made available to the programmer or that provide services to the community of machine users. The system software includes the *library* or libraries of subprograms in object form that may be called directly in a FORTRAN program; for example, the standard FORTRAN functions of page 65 and routines for handling the solution of commonly occurring problems in numerical analysis and statistics. In a large computing system, there would normally be hundreds or thousands of such routines in the library.

Translators for a variety of assembly, procedure-oriented, and special-purpose languages are also a part of the software for the system. The number of such programs is usually small, typically no more than 20 or 30 for even the largest computing installations. The FORTRAN, WATFOR, or WATFIV compilers would be included in this group of programs.

Finally, every large computing facility has a set of programs responsible for management of the hardware resources, maintenance of the users' accounts, and, in general, providing flexible user-oriented services, such as maintenance of files of data on secondary storage devices. This collection of programs constitutes the *operating system* for the machine and allows both novice and experienced programmers to make efficient use of the computing resources.

In this chapter, we discuss the functional characteristics of operating systems at a level of detail adequate for practical computing with procedure-oriented languages such as FORTRAN, WATFOR, and WATFIV. After a general discussion of the topic, we describe the hardware of the computing system currently in use at The University of Michigan and discuss some of the characteristics of the operating system called MTS (Michigan Terminal System) for this hardware. The chapter concludes with a short demonstration of a time-sharing system at a teletypewriter terminal.

We realize that the computer hardware and the operating system for the hardware at most installations will differ from those described here. There is, in fact, no single computing system/operating system combination that could be called *the* "standard." Virtually every large computing system can be viewed as unique; almost every facility has its own conventions for assigning access to the system, maintaining accounts, arranging decks for jobs, using the time-sharing service, etc. Since there seems little point in introducing yet another hypothetical computer, operating system, or command language for that operating system, we have elected to describe an existing system used regularly by more than 10,000 students, faculty, and staff at The University of Michigan. We would expect the instructor to supplement the material in this chapter with information about the computing facilities and operating system conventions at his own institution.

4.2 Operating Systems—Batch Processing

As recently as the mid-1950s, it was common practice for an individual programmer to control the gross processing of his program, such as the loading of cards and the initiation of execution, directly from the console of the computer. Typically, the programmer punched his machine-language program on cards and then carried out a sequence of tasks similar to the following:

1. The object program, preceded by another program called the *loader*, is positioned in the card reader. The loader is a program responsible for reading *other* programs into the memory.
2. The programmer enters one or more machine instructions directly at the console of the computer to initiate the reading of instructions in the *loader* into the machine's memory. Usually such loaders operate in bootstrap fashion, that is, the first few instructions, read as the result of the initial console instruction, are themselves input commands to read in additional instructions, in this case the object program.
3. Data cards for the object program are positioned in the card reader.
4. The programmer initiates execution of the object program directly from the console.
5. The executing program writes its results on some output device, usually a card punch or line printer.
6. If the programmer finds his results in error, he tries to debug the program by entering new instructions or replacing old instructions in the memory, directly from the console. He then initiates reexecution of the program. This process is repeated until the program is free of errors (hopefully!).

Such *open-shop* operation proved to be a very inefficient way of using a computer* because a substantial fraction of the potentially available computing time was wasted while the operator/programmer performed manual tasks, such as entering instructions at the console. In addition, program debugging from the console was usually not very effective because of the pressure of time and the necessity of working directly in the machine's language.

*Most small, inexpensive computers (*minicomputers*) are still operated directly by the user, although even these machines normally have simple monitors that allow automatic loading of programs. In this text, we are more concerned with large computing systems.

In the late 1950s, a new generation of computers, much faster than those previously available, came onto the commercial market. These machines were rather complex, with a variety of input/output devices, large memories, etc. It was clear that a professional staff would be required to operate the machine on a closed-shop basis to achieve reasonable throughput. (Here, *closed shop* indicates that the user writes his own programs, but the programs are run on the computer by a professional operator.) It was also clear that most of the tasks that had been done manually on the older computers would have to be performed automatically on the new machines, hopefully by the machines themselves.

Out of this necessity was born a whole new class of computer programs, usually called *the operating system*, *the executive system*, or *the monitor*, whose major purpose is to route other programs through the computing system, to take maximum advantage of the operating characteristics of input and output devices, fast and secondary storage, and the central processing unit. The major *system programming* tasks in the late 1950s and the early 1960s were related to the implementation of *batch processing systems*. These operating systems allow the sequential processing of a large number of programs (a *batch*) without external intervention by a human operator. The most flexible of these systems allow the user to take advantage of a wide variety of computing services such as the translation of different source languages into appropriate machine language and the execution of object programs with supplied data. The best of these systems allow each user complete flexibility in his choice of the available computer services.

How can this be achieved? As an example, we shall describe the behavior of a typical operating system, illustrating the sequence of operations that would be required to allow the automatic processing of many individual users' program decks (*jobs*) through a batch-processing system. If each programmer is to be allowed flexibility in his choice of services, and if he is not to be physically present at the computer at the time his program is executed, then the program deck must include not only *programs* that he wishes to translate and/or execute and any associated *data*, but also *commands* to the operating system itself. Thus, if a programmer wishes to have his FORTRAN program translated and executed on the computer, he must (1) indicate that he wishes to translate a FORTRAN program, (2) point out which of the cards in the deck contain the FORTRAN statements, (3) indicate that the object program should be loaded and executed, (4) specify which of the cards contain data for the object program, and (5) select the device on which the output is to be written (for example, which card punch or printer, etc.).

If the operating system is to keep its own accounting records, the deck must also include information such as name and account number. Another important characteristic of good operating systems is that they monitor the progress of activities that they initiate. For example, a programming error by one user must not be allowed to cause problems either for the other users or for the operating system itself. Limits on total computing time and total number of pages to be written or cards to be punched by each job (all to be specified by the user) must be enforced by the operating system as well.

A typical deck for compiling and executing a FORTRAN program in some batch processing system might appear as follows:

```
$SIGNON ISAAC NEWTON ACT507 TIME=1M PAGES=20 CARDS=0
$COMPILE FORTRAN
   {(FORTRAN program)
$ENDFILE
$LOAD OBJECT
$EXECUTE
   {(data cards for the FORTRAN program)
$ENDFILE
$SIGNOFF
```

Another user of the computing system who wished to compile and execute a program written in the ALGOL language might prepare his deck as follows:

```
$SIGNON JAMES WILKES ACT905 TIME=20S PAGES=35 CARDS=40
$COMPILE ALGOL
   {(ALGOL program)
$ENDFILE
$LOAD OBJECT
$EXECUTE
   {(data cards for the ALGOL program)
$ENDFILE
$SIGNOFF
```

Yet another user might have a deck containing an object program on punched cards that he wishes to load and execute. His deck sequence might appear as follows:

```
$SIGNON ALFRED NEUMAN ACT123 TIME=5.5S PAGES=10 CARDS=0
$LOAD OBJECT DECK
  {(object program on punched cards)
$ENDFILE
$EXECUTE
  {(data cards for the object program)
$ENDFILE
$SIGNOFF
```

In each case, the command lines to the operating system start with a special character ($). The first line contains information about the user: his name, his account number, how much computing time his job is expected to require, how many pages he expects his job to write on the printer, and how many cards will be punched by his job. Remaining cards contain program and/or data lines interspersed with appropriate commands to the operating system. Each user requests different services from the computing system, yet the structure of each deck is similar from the viewpoint of the operating system. Each command is followed by appropriate data.

The programs that constitute the operating system must themselves be available in object form and directly accessible to the central processor. Typically, the operating system programs are kept on auxiliary storage devices such as magnetic tapes or disk files. A possible configuration for a computer to be operated by a batch-processing system is shown in Fig. 4.1. It consists of a central processor, I/O units, and secondary storage devices to store copies of the operating-system programs, the translating programs such as the FORTRAN translator and an assembler, a library of frequently used subprograms (square root, sine, cosine, etc), and essential accounting information needed by the operating system to control access to the computer and to maintain up-to-date records. In addition, some secondary storage space will normally be needed by the translators to store temporarily the object programs as they are being generated. Although a separate storage device is shown for each of these functions in Fig. 4.1, all of this information might be stored on a single device, for example a single disk drive. The processing of a typical batch of individual jobs might occur as shown in Fig. 4.2. (The paragraph numbers correspond, roughly, to the numbers used in the figures.)

1. The operator assembles the card decks in order in the input hopper of card reader.
2. The operator keys in one or more instructions at the console to initiate reading of the command section of the operating system from auxiliary storage into the machine's memory. Such programs are usually self-loading, in that the first few instructions read in the remaining instructions, and then transfer machine control to the instructions just read in.
3. The computer begins execution of the command program or *monitor* responsible for interpreting command lines and bringing the required com-

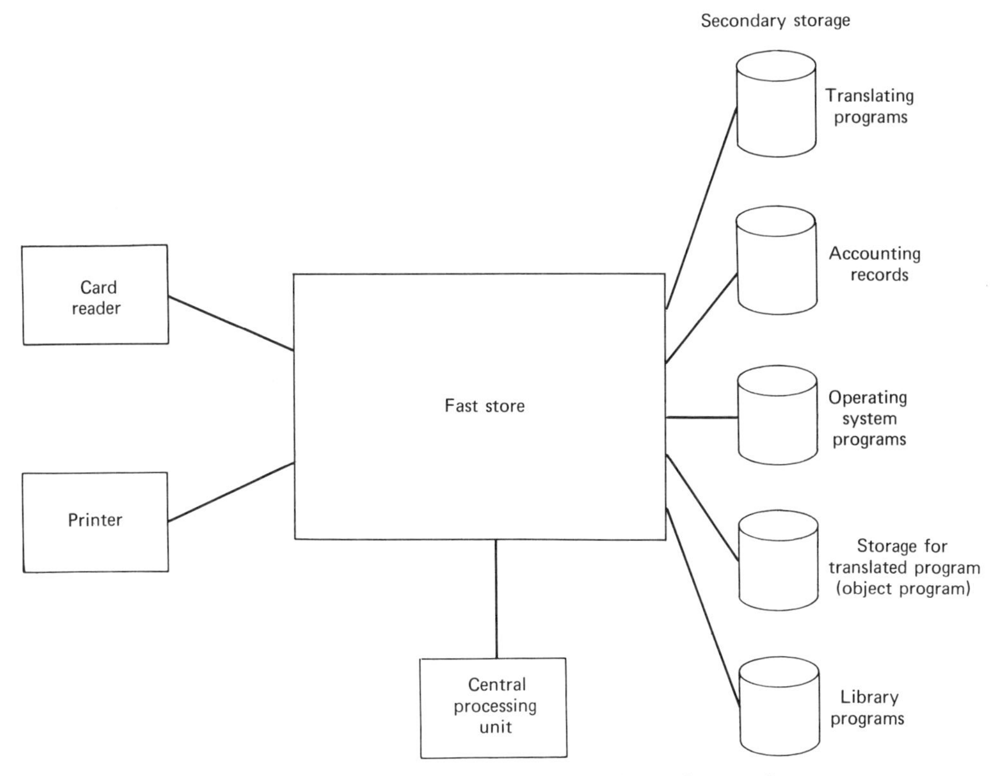

Figure 4.1 A possible configuration of a computer with a batch-processing system.

puting resources (for example, the system programs mentioned above) into action at the proper time. If the jobs are arranged as indicated above, the first card read in each deck will be a request for access to the system. The monitor can then scan through information in its accounting files in auxiliary memory to determine if the user should have access to machine services, that is, if his account is in order.

If service is to be denied, the monitor writes a copy of the sign-on request and a comment regarding the reason for the refusal. This lets the user know why his job was not run. The monitor then cycles through all the cards associated with the job and proceeds to process the next job in the batch.

7. Suppose that the request for services is approved. The system saves the specified page and card estimates for later reference, and then sets an internal *clock* to cause an interrupt (an automatic return to the monitor) should the job still be in process after the indicated computing time has elapsed.
8. The monitor reads the next card in the job deck; if the job were the first of the decks on page 205, this card would contain the characters: $COMPILE FORTRAN. The monitor would decode this string of characters and recognize that the FORTRAN translator is required. The monitor then reads a copy of the FORTRAN translating program (in machine-language form) from secondary storage into the fast memory, and transfers control of the machine to the newly stored FORTRAN translating program.
12. Next, the FORTRAN translating program reads the FORTRAN statement cards that follow the $COMPILE FORTRAN command. As the FORTRAN statements are read, the corresponding object code is produced

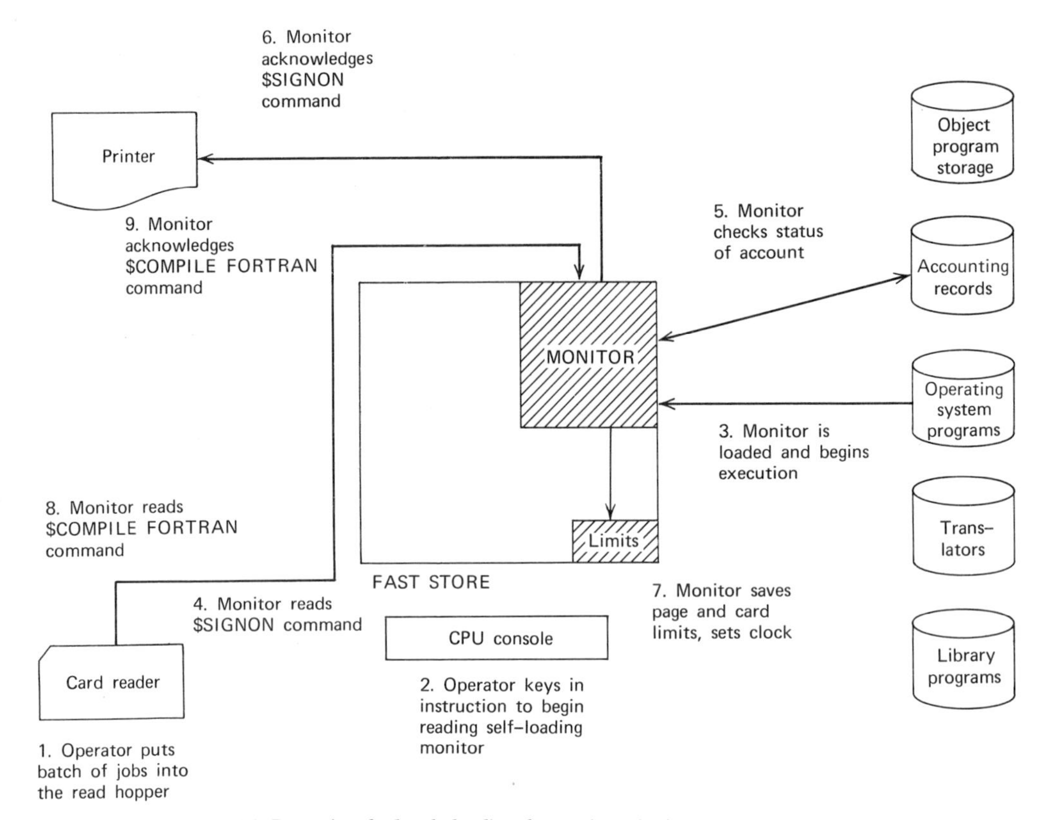

(a) Preparing the batch, loading the monitor, signing on (steps 1–9)

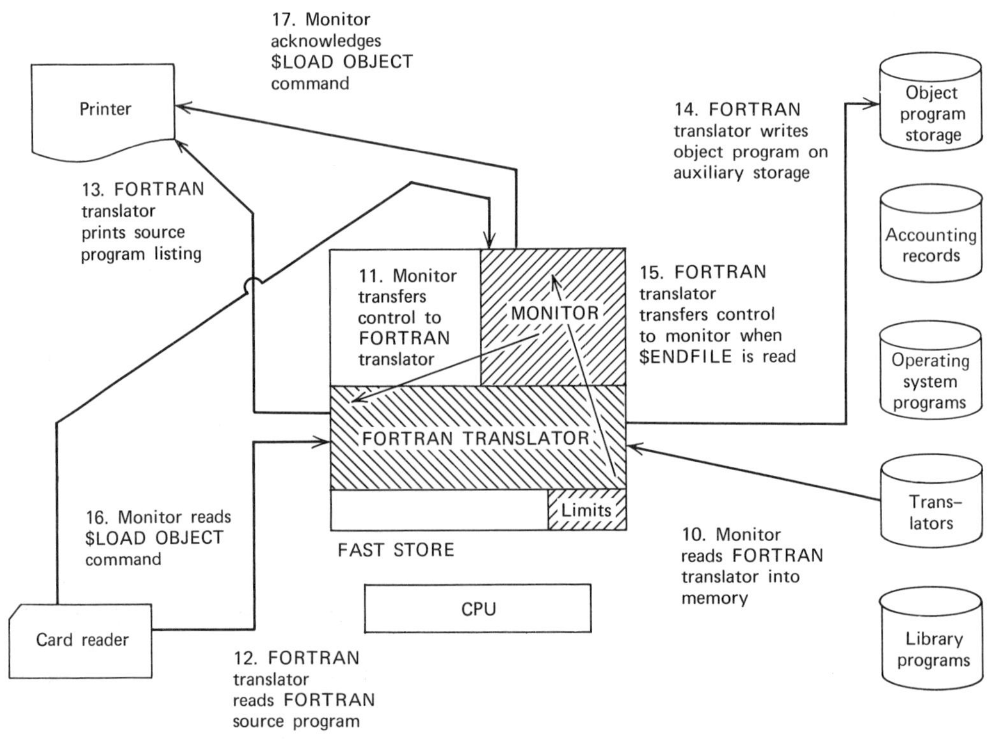

(b) Translating a FORTRAN program (steps 10–17)

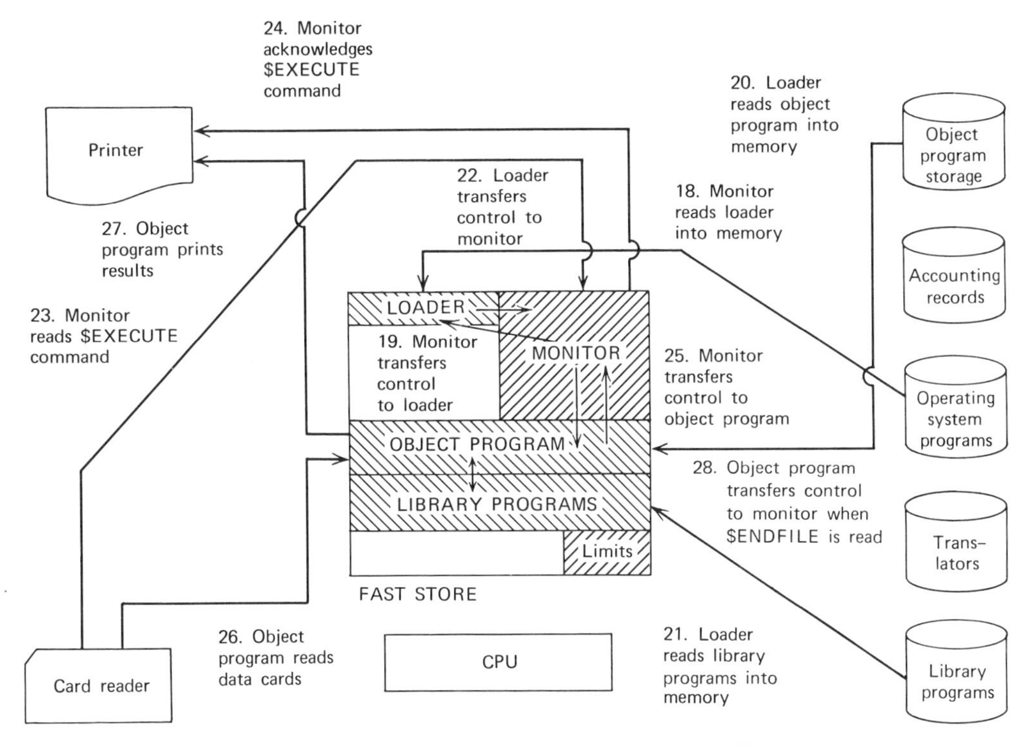

(c) Loading and executing a FORTRAN program (steps 18–28)

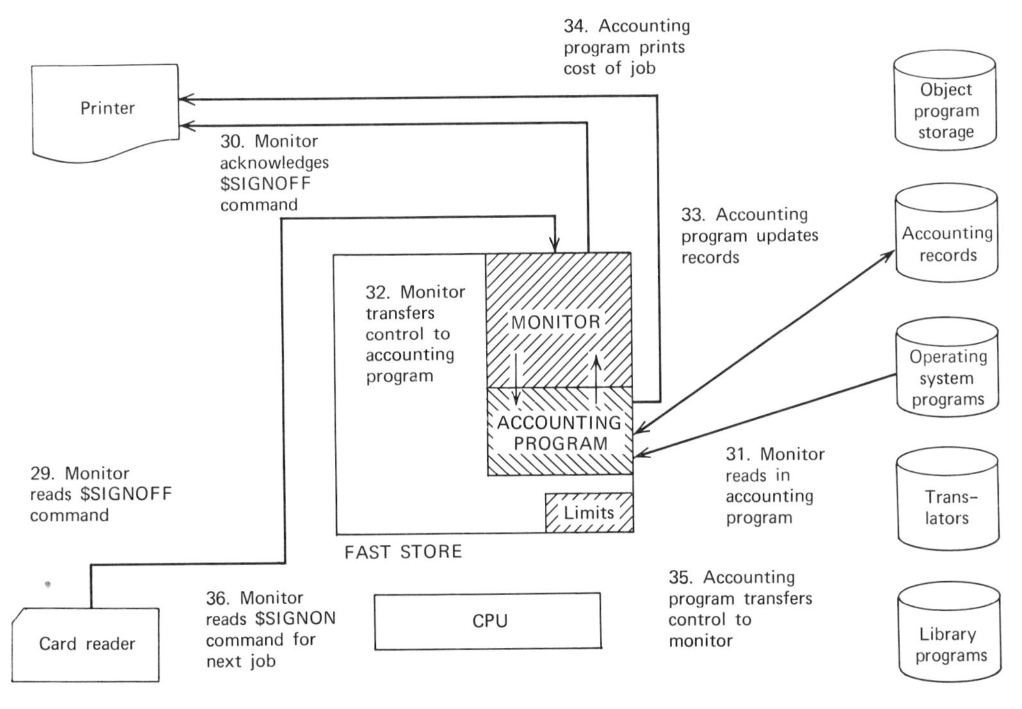

(d) Signing off, system accounting (steps 29–36)

Fig. 4.2 Processing a job in a batch-processing system.

and written onto some secondary-storage device, say a disk file. When the FORTRAN translator encounters the $ENDFILE command, the translation task is completed and the object program is available in secondary storage.

15. The FORTRAN translator can (a) transfer control directly back to the monitor program if it is still available in the memory or (b) initiate the loading of a new copy of the monitor from secondary storage just as the human operator did originally in step 1. The latter situation might occur if the memory is so small that the FORTRAN translator requires the fast storage space normally occupied by the monitor. (Many operating systems would not allow any encroachment on storage occupied by the most important parts of the operating system itself.)
16. The monitor reads the next command, $LOAD OBJECT. The monitor reads an appropriate loading program from the auxiliary storage into the memory and transfers control to it.
20. The loader reads the object program into the fast memory from secondary storage, brings in any required library programs, and returns control to the monitor.
23. The monitor reads the next command card, $EXECUTE and transfers control to the object program brought in by the loader.
26. During its execution, the object program reads data cards from the card reader, and produces written output on the line printer or card punch. As pages are written and cards punched, system programs responsible for output keep subtotals, comparing them with the user-specified limits.
28. Eventually, either the page, card, or time limits are exceeded, or the final $ENDFILE card is encountered. Appropriate comments are printed with the output to indicate the cause of termination, and the program in charge at the time (probably an input routine) returns control to the monitor if present in the memory, or initiates reading of a new copy of the monitor, if not.
29. The monitor reads the $SIGNOFF command and brings in the accounting program to update the system records and print accounting information for the job. When finished, the accounting program returns control to the monitor.
36. The monitor is now ready to examine the first card in the next job. This card should be another $SIGNON command, requesting access to the computing system. The whole process is repeated again, with variations depending on the services requested in the individual job.

As the various programs are executed, output for the user (for example, a listing of the FORTRAN program, information about the object program, and accounting records) is written on the printer along with the output produced by his object program.

The key to a successful operating system is the ability of one program to read another program from auxiliary storage and then to transfer control of the computer to it. If the program communication chain is not broken, intervention by a human operator is unnecessary.

4.3 Multiprogramming, Time-Sharing, and Multiprocessing

Introduction.

Before 1960, most computing machines were equipped with just one card reader, one card punch, and one printer, and were designed to perform only one task at a time. The memories were small, and normally only one person's object program could be stored in the fast memory at a given time, as illustrated in Fig. 4.3.

The execution sequence for most programs involves: (a) reading data, (b) com-

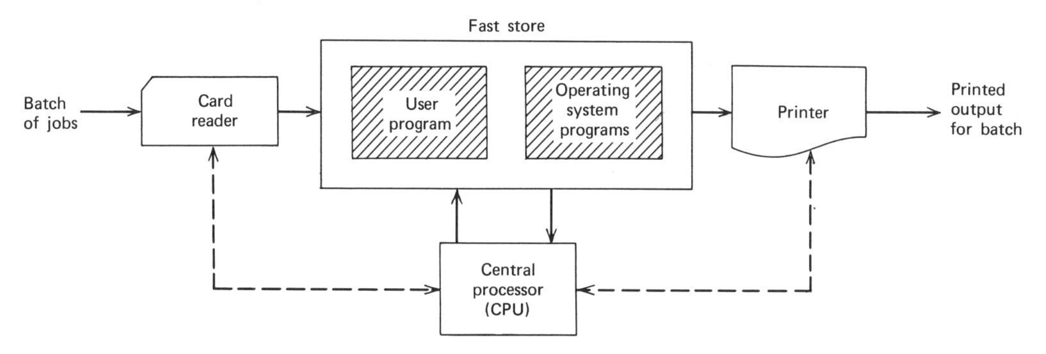

Figure 4.3 Typical early computer (circa 1958).

puting results, (c) printing results, and (d) returning to read another set of data. This cyclic pattern of read, compute, write, compute, read, compute, write, etc. is often repeated hundreds of times for a single program.

If the machine can perform just one task at a time (read, write, add, store, etc.), then the central processor (CPU) will be busy only during the compute segments of this cycle, as is illustrated by Fig. 4.4. Substantial idle time for the CPU will accumulate while the relatively slow input and output operations are being performed. Even on the fastest equipment available in 1960, one second was required to read just *five* cards or to print just *five* lines; yet central processors available in the late 1950s were capable of executing individual machine instructions at the rate of about *50,000* per second.

For the processing of many (probably most) programs, the active periods will total no more than a few percent of the elapsed time. Yet the throughput of the installation must be measured in terms of the elapsed time; unfortunately, the charges for using the computer are also based on elapsed time. Thus the central processor can be seen to have only the *potential* capability of processing 50,000 instructions per second, a potential that can be exploited only during the strictly computational phases of program execution. The *average* number of instructions processed per second, based on elapsed time, will be very substantially smaller than the potential number. Hopefully this average value will be large enough to justify the cost of the machine. The situation is very like that in the air transportation industry. The potential income from an air flight is measured in terms of the number of seats, while the actual income is measured by the number of seats filled with paying passengers. To operate profitably, the average load factor (the average fraction of total capacity actually used) must achieve some minimum value.

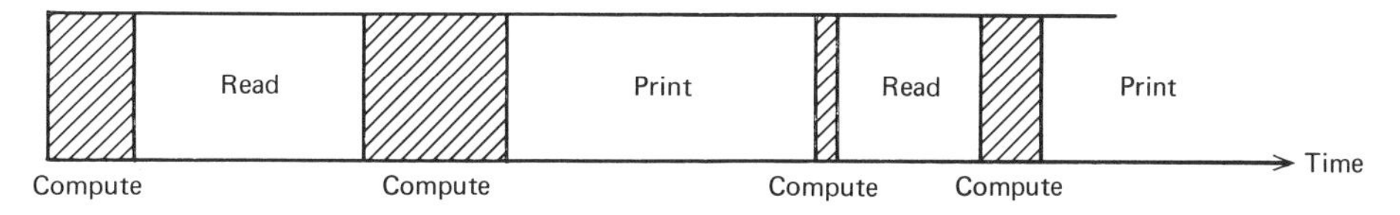

Figure 4.4 Central processor activity (crosshatched) and idleness (blank) during execution of a typical user program in a batch-processing environment.

Channels.

In the early 1960s, new computers were developed that allowed the input and output equipment and the central processor to operate *simultaneously*. The input and output tasks are handled by small communications computers called *channels*. The channels normally have only a small repertoire of machine instructions dealing with the transfer of information between the input/output equipment and the memory (and *vice versa*) and the operation of the input and output equipment. They are often programmable and usually have small memories, hence are rather special stored-program computers. (In some systems, a portion of the central memory is assigned to serve as the memory for the channels.) A read instruction encountered in the user's program causes the central processor to initiate the execution of a small input program by a channel. The central processor can then continue to process instructions, while the detailed commands to the card reader and the transfer of information from card reader to central memory are handled by the channel; when the channel is finished with its task it will normally have some way of signaling the central processor (for example, by interrupting the activities of the central processor momentarily, so that a special system program can be executed to update information about the status of the card reading task).

Thus a channel can be viewed as a slave computer to the central processor that handles the input and output tasks. This frees the central processor from the necessity of waiting in idleness while the slow input/output operations are being performed. Or does it? Presumably the basic cyclic behavior for the execution of a user program remains unchanged. Normally, computation cannot proceed until *after* all the necessary data have been read into the memory. Likewise, information to be printed must be retrieved from the appropriate memory locations *before* new data or results are stored in these locations.

These apparent timing problems can be solved in part by using regions of the fast memory as *buffers* between the executing user program and the slow input and output equipment (see Fig. 4.5). As soon as the object program is loaded, several card images are read directly into the input buffer. When a reading operation is called for in the user's program, a system program will be called that:

(a) Copies the requested data from the buffer-region card images into the memory locations assigned to those variables in the user's program; these operations are accomplished at central processor speeds, since only the fast memory and arithmetic unit are involved.

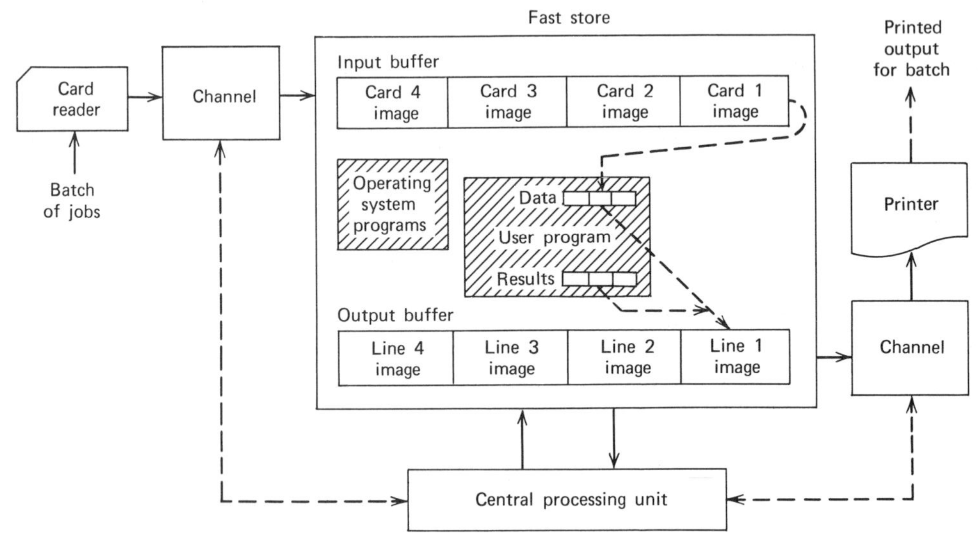

Figure 4.5 Buffered input and output in a computer equipped with channels.

(b) Shifts any "unread" card images to the front of the input buffer region.
(c) Initiates a program in a channel that will read new card images into the available "empty" locations in the input buffer.
(d) Transfers back to the user's program so that the central processor can continue executing the program at full speed.

In similar fashion, when a writing operation is encountered in the user's program, a system program is called that:

(a) Copies the output lines into the output buffer region of fast memory; as only the fast memory and arithmetic units are involved, this operation is accomplished at electronic speeds.
(b) Initiates a program in a channel that will transfer the buffered output line images to the printer.
(c) Transfers back to the user's program so that the central processor can continue execution at electronic speeds.

Thus it is possible that the card reader, printer, and central processor are all operating at the same time. Unfortunately, in many cases the buffers cannot be replenished (input) or dumped (output) rapidly enough to keep up with the computational capability of the central processor, and the processor will still experience considerable idle time. When this happens, the executing program is said to be "I/O bound."

Multiprogramming.

By the mid-1960s the memories of the larger machines were considerably expanded, so that it became possible to store instructions for more than one user's program in the fast store at the same time. In addition, several card readers and printers could be operated simultaneously, leading to the possibility that more than one batch of programs could be processed by a single processor, through careful switching of processor control from one user program to another.

Figure 4.6 shows the situation for a machine equipped with three printers and three card readers. Three streams of batch jobs are being read from the three card readers, and three object programs are loaded in the fast store. Program 1 is associated with batch stream 1 being processed from card reader 1; any printed output for this program will appear on printer 1. Similarly, programs 2 and 3 are associated with card readers and printers 2 and 3, respectively. Suppose that the

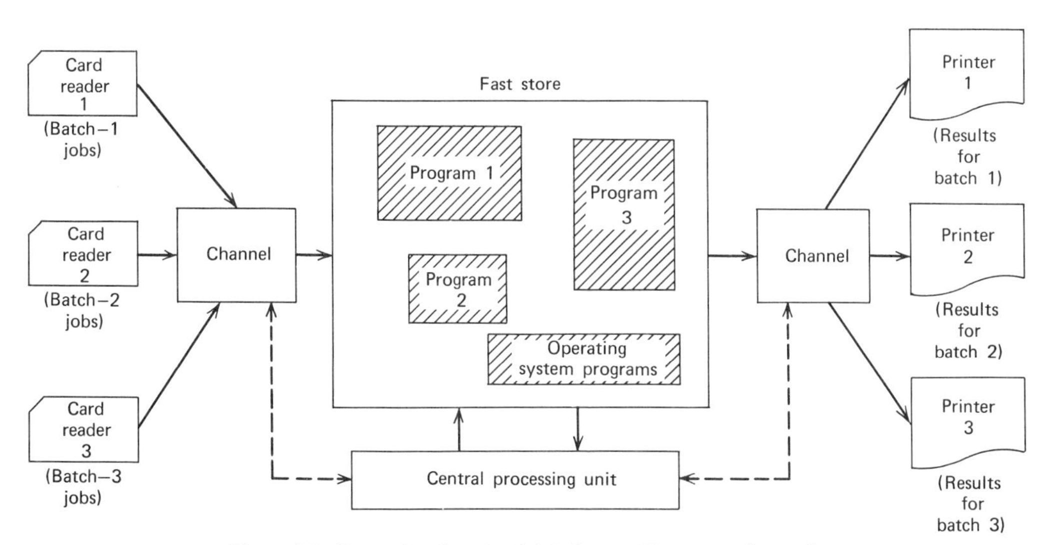

Figure 4.6 Processing three batch jobs in a multiprogramming mode.

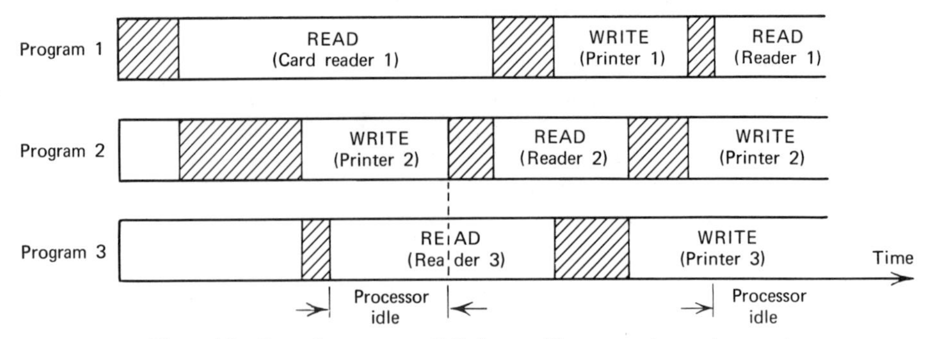

Figure 4.7 Central processor activity in a multiprogramming environment.

processor begins by executing instructions in program 1, as shown in Fig. 4.7. Let this program continue to have control of the processor until some input or output activity is required. Assume that a read operation is the first such instruction encountered. At this point, the CPU assigns the task of reading data for program 1 from card reader 1 to a channel, and immediately begins execution of program 2, leaving virtually no idle CPU time between the interruption for input activity in program 1 and the beginning of execution of program 2. Program 2 is in control of the CPU until a writing operation is encountered. The processor assigns a channel the task of writing output from program 2 on printer 2. Control is switched immediately to program 3, again leaving no idle time for the CPU.

Suppose that program 3 executes until a reading operation is encountered. The CPU assigns the task of reading data for program 3 from card reader 3 to a channel. Depending on the number of channels and their status (idle or busy), the reading operations may be initiated directly, or the task may be assigned to a queue awaiting execution whenever one of the channels finishes its current task. In any event, let us assume that the reading task for program 1 and the writing task for program 2 are unfinished. The processor must then remain idle until it receives a signal from one of the channels that the input or output operation of some program (Program 2 in Fig. 4.7) is completed. Processor control passes immediately to that program. If only a small number of programs is present in the fast memory, and if the processor is capable of executing millions of instructions per second, there is still likely to be a significant idle time for the processor, even if input and output buffers are used. Nevertheless, the processor is obviously being used more efficiently than in the configurations shown in Figures 4.3 and 4.5.

This approach to making more effective use of the available computing equipment, particularly of the fast memory and of the central processor, is called *multiprogramming*. Clearly, a multiprogramming operating system will be more complex than the simple batch-processing system described in Section 4.2. Note that whereas programs in the simple batch-processing environment are processed in *sequence*, more than one program is being run *concurrently*, though *not executed simultaneously*, in the multiprogramming environment. There is no guarantee that the first program to begin execution will be the first to finish. However, if the hardware is organized as shown in Fig. 4.6, then a program from any one batch must be completely processed before the processing of the *next* program in that batch can be started.

One way to improve the operating efficiency of a multiprogramming system is to increase the size of the fast memory, so that many user programs can be stored simultaneously. If the number of such programs is large enough (say 20 or 50 or 100), then there will almost always be at least one program awaiting execution by the processor. [If there is more than one program awaiting execution at any time, an execution *queue* can be set up by the operating system. The usual procedure is to assign positions in such a queue on a "first-come, first-served" basis; however, the position in the queue could as easily be assigned on a *priority* basis. The priority scheme might be determined by the operating system itself (based, for example, on the user's estimate of execution time required, or the amount of storage space occupied by the program); alternatively, the user might be given the option of choosing the priority level, by agreeing to pay a higher rate for faster service.] The CPU could then operate at essentially full capacity, with very little idle time. Such a system would probably require substantial input and output equipment, though the assignment of a separate card reader and printer to every program in the memory, as shown in Fig. 4.6, would hardly be practicable.

Spooling.

One way to avoid the notion of multiple batch streams, each with its own reader and printer, is to use some secondary storage device, usually a disk storage unit, as an intermediate repository for all incoming card images and outgoing line images for each job, as is illustrated in Fig. 4.8. Here, a copy of each job deck is read from a card reader into an assigned section of disk storage. *Any* card reader will do, since the job is not processed until *after* the card images for the entire job deck are available in secondary storage. The transfer of information from card reader to disk storage is accomplished by an operating system program and one or more channels. Once the complete copy of a job is available, the operating system can then assign to a channel the task of reading the card images from disk storage to fast memory. This scheme is practical because the card-to-fast-storage-to-disk storage reading and writing operations can take place concurrently with the execution of other user programs, so that the card reader can "get ahead of" the CPU, very like the tortoise and hare. In addition, the disk storage reading and writing operations can be executed much faster than reading from a card reader or writing on a printer directly.

Note that this approach has two distinct advantages over that illustrated in Fig. 4.6: (a) the number of active programs in the fast memory is no longer dependent on the number of card readers available, and (b) jobs can be *scheduled* for execution by the operating system. The latter advantage means that jobs need not be processed in the same order as they are submitted by the users at the card readers. Depending on the status of the system (the number of active programs already in the fast memory, the total fast storage space occupied, the fractional idleness of the processor during the preceding few seconds, etc.), the operating system can itself assign *priorities* for the entry of additional jobs into the fast memory. Thus, if a job requires only a small amount of fast storage and a short total calculational time, it might well be processed *before* a job demanding more of the computer's resources, even though it was submitted at the card reader *after* the more demanding job. Alternatively, the user might be allowed to influence the processing order, by agreeing to pay more for faster service, for example.

Output from the executing programs can be handled in similar fashion. Each job is assigned a section of disk storage for printed or punched information, as shown in Fig. 4.8. The successive line or card images for a job are written into its assigned disk area until the job is completely processed. Once the job is no longer active, that is, once the job has no object program in the fast memory still awaiting execution by the central processor, the printing and punching of the stored line and card images can be scheduled by the operating system, depending on the availability of the output equipment and the channels. (Some

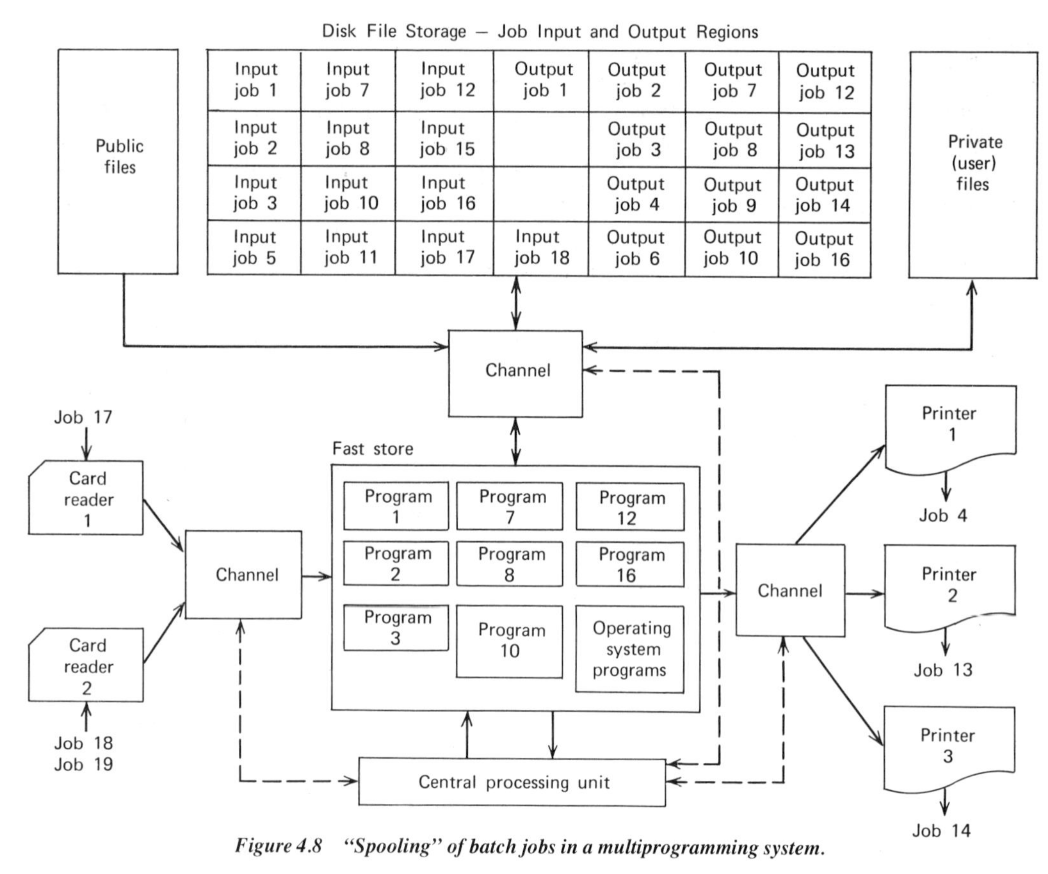

Figure 4.8 "Spooling" of batch jobs in a multiprogramming system.

systems use *multiplexor channels*, capable of communicating with several devices concurrently. Information to be transmitted between the fast memory and various input and output devices is processed in short bursts in "round-robin" fashion, giving the appearance that one channel is handling many devices simultaneously.)

Two important advantages of this operating scheme are: (a) it is not necessary to have one printer and one card punch unit for every executing job, and (b) the output from jobs can be scheduled for the output equipment that is actually available. The latter advantage means that priorities can be assigned for the printing and punching of results. For example, a job with rather little output can be scheduled for printing before a job with large amounts of output, even when execution of the "short" job is completed after that for the "long" job. Or the user might be given the option of paying more for faster overall processing of the job, in which case the operating system could place the associated output task near the front of the printing and/or punching queue.

In the situation illustrated in Fig. 4.8, there is a total of 19 jobs currently being processed. The status of each job is shown in Table 4.1.

Thus two jobs are being read into the job input regions of disk storage, and three jobs already present in their input regions of disk storage are awaiting loading of programs into the fast memory for subsequent execution. Eight active programs are in execution in the fast memory, sharing the services of the CPU; card input for these jobs is taken from the appropriate job input regions and printed output is written into the appropriate output regions of disk storage. Five jobs have completed the execution phase; the output of three of them is being printed, while the remaining two are in the printing

Table 4.1 Status of Jobs in the Multiprogramming System of Figure 4.8

Job Number	Current Status
1	In execution phase
2	In execution phase
3	In execution phase
4	Execution complete, output being printed on printer 1
5	Awaiting execution
6	Execution complete, awaiting printing of output
7	In execution phase
8	In execution phase
9	Execution complete, awaiting printing of output
10	In execution phase
11	Awaiting execution
12	In execution phase
13	Execution complete, output being printed on printer 2
14	Execution complete, output being printed on printer 3
15	Awaiting execution
16	In execution phase
17	Job being read on card reader 1
18	Job being read on card reader 2
19	Job awaiting reading on card reader 2

queue. One job is in the job reading queue at card reader 2. Note that the channels handle transmission of information to or from fast memory and one or more peripheral devices. Information is never passed directly from one device to another. Thus the reading of the job card images from the card reader to the job input region of disk storage is handled in two stages, card reader to fast memory, and fast memory to disk storage. The printing of information is handled similarly, from job output region of disk storage to fast memory, and from fast memory to printer.

As is shown in Fig. 4.8, not all input for a job need be included in the initially read job card deck. Some input might involve programs or data of a public nature, such as translating programs, library subprograms, or population census records, or of a private nature such as user-written programs or user data, already stored in files (see Section 4.7) on secondary storage devices, usually in disk memory or on magnetic tapes. Similarly, not all output from a job need be printed or punched; results might be written directly into a section of disk storage assigned to the user. Such output would not appear in the output region of disk memory assigned to the job.

The mode of operating a multiprogramming system illustrated in Fig. 4.8 is called *spooling*, and has greatly improved the performance and operating efficiency of computing systems. It allows the assignment of priorities to every phase of job processing. With a judicious selection of the hardware configuration, virtually all parts of the computing system can be operated at near capacity. Obviously, the hardware mix requires a careful balancing of the central processor speed, the sizes of the fast memory and disk storage, the number and transmission capacity of the channels, and the number and speeds of the input and output devices. And, of course, the proper selection of hardware ultimately depends on the number and nature of the jobs to be processed by the system.

Remote Batch Stations.

If the proper communication equipment is available, it is often practical to locate some of the card reading and printing equipment large distances (up to thousands of miles) from the central processor. These *remote batch stations* usually consist of a small computer, at least one card reader, printer, and card punch, possibly some channels, and a communications adapter that functions as a transmission control unit from the batch station. (Most small computers do not have independent channels; information is routed to and from the fast memory and the input and output devices by the central processor over high capacity communication buses.) One possible arrangement, using channels, is shown in Fig. 4.9. The communications terminal is interfaced with high-speed telephone, microwave, or earth satellite circuits through an electronic device called a *modem* (not shown in Fig. 4.9), which converts the digital bit strings from the terminal to circuit-compatible signals. At the central facility, another modem decodes the transmitted signals into digital bit patterns, which are then received by another communications terminal unit. The card images sent from the remote station are then fed from the communications terminal to a channel in the central computer for eventual assignment to job input regions of disk storage.

Once a complete job is available in its input region, it is then handled identically with jobs being read from card readers at the central facility. Similarly, once the execution of the remote job is completed, information in the job output region can be scheduled for transmission back to the remote batch station, where printed and punched results are produced for the user. For all practical purposes, the remote user has use of the full resources of the large central processor (including any available public and private stored files), even though he is physically quite far from the computing equipment.

In some cases, the remote station may itself be a fairly powerful computer and the remote station may be able to completely process small jobs locally, in addition to its major task of serving as remote input and output hardware for the central facility. And, if more than one large central facility is available, the remote station need not be tied permanently to any one of them, provided the appropriate communication circuits are created.

The latter arrangement leads to the possibility of creating a *computer network*, in which several large machines can be accessed from remote stations. Jobs requiring the resources of a particular large processor (for example, a job that requires data available only on a particular machine) might be routed to that processor; jobs that need no particular feature of any one processor might be routed to the one that is least busy at the time. Thus, one remote station might be transmitting jobs and receiving the output from jobs being processed on several different large machines at the same time. In addition, the network arrangement might allow a job being executed on one processor to use programs or data stored on another processor.

Time-Sharing.

An alternative path to the dissemination of computing services over large distances to individual users is called *time-sharing*. Here, the user typically has an inexpensive typewriterlike device, called a *teletypewriter* or *teletype*, located in his laboratory or office. Each terminal unit has a keyboard for input to the computing system and a printing element for output from the computing system. Devices that allow input and output of graphical information are also becoming quite common. Communication between user and central computer is normally made over low-capacity communication circuits, such as voicegrade telephone lines. The teletypewriter usually has no local computational capability, and is interfaced with the telephone system through a modem. The modem encodes typed information into appropriate signals, which are then transmitted through the telephone network to the central computing facility. There, another modem decodes the signals and passes them along to a communications terminal (transmission control unit). Information from the central facility to the user's terminal follows the reverse path.

Normally, the time-sharing user communicates directly with the operating system in the central processor; input and output regions of disk memory are

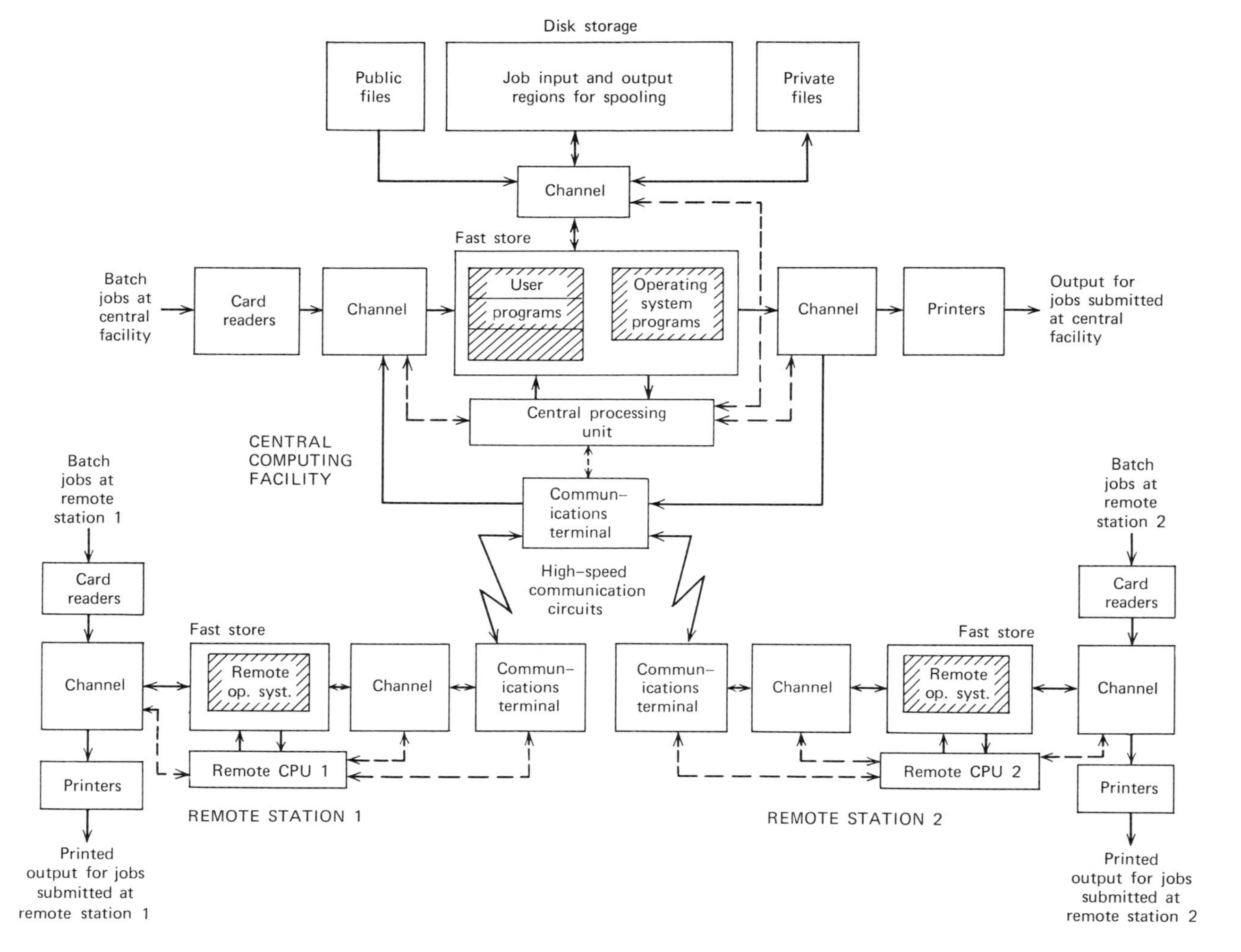

Figure 4.9 Central computing facility with two remote batch stations.

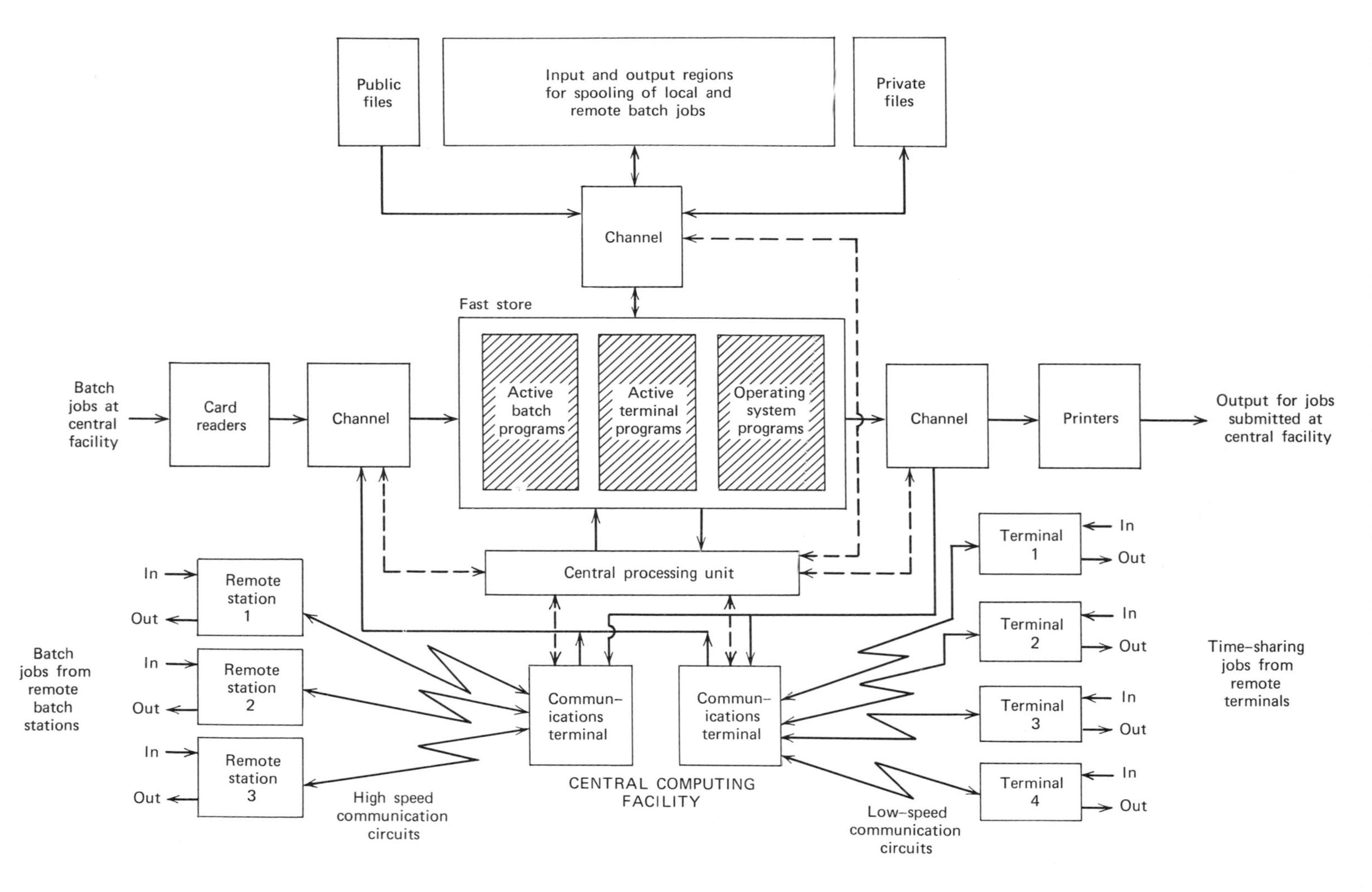

Figure 4.10 A multiprogramming system with both batch and time-sharing services.

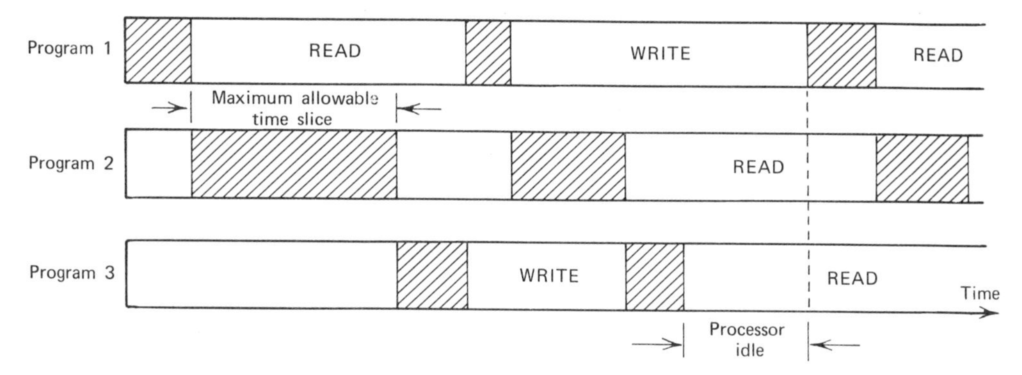

Figure 4.11 Central-processor activity in a time-sharing environment.

not assigned to accept the entire job input before execution or the entire job output before printing (punched card output is usually not available at the remote site, but punched paper tape attachments are quite common). In fact, the major thrust of the time-sharing service is to allow the user to *interact* with the operating system and with his executing programs *while* the job is being processed.

Figure 4.10 illustrates a time-sharing system that also allows the batch processing or spooling of jobs submitted on cards at the central facility or at remote batch stations. Thus, one way to implement a time-sharing system is to provide both interactive or *conversational* access from small remote terminals and the batch processing of jobs, all under control of a sophisticated multiprogramming system.

Whereas the principal objective of the strictly batch multiprogramming system is to keep the *processor* busy, a major objective of a time-sharing system must be to keep the *user* busy. In order to give the time-sharing users good *response time* (that is, a response within a period of 1 to 5 seconds), the standard multiprogramming scheme is usually altered so that an executing user program retains control over the processor until either: (a) processing is interrupted for a reading or writing operation, or (b) some maximum period of time called a *time slice* has passed. The situation for three active programs is illustrated in Fig. 4.11; execution of program 2 is suspended once after having control of the central processor for a full time slice. Note that this procedure precludes the possibility that one program can have exclusive processor control over an extended period of time. If, for example, the time slice is 50 milliseconds, then *at least* 20 users can be given some computing service each second. In fact, the number of programs actually served in a large time-sharing system with a 50-millisecond time slice would usually be much larger than 20, since most programs would be interrupted for reading or writing operations before the expiration of the maximum allowable time slice (recall that in 50 milliseconds, a large processor would be able to execute 50,000 or more individual machine instructions). Users are usually served in a "round-robin" fashion, although other priority schemes are also possible. In some systems, time-sharing jobs are always given priority over batch jobs, in which case the time-sharing jobs are said to be running in the "foreground" while the batch jobs are being run in the "background."

Multiprocessing.

The total computational capacity of a computing facility can be increased by either: (a) increasing the speed of the central processing unit, that is, by replacing the processor with a faster model, or (b) adding central processors to the system. When the latter alternative is chosen, the mode of operation is called *multiprocessing.* Typically, all of the processors execute programs from the *same* fast memory. Thus, as is illustrated in Fig. 4.12, if two processors are available, each can be executing a (usually different) program at the *same* time. If both processors are operated under a single multiprogramming system, then either processor can be assigned the task of executing any active program during its next time slice. Thus, a user program may be executed by processor 1 during one time slice and by processor 2 during some other time slice; this is analogous to solving some parts of a problem on one desk calculator and solving the other parts on another desk calculator. The user will, of course, be unaware of which processor is solving his problem, just as he is unaware of the assignment of particular time slices to his executing programs.

In the preceding paragraph, it was assumed that the two processors were identical. However, many alternative arrangements involving processors of different speeds and operating characteristics have also been employed in practical multiprocessing systems.

4.4 Storage Allocation

In its most general form, storage allocation is concerned with the management of memory at all levels, including the long-term storage of programs and data on secondary storage devices, the dynamic assignment of fast memory space to executing programs, and the efficient sharing of programs and data by all users of the computing system. On a large system, this problem is quite complex, and could itself be the subject of a textbook. Here, we shall restrict the discussion to some of the more important aspects of fast memory assignment for storage of executing object programs.

Absolute and Relocatable Code.

In the early machines, the main memory was the only memory. Object programs, either those written directly in the machine's language or produced by simple translators, were written in *absolute* code. "Absolute" implies that all addresses in the program, incuding those assigned to the variables and to the instructions themselves, are *fixed.* Thus a program in absolute code is always loaded into the *same* locations in the fast memory. Since program instructions are normally stored sequentially in memory, the code for a particular object program normally occupies a *block* of contiguous storage locations. For a program written in absolute code, this block is determined when the program is written, an arrangement

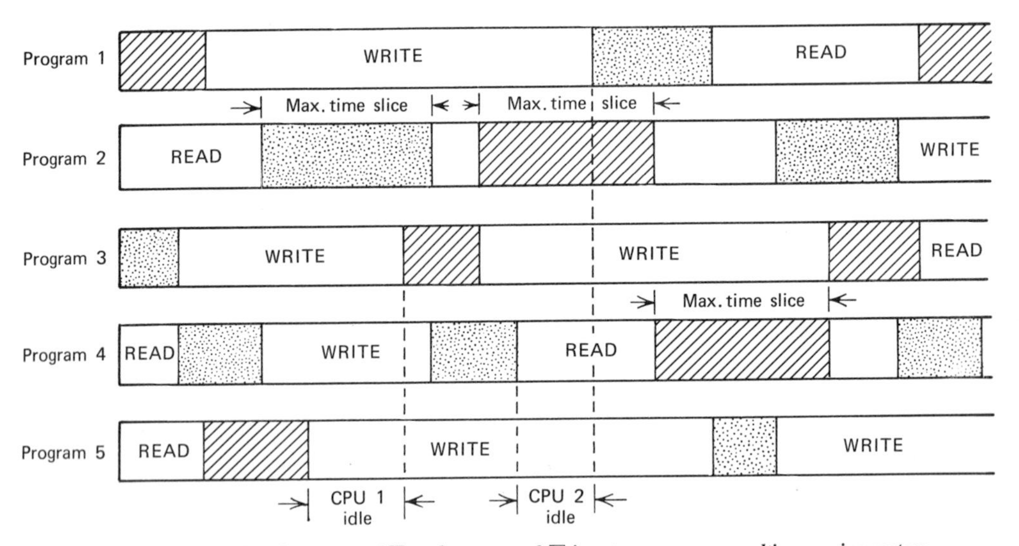

Figure 4.12 Activity of processor 1 ▨ and processor 2 ▩ in a two-processor multiprocessing system.

that is satisfactory only if the program is completely self-contained and is the only one present in the memory.

Programmers soon discovered that the writing and debugging of programs was greatly simplified by introducing *modularity* into program structures; hence the introduction of the subprogram concept in FORTRAN – II. This decomposition of large programs into a set of smaller, more manageable, *subprograms*, led rather quickly to the notion of a subprogram library containing routines to handle the most common functional evaluations (sine, square root, etc.) and numerical methods (solution of linear equations, for example). Clearly, if the object code for these subprograms is written using absolute addresses, serious addressing conflicts are certain to develop. For example, if the sine subprogram is written to be loaded only into a block of storage between address 1000 and address 1050, then no other subprogram that might be needed in the memory at the same time as the sine program may use addresses in this range. In addition, the user himself must carefully avoid using these locations in his program, if he expects to use the library sine function.

This problem of addressing conflict was solved with the introduction of *relocatable* object programs. In this widely used approach, every subprogram is written as if it were to be loaded as a block starting at the *same* location. Thus, all relocatable programs produced by the assembler or translator might be written with the first instruction assigned to address 0, the first storage location in the fast memory. When the programs are loaded into the memory, however, the loading program (usually called the *loader*) can assign the block of code to a block of storage starting at some *other* address; since many of the instructions will refer to addresses for data or other instructions, the loader must modify the address parts of such instructions to reflect the fact that the subprogram block has been relocated in the memory. Each subprogram, in turn, can be loaded starting with the first available memory location following already loaded subprograms, so that successive blocks of object code are tightly "packed" into the memory. Since the programmer will not know where the loader will store the various subprogram blocks, the *linking loader* must modify all instructions involving calls upon subprograms as well. Such loaders usually produce a *loading map*, listing the storage addresses occupied by each subprogram.

Overlays.

The introduction of procedure-oriented languages and the availability of extensive subprogram libraries made the generation of quite large object programs a rather simple matter; in many cases, the program may be larger than the total capacity of the fast memory. Since the central processor can execute machine instructions present in the fast memory only, such large programs must be processed in pieces. One solution to this problem, possible only when the machine has secondary storage devices such as disk files, is to partition the program into data regions and instruction regions. If the data regions are assigned locations in the fast memory, and the instruction regions can be partitioned into smaller blocks of instructions, then it may be possible to keep the data in the fast memory during the entire execution of the program, while successive blocks of the partitioned instructions are brought into the fast memory and executed by the processor. The blocks of instructions are usually called *overlays*; the part of the fast memory allocated for program instructions is overlaid with successive *core loads* of instructions. Great care must be taken in partitioning the instructions, so that the number of overlay operations is kept to a minimum; the efficiency of the processor can be drastically reduced if a large fraction of the computing time is spent in reading in the overlays. Because it is rather difficult to do this partitioning automatically, it is usually the responsibility of the individual programmer to determine the overlay strategy, an unsatisfactory situation for all but very experienced programmers (nevertheless, this approach is widely used.)

Multiprogramming operation considerably increases the complexity of the storage allocation problem. New object programs are introduced dynamically into a memory already containing many object programs of other users. If the user's program is to be loaded into a contiguous block of storage, then some unused block at least as large as the program must be found. This process of "squeezing in" new programs leads rather rapidly to a serious fragmentation of the fast memory, making it very difficult, for example, to find an unused block large enough for big programs. Several approaches to this problem are possible. Perhaps the simplest is to partition the fast memory into blocks of fixed size and then to store only one user's program in each *partition* at a time; effectively, this means that the machine has several separate fast memories, normally differing in size to allow for the processing of small programs without wasting storage space, while still allowing the running of large programs. Another possibility would be to relocate all users' program storage assignments dynamically to insure that no storage is "wasted" between storage blocks assigned to the different users; this solution to the problem may pose serious threats to the efficiency of the computing system, however, since a significant amount of the processor's computational capacity may be consumed in the relocation calculations. Still another approach would be to abandon the notion of a contiguous block of storage for each user; presumably the program could itself be broken into smaller blocks of fixed or variable size and "fitted in" the available memory space block by block. Again, however, the bookkeeping load on the operating system might considerably reduce the efficiency of the computing system.

Multiprogramming also leads to the possibility of having more than one user share the *same* program. For example, one copy of the FORTRAN translator could be executed by several users in multiprogramming operation, provided that: (1) the translator itself does not change as a result of its execution by a particular user—such unchanged programs are called *pure procedures*, and (2) any user-dependent data or results (for example, the user's source program or the translated object program) for different users are kept completely separate from one another. Usually such a pure procedure would be assigned a separate "working space" for each user.

Virtual Memory and the IBM 360/67 Computer.

Most of the recent solutions to the problem of automatically allocating storage in large computing systems are variations on the concept of the *one-level store* or *virtual* memory, first introduced on the ATLAS computer, built for Cambridge University (England) in the early 1960s. As of 1973, one of the most widely used commercially available machines employing a virtual memory is the IBM 360/67. The hardware characteristics for the dual processor version of this machine currently in use at the University of Michigan and the operating system for it, called MTS, are summarized in Sections 4.5 and 4.6, respectively. In the remainder of this section, we describe the implementation of the virtual memory concept in this machine.

The two IBM 360/67 processors execute instructions taken from the fast (magnetic core) memory; the core memory with somewhat more than 1.5 million bytes capacity is subdivided into *page frames* having a fixed size of 4096 bytes each. Thus, the main memory is subdivided into 384 page frames, each capable of holding about 1000 full (four byte) words. Three high-speed drum memories, each containing 900 page frames of 4096 bytes, operate as auxiliary memory to the central store. The processors *cannot* execute instructions from the drum memories. However, a page frame of information in the drum memory can be transmitted very rapidly to one of the page frames in the fast memory and *vice versa*, in an operation called *paging* or *page swapping* (when one page frame of fast memory is *paged out* while another page frame of drum memory is *paged in*). This arrangement means that a total of 3084 page frames of instructions and data can be present in the pooled fast and drum memories, even though no more than 384 of this total are directly accessible to the processors at any given time.

The operating system for the IBM 360/67 at The University of Michigan is called MTS (Michigan Terminal System). When an object program is waiting to be loaded into memory and executed, it is normally stored on disk-file memory. MTS, whose structure and use will be described in more detail in the latter sections of this chapter, reads a copy of the stored program, and formats the information into blocks of 4096 bytes, that is, into blocks of fixed size exactly equivalent to the page frame capacity of the fast and drum memories. Individual *pages* of the object program may be read into *any* available page frames in the fast memory without regard to order; that is, pages of program that follow in logical sequence need not be loaded into contiguous page frames of fast memory. Should there be fewer page frames of fast memory available than there are pages of object code, the additional pages of code are copied into available page frames of drum memory, again without any particular order required in the drum page-frame assignments. This arrangement of scattered pages of object program will be true for other computer users as well, and the processors will be operating in multiprogramming-multiprocessing mode.

The principal difference between this mode of operation and that shown in Figures 4.8 and 4.9 is that an individual user's object program is not stored in one contiguous block of fast memory; instead it is stored in page-frame size blocks in both fast and auxiliary memory. Only those pages of code that are being executed or accessed by a processor need be in the fast memory; hence it is quite possible that a single user program may be *larger* than the *entire capacity* of the fast memory. However, pages of code can be paged or swapped between the drum and fast memories, and the processor is capable of executing such a program without resort to *user-supplied* overlay strategies. The user will be unaware of the size limitations of fast memory (in the University of Michigan Computing System, pages of object code can be paged or swapped between disk storage and fast memory, should the drum memories become full, so the effective *virtual* memory available to a given user is enormous when compared with the physical capacity of the fast memory alone).

During the initial loading process, the operating system leaves at least one page of object code for each user in the fast memory, including the logically first page (that is, the page containing the first machine instruction in the program or the *entry point*). When the user is assigned his first time slice by the operating system, execution of his object program begins with instructions in this logically first page. If instructions in one page of the program make reference to instructions or data in another page of the program, and that other page is also in the fast memory, or if all of the instructions in one page have been executed and the logically next page of instructions is also in the fast memory, execution continues without interruption. However, if the executing program makes reference to instructions or data in a page *not* currently in the fast memory, or if all instructions in a given page have been executed and the logically next page of instructions is *not* in the fast memory, computation is suspended temporarily. The missing page (presumably available in one of the page frames of the drum memory) is located and the operating system initiates the reading (by a channel) of the page from a drum into any currently available page frame of fast memory. In the meantime, the central processor continues its multiprogramming time-sharing duties by executing instructions in other user's programs (see Fig. 4.13).

Once the missing page of a program has been copied into the fast memory, and the operating system assigns the program its next time slice, execution resumes and continues until: (1) an input operation is required, (2) an output operation is required, (3) a page not currently in the fast memory is referenced, (4) all instructions in one page have been executed and the logically next page of instructions is not in the fast memory, or (5) the assigned time slice has elapsed. At any instant in time, many users programs will be stored in the page frames of fast and drum memories. The processing sequence outlined above is applied to the programs of each of these users in turn.

The 2700-page capacity of the drum memories is much larger than the total capacity of the fast memory, so it will not,

Initial situation: Two users, A and B, are running in multiprogramming mode. All pages A1, A2, . . . , A8, and B1, B2, . . . , B7 of these programs are available on fast drum storage. The most recently active pages of programs A and B are resident in fast core memory. Assume that information about the location of pages in program A and B is maintained by MTS in the indicated page of fast memory, and that MTS and some associated operating system programs occupy certain pages of fast memory.

A4	Operating system programs	B7	B6
	A3	page location information	A5
A1	B5	A8	MTS
	B2	B1	A2

Fast memory

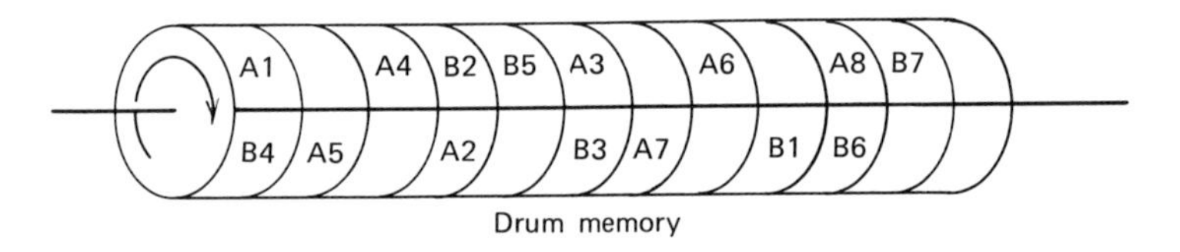

A new user, C, joins A and B as a competing customer for computing services. Assume that C requests that an object program with five pages be brought from disk storage and executed. When MTS assigns a time slice to user C, pages of code will be brought from disk to fast memory, and then written in any available pages on the drum.† Assume that copies of pages C1 and C5 are left in fast memory (users A and B will continue to have periodic control of the processor while the reading of program C proceeds). Once the entire program is read onto the drum, and the user is assigned a time slice, execution of program C will begin with page C1. During the page reading operations, MTS updates the page location information in the memory.

A4	Operating system programs	B7	B6
C1	A3	page location information	A5
A1	B5	A8	MTS
C5	B2	B1	A2

Fast memory

Disk memory

(† A page is not written onto drum storage until the fast memory page is needed, if at all.)

A1 C1 A4 B2 B5 A3 C3 A6 A8 B7

B4 A5 C2 A2 C4 B3 A7 B1 B6 C5

Drum memory

Figure 4.13 (a) Arrangement of pages of object code in fast memory and drum memory page frames in the IBM 360/67 computer.

in general, be possible to have all the pages for all users' programs in the fast memory at one time. What happens when the fast memory is full and a program references a page not in the fast memory? Some page in the fast memory must be *removed* to make room for the incoming page. An algorithm is available to choose the page to be removed; it is based on the length of time the page has remained in the fast memory without being accessed.

Certain hardware features simplify the selection and the mode of page replacement. Each page frame of fast storage has associated with it some additional storage bits for control purposes. One bit is called the *use* bit. Whenever information in the page frame is accessed, the bit is automatically set to 1. If the operating system resets all the use bits to 0 at specified time intervals, then the use bit can be used to establish which pages of object code have not been used since the last resetting operation.

Another bit, called the *change* bit, is set only when the content of a page frame has been changed. If the change bit of the page to be removed from the fast memory (the

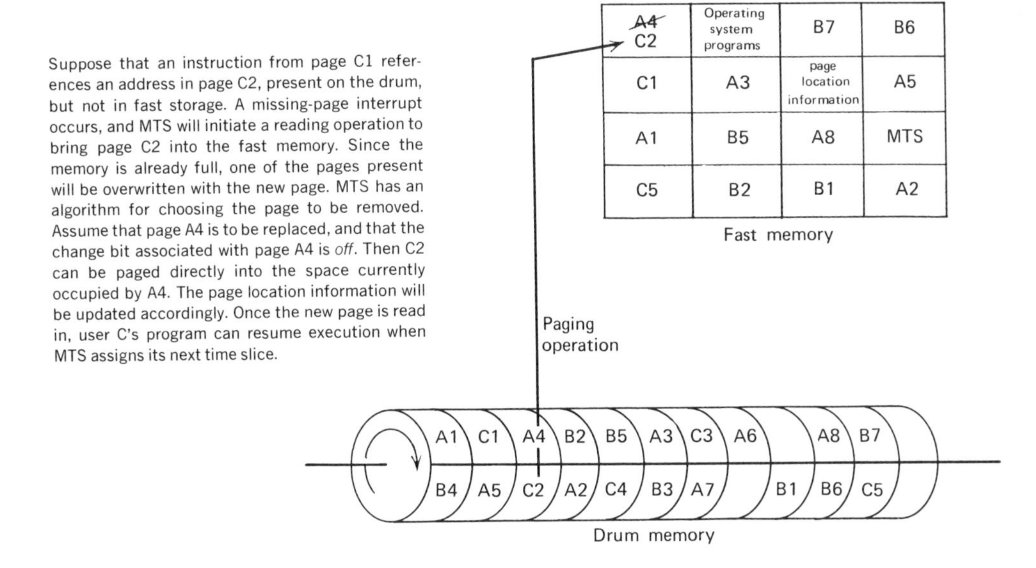

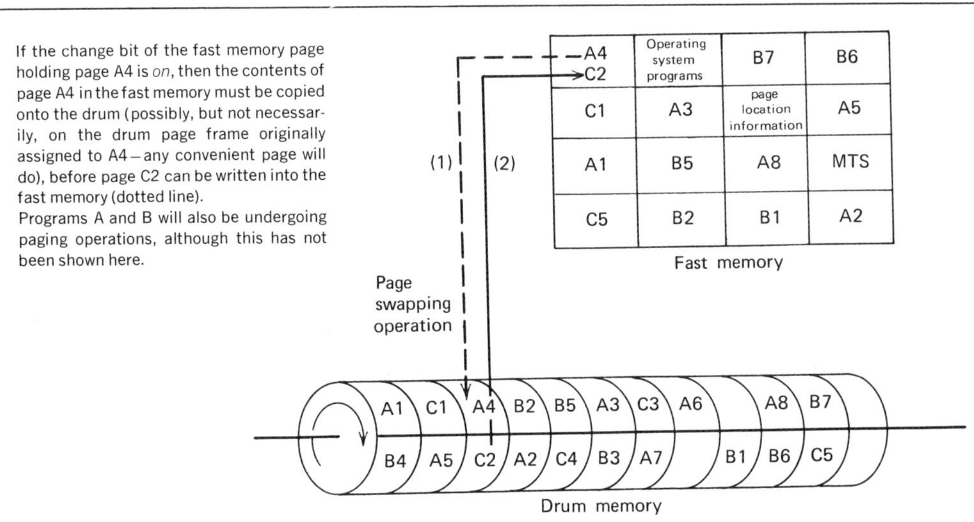

Figure 4.13 (b) Paging and page swapping operations in the IBM 360/67 computer.

outgoing page) is *off*, then the incoming page can be written directly over the fast memory page frame occupied by the outgoing page; a copy of the outgoing page will already be available in a page frame of drum memory, should it be needed again. This assumes that the outgoing page has already been written at least once on the drum; if the outgoing page has not yet been written on the drum, then it must, of course, be copied onto the drum before being overwritten. If the change bit of the outgoing page is *on*, then the page must be copied onto the drum before the associated page frame of fast memory is overwritten, to insure that an updated version of the page will be available for future use. Any convenient unassigned page frame in the drum store can be used, provided that the page location information is updated by the operating system. The process of transmitting pages of object program from drum page frames to fast memory page frames is called *paging*, while the two-way transmission of pages of object programs is called *page swapping* (see Fig. 4.13*b*).

The allocation of storage space in both the fast and drum memories is dynamic, that is, it changes with time. Normally, not all pages from a given program are present in the fast memory at one time, and pages that are present are scattered randomly throughout the memory. Because of the nature of the paging and page-swapping operations, the *same* program page may even be stored in *different* page frames of the memory at different times during the execution of a program. This arrangement has the distinct advantage of allowing very large programs to be executed from a relatively small fast store; the fast store is being used very efficiently because only the active parts (usually no more than a few pages at one time) of each user's program are present in the fast memory at any time. Since the machine is being multiprogrammed, this arrangement should also allow the storage of a sufficiently large number of user programs to keep the processor busy nearly all the time.

Unfortunately, there are serious disadvantages as well if all addresses in a page of object program must be recomputed each time a page is moved. While the relocatable code concept described on page 223 *could* be applied each time a page is moved, in fact, pages are moved so frequently in a busy system that the processor would be spending a substantial part of its computational capability simply computing addresses in relocated instructions if the problem is to be solved with computer software. And certainly, the programmer cannot be expected to handle the relocation problem himself, as he would be expected to do in the simple overlay strategy described earlier.

Fortunately, the virtual memory scheme overcomes the disadvantages of the preceding paragraph, while retaining all of the advantages of the decomposition of object programs into small blocks that are dynamically allocated storage in the fast memory when the processor has need of them. In this case, the addresses in object program instructions are viewed as *symbolic* only, spanning a *single level virtual addressing space*. That is, all programs are written without reference to the various levels of storage (e.g., disk, drum, magnetic core); the user writes his programs as if they were to be processed on a computer with a virtually unlimited fast memory without any secondary storage devices at all. There are, by definition, no overlay problems, at least for the *user*. *Each* user is assigned his *own* virtual memory; hence it is impossible for one user to refer to addresses in some other user's virtual memory. This approach leads automatically to a simple *storage protection* mechanism.

In the IBM 360/67 computer, the pages of object code for a particular user are assigned to *segments* containing up to 255 pages; as many as 4096 segments may be present in the virtual memory. Virtual addresses consist of 32 bits such that the segment number, page number within a segment, and byte number within a page (called the *displacement*) are given by the first 12, the next 8, and the last 12 bits of the address, respectively. Thus, theoretically, the virtual addressing space spans the range of addresses 0 through $2^{32}-1$. The virtual memory for this machine contains over *four billion* bytes.

These virtual addresses are *never* changed (for example, relocated) regardless of the hardware storage addresses of the locations where the program is stored. Thus there are *two* addressing spaces in a virtual memory machine, the *virtual addressing space*, and the *hardware addressing space* (usually called the *memory space*). Clearly, if the virtual addresses are never changed, then the movement of a page of object program from one page frame to another (for example, from a drum page frame to a fast memory page frame) does not require the operating system to carry out the burdensome task of relocation calculations. There is still the problem, of course, of relating a particular *virtual address* to the *true hardware address* associated with it at any given time. [In an analogy with a FORTRAN program, the user is perfectly happy to use the symbolic addresses (names) X and Y; when the processor is attempting to add X and Y, however, it must know the true hardware locations (say 5227 and 4126) corresponding to X and Y, so that the proper numbers will be used in the computation.] Thus, in a virtual memory machine, there must still be some mapping mechanism for relating the virtual addresses to the corresponding hardware addresses.

The address mapping function is handled primarily by special hardware (usually called the *address relocation hardware*); the operating system normally plays a role as well, by maintaining segment and page location information in the form of tables. In particular, in the IBM 360/67 computer, the operating system maintains, for *each* user: (1) one word of memory containing the first or *base* address of a *segment table*, and the *number of segments* in the program, (2) a *segment* table with three entries for each segment, the *number of pages* in the segment, the *base address* of a *page table* for the segment, and some *availability* bits to indicate if the segment is available in the computing system, and (3) a *page table* for each segment that contains two entries for each page, the *real* physical (hardware) *base address* in the fast memory where the page was last stored, and availability bits to indicate if the page is available in the fast memory or if it must be paged into the fast memory from the drum memory. Whenever a paging or page-swapping operation takes place, the information in the appropriate table is updated by the operating system.

When a user is assigned a time slice, the operating system loads the base address of that user's segment table into a special register in the processor called the *segment table base register*. Given this information, special hardware in the processor has full access to the chain of addressing information kept in the segment and page tables for the user, and the correspondence between the virtual addresses and true hardware addresses is automatically established without further resort to the operating system (hence, the processor's computational capability is not appreciably affected by the address mapping operations).

The heart of the address relocation hardware consists of eight special registers of *associative memory*, each containing the virtual memory base address and the corresponding true hardware base address for one "active" page in the user's program (a page in the fast memory that is being referenced by the central processor). These registers are loaded automatically from the information stored in the segment and page tables maintained by the operating system. When a virtual address is referenced in an instruction, the hardware effectively does a "table look-up" operation in the associative memory registers and establishes the correspondence between the virtual and real addresses; the real hardware address corresponding to a given virtual address is then transmitted to the central processor before the instruction is actually performed. The hardware details of this operation would be of little interest to the typical computer user, since the address relocation equipment functions automatically, and the user has no control over its performance.

In the preceding pages we have described just one implementation of the virtual memory concept in existing hardware. Other schemes are also possible, and are being incorporated into many of the newer digital computing systems. For example, in some systems, the storage frames have variable size, depending on the nature of the program; the storage and management of address information tables may be quite different from that described above. Nevertheless, the benefits to the programmer accruing from any virtual memory design promise to be very great indeed. The user is relieved almost completely of the memory management problem.

4.5 Hardware Configuration of an Existing Large-Scale Computing System

In this section we describe the hardware configuration of the computing system (as of 1973) that regularly serves more than 10,000 users from among the student body, faculty, and staff at four campuses of the University of Michigan (Main Campus and North Campus in Ann Arbor, and the Dearborn and Flint Campuses in Dearborn and Flint, Michigan, respectively). The equipment consists of central facilities located on the North Campus, and three remote batch stations, one each on the Main Campus, the Dearborn Campus, and the Flint Campus. It is a multiprocessing system with two identical IBM 360/67 central processors. The operating system multiprograms spooled batch jobs and time-shared jobs in a manner very like that illustrated in Fig. 4.10.

The principal hardware items, exclusive of the channels, a variety of input and output control units, and some of the communication equipment such as modems, are shown in Fig. 4.14 and listed below; some of the more important operating characteristics of each unit are included.

Central Computing Facility.

(1) *Two IBM 360/67 central processors.* Instruction times for each processor are of the order of one microsecond. The control unit of each processor partially decodes several instructions in advance of the one actually being executed. The instruction decoding operations are said to be *overlapped* and considerably improve the operating speed of the processors.

(2) *Six magnetic-core memory units*, usually called memory *boxes*. Each of the boxes is subdivided into 64 parts of equal capacity called *pages*. Each page contains 4096 bytes, or the equivalent of 1024 full words of memory. Thus there is a total of 384 pages or somewhat more than 1.5 million bytes of fast memory. The fetch and decoding time for an instruction word is about 750 nanoseconds (billionths of a second).

(3) *Three high-speed drum memories.* Each drum has a capacity of 900 pages of 4096 bytes each, giving a total drum store of 11,059,200 bytes. The central memory and auxiliary high-speed drum memory operate in tandem, as described in Section 4.4. Access time is about 5 milliseconds.

(4) *Twenty-eight disk drives*, arranged in groups of six or eight, each group having a separate controller (accessing device). Each drive holds one replaceable disk pack consisting of 20 surfaces. Each surface is divided into 400 circular tracks with 7294 bytes each (one byte equals eight bits). Thus each surface holds 2,917,600 bytes, and each drive 58,352,000 bytes. The total disk storage capacity is 1,633,856,000 bytes. Average disk access time is about 30 milliseconds.

(5) *Twelve magnetic tape drives.* Ten of these drives are equipped to read and write 9-track tapes (the standard tape for the IBM 360 system) and two are equipped to handle 7-track tapes (the standard tape for many other computing systems). The standard writing densities for 9-track tapes are 800 and 1600 bits per inch per track. The input/output transfer rate is 160,000 bytes per second for the higher density.

(6) *Two card readers.* One of these reads at a rate of 600 cards per minute, while the other reads at the rate of 1000 cards per minute. In addition to reading the standard punched card, the faster reader can read "mark-sense" cards having the same shape as the standard punched card. Information can be encoded on the mark-sense cards by marking appropriate squares with a lead pencil. Punched cards and mark-sense cards may be intermixed in an input job deck, so that users can prepare programs or make corrections to input decks without having access to keypunching equipment.

(7) *Three line printers.* Each unit has a 132-column printer carriage. One unit has a printing speed of 600 lines per minute, while the other two have capacities of 1000 lines per minute.

(8) *One read/punch unit.* This unit can read punched cards at the rate of 1000 per minute and punch cards at the rate of 300 per minute.

(9) *One high-speed paper-tape reader.* This unit reads 5-, 6-, 7-, or 8-channel punched paper tape at the rate of 300 characters per second. With appropriate conversion routines, paper tapes prepared by standard teletypewriter paper-tape attachments can be read.

(10) *One audio-response unit.* This device can generate audible sounds under computer control for transmission over the telephone network. It accepts input signals from the tone generators of a standard touch-tone telephone. By communicating with the operating system through the audio-response unit, any standard touch-tone telephone can become a remote terminal to the time-sharing system. With the exception of the printing of output (all output is spoken) any operation that can be performed from a teletypewriter terminal can be performed from a touch-tone telephone.

(11) *One communications terminal.* This unit can handle asynchronous signals (that is, signals that arrive at arbitrary times) from as many as 64 voice-grade telephone lines (up to 1200 bits per second) and 12 high-capacity lines (up to 9600 bits per second). It employs half-duplex transmissions, that is, it can either send or receive signals over a particular line but cannot both send and receive signals simultaneously. This unit is the principal communication device between the IBM 360/67 computing system and remote timesharing equipment such as teletypewriters, Datel, and IBM 2741 terminals. The high-speed lines serve as the communication links with small computers or batch terminals that allow the remote entry of batch jobs and the printing and punching of output for those jobs at remote locations (see below).

(12) *Data concentrators.* These devices are modified PDP-8 and PDP-11 computers that serve as sophisticated communications terminals operating in full-duplex mode, that is, they can send and receive signals over the same line simultaneously. The data concentrators communicate directly with the central processors (through channels) and can service a wide variety of remote terminal devices in time-sharing mode over 20 low-speed lines (300 bits per second) and 5 high-speed lines (2000 bits per second). These include the standard teletypewriterlike terminals, but a

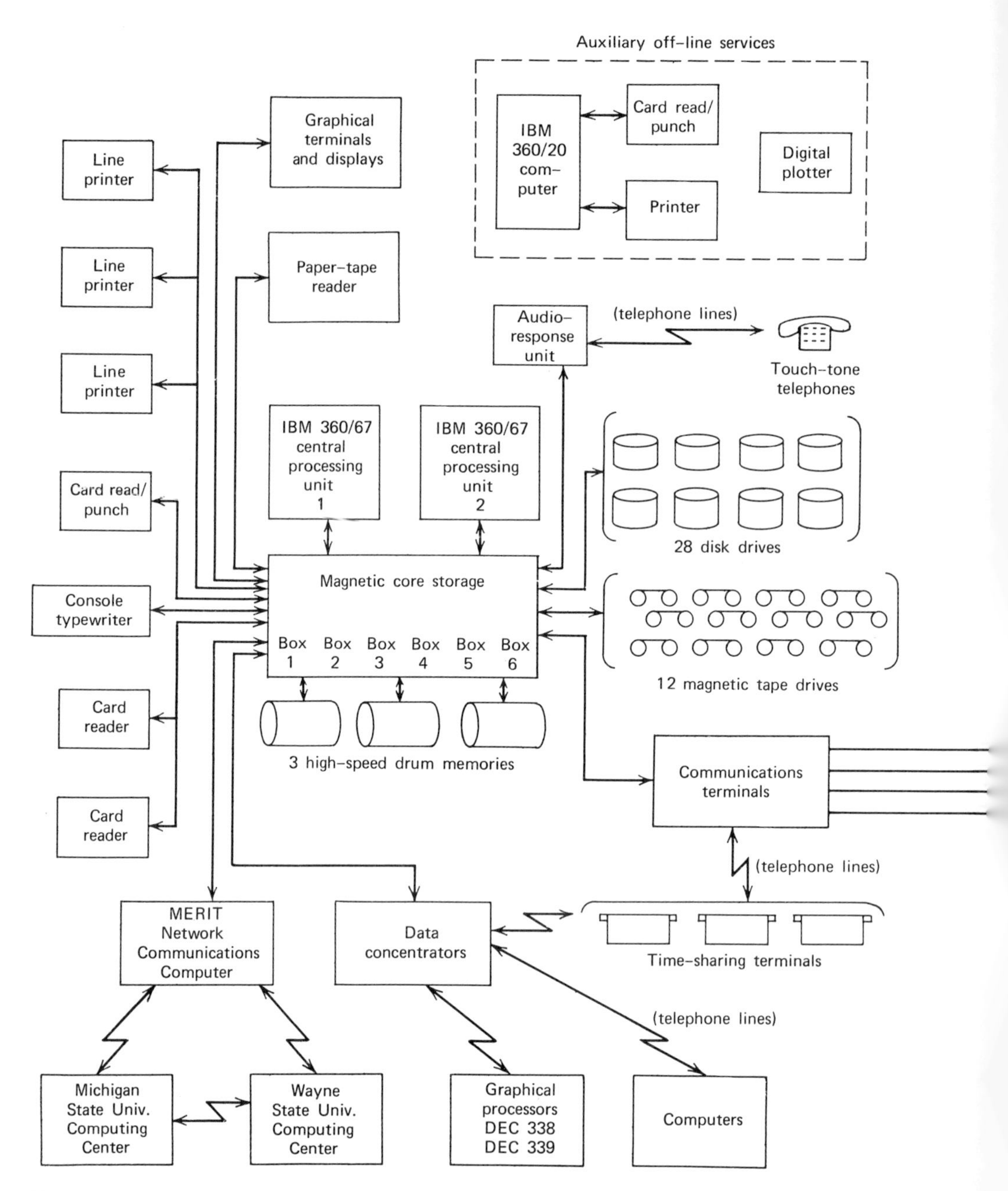

Figure 4.14 The University of Michigan Computing System (channels, modems, and various controllers not shown).

principal function of the data concentrators is to serve as an interface between other computers and the central facility. There are many "minicomputers" in university laboratories that can effectively use the resources of the central facility in accomplishing their tasks. Each of the data concentrators is equipped with a high-speed *paper-tape punch* with a punching speed of 60 characters per second.

(13) *Graphical display terminals.* Two terminals consist of a typewriterlike keyboard with a television screen for character display, and are used to

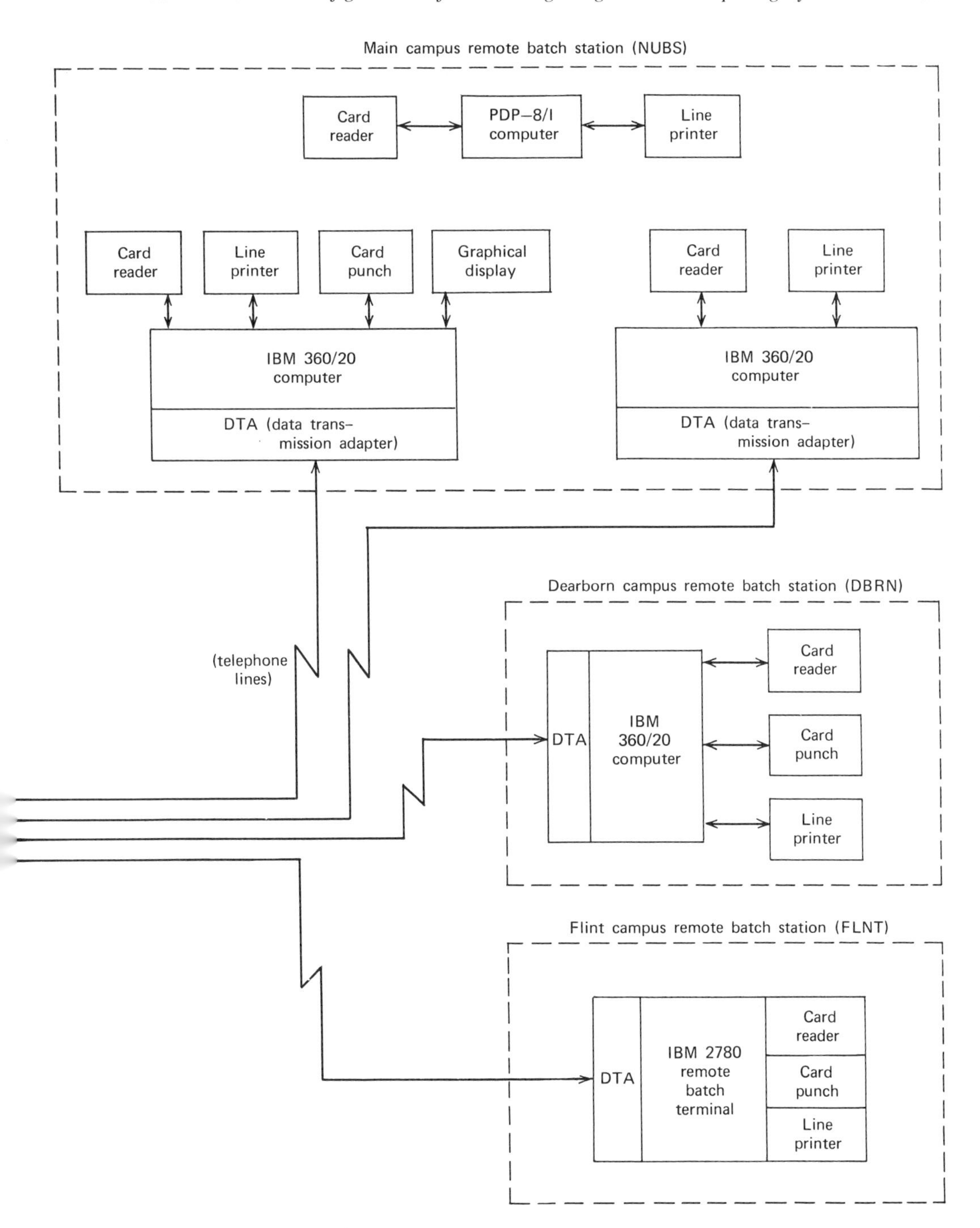

display the status of the system for operating personnel. In addition, the status of all batch jobs in the computing system is displayed on television screens in the Computing Center work areas.

(14) *One network communications computer.* The three major university computing facilities in Michigan have recently been linked together in a computer network called MERIT (Michigan Educational Research Information Triad). At the site of each central facility there is a network communications computer, a modification of the PDP-11

computer. A user at one university who wishes to use a program or data resource available only at one of the other university centers, is able to dial the time-sharing service of his own computer and request services from one of the other facilities. Since the hardware and software at each school is different, the associated communication computers translate some of the unique characteristics of each system into a uniform network format; the user will hopefully need to learn only minimal information about the software of another system in order to make use of its computing resources.

(15) *One IBM 360/20 computer.* This equipment does not communicate directly with the IBM 360/67 computing system, as does all the equipment described above. It is used primarily to prepare printed listings of punched-card decks and to reproduce and interpret (here, the character equivalents of previously punched cards are printed on the card) card decks.

(16) *Digital plotter.* This off-line device can record graphical information previously encoded in digital form on magnetic tape by the IBM 360/67 computer. Plots can be as wide as 30 inches and as long as desired.

(17) *Other equipment.* There is a substantial amount of equipment associated with the hardware (channels and controllers for the various input and output and peripheral storage devices) whose detailed operation would be of little interest to the typical computer user. There is also a significant investment in communications equipment, particularly in modems and other machinery associated with the commercial telephone system. In addition, there are many keypunches and teletypewriter terminals at the central facility.

Remote Batch Station – Main Campus.

(1) *Two IBM 360/20 computers.* Batch jobs may be submitted at the remote station for processing by the central computing facility. These computers serve only to format card images for transmission over high-speed telephone lines and to produce printed and punched output for card and line images transmitted from the central facility. The transmission rate from each computer is 7200 baud (bits per second, the equivalent of approximately 900 characters per second).

(2) *Two line printers.* Each printer has a carriage width of 132 columns and a printing speed of 1000 lines per minute.

(3) *Two card readers.* One reader has a capacity of 1000 cards per minute while the other has a capacity of 600 cards per minute.

(4) *One card punch.* The unit can punch cards at the rate of 300 per minute.

(5) *Television display.* The status of batch jobs being processed at the central facility is displayed continuously on television monitors in the user work areas.

(6) *One PDP-8/I computer.* This small computer is equipped with a card reader and printer, and is used only to prepare printed listings of punched card decks.

(7) *Other equipment.* There are several keypunches and teletypewriter terminals in the user work areas, and substantial amounts of communication equipment in the machine room.

Other Equipment on the Ann Arbor Campuses Provided by the Computing Center.

(1) *One DEC-338 and one DEC-339 graphical processors.* These two units, manufactured by the Digital Equipment Corporation, are stand-alone computers for manipulation of graphical information displayed on cathode

ray tubes. Each can also be operated as a time-sharing terminal of the central computing system through one of the data concentrators. Information can be entered by a variety of devices, including a light pen that allows the user to "draw" on the face of the televisionlike screen.

(2) *"Public" keypunches and time-sharing terminals.* Keypunches and teletypewriters for use of the computing community at the University are scattered throughout the most important classroom, office, and laboratory buildings. There are, of course, a large number of other small computers (14 of these can serve as "private" remote batch stations), keypunches, graphical displays, teletypewriters, etc. on campus that are not owned by the Computing Center, but that have access to the services of the computing system.

Remote Batch Station—Dearborn Campus.

The remote batch station at the Dearborn campus consists of one IBM 360/20 computer equipped with one card reader (600 cards per minute), one card punch (90 to 350 cards per minute), and one line printer (300 lines per minute). The maximum transmission rate to and from the central facility is 2000 bits per second.

Remote Batch Station—Flint Campus.

The remote batch station at the Flint campus is an IBM 2780 remote batch terminal. This self-contained card-reading, card-punching, and printing unit has a reading capacity of 150 cards per minute, a punching capacity of 90 to 350 cards per minute, and a printing capacity of 200 lines per minute. The maximum transmission rate to and from the central facility is 2000 bits per second.

4.6 The Operating System for the University of Michigan Computing System

In this section, we describe, in a rather general way, the organization of the operating system in use on the hardware described in Section 4.5. Although the system is usually called MTS (Michigan Terminal System), the principal supervisory program is a multiprogramming system called UMMPS (University of Michigan Multiprogramming System). It is responsible for scheduling the execution of the most important system programs on the two central processing units, in particular, a spooling program, named HASP (Houston Automatic Spooling Priority program), time-sharing monitors, named MTS, and several special programs that manage memory allocation and communication between fast memory and secondary-storage devices such as the disk files, high-speed drums, and magnetic tapes. From the user's viewpoint, the computer appears to be under direct control of a time-sharing monitor, MTS, since all commands in a user's job (even those in batch jobs) are eventually processed by MTS. Thus, for most purposes, MTS can be considered *the* operating system; the user will normally not be aware of the existence of either HASP or UMMPS.

Nevertheless, a brief description of the interaction between UMMPS, HASP, and MTS should help to explain the spooling of batch jobs and the relationships of batch and terminal (time-sharing) operations in the University of Michigan computing system. UMMPS is a multiprogramming, multiprocessing system that operates the two IBM 360/67 computers in "supervisor" mode, that is, it is able to execute certain "privileged" machine instructions, and is the only program that is allowed to perform the most critical system control functions, in particular: (a) scheduling all tasks to be run on both processors, (b) actually allowing a particular task to be run for a time interval (time slice) on one of the processors, (c) scheduling all requests for input and output, (d) initiating all channel activities, and (e) allocating and managing all storage, both virtual and real.

One of the tasks supervised by UMMPS is the spooling program, HASP. It is responsible for: (a) spooling batch jobs read from cards at the Computing Center and all of the remote batch stations, (b) assigning an execution priority to the incoming jobs, (c) initiating execution of the batch jobs, and (d) spooling the output from the batch jobs to printers and card punches at the Computing Center and remote batch stations. HASP does *not* control the execution of batch programs once they have been loaded into fast memory.

HASP creates an input region (file) and an output region (file) on disk storage for each batch job as described in Section 4.3 and as illustrated in Figures 4.8 and 4.9. Each job is then assigned an *execution priority*, determined by the user's estimate of the maximum execution time (total central-processor time) required.

Once the execution priority has been assigned, HASP is responsible for initiating execution of the job. From one to 20 batch jobs may be in execution at the same time. The exact number is determined by a rather complicated algorithm that takes into account the number of fast memory and drum memory pages occupied by active programs, and the fraction of available central-processor time being spent in the *wait state* (idle time).

HASP can be viewed as having 20 execution initiation ports for its queue of waiting jobs, which are positioned in the queue in descending order of priority and on a first-come first-served basis within each priority class. Each execution initiation port has a specific priority level at which it begins to accept jobs, so that not all ports accept jobs having the highest priority. This scheme ensures that low priority jobs will not be kept at the end of the queue indefinitely, simply because additional jobs with higher priority are entering the queue. Once a job has been introduced into execution through a particular port, no additional jobs can be initiated through that same port until the first job has completed the execution phase of processing.

When HASP initiates execution of a batch job, it requests UMMPS to create an MTS time-sharing monitor to supervise processing of the *batch* job in *time-sharing* mode, with the understanding that when MTS needs an input line it will read from the HASP input region rather than from a time-sharing terminal. Likewise, MTS will write any output lines or card images into the HASP output region for the job, rather than onto a terminal. Thus, in the Michigan system, there is little distinction between a batch job and a terminal job in the execution phase of processing.

Once the execution of a batch job is complete, HASP is responsible for scheduling and initiating the printing and punching of job output present in the output region on disk storage. Each job is assigned a *printing priority* based on the actual number of pages to be printed. Whenever a printer is free, the job with the current highest printing priority will be printed next. Within a given priority level, jobs are printed in the order of completion of execution.

Punched output, written into the output region of disk storage during the execution of the job, is punched after printing for the job is complete. There are no priorities assigned to the punching operation; jobs are punched in the order in which they finish printing.

So that the batch job programmer will be able to follow processing of his job, HASP displays the status of each job on televisionlike monitors located at the computing center. Under ideal circumstances, the *turnaround time* (total elapsed time from submission of the job at the input window to collection of the job output at the output window) for the processing of a short batch job will be about five minutes; under heavy loading conditions, particularly near the end of the semester, turnaround times may be considerably longer. Of course, if system malfunctions occur or if a part of the computing equipment is undergoing maintenance work, then system performance will be correspondingly degraded.

Each of the remote batch stations executes a remote HASP program which has four primary functions: (a) reading remote batch jobs from the card reader at the remote station, (b) communicating with the HASP spooling program operating in the central computer and transmitting the card images to the central facility over telephone lines, (c) accepting the output from the batch jobs transmitted from the central facility to the remote station by the HASP program executing in the central computer (central HASP maintains routing records so that output destined for remote stations is not printed and punched at the Computing Center), and (d) printing or punching the job output at the remote station.

Provided that a telephone line is available (see Section 4.9), a time-sharing user who dials the computer is assigned a copy of the time-sharing monitor, MTS. The tasks of this monitor are to interpret command lines for the job, to carry out the indicated orders, to read input data for executing programs, to write output lines from executing programs, etc. These time-sharing monitors have at their disposal many subsystems, such as those that actually handle the transmission of information to and from files (on disk storage, for example) and devices. Thus there are many MTS "operating systems," one for each terminal job and one for each batch job that is in its execution phase. In fact, HASP, the individual copies of MTS, and the major subsystem routines are processed as individual tasks in multiprogramming and multiprocessing mode by UMMPS. The various user-related jobs, each monitored by its own copy of MTS, are assigned equal priority by UMMPS, so that batch jobs and terminal jobs are treated identically during the execution phase of processing.

Since all commands for a user's job, either in batch or time-sharing mode, are eventually interpreted by a copy of the MTS time-sharing monitor, the entire computing system appears to be operated by MTS; thus the user needs to learn just one command language, that for MTS. Examples of MTS batch job decks have already appeared in Section 3.38, while an MTS time-sharing job is shown in Section 4.9.

4.7 Files

Virtually every large computing system, and in particular systems providing time-sharing services, allow the user to keep private information, such as programs and data, on secondary storage devices at the computing facility. Normally this information is saved on disk storage in named *files* associated with the user's account number. One of the major services provided by the operating system will deal with the creation, editing, manipulation, and removal of these files at the request of the user. Normally a file consists of a set of lines of information arranged in a particular order; a file might contain, for example, a FORTRAN source program, a FORTRAN object program, a set of data for a FORTRAN program, a FORTRAN compiler, a library of subprograms, the complete text of *Genesis*, medical histories of all patients in a hospital ward, etc.

In this section, we describe the characteristics of files in the University of Michigan computing system; other large facilities will almost certainly offer similar storage capabilities to their users. MTS recognizes two major types of file organization: (1) line files, and (2) sequential files. In a line file, lines are ordered by an associated *line number*; individual lines may be stored or retrieved in any order. Thus it is a simple matter to insert a new line between two lines already present in the file, or to replace or delete a line in a file. Sequential files also consist of lines, but these lines have no associated line numbers; lines are kept in the same sequence as originally written, so that insertion of a new line between two existing lines is not possible. Such files have many characteristics of information stored on magnetic tape, in that successive lines appear in fixed sequence on the storage medium. We shall describe only line files here, as these are the most useful to the typical FORTRAN programmer.

A line file consists of an ordered set of zero (if the file is empty) or more lines, each containing from 1 to 255 characters (bytes). The line number may be any number of the form *sddddd.ddd* ($s =$ sign, $d =$ digit) between -99999.999 and $+99999.999$. Lines are organized in *algebraic order* according to their line numbers. Most users choose to number the lines in their files with the small integers: 1, 2, 3, 4, etc.

Three classes of line files are available: (1) public, (2) private, and (3) temporary files. A *public file* usually contains a library program in object form. Contents of such files are available to *all* users of the computer, and can be loaded into the user's virtual memory and executed directly using the $RUN command, as described in Section 4.9. The *names* of all public files begin with an asterisk (*) followed by up to 15 characters, the last of which is not an asterisk. For example, the object code for the WATFIV compiler is available in a public file named *WATFIV. Public files can be created only by the Computing Center staff.

Users may create, write into, read from, empty, or destroy (remove from the system) *private files*, which are normally available only to the user who created them. Such files are given names of the user's choice consisting of up to 12 alphanumeric characters (for example, PROGRAM, DATA, JOE), and are stored more or less permanently (until the user wishes to remove them) on disk file storage. Each user is assigned a maximum amount of such storage space that he may keep permanently in the computing system. The smallest MTS file is automatically assigned two pages (about 8000 bytes) of storage, although the user may create a larger file if he wishes. In MTS, a user may, at his discretion, allow specified other users to have certain kinds of access to a given private file, in which case, the file is said to be *shared*. For example, a professor may make a program or data stored in one of his private files available to his students on a read-only basis (the student may make a copy of the file or read from the file, but not write into it).

In addition, a user may reference a *temporary* or *scratch file* during the processing of his job. Temporary files are named with two to nine characters, the first of which must be a minus sign. They may be manipulated just as private files are. However, a temporary file is available only from the time of its creation (the time it is first mentioned) until the job is completed. Thus a temporary file created during the processing of one job will *not* be available for reference by some future job.

The file management routines of MTS maintain an up-to-date catalog of all private files that have been created (but not yet destroyed) by each user of the computing facility. There is, of course, a charge for storage of such files (as of 1973, the cost was 0.0167¢ per page-hour, one page of storage for one hour).

In MTS, reference to the name of the file alone implies the whole file (all lines in the file). Thus the MTS command $LIST ALPHA will cause the complete contents of the private file named ALPHA to be listed on a printer or terminal. However, parts of files called *subfiles* may also be selected. For example, the MTS command $LIST ALPHA(1.7,55) will cause the contents of only those lines with line numbers 1.7 through 55 to be listed.

Subfiles may be joined together in any order to form a *concatenated* file. This can be done explicitly by joining the subfiles in the desired order with plus signs, for example,

```
JOE+ALPHA(5,12)+PROGRAM(7,7)
```

Thus it is a simple matter to gather together lines from any number of files in any order, allowing, for example, an executing object program to read data lines from a variety of files.

Devices are also given names by MTS, and are treated almost interchangeably with files, so that it is possible, for example, to have an executing program read some data from files and other data from devices, such as a time-sharing terminal; this capability is illustrated in the time-sharing example of Section 4.9.

4.8 Time-Sharing Terminals

MTS supports a wide variety of terminal devices for accessing the computing system in time-sharing mode. The most important are the various Bell teletypewriters (see Plate 10) and modifications of the IBM Selectric typewriter such as the IBM 2741 (see Plate 9) and Datel 30 terminals. The most widely used terminal for the many time-sharing systems in use is probably the Bell model 35KSR teletypewriter; the layouts of the keyboard and console control buttons and switches for this terminal are shown in Fig. 4.15. Basically, the keyboard has a standard typewriter format for the letters and digits. Since only the capital letters

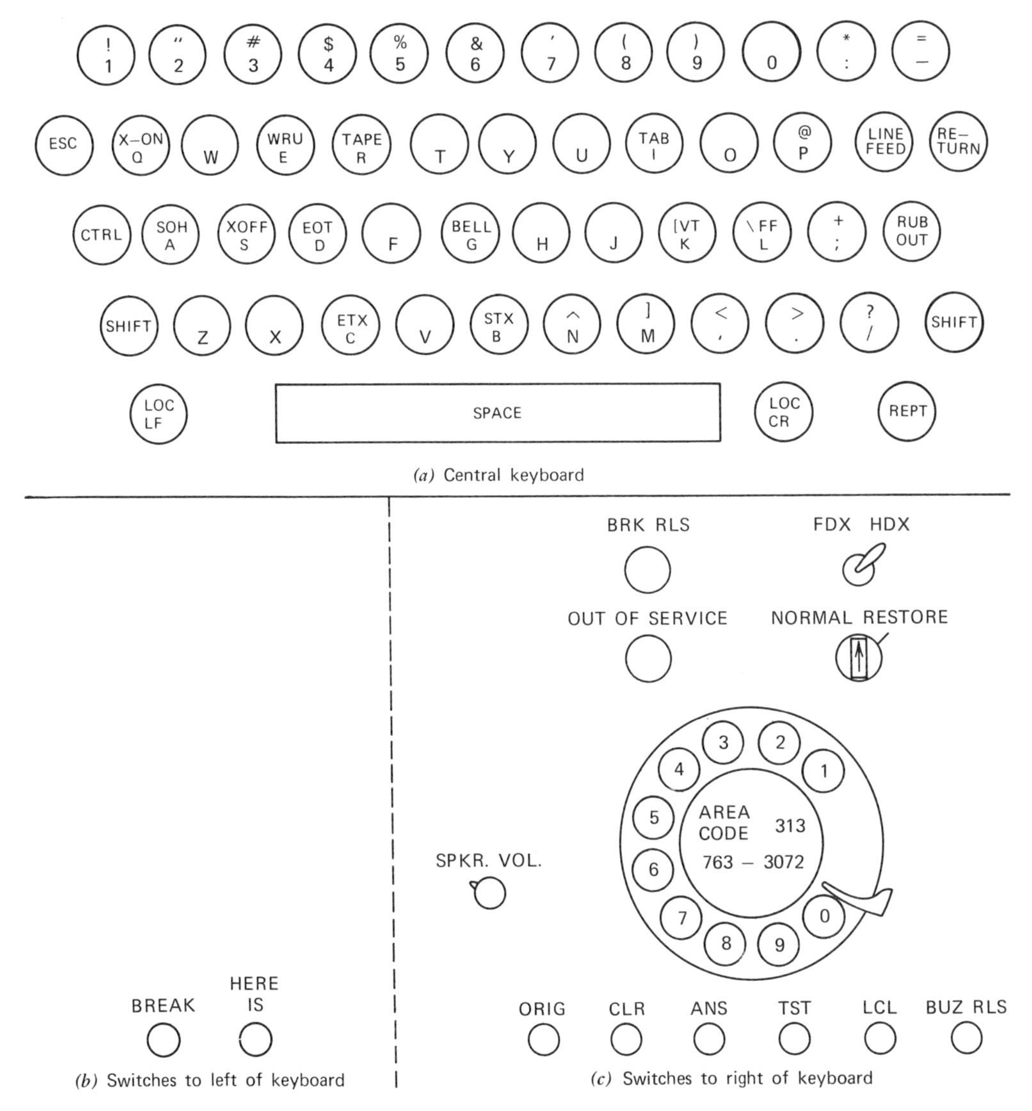

Figure 4.15 Model 35KSR teletypewriter keyboard, console buttons and switches (see also Plate 10).

are available, the shift function button is used exclusively to generate the special characters appearing above the digits and some of the letters.

In order to use the teletypewriter as a remote terminal of a time-sharing system, one must first establish a telephone connection between the unit and a communication terminal at the central computing facility. The teletypewriter is equipped with a speaker that sounds the exchange dial tone when the ORIG button is depressed. After the proper telephone number is dialed, the communication equipment at the central computer will answer, usually with a short beeping tone. The time-sharing operating system then takes over the signing-on protocol. Each teletypewriter is equipped with an "answer-back drum" that can be interrogated by the operating system, so that the teletypewriter may be uniquely identified. Depending on the particular communication terminal dialed at the computer, the teletypewriter will be operated in either *half-duplex* mode or *full-duplex* mode by the user's setting the FDX-HDX switch to the proper position. Here, *half-duplex* implies that only one-way transmissions are allowed, either from teletypewriter to computer or *vice versa*, while *full-duplex* indicates that messages may be traveling in both directions simultaneously. In any event, the operating system will normally issue prompting information to assist the programmer in signing on (in MTS, a prompting character, #, is printed at the left edge of the page whenever the user is expected to respond by typing at the keyboard).

Plate 9. Time-sharing terminal. This IBM 2741 unit is typical of many of the wide-carriage, typewriterlike units now on the market. (Courtesy IBM Corporation.)

Usually, anything typed on the keyboard will be printed automatically, so that the printed listing produced by the teletypewriter will contain a complete record of the time-sharing session, with messages from the computer and user responses appearing in chronological order. In virtually every time-sharing system, certain key combinations will be reserved for special editing functions, such as the deletion of the last typed character, deletion of the last typed line,

Plate 10. Teletypewriter terminal. This model 35KSR Bell Teletypewriter is among the most popular of the time-sharing terminals. Details of the keyboard arrangement are shown in Fig. 4.15. (Courtesy American Telephone and Telegraph Corporation.)

and the immediate interruption of job processing. In MTS the following are used:

1. *To delete the last character typed,* press the CTRL and H key simultaneously, a combination called "CONTROL-H." Nothing is printed on the paper, and the printing element is not backspaced. However, the input routines will automatically trim the previous character from the line. If the last *n* characters typed are to be deleted, then the CTRL key can be depressed while the H key is struck *n* times.
2. *To delete all characters already typed* in the current line, press the CTRL and N buttons. Nothing is printed as a result of the operation, and the printing element is not returned to the left margin of the carriage. Any additional characters typed will be entered into columns 1, 2, 3, . . . of the current line. If "CONTROL-N" is followed immediately by a carriage return (RETURN), then MTS responds by typing "LINE DELETED" and returns the carriage to the left margin.
3. *To interrupt any current processing action* and restore immediate contact with the operating system, MTS, press the BREAK button. This causes an *attention interrupt* to take place; whatever the current activity, such as the listing of a source program or data file or the execution of an object program, processing will be interrupted. MTS types a message, "ATTN!", and waits for the user to enter another command. This feature allows the user to have absolute control over the processing of his job.

Several other, though less essential, control functions are available from the teletypewriter as well, but are not included here.

4.9 An Example Time-Sharing Session at a Teletypewriter Terminal

This section illustrates the use of the time-sharing service provided by the

University of Michigan computing system. We attempt to illustrate the simplest features of the system, such as the creation of files, the entering of source program and data lines into files, the translation of source FORTRAN and WATFIV programs, and the execution of the generated object programs with user data. Aside from the specific form of the operating system (MTS) commands, similar features will be available in virtually any good time-sharing system. A brief explanation of each command will be provided as needed, as there is little point in describing the MTS command language in detail; the command language for other time-sharing services will differ from those shown here.

In the following example, everything that the user types at the terminal keyboard is underlined; all other characters appearing on the printed listing are written by the computer. The text is written as if *you* were sitting at the teletypewriter entering commands and data.

Signing On. Assuming that the dialing procedure has been followed, that a line is available to the communication terminal at the central computer, and that the computer is operating properly, MTS responds with the following lines:

```
MTS (LA00-0016)
WHO ARE YOU?
UMXXXXAAA
```

Here, the characters in the first line indicate which telephone communication lines have been assigned for transmissions between the teletypewriter and the central computer. The characters UMXXXXAAA were written by the answer-back equipment on the teletypewriter (see the previous section) in response to interrogation by the operating system. This code will be different for each teletypewriter.

Next, MTS types a pound sign (#) in the first column and waits. The appearance of # in column 1 of any line means that MTS is waiting for you to enter a line at the keyboard. Each user of the equipment has his own computing center user number that the operating system uses to control computer access, the nature and extent of allowed computing services (for example, whether a user may have access to both conversational and batch services, how much permanent file storage he may assign), and for maintaining accounting records. The first command must be a $SIGNON command with the appropriate user number; here, we assume that your user number is CCNO.

```
#SIGNON CCNO
```

Before MTS takes action on this or any other line typed at the terminal keyboard, you must depress the carriage-return button, RETURN (see Fig. 4.15). Throughout the following pages, each line that you type must be terminated by a carriage return.

After accepting this line, MTS checks the accounting records for your user number to determine that it is a valid number, that you are allowed access to the computing system through a remote terminal, and that there are funds remaining in the account. If any of these three criteria is not met, MTS types an appropriate comment at the terminal, and disconnects the telephone. Otherwise, MTS will request that you enter your user *password*, up to 12 characters known only to you and to the operating system. This password (which can be changed at any time, once the user gains access to the system using the proper old password) serves as a final key to computer access.

```
#ENTER USER PASSWORD
?
```

At this point, the user password may be entered at the keyboard *without* being printed on the paper, so that someone else could not find the password and user number together, should the paper be discarded. If the password is incorrect, MTS allows the user three tries before disconnecting the telephone line; if correct, MTS types the time when the account number was last used and the current time and date, and then returns with # in column 1 to await the next command.

```
#**LAST SIGNON WAS; 22:38.47   12-29-71
#  USER "CCNO" SIGNED ON AT  10:47.57 ON 01-03-72
```

Running Some Public Files. In MTS, an object program is loaded into virtual memory and execution is initiated with the $RUN command. Many object programs of a service nature are available in public files (see Section 4.7). Let us begin by running the program stored in the file *STATUS; this is an accounting program that writes out a brief summary of the status of the account.

```
#$RUN *STATUS
#EXECUTION BEGINS

 STATUS OF CCNO AT LAST SIGNOFF          USED       MAXIMUM      REMAINING

 CUMULATIVE CHARGE          ($)          291.52      850.00        558.48
 CURRENT DISK SPACE         (PAGES)          76         120            44
 CUMULATIVE TERMINAL TIME   (HR)          11.63

#EXECUTION TERMINATED
```

Here, the command $RUN *STATUS causes MTS to load the contents of the public file *STATUS into virtual memory. The message #EXECUTION BEGINS is written by MTS to indicate that the loading process is complete and that the processor is now executing the object program from *STATUS. The next lines are written by the program *STATUS as a result of examining the accounting records for the user number CCNO. User CCNO has a total of 76 pages (1 page = 4096 bytes) of private files saved in disk storage, and will be allowed to create additional files, provided that no more than 44 new pages of disk file space are needed. The message #EXECUTION TERMINATED is written by MTS to indicate that the execution of *STATUS is complete, and that the storage space allocated to the object program from *STATUS has been released for other programs.

We can find the names of the private files created under the user number CCNO by loading and executing the object program in a public file named *CATALOG, as follows:

```
#$RUN *CATALOG
#EXECUTION BEGINS
 PACERGO
 TRANS
 TRANST
 TRANSOBJ
 EQCALL
 PACER
 PACERDS
 PACER2
 PACERLIB
 USER CCNO HAS    9        FILE(S) WITH TOTAL SIZE OF      76          PAGES
#EXECUTION TERMINATED
```

More information about any of these files could be obtained by running another public file called *FILESNIFF (see page 247). The program *FILESNIFF examines the storage directory for private files for information about the specified file, such as the particular disk pack on which it is stored, and the number of lines in the file.

Next run the public file *USERS to determine the current loading of the computing system.

```
#$RUN *USERS
#EXECUTION BEGINS
  THERE ARE  53 TERMINAL USERS,  12 BATCH TASKS,  40 AVAILABLE LINES,
  AND 21 NON-MTS JOBS USING 1928 VIRTUAL PAGES AND 238 REAL PAGES.
  HARDWARE IS CPU'S P1, P2, STORAGE A, B, C, D, E, F, CCU'S 0, 1.
#EXECUTION TERMINATED
```

A total of 65 jobs is being processed by MTS; 53 are time-sharing jobs from remote terminals, and 12 are batch jobs submitted at the Computing Center or remote stations. The 21 additional jobs are major system communication and storage management tasks. Thus the multiprogramming system is processing a total of 86 active programs. Of the total virtual memory capacity of 3084 pages (900 each on three drums and 384 pages of core memory), 1928 were being used when *USERS was being executed; 238 of the 384 pages of the core memory (the "real" pages) were being used by active programs. All major hardware items are in operation, including the two central processors (P1 and P2), the six magnetic core memory boxes (storage A, B, C, D, E and F), and the two channel control units (CCU's 0 and 1). The latter units control operation of the channels (see Section 4.3).

Next, run the public file *TIME to determine the current clock time as follows:

```
#$RUN *TIME
#EXECUTION BEGINS
  CLOCK  10:51.21  DATE  01-03-72
#EXECUTION TERMINATED
```

Creating and Entering Lines into a File. Suppose that you wish to create a file named PROG, and to enter a source program in the FORTRAN-IV language into the new file. The first step is to bring PROG into existence using the $CREATE command.

```
#$CREATE PROG
#  FILE "PROG" HAS BEEN CREATED.
```

Since no parameters except the file name were specified in the $CREATE command, PROG will be a two-page line file assigned to disk file storage.

The $CREATE command causes the new file to become the *active* file. Any line entered at the terminal keyboard that begins with a line number will be entered directly into the file. If a line with that line number already exists in the file, then the new line will replace the old one. If no line with that line number is present in the file, the new line is inserted into the file at the appropriate location.

Usually we will want MTS to number the lines for us, starting with line number 1 and incrementing the line numbers by 1. Then the MTS command is:

```
#$NUMBER
```

If the numbering sequence were to begin with *n* and be incremented by *m*, then the command would be written as $NUMBER *n*, *m* or $NUMBER *n* *m*; for example, $NUMBER 10 5.

MTS will now supply the first line number followed by an underscore mark "_" and wait for us to enter the content of the first line. No action will be taken by MTS until we key in the carriage return for the line. Column 1 of the *content* of each line immediately follows the underscore mark, so that the first character typed will appear in column 1 of the line when it is saved in the file PROG. Suppose that we want to enter a simple FORTRAN source program into the file. The content of successive lines will be treated just as the content of a FORTRAN program on punched cards would be treated by one of the compilers. Unless special editing programs are to be used later, or unless the particular compiler allows source statements to be entered in free format, each FORTRAN statement must be entered according to the standard FORTRAN card format described in Section 3.13. Note that the line numbers in the file have no relationship to the FORTRAN statement numbers in the FORTRAN source program.

```
#        1_    1  READ (5,100,END=20,ERR=20)    X, Y
#        2_  100  FORMAT ( 2F10.1
#        3_       Z*X + Y
#        4_       WRITE (6,200)   X, Y, Z
#        5_  200  FORMAT (3F10.5)
#        6_       GO TO 1
#        7_   20  CALL EXIT
#        8_       END
#        9_$UNNUMBER
```

The MTS command $UNNUMBER causes the operating system to stop the automatic numbering of file lines.

Listing and Copying a File. Two commands, $LIST and $COPY, are available for listing the contents of a file, with and without the line numbers, respectively. We can now check the line numbers and the contents of PROG as follows:

```
#$LIST
>        1          1  READ (5,100,END=20,ERR=20)    X, Y
>        2        100  FORMAT ( 2F10.1
>        3             Z*X + Y
>        4             WRITE (6,200)   X, Y, Z
>        5        200  FORMAT (3F10.5)
>        6             GO TO 1
>        7         20  CALL EXIT
>        8             END
# END OF FILE
```

We could also copy the contents of the file. In this case, the line numbers are deleted and the first character in each line is used as a carriage-control character. Column 2 of the line content appears immediately after the MTS character >.

```
#$COPY
>    1  READ (5,100,END=20,ERR=20) X, Y
> 100   FORMAT ( 2F10.1
>       Z*X + Y
>       WRITE (6,200)   X, Y, Z
> 200   FORMAT (3F10.5)
>       GO TO 1
>  20   CALL EXIT
>       END
```

Now, suppose that we spot the error in line 2 (but not the one in line 3), and wish to correct it. Since PROG is still the active file, it suffices to type the line number, followed by the correct line content:

```
#2 100   FORMAT ( 2F10.2 )
```

To be assured that the line has replaced the original line 2, list the subfile consisting of lines 2, 3, and 4 in PROG.

```
#$LIST PROG(2,4)
>       2       100    FORMAT ( 2F10.2 )
>       3              Z*X + Y
>       4              WRITE(6,200) X,Y,Z
# END OF FILE
```

So far, we have only our FORTRAN program in the private file PROG. Next, let us create another private file named DATA, and use it for storing lines of input data values for the variables X and Y in the FORTRAN program, just as we would punch data cards for a batch run. If there are two sets of data, we might have:

```
#$CREATE DATA
#   FILE "DATA" HAS BEEN CREATED.
#$NUMBER 2,2
#        2_         6.97         4.32
#        4_        -6.97        69.0L
#        6_$UNNUMBER
#$LIST DATA
>        2               6.97         4.32
>        4              -6.97        69.0L
# END OF FILE
```

In this case, the initial line number and line increment are both two. Note that in line 4 of the file DATA, the character L was inadvertently used for the digit 1. Correct the line and relist the file.

```
#$NUMBER 4
#      4_      -6.97        69.01
#      5_$UNNUMBER
#$LIST DATA
>        2            6.97          4.32
>        4           -6.97         69.01
# END OF FILE
```

Compiling a FORTRAN Source Program. The IBM G-level FORTRAN-IV compiler is available in a public file named *FTN. We can compile the source program in the file PROG by loading and executing this compiler. Unless we specify otherwise on the $RUN command, the object program will automatically be written into a temporary file named −LOAD by *FTN.

```
#$RUN *FTN PAR=SOURCE=PROG
#EXECUTION BEGINS
       3.000              Z*X + Y
                          $
  01) SYNTAX
    1 SERIOUS ERRORS IN MAIN, NO OBJECT GENERATED.
```

Here, the command $RUN *FTN/PAR=SOURCE=PROG causes the *FTN compiler to be loaded and executed. The source program is read from the file PROG and the object program is written into the file −LOAD. However, since a serious error was encountered in the statement in line number 3, no object program was in fact generated or saved in −LOAD. When *FTN is run from a terminal, the source program is not printed on the terminal (unless specifically requested by an additional parameter in the $RUN command); only abbreviated compiler diagnostics (see Section 3.39 and Appendix D) and the offending FORTRAN source statements are listed.

The diagnostic comment produced indicates that the trouble is near the Z in the statement Z*X+Y (the location of "$" below the statement). Next, we can correct the line in the file and list the subfile PROG(2,4) to ensure that the corrected statement Z=X+Y has been entered properly. The file DATA is currently the active file; to make an already created file (PROG in this case) the active file, we use the $GET command

```
#$GET PROG
#READY.
#3         Z = X + Y
#$LIST PROG(2,4)
>        2        100    FORMAT ( 2F10.2 )
>        3               Z = X + Y
>        4               WRITE (6,200)   X, Y,
# END OF FILE
```

Now, compile the corrected program.

```
#$RUN *FTN PAR=SOURCE=PROG
#EXECUTION BEGINS
  NO ERRORS IN MAIN
```

There are no language errors in the source program, and the object program is stored in the file −LOAD in disk storage.

Running the Object Program. The compilation appears to be successful, so the object program in the temporary file −LOAD can be loaded and executed using the $RUN command. Let input unit 5, used in the READ statement of the original FORTRAN program, be assigned to the file DATA containing the two sets of data. Let the output from the program, produced by WRITE statements in the original program referencing output unit 6, be written into the temporary file −OUT. Then the appropriate command and printed output at the terminal are:

```
#$RUN -LOAD 5=DATA 6=-OUT
#EXECUTION BEGINS
#EXECUTION TERMINATED
```

The comments "#EXECUTION BEGINS" and "#EXECUTION TERMINATED", written by MTS, refer to the execution of the object program in −LOAD. The output from the program should be available in the file −OUT. The file −OUT should be copied rather than listed if the first character of each line (a blank in this case) is to be correctly interpreted as the carriage control character.

```
#$COPY -OUT
>   6.97000    4.32000  11.29000
>  -6.97000   69.00999  62.03999
```

Except for some rounding error in the second line, the results appear to be correct and conform to the specified format (3F10.5).

Next, let us write the content of the file DATA into the file PROG, following the FORTRAN source program, and then list PROG to insure that the data lines have been entered properly.

```
#$COPY DATA PROG(LAST+1)
#$LIST PROG
>       1         1  READ (5,100,END=20,ERR=20)    X, Y
>       2       100  FORMAT ( 2F10.2 )
>       3            Z = X + Y
>       4            WRITE (6,200)  X, Y, Z
>       5       200  FORMAT (3F10.5)
>       6            GO TO 1
>       7        20  CALL EXIT
>       8            END
>       9            6.97       4.32
>      10           -6.97      69.01
# END OF FILE
```

Note that the last line in the file may be referred to as LAST, so that LAST+1 was a line with the line number 9 when the $COPY command was executed.

Now, execute the public file *FILE-SNIFF with the file name PROG as the parameter.

```
#$RUN *FILESNIFF PAR=PROG
#EXECUTION BEGINS
    FILE=PROG
      LOC= DISK
      TYPE= LINE
      VOLUME=MTS002, NO. EXTENTS=01
      NO. LINES=00010      MAX. LINE LENGTH=00039
      NO. PAGES=(0002,0002), SIZE PARS=(000146,000146).
      PERMIT CODE = NONE
#EXECUTION TERMINATED
```

The output from *FILESNIFF indicates that PROG is a two-page line file saved on disk storage on the drive with disk pack number MTS002. The file is in one contiguous area on the disk file, is available only to the programmer who created it, and contains ten lines with a maximum line length of 39 characters.

Run *CATALOG again to ensure that the two new files PROG and DATA have been cataloged under your user number.

```
#$RUN *CATALOG
#EXECUTION BEGINS
 PACERGO
 TRANS
 TRANST
 TRANSOBJ
 EQCALL
 PACER
 PACERDS
 PACER2
 PACERLIB
 PROG
 DATA
 USER CCNO HAS  11     FILE(S) WITH TOTAL SIZE OF    80        PAGES
#EXECUTION TERMINATED
```

Now, destroy the file DATA, since we no longer need it (the original content of DATA has already been written into the file PROG).

```
#$DESTROY DATA
#FILE "DATA" IS TO BE DESTROYED.  PLEASE CONFIRM.
?OK
#DONE.
```

Since the file DATA has been destroyed, any attempt to reference it should be refused by MTS. For example:

```
#$GET DATA
#FILE DOES NOT EXIST.
```

Let us run the object program (still present in the file −LOAD) with the data in the subfile PROG(9,10) and have the results printed directly at the terminal.

```
#$RUN -LOAD 5=PROG(9,10) 6=*SINK*
#EXECUTION BEGINS
    6.97000    4.32000   11.29000
   -6.97000   69.00999   62.03999
#EXECUTION TERMINATED
```

In this case, the input device 5 is equated with the subfile PROG(9,10), so that the data will be read from lines 9 and 10 of PROG. Device 6 is equated with *SINK*, the MTS name for the printing element of the terminal. Hence, any output produced by the WRITE statement in the FORTRAN source program is written directly on the terminal, rather than into a file as in the preceding example.

To illustrate the point that concatenated files may be referenced as logical input/output units, consider the input data to consist of (a) the data in line 9 of file PROG, (b) any number of data lines entered at the keyboard of the terminal, and (c) the data in line 10 of the file PROG. In MTS, the terminal keyboard is given the name *SOURCE*.

Here, the first line following the comment "#EXECUTION BEGINS" contains the results for the data from line 9 of PROG. The next four lines are alternately data lines entered at the terminal and the corresponding output from the program. Any number of data lines could be entered at the keyboard (*SOURCE*); to indicate that MTS should proceed to the next subfile of the concatenated input file, the user enters the command $ENDFILE. The line preceding the comment "#EXECUTION TERMINATED" is the program output for the data taken from line 10 of the file PROG.

Running a WATFIV Program. Suppose that we wish to enter and debug a WATFIV FORTRAN program at a terminal that consists of a small main program that reads N and an array of numbers X(1), . . . , X(N), calls upon a function AVG(N,X) that calculates the average of X(1), . . . , X(N), writes out the data and the computed average, and then repeats the process for any number of data sets.

Let us create a file named WPROG, enter the program using the automatic numbering feature, and then list the contents of the file (see top of next page).

The file containing the source WATFIV program for the WATFIV compiler, is called SCARDS in MTS, and must contain any commands for the compiler ($COMPILE, $DATA, $STOP—see Section 3.38), and, if unformatted READ statements are used, the lines of input data for the object program. To insert a single dollar sign as the content of column 1 of a

```
#$RUN -LOAD 5=PROG(9,9)+*SOURCE*+PROG(10,10)
#EXECUTION BEGINS
    6.97000    4.32000   11.29000
      42.01      -3.55
   42.00999   -3.55000   38.45999
       2.00       3.00
    2.00000    3.00000    5.00000
  $ENDFILE
   -6.97000   69.00999   62.03999
#EXECUTION TERMINATED
```

```
#$CREATE WPROG
# FILE "WPROG" HAS BEEN CREATED
#$NUMBER 10,10
#     10_  1  READ, N, (X(I), I=1,N)
#     20_     PRINT, N, AVG(N,X), (X(I), I=1,N)
#     30_     GO TO 2
#     40_     END
#     50_     FUNCTION AVG(N,X)
#     60_     DIMENSION X(20)
#     70_     AVG = 0.0.
#     80_     DO 2  I=1,N
#     90_  2  AVG = AVG/N
#    100_     RETURN
#    110_     END
#    120_$UNNUMBER
```

```
#$LIST WPROG
>        10        1  READ, N, (X(I), I=1,N)
>        20           PRINT, N, AVG(N,X), (X(I), I=1,N)
>        30           GO TO 2
>        40           END
>        50           FUNCTION AVG(N,X)
>        60           DIMENSION X(20)
>        70           AVG = 0.0.
>        80           DO 2  I=1,N
>        90        2  AVG = AVG/N
>       100           RETURN
>       110           END
# END OF FILE
```

line, it is necessary to enter *two* dollar signs in sequence (this avoids the problem of interpreting such lines as MTS commands, which normally begin with a single dollar sign). Since WPROG is the current active file, the compiler commands and two illustrative data sets can be entered directly into WPROG.

```
#1$$COMPILE LIST
#$NUMBER 200,5
#     200_$$DATA
#     205_2,1.5,3.0
#     210_3 1. 2. 3.
#     215_$$STOP
#     220_$UNNUMBER
```

In the data set entered into line 205, the optional comma has been used as a delimiter between successive data items; blanks have been used to delimit the successive entries in the data entered into line 210. The parameter LIST in the $COMPILE command forces the listing of the source program on the output device called SPRINT at compilation time (otherwise only statements containing errors and appropriate diagnostics are printed).

Now, list the complete WATFIV source file in WPROG.

```
#$LIST WPROG
>         1     $COMPILE LIST
>        10        1  READ, N, (X(I), I=1,N)
>        20           PRINT, N, AVG(N,X), (X(I), I=1,N)
>        30           GO TO 2
>        40           END
>        50           FUNCTION AVG(N,X)
>        60           DIMENSION X(20)
>        70           AVG = 0.0.
>        80           DO 2  I=1,N
>        90        2  AVG = AVG/N
>       100           RETURN
>       110           END
>       200     $DATA
>       205     2,1.5,3.0
>       210     3 1. 2. 3.
>       215     $STOP
# END OF FILE
```

The source file appears to be in order for processing by the WATFIV compiler, available in a public file named *WATFIV. *WATFIV reads the source program from a logical input device called SCARDS and writes its source program listings, any diagnostics, and the results of object program execution that are written with the PRINT statement, on a logical output device called SPRINT. In this case, SCARDS will be equivalent to the source file WPROG; let the printed output be written into a temporary file named —OUT, and then copy the contents of —OUT to the terminal.

The WATFIV compiler numbers each statement; normally these numbers will differ from both the line numbers in the source file and any FORTRAN statement numbers in the WATFIV program. In WATFIV parlance, these numbers are called ISN's (internal statement numbers) and are used in conjunction with some of the diagnostic messages. For example, the WATFIV statement with the FORTRAN statement number 2 has line number 90 in the file WPROG, and has been assigned the internal statement number 9 by the compiler.

```
#$RUN *WATFIV SCARDS=WPROG SPRINT=-OUT
#EXECUTION BEGINS

  EXECUTION SUPPRESSED...
#EXECUTION TERMINATED
#$COPY -OUT
> $COMPILE LIST
>       1        1  READ, N, (X(I), I=1,N)
> ***ERROR***   INVALID ELEMENT IN INPUT LIST OR DATA LIST
>       2           PRINT, N, AVG(N,X), (X(I), I=1,N)
> *EXTENSION*  OTHER COMPILERS MAY NOT ALLOW EXPRESSIONS IN OUTPUT LISTS
> *EXTENSION*  OTHER COMPILERS MAY NOT ALLOW EXPRESSIONS IN OUTPUT LISTS
>       3           GO TO 2
>       4           END
> ***ERROR***  MISSING STATEMENT NUMBER       2 USED IN LINE       3

>       5           FUNCTION AVG(N,X)
>       6           DIMENSION X(20)
>       7           AVG = 0.0.
> ***ERROR***   ILLEGAL USE OF DECIMAL POINT.UNEXPECTED . BEFORE END-OF-STATEMENT
>       8           DO 2  I=1,N
>       9        2  AVG = AVG/N
>      10           RETURN
>      11           END
> ***ERROR***  SUBPROGRAM X       USED IN LINE       1 IS MISSING
```

The diagnostic error messages (see Section 3.39 and Appendix F) can be interpreted as follows:

ERROR INVALID ELEMENT IN INPUT LIST OR DATA LIST

This is error number IO-2 from Appendix F. Since X has not been dimensioned prior to the READ statement, X is thought to be the name of a function. It is invalid to read a data value for a function reference. This error comment will be eliminated once the array X is properly dimensioned. Hence, the READ statement need *not* be changed.

EXTENSION OTHER COMPILERS MAY NOT ALLOW
EXPRESSIONS IN OUTPUT LISTS

This comment (message IO-3 in Appendix F) appears twice following the PRINT statement, once to indicate that the function reference AVG(N,X) appears, and the second time to indicate that the (assumed) function reference X(I) appears. These are allowable WATFIV extensions to IBM FORTRAN, and hence are flagged with extension rather than error messages. Once X is properly dimensioned, the second of these comments will not appear.

```
***ERROR***  MISSING STATEMENT NUMBER 2 USED IN LINE 3
```

The statement with internal statement number 3 (GO TO 2) refers to a WATFIV statement with FORTRAN statement number 2, but no such statement number appears in the main program. The offending statement should be GO TO 1. This error message (ST-0 in Appendix F) indicates that the error is serious enough to prevent compilation of a complete object program.

```
***ERROR***  ILLEGAL USE OF DECIMAL POINT. UNEXPECTED
             .  BEFORE END-OF-STATEMENT
```

This error message (CN-5 in Appendix F) appears because the constant 0.0. in the statement with internal statement number 7 contains two decimal points.

```
***ERROR***  SUBPROGRAM X USED IN LINE 1 IS MISSING
```

Because X was not dimensioned in the main program, it is assumed to be a function by the WATFIV compiler. Since no function named X was supplied as part of the source file or was found in the WATFIV subprogram library, X is termed a "missing" function. This comment (SR-0 in Appendix F) will be eliminated once X is properly dimensioned.

Make the indicated corrections to the file WPROG, empty the file −OUT to remove the current content, call *WATFIV to compile and execute the program as before, and copy the output file −OUT.

```
#1.5        DIMENSION X(20)
#30         GO TO 1
#70         AVG = 0.0
#$EMPTY -OUT
#DONE.


#$RUN *WATFIV SCARDS=WPROG SPRINT=-OUT
#EXECUTION BEGINS

   EXECUTION BEGINS...
#EXECUTION TERMINATED


#$COPY -OUT
> $COMPILE LIST
>       1           DIMENSION X(20)
>       2       1   READ, N, (X(I), I=1,N)
>       3           PRINT, N, AVG(N,X), (X(I), I=1,N)
> *EXTENSION*  OTHER COMPILERS MAY NOT ALLOW EXPRESSIONS IN OUTPUT LISTS
>       4           GO TO 1
>       5           END

>       6           FUNCTION AVG(N,X)
>       7           DIMENSION X(20)
>       8           AVG = 0.0
>       9           DO 2  I=1,N
>      10       2   AVG = AVG/N
>      11           RETURN
>      12           END
>              2     0.0000000E 00     0.1500000E 01     0.3000000E 01
>              3     0.0000000E 00     0.1000000E 01     0.2000000E 01     0.3000000E 01
> **ENDFILE**  END OF FILE ENCOUNTERED ON UNIT SCARDS    (IBM CODE IHC217)

> PROGRAM WAS AT LINE     2 IN ROUTINE M/PROG WHEN TERMINATION OCCURRED
```

This time there are no compilation errors, but the calculated results are incorrect; the program contains one or more *logical* errors, but no syntactical errors. The message

```
**ENDFILE**  END OF FILE ENCOUNTERED (IBM CODE IHC217)
```

indicates that all the data for the program have been processed, as described in Section 3.22. The program terminated with the READ statement with internal statement number 2 in the main program (M/PROG) when the $STOP compiler command was encountered in the file WPROG.

The remaining errors are both in the function AVG. Line 90 in the file WPROG is in error, and an additional statement is needed between the statements in lines 90 and 100. Making these insertions, listing the modified source file WPROG, and calling on the compiler *WATFIV to compile the new source program and to print the output directly on the terminal (SPRINT defaults to *SINK* if unassigned) leads to the following:

```
#90    2  AVG = AVG + X(I)
#95       AVG = AVG/N
#$LIST WPROG
>        1        $COMPILE LIST
>        1.5            DIMENSION X(20)
>       10            1 READ, N, (X(I), I=1,N)
>       20              PRINT, N, AVG(N,X), (X(I), I=1,N)
>       30              GO TO 1
>       40              END
>       50              FUNCTION AVG(N,X)
>       60              DIMENSION X(20)
>       70              AVG = 0.0
>       80              DO 2  I=1,N
>       90            2 AVG = AVG + X(I)
>       95              AVG = AVG/N
>      100              RETURN
>      110              END
>      200        $DATA
>      205        2,1.5,3.0
>      210        3 1. 2. 3.
>      215        $STOP
# END OF FILE

#$RUN *WATFIV SCARDS=WPROG
#EXECUTION BEGINS
  $COMPILE LIST
       1            DIMENSION X(20)
       2          1 READ, N, (X(I), I=1,N)
       3            PRINT, N, AVG(N,X), (X(I), I=1,N)
  *EXTENSION*  OTHER COMPILERS MAY NOT ALLOW EXPRESSIONS IN OUTPUT LISTS
       4            GO TO 1
       5            END

       6            FUNCTION AVG(N,X)
       7            DIMENSION X(20)
       8            AVG = 0.0
       9            DO 2  I=1,N
      10          2 AVG = AVG + X(I)
      11            AVG = AVG/N
      12            RETURN
      13            END

  EXECUTION BEGINS...
                2    0.2250000E 01    0.1500000E 01    0.3000000E 01
                3    0.2000000E 01    0.1000000E 01    0.2000000E 01    0.3000000E 01
  **ENDFILE**  END OF FILE ENCOUNTERED ON UNIT SCARDS   (IBM CODE IHC217)

  PROGRAM WAS AT LINE      2 IN ROUTINE M/PROG WHEN TERMINATION OCCURRED
#EXECUTION TERMINATED
```

This time the results are correct.

Signing Off. To terminate processing of the time-sharing job, the MTS command is $SIGNOFF.

```
#$SIGNOFF
#OFF AT 11:27.44      01-03-72
#ELAPSED TIME         39.784 MIN.        $1.98
#CPU TIME USED        34.563 SEC.        $3.00
#CPU STOR VMI         15.7   PAGE-MIN.    $.82
#WAIT STOR VMI         5.525 PAGE-HR.
#DRUM READS           76
#APPROX. COST OF THIS RUN IS     $5.79
#DISK STORAGE       1676.116 PAGE-HR.     $.50
#APPROX. REMAINING BALANCE:    $551.41
```

Here, "ELAPSED TIME" is the time between the execution of the $SIGNON and $SIGNOFF commands. "CPU TIME USED" is the sum of all the time slices during which object programs associated with the job were being executed by one of the processors; this includes the running of system programs such as the file-editing routines, loading programs, etc. "CPU STOR VMI" or "processor control virtual memory integral" is a measure of storage occupancy, called *'tenancy'* in MTS. This is the sum of the product of the virtual memory occupancy in pages times the length of each time slice assigned the job, expressed in units called page-minutes; thus it is a measure of the fast memory occupied while the job is in control of one of the processors. The "WAIT STOR VMI" is a similar integral computed during periods when the job is *not* in control of one of the processors (there is no charge for this kind of tenancy). "DRUM READS" is the number of paging operations (see Section 4.4) occurring between the drum memories and the fast memory during execution of the job.

The total charge of $5.79 covers all activity at the terminal during execution of the job, and is typical of the $10.00 to $12.00 per hour for running time-shared jobs consisting of a mixture of file-editing tasks, compilations, and short debugging runs. "DISK STORAGE" is the charge for storage of all private files stored under the user number incurred since the termination of the last job run under the same number; one page of file storage saved for one hour incurs a secondary storage tenancy of one page-hour. "APPROX. REMAINING BALANCE" is the account balance after deduction of the charge for the job and for the storage of private files.

As of early 1973, the charges for running a time-sharing job on the Michigan computer are:

Terminal connection time:	$ 3.30/hour
Central processing time:	$ 0.079/sec
CPU storage tenancy:	$ 0.047/page-min
Disk storage:	$ 0.000167/page-hour

CHAPTER 5

Solution of Single Equations

5.1 General Remarks on Numerical Methods

The techniques known as *numerical methods* deal with the quantitative solutions of problems that have been formulated algebraically, but whose solutions in straightforward algebraic terms are either difficult, tedious, or impossible.

Consider the general quadratic equation

$$ax^2 + bx + c = 0, \qquad (5.1)$$

which has the algebraic solutions

$$x_1 = \frac{-b + \sqrt{b^2 - 4ac}}{2a}, \qquad x_2 = \frac{-b - \sqrt{b^2 - 4ac}}{2a}. \qquad (5.2)$$

If $a = 1$, $b = -1$, and $c = -6$, the mere substitution of these numbers into (5.2), giving $x_1 = 3$ and $x_2 = -2$, would constitute a numerical *solution* of (5.1). It would hardly qualify as a numerical *method*, because little more than the evaluation of a formula is involved (*assuming* that a procedure for taking square roots is already available).

But what if we are confronted with the equation

$$a \sin^2 x + bx + c = 0, \qquad (5.3)$$

and are asked to find a numerical solution for the same values of a, b, and c as above? Since there is no ready-made algebraic solution—and none is readily apparent—we must attempt to devise a procedure of our own, which leads us immediately into the realm of *numerical methods*.

Numerical methods can be applied to a much wider variety of problems than the solution of a single algebraic equation such as (5.3). However, we choose to introduce the various techniques as simply as possible, and the next few sections will involve only the solution of *single* equations in a single variable. In subsequent chapters, we shall study a greater range of topics, including the solution of systems of *simultaneous* equations, the approximation of functions, the evaluation of integrals, the solution of ordinary differential equations, and certain optimization techniques.

Numerical methods usually possess the following characteristics:

(a) They frequently yield only an *approximation* to the exact solution of a problem. However, this approximation can be *refined* if we are prepared to expend more computational effort in order to obtain better accuracy.

(b) They are conceptually fairly simple, not involving an elaborate knowledge of mathematics, and can be expressed concisely in algorithmic form.

(c) They are readily adaptable to implementation on a digital computer.

(d) They occasionally fail—this point will be illustrated when discussing the individual methods.

5.2 Single Equations

Single equations in a single variable or unknown, such as x, can always be expressed as

$$f(x) = 0, \qquad (5.4)$$

where $f(x)$ is the function that results from taking all terms over to the left-hand side. Thus, an equation such as $x^2 = 2x^3 - 3$ can be rewritten as $f(x) = x^2 - 2x^3 + 3 = 0$, or as $f(x) = x - 2x^2 + 3/x = 0$, and so on. We shall be concerned with finding the *root* (or roots) of (5.4)—that is, finding the numerical value (or values) of x that satisfy the equation. A root of the equation $f(x) = 0$ is also called a *zero* of the function $f(x)$. Even though $f(x) = 0$ may have several roots, we may be interested in evaluating only a particular one of them, depending on the problem at hand.

Typical situations are shown graphically in Fig. 5.1; these could involve: (a) a single root $x = \alpha$, (b) two roots $x = \alpha_1$ and $x = \alpha_2$, and (c) no real root.

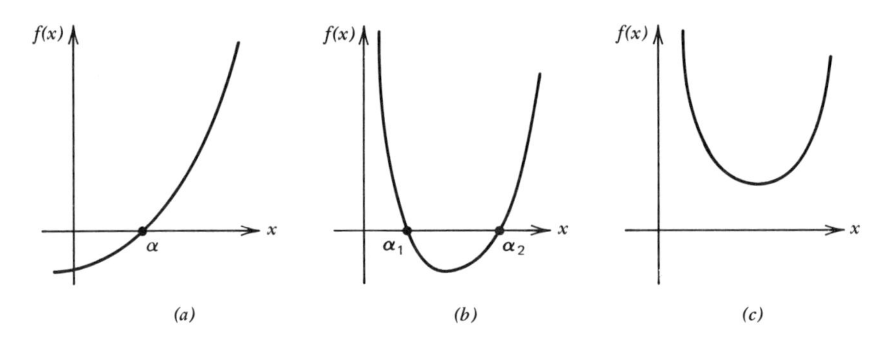

Figure 5.1 Roots of $f(x) = 0$.

5.3 Method of Successive Substitutions

We can attempt to solve $f(x) = 0$ by rewriting it in the form

$$x = F(x), \qquad (5.5)$$

where $F(x)$ is a new function of x. For example, the equation $f(x) = x^2 + 3e^{-x} - 7.2 = 0$ can be reexpressed in a variety of ways, such as

$$x = \sqrt{7.2 - 3e^{-x}},$$

$$x = x^2 + 3e^{-x} - 7.2 + x,$$

$$x = -\ln\left(2.4 - \frac{x^2}{3}\right),$$

and so on. Suppose that an initial or first approximation, x_1, is available for a root of (5.5). Evaluation of the right-hand side of (5.5) using x_1 would then provide a second (but not necessarily better) approximation x_2 to the root:

$$x_2 = F(x_1).$$

The procedure can clearly be repeated, giving

$$x_3 = F(x_2),$$

$$x_4 = F(x_3), \text{ etc.}$$

The general algorithm, which is iterative, is

$$x_k = F(x_{k-1}), \qquad k = 2, 3, \ldots \qquad (5.6)$$

Under certain conditions, demonstrated below, the values x_k will change less and less from one iteration to the next, and for all practical purposes the sequence will eventually keep on producing the same number for k sufficiently large. In other words, x_k will have *converged* to a value (α, for example) that satisfies the equation $x = F(x)$ and is therefore also a root of the original equation $f(x) = 0$.

The above procedure, known as the *method of successive substitutions*, is illustrated in Fig. 5.2*a*, which shows how the successive values of x might approach the root α. The staircase-type construction is based on: (a) the curve $F(x)$ versus x, and (b) a line that passes through the origin and is inclined at 45° to the x axis. The reader should convince himself that this graphical interpretation does indeed correspond to equation (5.6). A similar construction holds for an initial approximation that lies below the root (that is, $x_1 < \alpha$).

Another possible situation is illustrated in Fig. 5.2*b*; even though x_1 is quite close to α, each iteration takes us further *away*

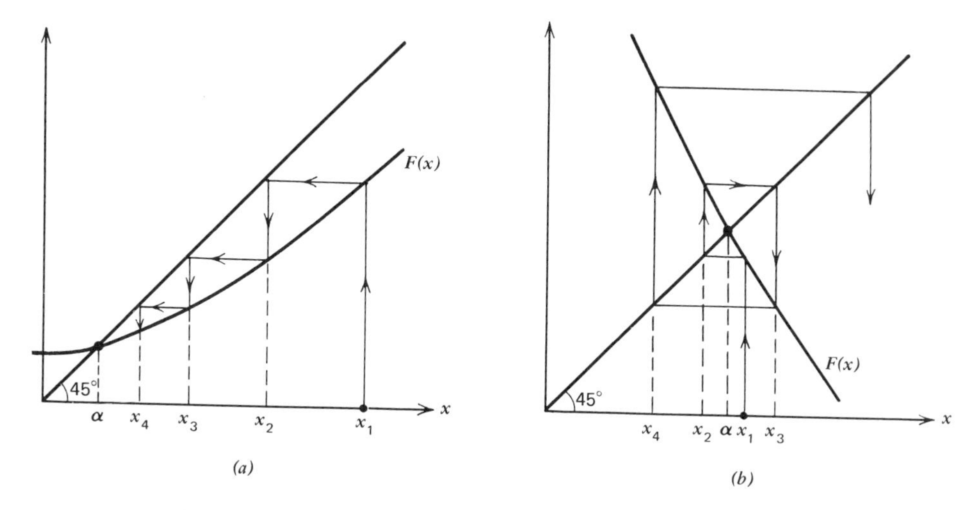

Figure 5.2 Method of successive substitutions.

from the root, and the method does not converge (it *diverges*) in this case. The reader can easily verify by geometrical construction that the steepness of the $F(x)$ curve is the key factor in determining convergence or divergence. In fact, convergence is guaranteed if the magnitude of the gradient or slope of $F(x)$ is less than one everywhere in the region between the initial approximation and the root. Thus the *convergence criterion* is

$$\left|\frac{dF}{dx}\right| < 1 \qquad \text{for} \qquad |x-\alpha| \leqslant |x_1-\alpha|. \tag{5.7}$$

The quantity $dF(\alpha)/dx$ is called the *asymptotic convergence factor*. In general, if the asymptotic convergence factor is near zero, convergence will be rapid, whereas if it is near one in absolute value, convergence will be slow.

The method of successive substitutions is also readily illustrated by the flow diagram shown in Fig. 5.3. Note that provision is made for accepting an initial value x_1 for x. The calculations are terminated here when either of two conditions is satisfied:

(1) A prescribed number, n, of iterations has been performed. (If convergence is not occurring within a reasonable number of steps, useless calculations are thereby avoided.)
(2) The computed value of x does not change fractionally from one iteration to the next by more than a prescribed small value ε, such as $\varepsilon = 0.001$, or

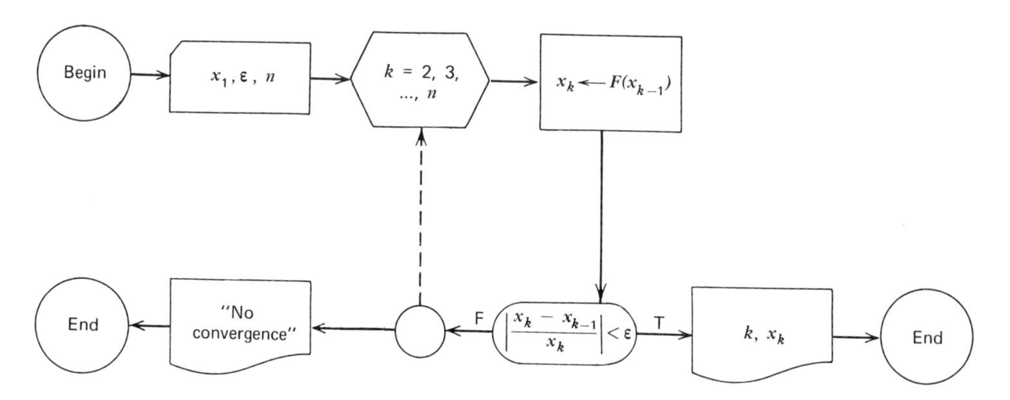

Figure 5.3 Flow diagram for method of successive substitutions.

even less. (While this does not *guarantee* that we are close to the root, the odds are strongly in favor of being so.) Note that this test cannot be used when the root is zero or close to zero; in this case, a comparison of $|x_k - x_{k-1}|$ against ε would be appropriate. Another alternative test would be to compare $|f(x_k)|$ against ε.

If the flow diagram of Fig. 5.3 were translated into a computer program, n storage locations would be implied for x_1 through x_n. Since there is little point in saving every such approximation except the most recent, we can economize greatly on storage by abandoning the subscripts and using the equivalent form shown in Fig. 5.4. Note that although x always contains the most recent estimate of the root, a second variable, x^*, is needed for saving temporarily the immediately preceding value before making the convergence test.

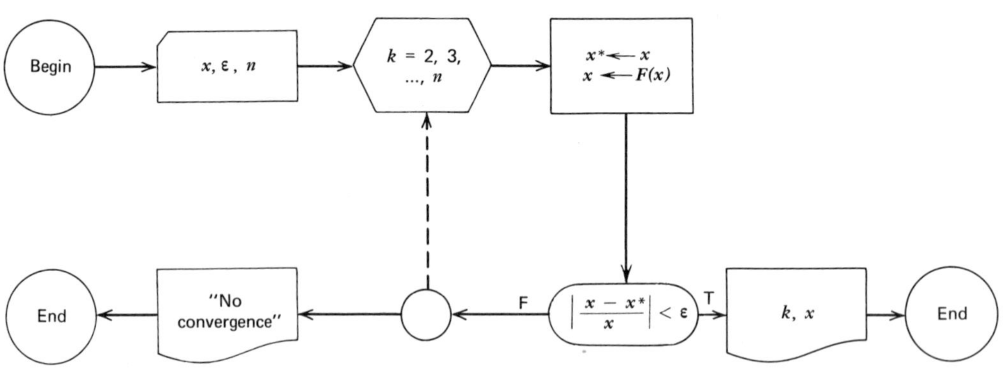

Figure 5.4 Alternative flow diagram for method of successive substitutions.

The method of successive substitutions has the merit of simplicity, but the rate of convergence is usually appreciably slower than that resulting from Newton's method (see Section 5.4). It also has the disadvantage that success hinges on finding a suitable rearrangement of the original equation (in this connection, see Problem 5.17).

EXAMPLE 5.1

Van der Waals' equation of state for an imperfect gas is

$$\left(P+\frac{a}{v^2}\right)(v-b) = RT, \qquad (5.8)$$

where P is the absolute pressure (atm), v is the molal volume (liters/g mole), T is the absolute temperature (°K), R is the gas constant (0.082054 liter atm/g mole °K), and a and b are constants that depend on the particular gas. If $P = 200$ atm and $T = 500$°K, evaluate the molal volume of carbon dioxide, for which [10] $a = 3.592$ liter² atm/(g mole)² and $b = 0.04267$ liter/g mole.

A rearrangement of Van der Waals' equation that conforms to (5.5) is

$$v = \frac{RT}{P+\dfrac{a}{v^2}} + b, \qquad (5.9)$$

that is,

$$v = \frac{41.027}{200+3.592/v^2} + 0.04267 = F(v). \qquad (5.10)$$

As a plausible initial approximation, we take the value of v that would be predicted by the ideal gas law ($a = 0, b = 0$); that is, $v_1 = 41.027/200 = 0.205135$. From (5.10), the second approximation becomes

$$v_2 = F(v_1) = F(0.205135) \doteq 0.186443.$$

Likewise, the third approximation is

$$v_3 = F(v_2) = F(0.186443) \doteq 0.177923.$$

The result of 20 such iterations is shown in Table 5.1. Although the computations have been performed using about ten-figure accuracy on a desk calculator, only the first six significant digits are given in the table. Even this degree of precision is much more than would actually be warranted by Van der Waals' equation, which is itself only an approximation to the physical behavior of gases. However, we have retained the extra digits to help illustrate the convergence of the *method.*

Table 5.1 Molal Volume at Each Iteration

Iteration number, k	Molal volume, v_k	Iteration number, k	Molal volume, v_k
1	0.205135	11	0.168191
2	0.186443	12	0.168143
3	0.177923	13	0.168115
4	0.173551	14	0.168099
5	0.171178	15	0.168090
6	0.169852	16	0.168084
7	0.169098	17	0.168081
8	0.168667	18	0.168079
9	0.168418	19	0.168078
10	0.168274	20	0.168078

EXAMPLE 5.2

SOLUTION OF VAN DER WAALS' EQUATION

Introduction and Problem Statement

Read Example 5.1, and then write a computer program that will implement the method. The intention is to introduce the reader to the computer application of numerical methods by programming an algorithm that has already been performed by hand.

Method of Solution

The key step of the algorithm is taken from equation (5.9):

$$v \leftarrow \frac{RT}{P + a/v^2} + b.$$

The procedure is patterned closely on the flow diagram of Fig. 5.4, which obviates the need for subscripts and thereby economizes on storage. Note the following additional features in the flow diagram below:

(a) The initial approximation is not read as data, but is estimated (as in Example 5.1) from the ideal gas law: $v = RT/P$.

(b) Values for a, b, P, T, R, and *set* (a run identification number) are included in the data, so that the program could be used for a variety of gases under various conditions.

(c) For additional clarity, the output includes a reprint of the input data, plus the values of the molal volume at each iteration.

Flow Diagram

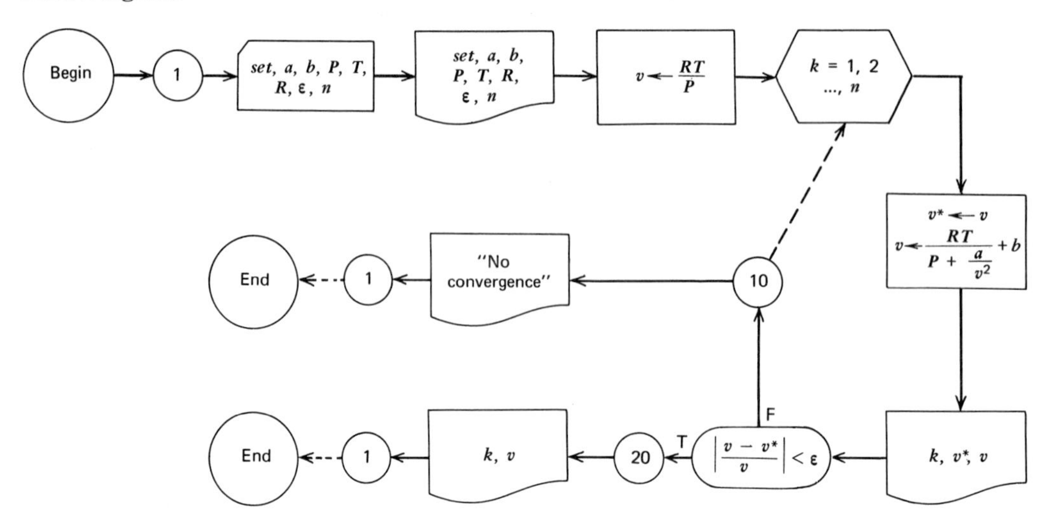

FORTRAN Implementation

List of Principal Variables

Program Symbol	Definition
A, B	Constants a, liter^2 atm/(g mole)^2, and b, liter/g mole, in the Van der Waals' equation.
EPS	Tolerance, ε, used in convergence test.
K	Iteration counter, k.
N	Maximum allowable number of iterations, n.
P	Absolute pressure, P, atm.
R	Gas constant, R, 0.082054 liter atm/g mole °K.
SET	Integer run identification number, *set*.
T	Absolute temperature, T, °K.
V	Latest estimate of molal volume, v, liter/g mole.
VSTAR	Temporary storage for previous estimate of molal volume, v^*.

The following program achieves format-free input by using a NAMELIST declaration in conjunction with an appropriate input statement. The abbreviated second set of data emphasizes that the entire list need not be read; those variables already assigned values in the first set will automatically retain their previous values. The DO loop below ends with a CONTINUE statement; it would be illegal to delete this statement and move the statement number 10 up to the previous IF statement because FORTRAN does not allow a DO loop to terminate in a statement that involves a transfer. This would be legal in WATFIV, however (see Section 3.17).

Program Listing

```
C         DIGITAL COMPUTING                    EXAMPLE 5.2
C         SOLUTION OF VAN DER WAALS'S EQUATION BY SUCCESSIVE SUBSTITUTION
C
      INTEGER SET
      NAMELIST /DATA/ A, B, P, T, R, EPS, SET, N
C
C     ..... PRINT HEADING, READ & CHECK INPUT DATA .....
      WRITE (6,200)
    1 READ (5,DATA,END=999,ERR=1)
      WRITE (6,201) SET, A, B, P, T, R, EPS, N
C
C     ..... PERFORM SUCCESSIVE SUBSTITUTION ITERATIONS .....
      WRITE (6,202)
      V = R*T/P
      DO 10  K = 1, N
      VSTAR = V
      V = R*T/(P + A/V**2) + B
      WRITE (6,203) K, VSTAR, V
C
C     ..... TEST FOR CONVERGENCE .....
      IF (ABS((V - VSTAR)/V) .LT. EPS)  GO TO 20
   10 CONTINUE
C
C     ..... PRINT OUT NO CONVERGENCE MESSAGE .....
      WRITE (6,204)
      GO TO 1
C
C     ..... PRINT OUT FINAL CONVERGED SOLUTION .....
   20 WRITE (6,205) K, V
      GO TO 1
C
  999 CALL EXIT
C
C     ..... FORMATS FOR OUTPUT STATEMENTS .....
  200 FORMAT ('1DIGITAL COMPUTING                  EXAMPLE 5.2'/ ' SOLUTIO
     1N OF VAN DER WAALS''S EQUATION BY SUCCESSIVE SUBSTITUTIONS.')
  201 FORMAT ('0'/'0THE INPUT VALUES FOR DATA SET NO.', I3, ' ARE'/
     1        '0    A     =', F9.3 /'     B     =', F11.5/
     2        '     P     =', F7.1 /'     T     =', F7.1 /
     3        '     R     =', F12.6/'     EPS   =', E11.1/
     4        '     N     =',I5)
  202 FORMAT ('0VALUES DURING SUCCESSIVE ITERATIONS ARE'/
     1        '0    K        VSTAR           V'/ ' ')
  203 FORMAT (I6, 2F11.6)
  204 FORMAT ('0NO CONVERGENCE')
  205 FORMAT ('0SOLUTION CONVERGED, WITH'/
     1        '0    K     =' I5/ '     V     =', F12.6)
C
      END
```

Data

```
&DATA A = 3.592, B = 0.04267, P = 200., T = 500., R = 0.082054,
EPS = 1.0E-4, N = 20, SET = 1 &END
&DATA EPS = 1.0E-4, N = 10, SET = 2 &END
```

Computer Output

```
DIGITAL COMPUTING                    EXAMPLE 5.2
SOLUTION OF VAN DER WAALS'S EQUATION BY SUCCESSIVE SUBSTITUTIONS.


THE INPUT VALUES FOR DATA SET NO.  1 ARE

     A      =     3.592
     B      =     0.04267
     P      =   200.0
     T      =   500.0
     R      =     0.082054
     EPS    =     0.1E-03
     N      =    20

VALUES DURING SUCCESSIVE ITERATIONS ARE

     K      VSTAR          V

     1   0.205135     0.186442
     2   0.186442     0.177923
     3   0.177923     0.173551
     4   0.173551     0.171178
     5   0.171178     0.169851
     6   0.169851     0.169098
     7   0.169098     0.168666
     8   0.168666     0.168418
     9   0.168418     0.168274
    10   0.168274     0.168191
    11   0.168191     0.168143
    12   0.168143     0.168115
    13   0.168115     0.168099

SOLUTION CONVERGED, WITH

     K      =    13
     V      =     0.168099


THE INPUT VALUES FOR DATA SET NO.  2 ARE

     A      =     3.592
     B      =     0.04267
     P      =   200.0
     T      =   500.0
     R      =     0.082054
     EPS    =     0.1E-03
     N      =    10

VALUES DURING SUCCESSIVE ITERATIONS ARE

     K      VSTAR          V

     1   0.205135     0.186442
     2   0.186442     0.177923
     3   0.177923     0.173551
     4   0.173551     0.171178
     5   0.171178     0.169851
     6   0.169851     0.169098
     7   0.169098     0.168666
     8   0.168666     0.168418
     9   0.168418     0.168274
    10   0.168274     0.168191

NO CONVERGENCE
```

Discussion of Results

The computations are for carbon dioxide at 200 atm and 500°K—the same as in Example 5.1. The main modification is the incorporation of a convergence test: both data sets 1 and 2 seek less than a 10^{-4} fractional change in the computed value of v. Convergence within this criterion is achieved for the first data set (maximum allowable number of iterations is $n = 20$) but not for the second ($n = 10$), even though the last computed value of $v = 0.168191$ would be acceptable for most purposes. This emphasizes that the values assigned to ε and n are somewhat arbitrary, and the programmer must exercise judgment, based on his experience, as to what he thinks is reasonable.

The values computed in Example 5.2 are not identical with those of Example 5.1 (examine the results for $k = 2$, 6, and 8). The reason is that the desk calculator used in Example 5.1 was capable of ten-digit accuracy; the digital computer used in Example 5.2 is capable of no more than seven-digit accuracy in standard (single-precision) arithmetic.

Note that the above program could be reduced to about half its length by omitting the comment cards and using less elaborate output formats. However, we firmly believe that such embellishments are worthwhile for any program that is to have more than a transient value.

5.4 Newton's Method

Newton's method, illustrated in Fig. 5.5, adopts the following procedure for approximating a root $x = \alpha$ of $f(x) = 0$. Start at point A on the x axis, corresponding to an initial approximation x_1, and

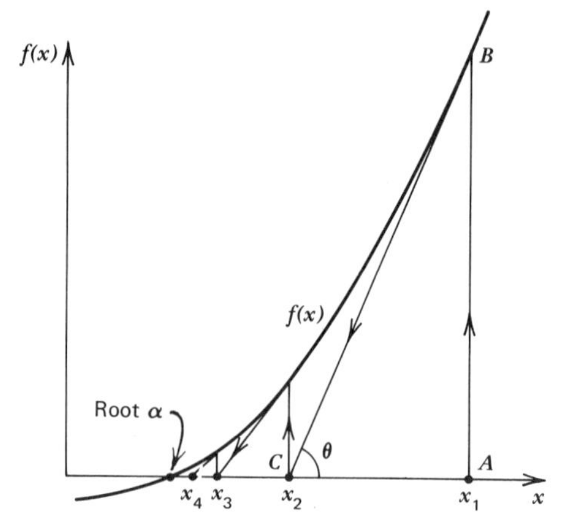

Figure 5.5 Newton's method.

erect a perpendicular to meet the curve $f(x)$ at B. Draw the tangent to the curve at B and project to meet the x axis at C, thus giving a second approximation x_2. Repeat the procedure to give subsequent approximations x_3, x_4, etc. Hopefully, the process will eventually converge to the root $x = \alpha$.

To obtain an algorithm in algebraic terms, first note that $AB = f(x_1)$ and $AC = x_1 - x_2$, so that

$$\tan\theta = \frac{AB}{AC} = \frac{f(x_1)}{x_1 - x_2}.$$

But $\tan\theta$ is also the slope of the tangent BC at B. That is,

$$\tan\theta = \left(\frac{df}{dx}\right)_{x=x_1} = \frac{f(x_1)}{x_1 - x_2}.$$

Abbreviating the derivative as $f'(x_1)$, we find that

$$x_2 = x_1 - \frac{f(x_1)}{f'(x_1)}.$$

Similarly,

$$x_3 = x_2 - \frac{f(x_2)}{f'(x_2)},$$

and, in general,

$$x_k = x_{k-1} - \frac{f(x_{k-1})}{f'(x_{k-1})}, \qquad k = 2, 3, \ldots. \tag{5.11}$$

Alternatively, we can dispense with subscripts and express the algorithm for Newton's method as:

$$x \leftarrow x - \frac{f(x)}{f'(x)}. \tag{5.12}$$

Here, the dotted arrow emphasizes that as soon as a new value of x is computed, it is immediately cycled back into the right-hand side, generating yet another value of x, and so on.

Note that Newton's method has the general form of (5.6) with $F(x_{k-1}) = x_{k-1} - [f(x_{k-1})/f'(x_{k-1})]$. The corresponding asymptotic convergence factor is thus $f(\alpha)f''(\alpha)/[f'(\alpha)]^2 = 0$ for simple roots, in which case $f'(\alpha) \neq 0$. Hence, at the price of having to compute a derivative value at each iteration, we would expect Newton's method to converge fairly rapidly to a root of $f(x) = 0$ in the majority of problems. However, in common with most numerical methods, it may fail occasionally in certain cases. A possible initial oscillation followed by a jump away from a root is illustrated in Fig. 5.6. Note, however, that the method *would* have converged in this case if the initial approximation had been somewhat closer to the root.

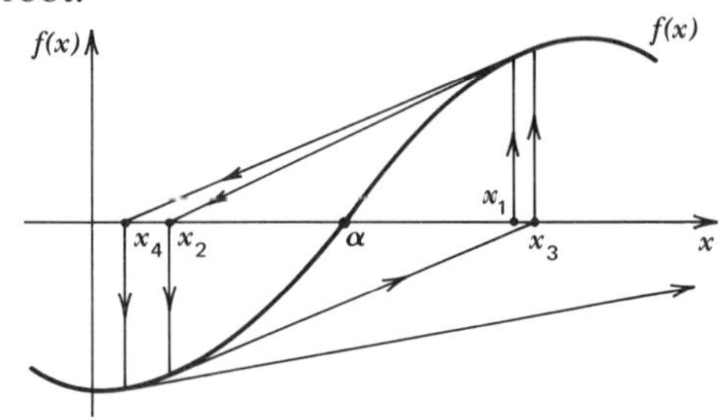

Figure 5.6 Failure of Newton's method.

EXAMPLE 5.3

Use Newton's method to generate an algorithm for finding the square root x of a positive number a:

$$x = \sqrt{a},$$

that is,

$$f(x) = x^2 - a = 0. \qquad (5.13)$$

From equation (5.13), by differentiation,

$$\frac{df}{dx} = f'(x) = 2x,$$

so that the algorithm of (5.12) becomes

$$x \leftarrow x - \frac{x^2 - a}{2x} = \frac{x^2 + a}{2x}.$$

That is,

$$x \leftarrow \frac{1}{2}\left(x + \frac{a}{x}\right). \qquad (5.14)$$

The algorithm of (5.14) is incorporated into the flow diagram of Fig. 5.7, which also includes provision for terminating the calculations if either: (a) a specified number of iterations, n, has been performed, or (b) the fractional change in x from one step to the next falls below a small specified tolerance ε.

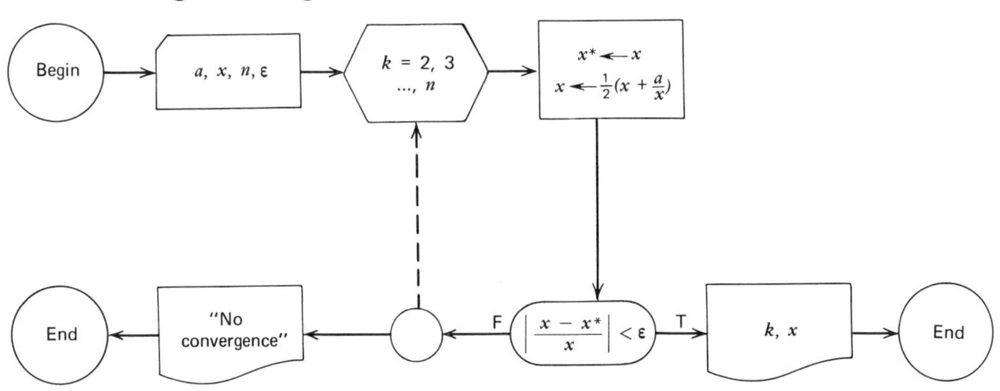

Figure 5.7 Flow diagram for finding $\sqrt{a}$ by Newton's method.

Now consider specifically $a = 100$, so that (5.14) becomes

$$x \leftarrow \frac{1}{2}\left(x + \frac{100}{x}\right).$$

We must also make an initial approximation or "starting guess" for x. Suppose we make the rather poor guess that $x \doteq a = 100$. The next value for x generated by the algorithm is

$$x \leftarrow \frac{1}{2}\left(100 + \frac{100}{100}\right) = 50.5.$$

Working to six decimal places, with $n = 10$ and $\varepsilon = 0.001$, subsequent iterations along the lines of Fig. 5.7 generate the following sequence:

Iteration Number, k	Approximation to Root, x
1	100
2	50.5
3	26.240099
4	15.025530
5	10.840435
6	10.032579
7	10.000053
8	10.000000

5.5 Half-Interval Method

Another technique with a simple graphical explanation is illustrated in Fig. 5.8.

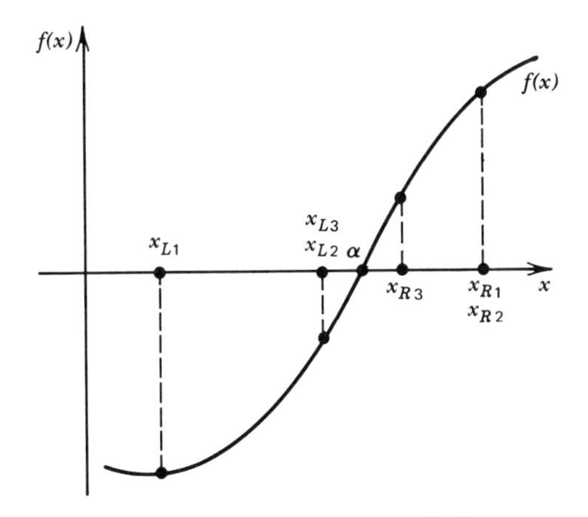

Figure 5.8 Half-interval method.

It gives a root of $f(x) = 0$ if starting values x_{L1} and x_{R1} are known such that $f(x_{L1})$ and $f(x_{R1})$ are opposite in sign. For continuous functions, the number $f((x_{L1}+x_{R1})/2)$, being the value of the function at the half-way point $x = (x_{L1}+x_{R1})/2$, will either be zero, or have the sign of $f(x_{L1})$, or have the sign of $f(x_{R1})$. In the event that the value is not zero, a second pair x_{L2} and x_{R2} can be chosen from the three numbers x_{L1}, x_{R1}, and $(x_{L1}+x_{R1})/2$ so that $f(x_{L2})$ and $f(x_{R2})$ are opposite in sign, while $|x_{L2}-x_{R2}| = |x_{L1}-x_{R1}|/2$. Continuing in this manner for several steps, we can obtain a much-reduced interval within which a root α must lie. Note that the method will yield only *one* root even though there may conceivably be 3, 5, 7, ... roots within the starting interval.

Because each new application of the iterative scheme reduces by half the length of the interval in x known to contain α, this procedure is called the *half-interval method.* Note that since the interval of uncertainty is always known, we can specify *a priori* the number of iterations required to locate the root within a prescribed tolerance. If Δ is the starting interval, then the number n of interval-halving operations required to reduce the interval of uncertainty to within δ can readily be shown to equal

$$n = \frac{\ln(\Delta/\delta)}{\ln 2}, \tag{5.15}$$

rounded, if necessary, up to the next higher integer.

5.6 Regula-Falsi Method

The final technique to be discussed here for finding a root of $f(x) = 0$ is the *regula-falsi method*, also known as *linear inverse interpolation.* Referring to Fig. 5.9, let x_{L1} and x_{R1} be numbers such that $f(x_{L1})$ and $f(x_{R1})$ are opposite in sign. Let x_2 be the abscissa of the point of intersection of the

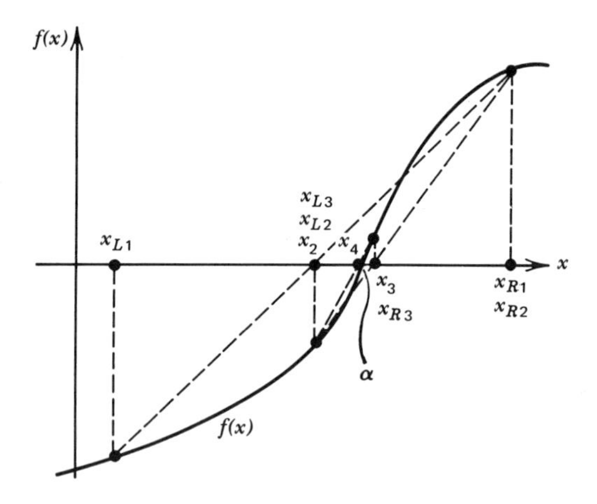

Figure 5.9 Regula-falsi method.

x axis and the chord joining the points $(x_{L1}, f(x_{L1}))$ and $(x_{R1}, f(x_{R1}))$; from geometrical considerations (similar triangles), we have

$$\frac{x_2 - x_{L1}}{x_{R1} - x_2} = \frac{-f(x_{L1})}{f(x_{R1})}.$$

That is,

$$x_2 = \frac{x_{L1}f(x_{R1}) - x_{R1}f(x_{L1})}{f(x_{R1}) - f(x_{L1})}. \tag{5.16}$$

If $f(x_2) = 0$, the process terminates with a zero of $f(x)$. If $f(x_2)$ has the same sign as $f(x_{R1})$, the next step is to choose $x_{L2} = x_{L1}$ and $x_{R2} = x_2$. On the other hand, if $f(x_2)$ has the same sign as $f(x_{L1})$, the next step is to choose $x_{L2} = x_2$ and $x_{R2} = x_{R1}$. The process is continued in the same manner, creating the sequence of pairs (x_{Lk}, x_{Rk}) and the accompanying number x_k, which will hopefully converge to the root α.

EXAMPLE 5.4

PERFORMANCE OF CENTRIFUGAL PUMP (*REGULA-FALSI* METHOD)

Introduction

As shown in Fig. 5.4.1, a centrifugal pump is used for transferring a liquid from one tank to another, with both tanks at the same level. The pump raises the pressure of the liquid from p_1 (atmospheric pressure) to p_2, but this pressure is gradually lost because of friction inside the long pipe, and, at the exit, p_3 is back down to atmospheric pressure.

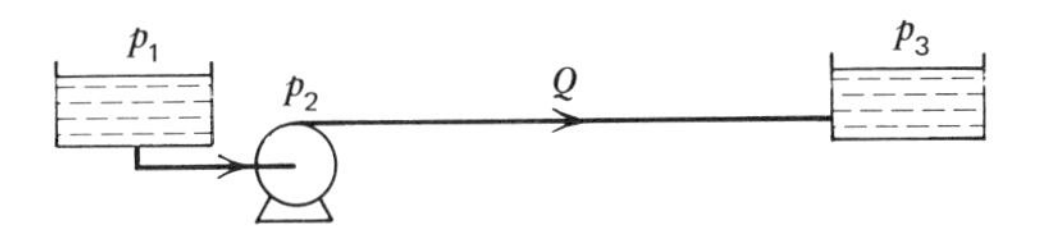

Figure 5.4.1 Pump and pipe installation.

The pressure rise in psi across the pump is given approximately by the empirical relation

$$p_2 - p_1 = a - bQ^{1.5}, \qquad (5.4.1)$$

where a and b are constants that depend on the particular pump being used, and Q is the flow rate in gpm. Note the interpretation of (5.4.1): for $Q = 0$ (no flow), the pressure increase $p_2 - p_1$ equals a (known as the shut-off pressure); for larger values of Q, $p_2 - p_1$ falls off (and eventually becomes zero for Q sufficiently large).

Also, the pressure drop in the horizontal pipe is given by the *Fanning* equation (see Kay [2], for example):

$$p_2 - p_3 = \frac{1}{2} f_M \rho u_m^2 \frac{L}{D}. \qquad (5.4.2)$$

Here, f_M is the dimensionless *Moody* friction factor, ρ is the liquid density, u_m is the mean velocity, and L and D are the length and diameter of the pipe, respectively. Since the volumetric flow rate is $Q = (\pi D^2/4)u_m$, equation (5.4.2) becomes

$$p_2 - p_3 = \frac{8 f_M \rho Q^2 L}{\pi^2 D^5}. \qquad (5.4.3)$$

All quantities in equation (5.4.3) must be in consistent units. However, if p_2 and p_3 are expressed in psi (lb_f/sq in.), ρ in lb_m/cu ft, Q in gpm (gallons/min), L in ft, and D in inches, we obtain

$$p_2 - p_3 = 2.16 \times 10^{-4} \frac{f_M \rho L Q^2}{D^5}. \qquad (5.4.4)$$

For the present purposes, the Moody friction factor f_M is treated as a constant although it really depends somewhat on the pipe roughness and on the Reynolds number, $Re = \rho u_m D/\mu$, where μ is the fluid viscosity.

Problem Statement

Write a program whose input will include values for a, b, ρ, L, D, f_M, ε (a tolerance used in convergence testing), and n (maximum number of iterations), and that will use the *regula-falsi* method to compute the flow rate Q and the intermediate pressure p_2. If the method fails to converge, print a message to that effect. Use the following two sets of test data:

Set 1: $D = 1.049$ in., $L = 50$ ft, $\rho = 51.4$ lb_m/cu ft (kerosene), $f_M = 0.032$, $a = 16.7$ psi, and $b = 0.052$ psi/(gpm)$^{1.5}$.

Set 2: D = 2.469, $L = 210.6$, $\rho = 62.4$ (water), $f_M = 0.026$, $a = 38.5$, and $b = 0.0296$.

Select values for ε and n that seem appropriate.

Method of Solution

Eliminating the intermediate pressure p_2 from equations (5.4.1) and (5.4.4), we have

$$p_1 + a - bQ^{1.5} = p_3 + cQ^2, \qquad (5.4.5)$$

where

$$c = 2.16 \times 10^{-4} \frac{f_M \rho L}{D^5}. \qquad (5.4.6)$$

But $p_1 = p_3 = 0$ psig (gauge pressure), since they are both at atmospheric pressure, so that (5.4.5) becomes

$$f(Q) = a - bQ^{1.5} - cQ^2 = 0. \quad (5.4.7)$$

Since a, b, and c are known, (5.4.7) can readily be solved by the *regula-falsi* method of Section 5.6. Let Q_L and Q_R denote the values of Q at the left and right ends of the current interval; also define $f_{QL} = f(Q_L)$ and $f_{QR} = f(Q_R)$. A suitable starting interval within which the root Q of (5.4.7) must lie is easily delineated: $Q_L = 0$ since there must be some positive flow, but this cannot exceed $Q_R = (a/c)^{1/2}$ because this would mean that the maximum possible pump pressure was completely consumed by pipe friction.

There is no need to retain subscripts for each iteration, so (5.16) becomes

$$Q = \frac{Q_L f_{QR} - Q_R f_{QL}}{f_{QR} - f_{QL}}. \quad (5.4.8)$$

The iterations are terminated when either: (a) n iterations (50, for example) have been performed, or (b) $|f(Q)|$ falls below ε (10^{-3}, for example). Once Q has been found, p_2 is obtained from equation (5.4.1).

Flow Diagram

(Define function $f(Q) = a - bQ^{1.5} - cQ^2$)

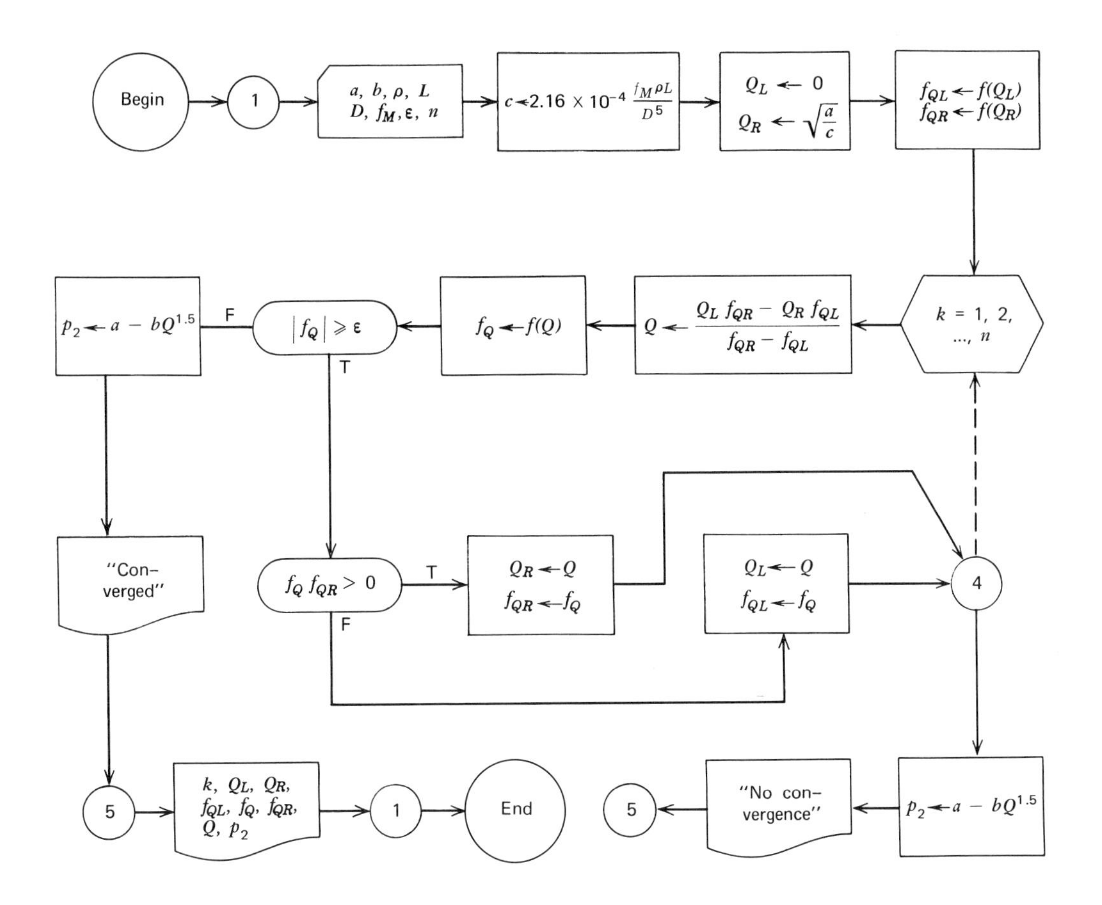

FORTRAN Implementation

List of Principal Variables

Program Symbol	Definition
A, B	Constants, a, b, for centrifugal pump in equation (5.4.1).
C	Constant, c, defined by equation (5.4.6).
D, L	Diameter, D (in.), and length, L (ft), of pipe, respectively.
EPS	Tolerance, ε, used in convergence testing.
F	Statement function representing equation (5.4.7), $f(Q) = a - bQ^{1.5} - cQ^2$.
FM	Moody friction factor, f_M.
FQ	Functional value, $f(Q)$.
FQL, FQR	Functional values, f_{QL} and f_{QR}.
K	Counter, k, for *regula-falsi* iteration.
N	Maximum allowable number, n, of *regula-falsi* iterations.
P2	Intermediate pressure, p_2, psig.
Q	New estimate of root obtained from (5.4.8).
QL, QR	Values, Q_L and Q_R, at left and right-hand ends of current interval.
RHO	Liquid density, ρ, lb_m/cu ft.

Program Listing

```
C         DIGITAL COMPUTING                          EXAMPLE 5.4
C         PERFORMANCE OF CENTRIFUGAL PUMP (REGULA FALSI METHOD).
C
      REAL  L
C
C
C     ..... DEFINE PUMP FUNCTION .....
      F(Q) = A - B*Q**1.5 - C*Q**2
      WRITE (6,200)
C
C     ..... READ AND PRINT OUT INPUT DATA AND CONSTANT C .....
    1 READ (5,100,END=999,ERR=1)   A, B, RHO, L, D, FM, EPS, N
      C = 2.16E-4*FM*RHO*L/D**5
      WRITE (6,201)   A, B, RHO, L, D, FM, EPS, N, C
C
C     ..... COMPUTE AND PRINT STARTING RANGE FOR Q AND F(Q) .....
      QL = 0.
      QR = SQRT(A/C)
      FQL = F(QL)
      FQR = F(QR)
      K = 0
      WRITE (6,202) K, QL, QR, FQL, FQR
C
C     ..... REGULA FALSI ITERATIONS, COMPUTE NEW VALUE FOR Q .....
      DO 4   K=1,N
      Q = (QL*FQR - QR*FQL)/(FQR - FQL)
      FQ = F(Q)
C
C     ..... IF FUNCTIONAL VALUE LESS THAN EPS, METHOD HAS CONVERGED .....
      IF ( ABS(FQ) .GE. EPS )   GO TO 2
C
C     ..... COMPUTE P2 AND PRINT RESULTS FOR OK CONVERGENCE .....
      P2 = A - B*Q**1.5
      WRITE (6,203)
      GO TO 5
C
C     ..... IF NOT, NEW Q WILL BECOME NEW QL OR NEW QR .....
    2 IF ( FQ*FQR .GT. 0. )   GO TO 3
      QL = Q
      FQL = F(Q)
      GO TO 4
    3 QR = Q
      FQR = F(Q)
    4 WRITE (6,202) K, QL, QR, FQL, FQR
C
C     ..... COMPUTE P2 AND PRINT RESULTS FOR NO CONVERGENCE .....
      P2 = A - B*Q**1.5
      WRITE (6,204)
    5 WRITE (6,205)   K, QL, QR, FQL, FQ, FQR, Q, P2
      GO TO 1
C
  999 CALL EXIT
C
C     ..... FORMATS FOR INPUT AND OUTPUT STATEMENTS .....
  100 FORMAT ( 3X, F6.1, 9X, F7.4, 11X, F5.1, 9X, F7.1 /  3X, F8.3,
     1   7X, F7.4, 11X, E8.1, 6X, I5 )
  200 FORMAT ('1DIGITAL COMPUTING                     EXAMPLE 5.4'/' PERFORMA
     1NCE OF CENTRIFUGAL PUMP (REGULA FALSI METHOD).')
  201 FORMAT ('0'/'0THE INPUT DATA AND THE VALUE OF C ARE'/
     1       '0     A      =', F7.1 /'     B      =', F10.4/
     2       '      RHO    =', F7.1 /'     L      =', F7.1 /
     3       '      D      =', F9.3 /'     FM     =', F10.4/
     4       '      EPS    =', F11.5/'     N      =', I5   /
     5       '      C      =', F12.6/'0VALUES DURING SUCCESSIVE ITERATION
     6S ARE'/ '0    K       QL           QR           FQL          FQR'/ ' ')
```

Program Listing (*continued*)

```
  202  FORMAT (I6, 2F11.3, 2F12.5)
  203  FORMAT ('0PROCEDURE CONVERGED WITHIN SPECIFIED TOLERANCE, GIVING')
  204  FORMAT ( '0NO CONVERGENCE WITHIN SPECIFIED NUMBER OF ITERATIONS' /
      1    ' CURRENT VALUES ARE' )
  205  FORMAT ( '0      K      =' , I5 / '       QL      =' , F10.4 /
      1    '      QR      =' , F10.4 / '      FQL     =' , F11.5 /
      2    '      FQ      =' , F11.5 / '      FQR     =', F11.5 /
      3    '      Q       =' , F10.4 / '      P2      =', F8.2 )
C
       END
```

Data

```
A =  16.7,     B = 0.0520,     RHO = 51.4,       L =   50.0,
D =   1.049,  FM = 0.0320,     EPS =   1.0E-3,   N =   50
A =  38.5,     B = 0.0296,     RHO = 62.4,       L =  210.6,
D =   2.469,  FM = 0.0260,     EPS =   1.0E-3,   N =   50
```

Computer Output

```
DIGITAL COMPUTING                        EXAMPLE 5.4
PERFORMANCE OF CENTRIFUGAL PUMP (REGULA FALSI METHOD).

THE INPUT DATA AND THE VALUE OF C ARE

     A      =   16.7
     B      =    0.0520
     RHO    =   51.4
     L      =   50.0
     D      =    1.049
     FM     =    0.0320
     EPS    =    0.00100
     N      =   50
     C      =    0.013985

VALUES DURING SUCCESSIVE ITERATIONS ARE'

     K        QL          QR          FQL          FQR

     0        0.0       34.556      16.70000    -10.56317
     1       21.167     34.556       5.36979    -10.56317
     2       25.680     34.556       0.71069    -10.56317
     3       26.239     34.556       0.08206    -10.56317
     4       26.303     34.556       0.00934    -10.56317
     5       26.311     34.556       0.00112    -10.56317

PROCEDURE CONVERGED WITHIN SPECIFIED TOLERANCE, GIVING

     K      =     6
     QL     =    26.3107
     QR     =    34.5564
     FQL    =     0.00112
     FQ     =     0.00016
     FQR    =   -10.56317
     Q      =    26.3115
     P2     =     9.68

THE INPUT DATA AND THE VALUE OF C ARE

     A      =   38.5
     B      =    0.0296
     RHO    =   62.4
     L      =  210.6
     D      =    2.469
     FM     =    0.0260
     EPS    =    0.00100
     N      =   50
     C      =    0.000804

VALUES DURING SUCCESSIVE ITERATIONS ARE
     K        QL          QR          FQL          FQR

     0        0.0      218.775      38.50000    -95.78302
     1       62.725    218.775      20.63081    -95.78302
     2       90.380    218.775       6.49636    -95.78302
     3       98.535    218.775       1.73828    -95.78302
     4      100.678    218.775       0.44517    -95.78302
     5      101.224    218.775       0.11272    -95.78302
     6      101.362    218.775       0.02848    -95.78302
     7      101.397    218.775       0.00717    -95.78302
     8      101.406    218.775       0.00186    -95.78302

PROCEDURE CONVERGED WITHIN SPECIFIED TOLERANCE, GIVING

     K      =     9
     QL     =   101.4061
     QR     =   218.7752
     FQL    =     0.00186
     FQ     =     0.00044
     FQR    =   -95.78302
     Q      =   101.4084
     P2     =     8.27
```

Discussion of Results

The program worked satisfactorily, requiring only $k = 6$ iterations for the first data set and nine iterations for the second. In both cases, because of the particular nature of $f(Q)$, the right-hand end of the interval, Q_R, always maintains its initial value.

The pressure drops in the pipe are comparable ($p_2 = 9.68$ and 8.27). Even though both the length and flow rate for the second data set are roughly four times those for the first data set, the effect of the increased diameter (which appears as D^5 in equation (5.4.4)) is clearly evidenced.

Problems

5.1 Why does the procedure in Example 5.1 apparently converge?

5.2 If Example 5.1 were patterned after Fig. 5.3 with $\varepsilon = 0.001$ and $n = 15$, what would be the displayed output? What if $\varepsilon = 0.0001$ and $n = 10$?

5.3 Make another rearrangement of Van der Waals' equation (5.8), again along the lines $v = F(v)$. Investigate the first few values in the resulting sequence with the same initial value $v_1 = 0.205135$.

5.4 Draw diagrams analogous to Fig. 5.2 that illustrate: (a) successful convergence, with $dF/dx = F'(x)$ negative and $x_1 < \alpha$, (b) no convergence, with $F'(x)$ positive and $x_1 > \alpha$.

5.5 The equation $x^3 + x^2 = 100$ is to be solved by the method of successive substitutions. If a root is known to be approximately $x \doteq 4.5$, would you expect the algorithm $x \leftarrow (100 - x^2)^{1/3}$ to converge? What about $x \leftarrow (100 - x^3)^{1/2}$?

5.6 Is it possible for the method of successive substitutions to converge even though (5.7) is violated—that is, even though $|dF/dx|$ is not less than 1 everywhere between the initial approximation and the root?

5.7 Develop, but do not implement, a Newton's method algorithm for solving (5.8), Van der Waals' equation.

5.8 Draw up a table that shows the relative advantages and disadvantages of the following methods for solving $f(x) = 0$: (a) successive substitutions, (b) Newton's, (c) half interval, and (d) *regula falsi*.

5.9 Develop a simple algorithm for finding the cube root x of a number a. Within slide-rule accuracy, check the algorithm by trying five iterations for $a = 27$ and also for $a = -27$.

5.10 Consider the following procedure for estimating $\sqrt{a}$. Suppose that an initial approximation x_1 is already available and that it differs from $\sqrt{a}$ by a small amount δ:

$$x_1 + \delta = \sqrt{a}.$$

Now square both sides, ignoring the term δ^2 as being small, and solve for δ. What is then the formula for the second approximation x_2? Comment.

5.11 In the iterative techniques for solving $f(x) = 0$, the error between the kth approximation x_k and the root α is $\delta_k = x_k - \alpha$. Based on the numerical values computed in Examples 5.1 and 5.3, construct plots of the ratios δ_{k+1}/δ_k and δ_{k+1}/δ_k^2 against k. What conclusions, if any, can you draw concerning the rates of convergence of (a) the method of successive substitutions and (b) Newton's method?

5.12 Draw diagrams, similar to Fig. 5.5, that illustrate Newton's method for the cases: (a) $x_1 < \alpha, f'(x) > 0$, and (b) $x_1 < \alpha, f'(x) < 0$.

5.13 Draw a diagram showing what might happen if Newton's method were applied to the situation shown in Fig. 5.1*c*, where $f(x)$ does not cross the x axis.

5.14 Verify formula (5.15) for the number of iterations needed in the half-interval method.

5.15 Consider the algorithm of (5.6), $x_k = F(x_{k-1})$, for attempting to compute a root $x = \alpha$ of $f(x) = 0$. Although this algorithm was developed first for the method of successive substitutions, we have already remarked that it also represents Newton's method if $F(x)$ is defined as $F(x) = x - [f(x)/f'(x)]$.

Let δ_k be the discrepancy or error of the kth iterate, x_k, from the root, α, so that $x_k = \alpha + \delta_k$. Assuming that $F(x)$ is suitably differentiable, expand $F(x)$ in Taylor's series (see Section 6.9) about α and show that

$$\delta_k = \delta_{k-1}F'(\alpha) + \frac{\delta_{k-1}^2 F''(\alpha)}{2!} + \frac{\delta_{k-1}^3 F'''(\alpha)}{3!} + \cdots$$

Investigate the relation, if any, between this result and the convergence criterion of equation (5.7) for the method of successive substitutions.

5.16 The *order* of the iteration (5.6), $x_k = F(x_{k-1})$, is defined as the order of the lowest nonzero derivative of $F(x)$ at the root $x=\alpha$. Show that for simple roots (for which $f'(\alpha) \neq 0$), Newton's method is a second-order or quadratic process, whereas for multiple roots (for which $f'(\alpha)=0$) it is a first-order process.

5.17 Suppose the equation $f(x)=0$ has been rewritten as $x=F(x)$, appropriate for an attempted solution by the method of successive substitutions. Show that it may also be rewritten in the modified form

$$x=(1-c)x+cF(x),$$

where c is an arbitrary constant.

Sketch the function $f(x)=x^3-3x+1$. Show that the rearrangement $x=F(x)=(x^3+1)/3$ would be suitable for determining only one of the roots of $f(x)=0$, but that the above modification with $c=-1/2$ could be used for the other two.

It is appropriate to mention *Wegstein's method* here. Suppose that the rearrangement $x=F(x)$ has been made, that the functional evaluations $F(x_1)$ and $F(x_2)$ have been made at two arbitrary starting points x_1 and x_2, and that a preliminary estimate $s=(F(x_2)-F(x_1))/(x_2-x_1)$ of the slope yields $|s|>1$. Draw a straight line through $(x_1, F(x_1))$ and $(x_2, F(x_2))$ and let it intersect the 45° line of Fig. 5.2 (b) at the point (x_3, x_3). Show that Wegstein's method amounts to the following successive substitution algorithm:

$$x_3=(1-c)x_2+cF(x_2),$$

in which $c=1/(1-s)$. Note that by using x_3 and x_2, a new estimate x_4 can be obtained, and so on. For typical situations, investigate graphically whether or not convergence is likely to occur.

5.18 Write a function, named SQRT, that returns as its value the square root of a real positive argument. Newton's method is suggested (see Example 5.3). Anticipate that typical references to the function will be

```
X = SQRT (A)
DISCR = SQRT(B*B - 4.*A*C)
```

5.19 Devise a procedure, based on one of the standard equation-solving techniques, for evaluating $a^{1/n}$. Assume that a is a real positive number, that n is a positive integer, and that the real positive nth root is required. Write a function, named ROOT, that returns such a root as its value. Anticipate that a typical reference to the function will be

```
X = ROOT (A, N)
```

in which the arguments have obvious interpretations.

5.20 Devise two or more practical schemes for evaluating, within a prescribed tolerance ε, all the zeros of the nth-degree Legendre polynomial $p_n(x)$, for which the general recursion relation is given in equation (7.23). Use should be made of the fact that all the roots of $p_n(x)=0$ lie between -1 and 1. Write a program that implements one of these methods. Your program should automatically generate the necessary coefficients of the appropriate Legendre polynomial, according to (7.23). Check your computed values for $n=2, 3, 4$, and 5 with those given in Table 7.1.

5.21 The Beattie-Bridgeman equation of state for an imperfect gas is

$$P=\frac{RT}{v}+\frac{\beta}{v^2}+\frac{\gamma}{v^3}+\frac{\delta}{v^4},$$

where P is the absolute pressure (atm), v is the molal volume (liter/g mole), T is the absolute temperature (°K), and R is the gas constant (0.082054 liter atm/g mole °K). The second, third, and fourth terms on the right-hand side of the above equation may be viewed as corrections of the ideal gas law, $P=RT/v$.

The parameters β, γ, and δ are defined by

$$\beta=RTB_0-A_0-\frac{Rc}{T^2},$$

$$\gamma=-RTB_0b+A_0a-\frac{RcB_0}{T^2},$$

$$\delta=\frac{RbcB_0}{T^2},$$

where A_0, B_0, a, b, and c are known constants for each particular gas.

Write a computer program that uses a technique such as Newton's method to solve the Beattie-Bridgeman equation for the molal volume v, given values for P, T, R, A_0, B_0, a, b, and c. Compute also the compressibility factor, $z = Pv/RT$, whose deviation from 1 gives an indication of the nonideality of the gas behavior.

Suggested Test Data

Compute and print the compressibility factors at $P = 20, 40, \ldots, 200$ atm for helium, carbon dioxide, and methane at 50°C and 150°C, and for ammonia at 150°C. The relevant Beattie-Bridgeman constants in units of °K (= °C + 273.16), g mole, liter, atm, are [32]:

Gas	A_0	B_0	a	b	$10^{-4} \times c$
Helium	0.0216	0.01400	0.05984	0.0	0.0040
Carbon dioxide	5.0065	0.10476	0.07132	0.07235	66.00
Methane	2.2769	0.05587	0.01855	−0.01587	12.83
Ammonia	2.3930	0.03415	0.17031	0.19112	476.87

5.22 *Introduction.* This problem illustrates: (a) Newton's method for solving equations, and (b) the use of variables for storing nonnumerical information.

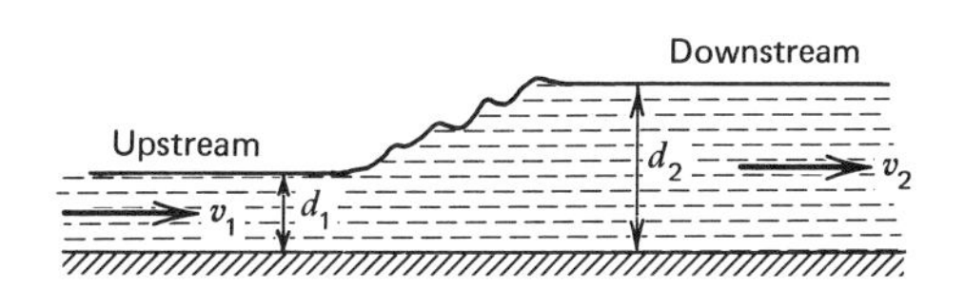

Figure P5.22.

Figure P5.22 shows a *hydraulic jump*, which occurs when a rapidly moving stream of liquid in a channel suddenly increases its depth, with an accompanying loss of energy due to turbulence. The phenomenon is important because the more tranquil and deeper stream tends to minimize erosion effects.

A dimensionless group F, known as the Froude number, is defined as

$$F = \frac{v^2}{gd},$$

where v = liquid velocity (ft/sec), d = depth (ft), and g = gravitational acceleration (32.2 ft/sec^2). On the basis of mass and momentum balances, it can be shown (see Streeter [21], for example) that the following relation holds between the upstream and downstream Froude numbers, F_1 and F_2, respectively:

$$(\sqrt{1+8F_1} - 1)^3 = 8F_1/F_2.$$

It can be shown that the jump can only occur if $F_1 > 1$ (in which event $F_2 < 1$). The following relation also holds:

$$F_1 d_1^3 = F_2 d_2^3.$$

Problem Statement. Write a program that accepts data using the following statement:

```
1 READ (5, 100, END=999) NAME, F, D, EPS, ITMAX
```

In the accompanying format, an A4 (4-column alphanumeric) field should be used for NAME, which is therefore understood to contain a string of four characters, including possible blanks.

The program should investigate the contents of NAME, and take the following action:

(a) If NAME contains the character string F1D1, then F and D should be interpreted as the upstream Froude number and depth, F_1 and d_1, respectively. The program should compute the remaining variables and print values for NAME, F1, V1, D1, F2, V2, D2 (with obvious meanings) before returning to read another set of data.

(b) If NAME contains the character string F2D2, then F and D should be interpreted as the downstream values, F_2 and d_2, respectively. The same set of values should be printed as in (a). However, Newton's method or similar should now be used to solve for F_1, in which case EPS is a convergence tolerance and ITMAX is an upper limit on the number of iterations to be used.

(c) If NAME contains neither F1D1 nor F2D2, then its content followed by the message BAD DATA should be printed, before returning to read another set of data. In this case, no calculations should be attempted.

Data. Test your program with the following sets of data, but also try a few sets of your own choice:

NAME	F	D	EPS	ITMAX
F2D2	0.25	10.0	10^{-5}	10
F2D2	0.50	10.0	10^{-5}	10
F2D2	0.75	10.0	10^{-5}	10
F1D1	22.35	5.0	10^{-5}	10
F2D2	0.0938	31.03	10^{-5}	10
F1D2	22.35	31.03	10^{-5}	10
F2D2	0.0938	31.03	10^{-15}	5

5.23 For the isentropic flow of a perfect gas from a reservoir through a converging-diverging nozzle, operating with sonic velocity at the throat, it may be shown (see Kay [2]) that

$$\frac{A_t^2}{A^2} = \left(\frac{\gamma+1}{2}\right)^{((\gamma+1)/(\gamma-1))}\left(\frac{2}{\gamma-1}\right)$$
$$\times\left[\left(\frac{P}{P_1}\right)^{2/\gamma} - \left(\frac{P}{P_1}\right)^{(\gamma+1)/\gamma}\right].$$

Here, P is the pressure at a point where the cross-sectional area of the nozzle is A, P_1 is the reservoir pressure, A_t is the throat area, and γ is the ratio of the specific heat at constant pressure to that at constant volume.

If A_t, γ, P_1, and A ($> A_t$) are known, devise a scheme for computing the *two* possible pressures P that satisfy the above equation. Implement your method on the computer; suggested test data are: $A_t =$ 0.1 sq ft, $\gamma = 1.41$, $P_1 = 100$ psia, and $A = 0.12$ sq ft.

5.24

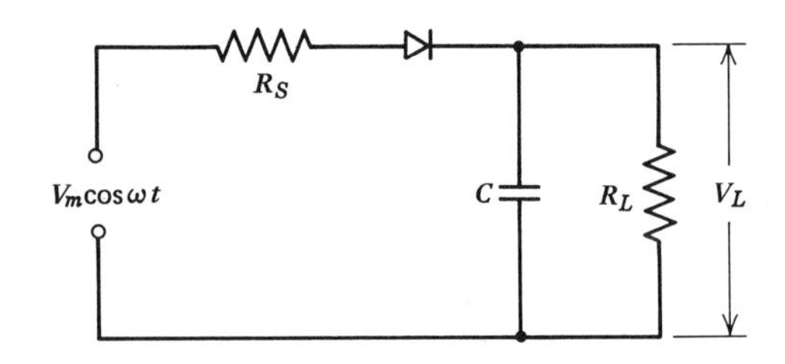

Figure P5.24.

Fig. P5.24 shows a load R_L that is supplied from an ac source of peak voltage V_m through a resistor R_S in series with a diode; the capacitor is sufficiently large to eliminate any ripple in the resulting dc voltage V_L across the load. The following approximate relation can be established between the two ratios V_L/V_m and R_S/R_L (see page 230 of Gray and Searle [24], for example):

$$\frac{V_L}{V_m} = 1 - \left(\frac{\pi V_L R_S}{\sqrt{2}V_m R_L}\right)^{2/3}.$$

Develop a numerical method, and implement it either by hand or on the computer, that will compute the voltage ratio V_L/V_m for specified values of R_S/R_L (such as 0.01, 0.1, 1, 10, for example). Plot V_L/V_m against R_S/R_L.

5.25 A more accurate treatment of the situation discussed in Problem 5.24 gives the following two equations:

$$\frac{V_L}{V_m} = \cos \omega\tau,$$

$$\sin \omega\tau - \omega\tau\frac{V_L}{V_m} = \frac{\pi V_L R_S}{V_m R_L}.$$

(These equations result from the facts that: (a) during a representative cycle $-\pi/\omega \leqslant t \leqslant \pi/\omega$, the diode conducts only during an interval $-\tau < t < \tau$, during which $V_m \cos \omega t > V_L$, and (b) the integrated current passing through the diode during this conduction period equals the total charge passing through the load over the whole cycle.)

(a) Referring to Section 6.9 if necessary, show that for small $\omega\tau$, these more accurate equations may be approximated by the single equation given in Problem 5.24.

(b) Based on the above two equations, develop a numerical method, and implement it on the computer, that will compute $\omega\tau$ and V_L/V_m for specified values of R_S/R_L (such as 0.01, 0.1, 1, and 10). Plot V_L/V_m against R_S/R_L. Compare with the results of Problem 5.24, if available.

5.26 For the 6J5 triode, the anode or plate current i_A (amp) is related to the grid potential v_G and anode potential v_A (both in volts relative to the cathode) by the following correlation, approximately valid for the ranges $-25 \leq v_G \leq 0$, $0 \leq v_A \leq 450$:

(a) if $15.0v_G + v_A > 0$, then

$$i_A = 10^{-3}e^m(15.0v_G + v_A)^n,$$

in which

$$m = -3.30 + 0.633v_G + 0.0147v_G^2,$$

$$n = 1.25 - 0.1135v_G - 0.00295v_G^2;$$

(b) if $15.0v_G + v_A \leq 0$, then $i_A = 0$.

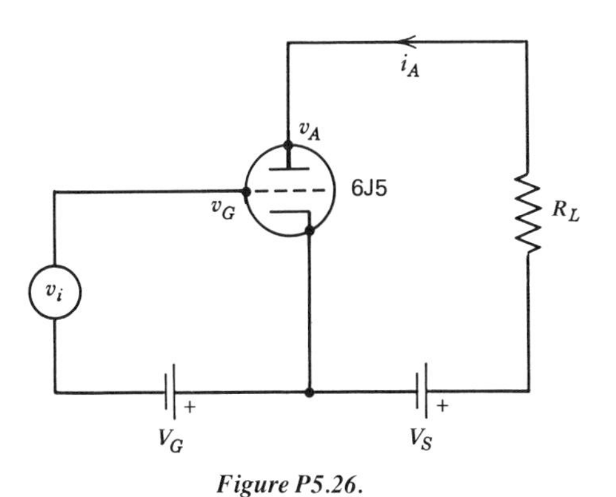

Figure P5.26.

The 6J5 triode shown in Fig. P5.26 is used for amplifying a variable input signal v_i volts. If the high-tension supply is constant at V_S volts, Ohm's law applied across the output load R_L ohms gives:

$$R_L i_A = V_S - v_A.$$

Also, the grid potential v_G at any instant is the sum of the input signal v_i and the constant grid bias V_G:

$$v_G = V_G + v_i.$$

In determining the performance of the amplifier, we first need to determine the *quiescent point*, which amounts to finding the anode current i_{Aq} and potential v_{Aq} when the input signal is zero—that is, when $v_G = V_G$. If any finite input signal v_i results in an anode potential of v_A, the corresponding *amplifier gain A* is then given by

$$A = \frac{v_A - v_{Aq}}{v_i}.$$

Write a program that will accept values for V_S, V_G, R_L, n, and Δv_i as data, and that will proceed to compute and print values for: (a) the quiescent-point values i_{Aq} and v_{Aq}, and (b) the amplifier gain for all values of $v_i = \pm\Delta v_i, \pm 2\Delta v_i, \ldots, \pm n\Delta v_i$.

Suggested Test Data

Investigate $R_L = 1500$, 8200, and 22,000 ohms, with $\Delta v_i = 1$ volt throughout, for each of (a) $V_S = 250$, $V_G = -5$ volts, $n = 5$, (b) $V_S = 250$, $V_G = -10$, $n = 10$, and (c) $V_S = 400$, $V_G = -10$, $n = 10$.

5.27 A long-range ballistic missile is fired with launch velocity v and angle of inclination α to the horizontal from a point A on the earth's surface. The intended target is at point B, also on the earth's surface, and is such that AB subtends an angle ϕ at the center of the earth; that is, the range measured along the surface is $R\phi$, where R is the radius of the earth.

Neglecting atmospheric resistance and rotation of the earth, it may be shown [22] that

$$\tan(\phi/2) = \frac{\sin\alpha\cos\alpha}{(K/Rv^2) - \cos^2\alpha},$$

where $K = GM$ is the product of the gravitational constant and the mass of the earth. (K also equals gR^2, where g is the gravitational acceleration at the earth's surface.)

Write a program that accepts values for ϕ and Rv^2/K as data and that proceeds to compute an appropriate value (if such exists) for the angle α. Note that if Rv^2/K is too small, the target may be unattainable, whereas if it is too large, the missile may escape completely from the vicinity of the earth.

Suggested Test Data

ϕ:	80°	80°	80°	160°	160°	160°	160°
Rv^2/K:	1	0.8	0.4	1	1.5	2	2.5

5.28

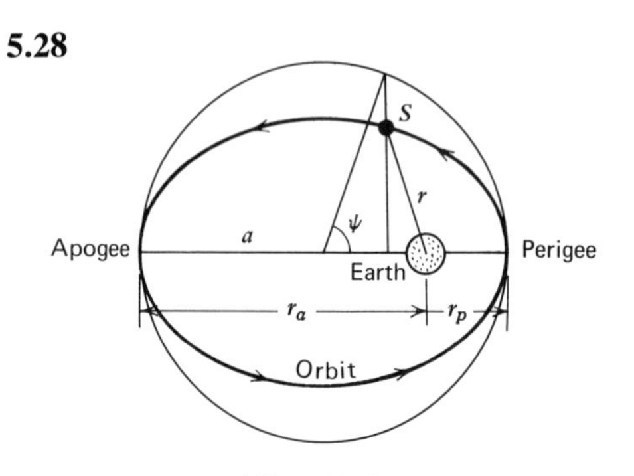

Figure P5.28.

The orbit of an earth satellite S is an ellipse with semimajor axis a, eccentricity e, perigee distance r_p, and apogee distance r_a. Using an auxiliary circle of radius a shown in Fig. P5.28, an angle ψ radians, called the *eccentric anomally*, serves to define the position of the satellite at any instant.

The eccentric anomally at a time t after the perigee obeys Kepler's law [22]:

$$t\sqrt{\frac{K}{a^3}} = \psi - e\sin\psi,$$

in which $K = GM = 1.407 \times 10^{16}\ \text{ft}^3/\text{sec}^2$ (where G is the gravitational constant and M is the earth's mass). It can also be shown that the earth/satellite distance r and the satellite speed v are given by

$$r = a(1 - e\cos\psi),$$

$$\frac{v^2}{K} = \frac{2}{r} - \frac{1}{a}.$$

Consistent units are assumed in these equations.

Write a program that will accept values for r_p, r_a, and an integer n as data, and proceed to compute and print values for e, a (miles), and T (hr, the time for one complete orbit). The program should then compute and print values for ψ (degrees), r (miles), and v (miles/hr) at n equally spaced time intervals between perigee and apogee (that is, $n+1$ sets of values, including perigee and apogee).

Suggested Test Data

r_p (miles):	4,346	4,157	4,300
r_a (miles):	4,664	30,400	4,300
n	10	10	10

5.29 A spherical pocket of high-pressure gas, initially of radius r_0 and pressure p_0, expands radially outwards in an adiabatic submarine explosion. For the special case of a gas with $\gamma = 4/3$ (ratio of specific heat at constant pressure to that at constant volume), the radius r at any subsequent time t is given by [3]:

$$\frac{t}{r_0}\sqrt{\frac{p_0}{\rho}} = \left(1 + \frac{2}{3}\alpha + \frac{1}{5}\alpha^2\right)(2\alpha)^{1/2},$$

in which $\alpha = (r/r_0) - 1$, ρ is the density of water, and consistent units are assumed. During the adiabatic expansion, the gas pressure is given by $p/p_0 = (r_0/r)^{3\gamma}$.

Develop a procedure for computing the pressure and radius of the gas at any time. Suggested test data (not in consistent units): $p_0 = 10^4$ lb$_f$/sq in., $\rho = 64$ lb$_m$/cu ft, $r_0 = 1$ ft, $t = 0.5$, 1, 2, 3, 5, and 10 milliseconds.

5.30 F moles/hr of an n-component natural gas stream are introduced as feed to the flash-vaporization tank shown in Fig. P5.30. The resulting vapor and liquid streams are withdrawn at the rates of V and L moles/hr, respectively. The mole fractions of the components in the feed, vapor, and liquid streams are designated by z_i, y_i, and x_i, respectively ($i = 1, 2, \ldots, n$).

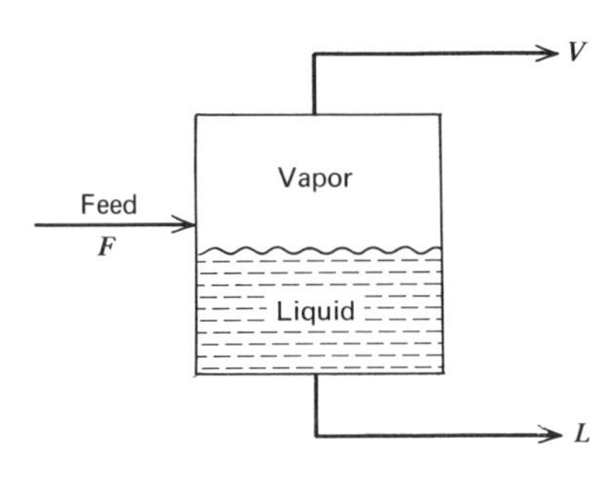

Figure P5.30.

Assuming vapor/liquid equilibrium and steady-state operation, we have:

$$\begin{aligned} &\textit{Overall Balance} && F = L + V, \\ &\textit{Individual Balance} && z_i F = x_i L + y_i V, \\ &\textit{Equilibrium Relation} && K_i = y_i/x_i, \end{aligned} \quad \Big\} i = 1, 2, \ldots, n.$$

Here, K_i is the equilibrium constant for the ith component at the prevailing temperature and pressure in the tank. From these equations and the fact that $\Sigma_{i=1}^{n} x_i = \Sigma_{i=1}^{n} y_i = 1$, show that

$$\sum_{i=1}^{n} \frac{z_i(K_i - 1)}{V(K_i - 1) + F} = 0.$$

Write a program that reads values for F, the z_i, and the K_i as data, and that uses Newton's method to solve the last equation above for V. The program should also compute the values of L, the x_i, and the y_i by using the first three equations given above.

The following test data [4] relate to the flashing of a natural gas stream at 1600 psia and 120°F:

Component	i	z_i	K_i
Carbon dioxide	1	0.0046	1.65
Methane	2	0.8345	3.09
Ethane	3	0.0381	0.72
Propane	4	0.0163	0.39
Isobutane	5	0.0050	0.21
n-Butane	6	0.0074	0.175
Pentanes	7	0.0287	0.093
Hexanes	8	0.0220	0.065
Heptanes +	9	0.0434	0.036
		1.0000	

Assume that $F = 1000$ moles/hr. A small tolerance ε and an upper limit on the number of iterations should also be read as data. What would be a good value V_1 for starting the iteration?

5.31 For turbulent flow of a fluid in a *smooth* pipe, the following relation exists between the Fanning friction factor c_f (one quarter of the Moody friction factor f_M) and the Reynolds number Re (see, for example, Kay [2]):

$$\sqrt{\frac{1}{c_f}} = -0.4 + 1.74 \ln (Re\sqrt{c_f}).$$

Compute c_f for $Re = 10^4$, 10^5, and 10^6.

5.32 For steady flow of an incompressible fluid in a pipe of length L feet and inside diameter D inches, the pressure drop Δp in lb_f/sq in. is given by

$$\Delta p = \frac{f_M \rho u_m^2 L}{24 g_c D},$$

where ρ is the fluid density (lb_m/cu ft), u_m is the mean fluid velocity (ft/sec), and f_M is the Moody friction factor (dimensionless). For the indicated units, the conversion factor g_c has the value 32.2 lb_m ft/lb_f sec^2. The friction factor is a function of the pipe roughness ε (in.) and the Reynolds number Re, where

$$Re = \frac{\rho u_m D}{12\mu}.$$

Here, μ is the fluid viscosity (lb_m/ft sec). For $Re \leq 2000$, which corresponds to laminar flow, the friction factor is given by $f_M = 64/Re$, whereas for $Re > 2000$ (turbulent flow) it is given by the solution of the *Colebrook equation*

$$\frac{1}{\sqrt{f_M}} = -2 \log_{10}\left[\frac{\varepsilon}{3.7D} + \frac{2.51}{Re\sqrt{f_M}}\right].$$

A good starting guess for the iterative

solution of the Colebrook equation may be found from the *Blasius equation*,

$$f_M = 0.316\, Re^{-0.25},$$

appropriate for turbulent flow, up to about $Re = 10^5$, in smooth pipes ($\varepsilon = 0$).

Write a function, named PDROP, that could be called with the statement

```
DELTAP= PDROP(Q, D, L, RHO, MU, E)
```

where the value of PDROP is the pressure drop in lb_f/sq in. for a flow rate of Q gal/min of a fluid of density RHO and viscosity MU through a pipe of length L, inside diameter D, and roughness E.

To test PDROP, write a main program that reads values for Q, D, L, RHO, MU, and E, calls on PDROP to compute the pressure drop, prints the data and result, and returns to read another data set.

Suggested Test Data

	Set 1	Set 2
Q, gal/min	170	4
D, in.	3.068	0.622
L, ft	10,000	100
ρ, lb_m/cu ft	62.4	80.2
μ, lb_m/ft sec	0.000688	0.0535
ε, in.	0.002	0.0005

5.33 Rework Example 5.4, now allowing for a variation of the Moody friction factor f_M with pipe roughness and Reynolds number, as in Problem 5.32. Use the same two sets of test data, and assume in both cases that the pipe roughness is 0.0005 ft, corresponding to galvanized iron pipe. Appropriate viscosities (at 68°F, in centipoise) are: $\mu = 2.46$ (kerosene) and 1.005 (water). Note that 1 centipoise = 0.000673 lb_m/ft sec. For each data set, the program should print values for f_M and the Reynolds number.

5.34 A semiinfinite medium is at a uniform initial temperature T_0. For $t > 0$, a constant heat-flux density q is maintained into the medium at the surface $x = 0$. If the thermal conductivity and thermal diffusivity of the medium are k and α, respectively, it can be shown (see Carslaw and Jaeger [5], for example) that the resulting temperature $T = T(x, t)$ is given by

$$T = T_0 + \frac{q}{k}\left[2\sqrt{\frac{\alpha t}{\pi}}\, e^{-x^2/4\alpha t} - x \operatorname{erfc}\frac{x}{2\sqrt{\alpha t}}\right].$$

If all other values are given, devise a scheme for finding the time t taken for the temperature at a distance x to reach a preassigned value T^*. Implement your method on the computer; suggested test data are: $T_0 = 70$°F, $q = 300$ BTU/hr sq ft, $k = 1.0$ BTU/hr ft °F, $\alpha = 0.04$ sq ft/hr, $x = 1$ ft, and $T^* = 120$°F. *Note:* the *complementary error function* erfc z is defined as $1 - \operatorname{erf} z$, where the *error function* itself is the integral

$$\operatorname{erf} z = \frac{2}{\sqrt{\pi}} \int_0^z e^{-\beta^2}\, d\beta,$$

which can be obtained by standard numerical integration techniques (see Problem 7.29) or by calling directly on the library function ERF.

5.35 A bare vertical wall of a combustion chamber containing hot gases is exposed to the surrounding air. Heat is lost at a rate q BTU/hr sq ft by conduction through the wall and by subsequent radiation and convection to the surroundings, assumed to behave as a blackbody radiator. Let T_g, T_w, and T_a denote the temperatures of the gases, the exposed surface of the wall, and the air, respectively. If σ = the Stefan-Boltzmann constant, 0.171×10^{-8} BTU/hr sq ft °R^4, and ε, t, and k denote the emissivity, thickness, and thermal conductivity, respectively, of the wall, we have:

$$q = \frac{k(T_g - T_w)}{t} = \varepsilon\sigma(T_{wR}^4 - T_{aR}^4) + h(T_w - T_a).$$

The extra subscript R emphasizes that the absolute or Rankine temperature must be used in the radiation term (°R = °F + 460°). The convectional heat transfer coefficient h, BTU/hr sq ft °F, is given by the correlation $h = 0.21(T_w - T_a)^{1/3}$, suggested by Rohsenow and Choi [6].

Assuming that T_g, T_a, ε, k, and t are specified, rearrange the above relations to give a single equation in the outside wall temperature T_w. Compute T_w for the following test data: $T_a = 100$°F, $t = 0.0625$ ft, with (a) $T_g = 2100$°F, $k = 1.8$ (fused alumina) BTU/hr ft °F, $\varepsilon = 0.39$, (b) $T_g = 1100$, $k = 25.9$ (steel), $\varepsilon = 0.14$ (freshly polished) and $\varepsilon = 0.79$ (oxidized). In each case, also compute q and the relative importance of radiation and convection as mechanisms for transferring heat from the hot wall to the air.

5.36 A vertical mast of length L has Young's modulus E and a weight w per unit length; its second moment of area is I. Timoshenko [7] shows that the mast will just begin to buckle under its own weight when $\beta = 4wL^3/9EI$ is the smallest root of

$$1 + \sum_{n=1}^{\infty} c_n \beta^n = 0.$$

The first coefficient is $c_1 = -3/8$, and the subsequent ones are given by the recursion relation

$$c_n = -\frac{3c_{n-1}}{4n(3n-1)}.$$

Determine the appropriate value of β.

5.37 Methane and excess moist air are fed continuously to a torch. Write a program that will compute, within 5°F, the adiabatic flame temperature T^* for complete combustion according to the reaction

$$CH_4 + 2O_2 = CO_2 + 2H_2O.$$

The data for the program should include the table of thermal properties given below, together with values for T_m and T_a (the incoming methane and air temperatures, respectively, °F), p (the percentage excess air over that theoretically required), and w (lb moles water vapor per lb mole of incoming air on a dry basis). Assume that dry air contains 79 mole % nitrogen and 21 mole % oxygen.

The heat capacities for the five components present can be computed as functions of temperature from the general relation

$$c_{p_i} = a_i + b_i T_k + c_i T_k^2 + d_i/T_k^2 \quad \text{cal/g mole °K},$$

where $i = 1, 2, 3, 4$, and 5 for CH_4, O_2, N_2, H_2O (vapor), and CO_2, respectively, and T_k is in °K. The following table [8] also shows the standard heat of formation at 298°K, $\Delta H^f_{i,298}$, cal/g mole, for each component.

i	a_i	b_i	c_i	d_i	$\Delta H^f_{i,298}$
1	5.34	0.0115	0.0	0.0	−17889.
2	8.27	0.000258	0.0	−187700.	0.0
3	6.50	0.00100	0.0	0.0	0.0
4	8.22	0.00015	1.34×10^{-6}	0.0	−57798.
5	10.34	0.00274	0.0	−195500.	−94052.

Suggested Test Data

$T_m = T_a = 60$°F, $p = 0$ to 100% in steps of 10%, $w = 0$ and 0.015.

5.38 The intensity q of blackbody radiation is given as a function of wavelength λ in equation (7.2.1) on page 330. For a given surface temperature T, devise a scheme for determining the wavelength λ_{max} for which the radiant energy is the most intense—that is, find λ corresponding to $dq/d\lambda = 0$. Write a program that implements the scheme. The input data should consist of values for T, such as 1000, 2000, 3000, and 4000°K; the output should consist of printed values for λ_{max} and the corresponding value of q. Verify numerically that *Wien's displacement law*, $\lambda_{max}T =$ constant, is obeyed.

5.39 The isothermal irreversible second-order constant volume reaction $A+B \rightarrow C+D$ has a reaction rate constant k. A volumetric flow rate v of a solution containing equal inlet concentrations a_0 each of A and B is fed to two CSTRs (continuous stirred tank reactors) in series, each of volume V. Denoting the exit concentrations of A from the first and second CSTRs by a_1 and a_2, respectively, rate balances give:

$$ka_1^2 = \frac{v(a_0 - a_1)}{V},$$

$$ka_2^2 = \frac{v(a_1 - a_2)}{V}.$$

If $k = 0.075$ liter/g mole min, $v = 30$ liter/min, $a_0 = 1.6$ g moles/liter, and the final conversion is 80% (that is, $a_2/a_0 = 0.2$), determine the necessary volume V (liter) of each reactor. For an extension to n CSTRs in series, see Problem 8.25.

5.40 Consider a small beam or reed of breadth b, length L, thickness t, cross-sectional area $A = bt$, second moment of area $I = bt^3/12$, Young's modulus E, and density ρ. It can be shown (see Wylie [9], for example) that the circular frequencies ω of natural vibrations of the reed are given by

$$\omega = \left(\frac{EI}{A\rho}\right)^{1/2} p^2,$$

where p is a root of

$$\cosh pL \cos pL = -1.$$

Write a program that will determine the first ten positive roots of the last equation and hence the corresponding values of ω. Suggested data, for a light-alloy reed: $E = 10^7$ lb$_f$/sq in., $L = 10$ in., $\rho = 170$ lb$_m$/cu ft, $b = 0.5$ in., and $t = 0.02$ in. The conversion 1 lb$_f = 32.2$ lb$_m$/ft sec^2 will be needed.

5.41 *Introduction.* Consider a spaceship propelled by a continuously operating low-thrust electric rocket motor that emits gas with an exhaust velocity of c meters/sec. Discounting gravitational effects, the following equation can be shown to give ϕ, the ratio of the final payload mass m_p to the initial mass m_0:

$$\phi = \frac{m_p}{m_0} = e^{-\Delta v/c} - \frac{\alpha c^2}{2t_f}(1 - e^{-\Delta v/c}).$$

Here, $t_f =$ flight time for the mission, sec,
$\alpha =$ specific mass of rocket power source, kg/watt,
$\Delta v =$ velocity increment for the spaceship, meters/sec, during the time t_f.

Representative characteristics of four types of nuclear/electric rocket motors are given in the following table:

Motor Type	Exhaust Velocity c, meters/sec
Nuclear	5,000– 10,000
Arcjet	7,000– 25,000
Electromagnetic	25,000–100,000
Ion	75,000–600,000

If t_f, α, and Δv are specified, a proper choice of motor will permit an exhaust velocity to be selected that maximizes the payload ratio ϕ. This can be achieved by setting $d\phi/dc = 0$, and using a standard equation-solving technique to find the corresponding optimum exhaust velocity c_{opt}.

Problem Statement

Write a program whose data include values for t_f, α, and Δv, and that computes and prints values for the following:

(a) The optimum exhaust velocity, c_{opt}.

(b) The corresponding maximum payload ratio, ϕ_{opt}.

(c) The payload ratio if c is chosen to be (i) $c_{opt}/5$, and (ii) $5c_{opt}$.

(d) The name of the type of motor that is best suited to the mission.

Suggested Test Data

(a) Moon flight, with $t_f = 3$ *days*, $\alpha = 0.005$ kg/watt, $\Delta v = 6900$ m/sec, and (b) Mars flight, with $t_f = 400$ days, $\alpha = 0.001$ kg/watt, $\Delta v = 5700$ m/sec.

CHAPTER 6
Numerical Approximation

6.1 Introduction

In this chapter, we shall study ways for constructing approximations $g(x)$ to different functions $f(x)$, so that $g(x) \doteq f(x)$. There are various reasons why we might wish to do this. As an introduction, we present three commonly occurring situations.

1. Tabulated Functional Values. First, $f(x)$ may be supplied as a series of tabulated values at a limited number of values for x, also known as *base points*:

Base Points	*Functional Values*
x_1	$f(x_1)$
x_2	$f(x_2)$
$\vdots$	$\vdots$
x_i	$f(x_i)$
$\vdots$	$\vdots$
x_n	$f(x_n)$

We shall assume that the above values have been determined with considerable accuracy—that is, we can trust the tabulated values almost completely.

The tabulated values are also illustrated in Fig. 6.1; the broken line emphasizes that although $f(x)$ is likely to be a continuous function of x, its exact variation between the tabulated values is probably unknown. Such a situation frequently

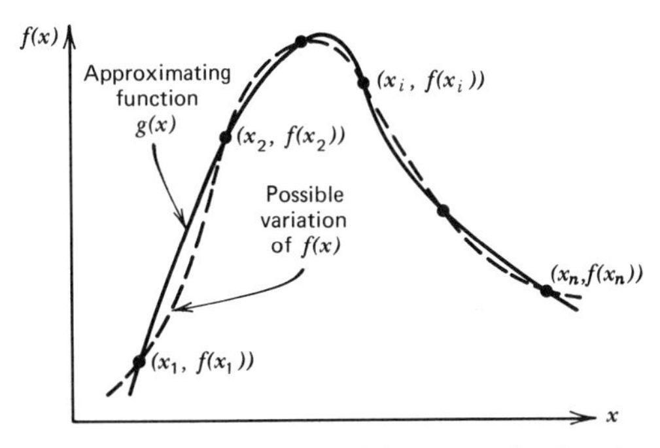

Figure 6.1 Approximation of f(x) by another function g(x).

arises in dealing with physical properties in engineering design problems. For example, the viscosity, $f(x)$, of a lubricating oil might be tabulated at different temperatures, x; the refractive index of borosilicate crown glass might be available at different wavelengths of light; the boiling point of a hydrocarbon such as octane might be specified at various pressures; or the attenuation of a band-pass filter might be given at certain frequencies.

The question now arises: If a certain value of x (pressure, for example) is given, what is approximately the corresponding value of $f(x)$ (the boiling point of octane, for example)? Clearly, the question cannot be answered precisely because there is no guarantee as to how the function behaves *between* the given base points. However, if we are prepared to make one or two intelligent assumptions, we can develop techniques that will be useful in the majority of cases for predicting or approximating the functional behavior of $f(x)$ between the base points. For example, the full curve in Fig. 6.1 represents a possible approximating function, $g(x)$, that at least reproduces the base-point values exactly, but probably does not duplicate $f(x)$ exactly between the base points. Some techniques for generating suitable approximating functions will be discussed in more detail in Sections 6.3, 6.4, and 6.5 (dealing with *interpolating polynomials*) and Section 6.7 (on *spline functions*).

2. Experimentally Determined Values. Second, we may be given a set of m pairs of experimentally determined points, (x_i, y_i), $i = 1, 2, \ldots, m$, in which we recognize that the values are not completely accurate, being subject to random error or noise that unavoidably creeps more or less into all experimental measurements. For example: (a) the y_i might represent the fuel consumption of a ship observed at different cruising speeds x_i, or (b) the y_i might represent the observed intensity of a shortwave radio signal at different distances from the transmitter. In both cases, there is little chance that the results could be duplicated with complete accuracy from one day to the next: prevailing ocean currents in (a) and atmospheric disturbances in (b) could easily influence the measured values. What we could hope to establish is something along the lines that, *on the average*, the fuel consumption y of the ship is a definite function, $y = y(x)$, of its speed x and that, *on the average*, the intensity y of the radio signal is a definite function, $y = y(x)$, of the distance x from the transmitter. The situation is illustrated in Fig. 6.2. The full curve shows a possible

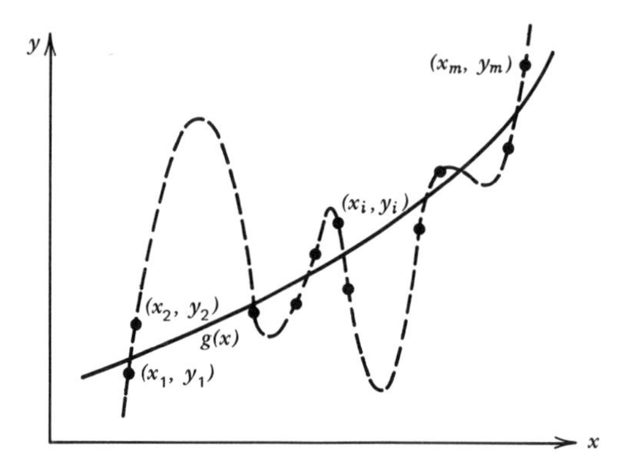

Figure 6.2 Approximation of experimentally determined points.

approximating function $g(x)$ that represents the general trend we are seeking; the dotted line indicates that disaster may strike if we blindly pretend that the data lie completely on a single curve. The *least-squares* method for determining a suitable approximating function will be discussed in Section 6.8.

3. Taylor's Expansion of a Known Function. Even if $f(x)$ is known analytically ($f(x) = \ln x$ or $\operatorname{erf} x$, for example), it is sometimes desirable to approximate it by another function $g(x)$ that is simpler to evaluate or manipulate. Section 6.9 presents Taylor's expansion for determining polynomial approximations $g(x)$ to any specified function $f(x)$, provided it is continuous and suitably differentiable.

6.2 Polynomials

In this chapter, we shall make extensive use of *polynomials* for the approximation of functions. The general nth-degree polynomial in a variable x is designated as $p_n(x)$, and is defined by

$$p_n(x) = a_0 + a_1x + a_2x^2 + \cdots + a_nx^n, \quad (6.1)$$

where the *coefficients* $a_0, a_1, a_2, \ldots, a_n$ are arbitrary, as yet. Examples of zeroth-, first-, and second-degree polynomials (with $a_0 = 3$, $a_1 = 1$, and $a_2 = -1/2$, for example) are:

$$p_0(x) = 3,$$

$$p_1(x) = 3 + x,$$

$$p_2(x) = 3 + x - \frac{1}{2}x^2.$$

From Fig. 6.3, note that $p_0(x)$ represents a straight line parallel to the x axis, $p_1(x)$ represents a straight line with a nonzero slope, and $p_2(x)$ represents a curve with a single maximum (or minimum).

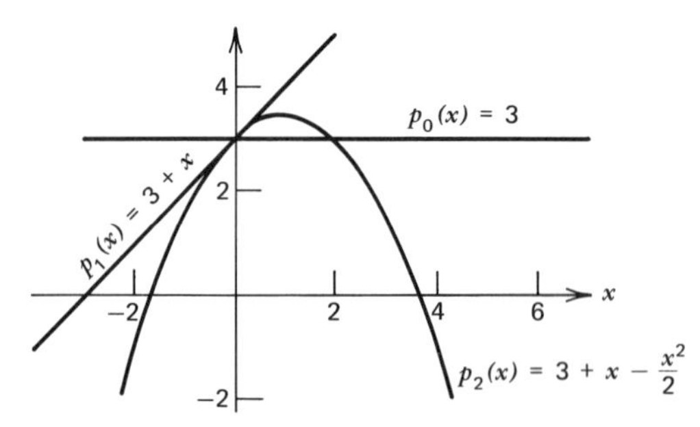

Figure 6.3 Examples of polynomials.

The evaluation of polynomials is of some interest—that is, the process of substituting numbers for x and the coefficients $a_0, a_1, \ldots, a_n$, and obtaining a value for $p_n(x)$. By first noting that (6.1) can be reexpressed as

$$p_n(x) = \sum_{i=0}^{n} a_i x^i, \quad (6.2)$$

we can readily construct the algorithm given in Fig. 6.4, in which P eventually contains the final value of the polynomial.

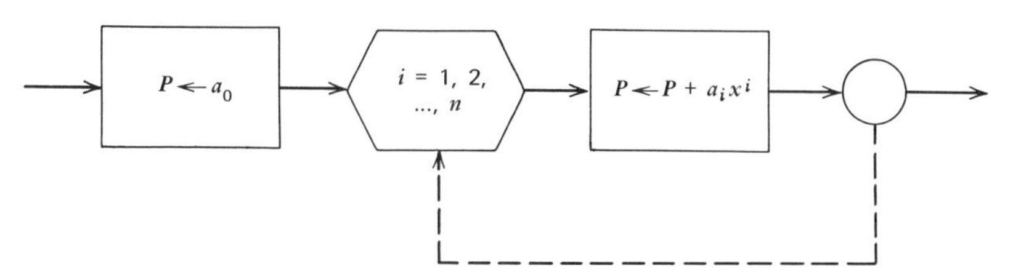

Figure 6.4 Evaluation of $p_n(x)$.

This procedure is not quite so readily translated into FORTRAN, because: (a) a zeroth vector element, such as A(0), is not allowed, and (b) the iteration or DO loop is performed at least once, even when the upper limit of the index is smaller than the initial value of the index (in particular, the case of a zeroth-degree polynomial, with $n = 0$, will need special treatment). Both of these difficulties can be overcome, however, by: (a) storing the coefficients $a_0, a_1, \ldots, a_n$ in elements A(1), A(2), . . . , A(N + 1) of the FORTRAN vector A, and (b) performing a test that will bypass the DO loop for the special case of $n = 0$. Assuming that N and X have been assigned the values of n and x, respectively, the necessary statements are:

```
      P = A(1)
      IF (N .EQ. 0)  GO TO 2
      DO 1  I = 1, N
    1 P = P + A(I + 1)*X**I
    2 (next statement)
```

Unless n is quite small, the above procedure is an inefficient way of evaluating a polynomial. By inspection of (6.1), and assuming that the exponentiation is being performed by repeated multiplication (the usual case for small n), observe that $n \times (n+1)/2$ multiplications and n additions are required to evaluate $p_n(x)$. But now consider the *nested* form:

$$p_n(x) = a_0 + x(a_1 + x(a_2 + \cdots + x(a_{n-1} + xa_n) \ldots)). \qquad (6.3)$$

By counting the number of multiplication and addition steps needed in (6.3), the reader should discover that it is considerably more efficient than (6.1), especially if n is large. The procedure described by (6.3) is known as *Horner's rule*, and forms the basis for Problem 6.1. Since both the nested and regular forms of the polynomial can be evaluated by the digital computer in a tiny fraction of a second, the greater computational efficiency afforded by the nested form will scarcely be noticed if the polynomial has to be evaluated a few times only. However, except for beginners, we recommend that the programmer should try to write efficient programs, unless in the process the statements become so hopelessly contorted that their meaning is largely lost. The reason is fairly obvious: the engineer, scientist, or applied mathematician will sooner or later be writing long programs that will take minutes or even hours to run. In such cases, even a 20% saving in total execution time, for example, could mean substantial savings in total running cost.

6.3 Linear Interpolation

We now reconsider the problem stated in the first part of Section 6.1—that of finding a suitable approximating function for the tabulated values $(x_i, f(x_i))$, $i = 1, 2, \ldots, n$. To a first approximation, we can draw straight-line segments between the successive points, as indicated in Fig. 6.5. One such segment is shown in Fig. 6.6; let the equation of this *linear* or *first-degree interpolating polynomial* be

$$p_1(x) = a_0 + a_1 x. \qquad (6.4)$$

The coefficients a_0 and a_1 can easily be obtained by noting that $p_1(x)$ passes through the base-points A and B, yielding

$$\begin{aligned} f(x_i) &= a_0 + a_1 x_i, \\ f(x_{i+1}) &= a_0 + a_1 x_{i+1}. \end{aligned} \qquad (6.5)$$

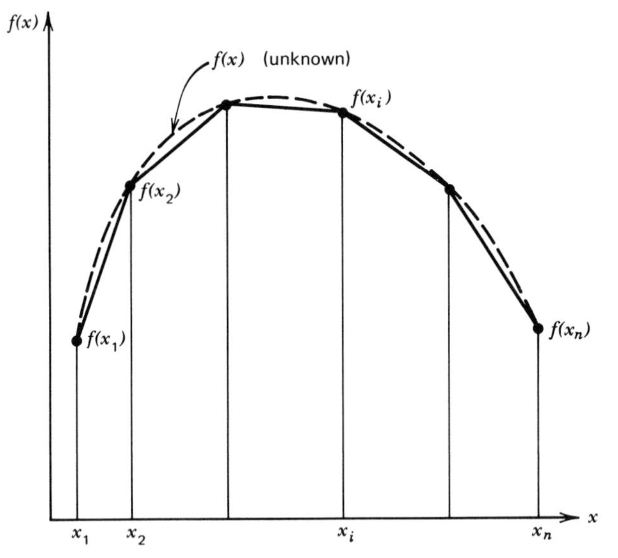

Figure 6.5 Straight-line segments between base points.

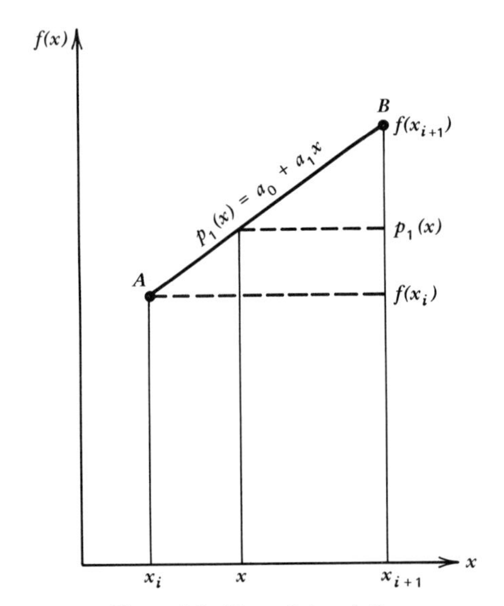

Figure 6.6 Linear interpolation.

These simultaneous equations have the solution

$$a_0 = \frac{x_{i+1}f(x_i) - x_i f(x_{i+1})}{x_{i+1} - x_i},$$
$$a_1 = \frac{f(x_{i+1}) - f(x_i)}{x_{i+1} - x_i}. \tag{6.6}$$

Substitution of a_0 and a_1 from (6.6) into (6.4), rearranging, and recognizing that $p_1(x)$ is an approximation to $f(x)$, we obtain:

$$f(x) \doteq p_1(x)$$
$$= f(x_i) + (x - x_i)\frac{f(x_{i+1}) - f(x_i)}{x_{i+1} - x_i} \tag{6.7}$$

$$= \frac{x - x_{i+1}}{x_i - x_{i+1}} f(x_i) + \frac{x - x_i}{x_{i+1} - x_i} f(x_{i+1}). \tag{6.8}$$

Equation (6.7) or (6.8) can be used for *estimating* values of $f(x)$ corresponding to *any* value of x. However, as shown in Fig. 6.7, the accuracy of the approximation is likely to be better if x lies *between* the base points A and B rather than *beyond* either of them (in which case we call the process linear *extrapolation*).

EXAMPLE 6.1

The following values are available [8] for the viscosity (μ) of 60% sucrose solution in water at various temperatures (T):

i	T_i (°C)	μ_i (centipoise)
1	10	113.9
2	20	56.7
3	30	34.01
4	40	21.30

Estimate μ at $T = 35$°C and at $T = 37$°C.

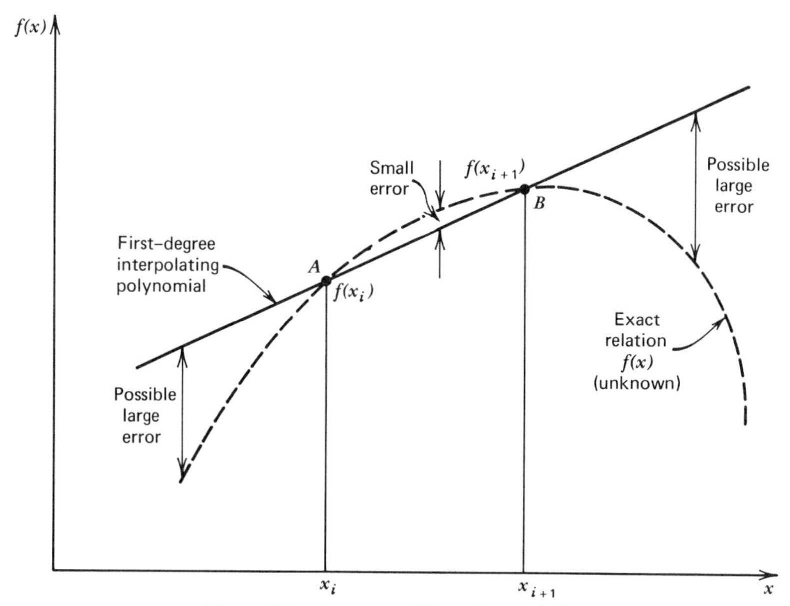

Figure 6.7 Accuracy of linear interpolation.

Since the given temperatures lie between $T_3 = 30$ and $T_4 = 40$, it is appropriate (and more accurate) to use the first-degree interpolating polynomial based on these points—and not, for example, that based on T_1 and T_2. Corresponding to equation (6.7), we have, with $i = 3$, replacing x by T and $f(x)$ by $\mu(T)$:

$$\mu \doteq \mu_3 + (T - T_3)\frac{\mu_4 - \mu_3}{T_4 - T_3}$$

$$= 34.01 + (T - 30)\frac{21.30 - 34.01}{40 - 30}.$$

That is,

$$\mu \doteq 72.14 \times 1.271\, T. \tag{6.9}$$

Substitution of $T = 35°C$ and $37°C$ into (6.9) gives estimated viscosities $\mu(35) \doteq 27.65$ and $\mu(37) \doteq 25.11$ centipoise. The actual value at 35°C, taken from a fuller version of the above table, is $\mu(35) = 26.62$ centipoise. The main reason for the discrepancy is that the *linear* interpolation has failed to account for the *curvature* in the actual relation between μ and T.

EXAMPLE 6.2

VISCOSITIES BY TABLE LOOK-UP AND LINEAR INTERPOLATION

Problem Statement

Consider n pairs of tabulated values (x_i, f_i), $i = 1, 2, \ldots, n$, with the x_i arranged in ascending algebraic order. (It is convenient when writing programs to suppose that the usual functional values $f(x_i)$, $i = 1, 2, \ldots, n$, have been stored as $f_1, f_2, \ldots, f_n$—that is, as successive elements of a vector f.) Given an interpolant value *xval*, write a function named TLULIN that performs a *table look-up procedure* to find those base points that lie nearest to *xval*, and then uses linear interpolation based on these base points to estimate the corresponding functional value *fval* $(= f(xval))$. Write a main program that calls on TLULIN to interpolate in Table 6.2.1 [8].

Table 6.2.1 Viscosities of 60% Sucrose Solution

i	T_i (°C)	μ_i (centipoise)
1	10	113.90
2	20	56.70
3	30	34.01
4	40	21.30
5	50	14.06
6	60	9.87
7	70	7.18
8	80	5.42
9	90	4.17

Use the following interpolant values: $T^* = 3.6, 10, 15, 35, 37, 41.37, 45, 50, 75, 80, 85, 90$, and 97.8°C. (Here, an asterisk has been appended to T for the same reason that x has been renamed *xval*—to avoid confusion with a vector of the same name.)

Method of Solution

The table look-up operation is best understood by considering the usual n pairs of tabulated values, with $x_1, x_2, \ldots, x_n$ arranged in ascending order of magnitude.

Case A:	$xval \rightarrow$		
		x_1	f_1
		x_2	f_2
		$\vdots$	$\vdots$
		x_i	f_i
Case B:	$xval \rightarrow$		
		x_{i+1}	f_{i+1}
		$\vdots$	$\vdots$
		x_{n-1}	f_{n-1}
		x_n	f_n
Case C:	$xval \rightarrow$		

Given *xval*, the main problem is to find the two base points that are most appropriate for use in the linear-interpolation formula, (6.7). If possible, base points x_i and x_{i+1} will be chosen so that $x_i \leqslant xval < x_{i+1}$, as in case B above. However, if *xval* lies outside the range of the table, extrapolation will be needed; thus case A will use x_1 and x_2 (the best available), and case C will employ x_{n-1} and x_n. The algorithm shown in the flow diagram uses a counter $i = 1, 2, \ldots, n-1$ to hunt through the table until *xval* is found to be smaller than the entry x_{i+1}; the appropriate base points are then x_i and x_{i+1}. If this never occurs, case C prevails and the base points are x_{n-1} and x_n.

An examination of Table 6.2.1 shows that the viscosity is nowhere near a linear function of temperature—that is, a plot of the μ_i against the T_i does not yield approximately a straight line. Hence, for a given temperature T^*, the error in the interpolated viscosity μ^* is probably appreciable. However, it is known [8] that the *logarithm* of viscosity *is* approximately a linear function of the *logarithm* of *absolute* temperature. Thus, in addition to interpolating in a table of μ_i versus T_i to produce a μ^* corresponding to a given T^*, we shall also employ the function TLULIN to interpolate in a table of ν_i

($= \ln \mu_i$) versus s_i ($= \ln (T_i + 273.16)$) to produce ν^* corresponding to a given s^* ($= \ln (T^* + 273.16)$). A second (and probably more accurate) estimate of the viscosity is then given by $\mu' = e^{\nu^*}$.

The function TLULIN is so arranged that the logical variable *flag* is set to T (true) if extrapolation is needed; if this is the case, an appropriate warning message is displayed in the main flow diagram.

Flow Diagram

Main (T_1 through T_9 and μ_1 through μ_9 are preset with the tabulated temperatures and corresponding viscosities of Table 6.2.1)

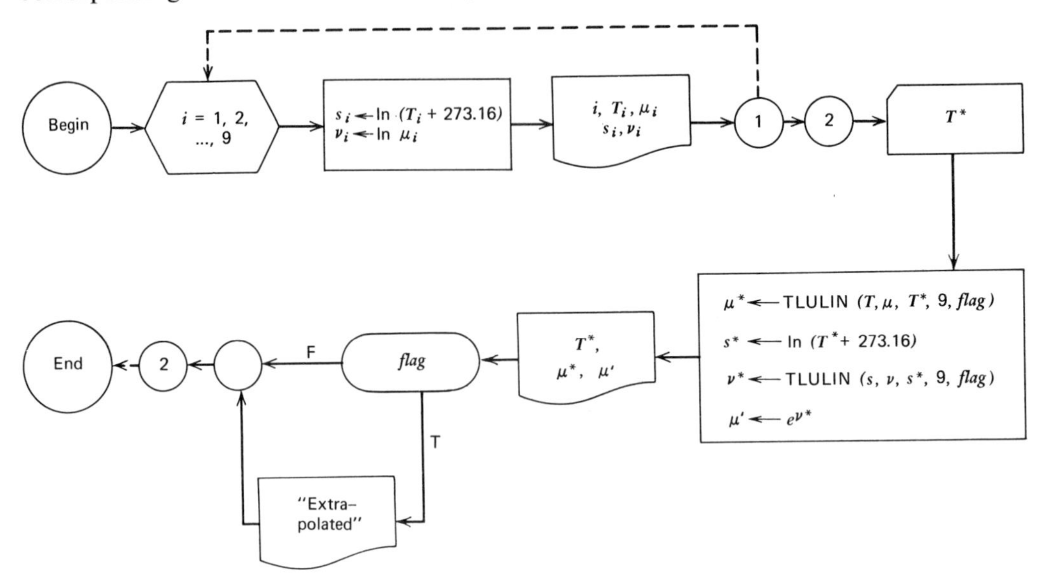

Function TLULIN $(x, f, xval, n, flag)$

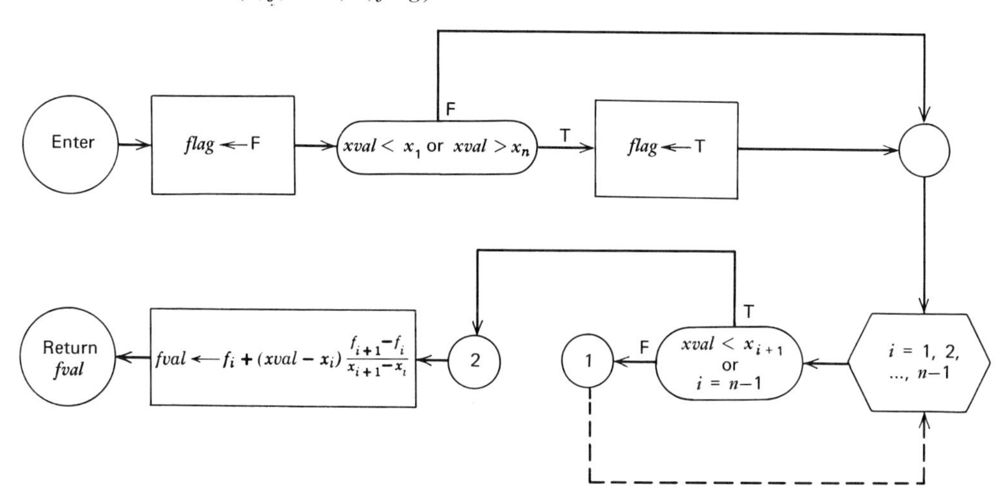

FORTRAN Implementation

List of Principal Variables

Program Symbol	Definition
(Main)	
FLAG	Logical variable, returned by TLULIN as .TRUE. if extrapolation is needed, and .FALSE. otherwise.
MU, NU	Vectors of tabulated viscosities μ_i (centipoise) and their logarithms ν_i.
MUSTAR, MULPRIM	Values of μ corresponding to T^*, obtained by direct and logarithmic interpolation, respectively.
NUSTAR	Interpolated value of ν corresponding to s^*.
SSTAR	s^*, logarithm of absolute temperature, $\ln(T^* + 273.16)$.
S	Vector of the s_i, the logarithms of absolute temperatures.
T	Vector of temperatures T_i, °C.
TSTAR	Temperature T^*, °C, at which the viscosity is required.
(Function TLULIN)	
F	Vector of functional values f_i.
N	Total number of tabulated values, n.
X	Vector of tabulated values x_i.
XVAL	Interpolant value, $xval$.

Program Listing

Main Program

```
C        DIGITAL COMPUTING                    EXAMPLE 6.2
C        VISCOSITIES BY TABLE LOOK-UP AND LINEAR INTERPOLATION.
C
      LOGICAL  FLAG
      REAL  MU, NU, MUPRIM, MUSTAR, NUSTAR
      DIMENSION  MU(9), NU(9), S(9), T(9)
C
C     ..... PRESET TEMPERATURES AND VISCOSITIES .....
      DATA  T / 10., 20., 30., 40., 50., 60., 70., 80., 90. /
      DATA  MU / 113.9,56.7,34.01,21.30,14.06,9.87,7.18,5.42,4.17 /
C
C     ..... PRINT TITLE AND CHECK TABULATED TEMPERATURES
C           AND VISCOSITIES AND THEIR LOGARITHMS .....
      WRITE (6,200)
      DO 1   I=1,9
      S(I) = ALOG(T(I) + 273.16)
      NU(I) = ALOG(MU(I))
    1 WRITE (6,201)    I, T(I), MU(I), S(I), NU(I)
      WRITE (6,202)
C
C     ..... READ VALUE FOR TEMPERATURE .....
    2 READ (5,100,END=999,ERR=2)   TSTAR
C
C     ..... USE TLULIN TO GET VISCOSITY BY DIRECT INTERPOLATION .....
      MUSTAR = TLULIN( T, MU, TSTAR, 9, FLAG )
C
C     ..... USE TLULIN TO GET VISCOSITY, INTERPOLATING VIA LOGS .....
      SSTAR = ALOG(TSTAR + 273.16)
      NUSTAR = TLULIN( S, NU, SSTAR, 9, FLAG )
      MUPRIM = EXP(NUSTAR)
C
C     ..... PRINT RESULTS, AND RETURN TO GET NEW TEMPERATURE .....
      WRITE (6,203)    TSTAR, MUSTAR, MUPRIM
      IF ( FLAG )   WRITE (6,204)
      GO TO 2
C
  999 CALL EXIT
C
C     ..... FORMATS FOR INPUT AND OUTPUT STATEMENTS .....
  100 FORMAT ( 8X, F6.2 )
  200 FORMAT ( '1DIGITAL COMPUTING                    EXAMPLE 6.2'/
     1   ' VISCOSITIES BY TABLE LOOK-UP AND LINEAR INTERPOLATION'/
     2   '0THE TABULATED TEMPERATURES, VISCOSITIES, AND THEIR LOGARITHMS
     3 ARE'/ '0', 5X, 'I', 10X, 'T(I)', 10X, 'MU(I)', 10X, 'S(I)', 11X,
     4   'NU(I)' / ' ' )
  201 FORMAT ( ' ', I6, F14.1, F15.2, 2F15.4 )
  202 FORMAT ( '0' / '0THE RESULTS GIVEN BY TLULIN FOR THE VARIOUS TEMPE
     1RATURES ARE' / '0', 15X, 'TSTAR', 9X, 'MUSTAR', 9X, 'MUPRIM' /' ')
  203 FORMAT ( ' ', 5X, 3F15.2 )
  204 FORMAT ( '+', 55X, '(EXTRAPOLATED)' )
C
      END
```

Function TLULIN

```
      FUNCTION TLULIN (X, F, XVAL, N, FLAG)
C
C       FUNCTION THAT PERFORMS TABLE LOOK-UP AND LINEAR INTERPOLATION
C       ON THE N TABULATED PAIRS OF VALUES X(1) ... X(N) AND FX(1) ...
C       FX(N).   GIVEN X = XVAL, THE CORRESPONDING FXVAL IS RETURNED.
C
      LOGICAL  FLAG
      DIMENSION  X(1), F(1)
C
      FLAG = .FALSE.
C
C     ..... SET FLAG TO TRUE IF EXTRAPOLATION IS NEEDED .....
      IF ( XVAL.LT.X(1) .OR. XVAL.GT.X(N) )   FLAG = .TRUE.
C
C     ..... PERFORM TABLE LOOK-UP TO SEE WHERE XVAL FITS .....
      NM1 = N - 1
      DO 1   I=1,NM1
      IF ( XVAL.LT.X(I+1) .OR. I.EQ.NM1 )   GO TO 2
    1 CONTINUE
C
C     ..... PERFORM LINEAR INTERPOLATION ON ITH AND (I+1)TH ENTRIES .....
    2 TLULIN = F(I) + (XVAL - X(I))*(F(I+1) - F(I))/(X(I+1) - X(I))
      RETURN
C
      END
```

Data

```
TSTAR =   3.60
TSTAR =  10.00
TSTAR =  15.00
TSTAR =  35.00
TSTAR =  37.00
TSTAR =  41.37
TSTAR =  45.00
TSTAR =  50.00
TSTAR =  75.00
TSTAR =  80.00
TSTAR =  85.00
TSTAR =  90.00
TSTAR =  97.80
```

Computer Output

```
DIGITAL COMPUTING                         EXAMPLE 6.2
VISCOSITIES BY TABLE LOOK-UP AND LINEAR INTERPOLATION

THE TABULATED TEMPERATURES, VISCOSITIES, AND THEIR LOGARITHMS ARE

        I           T(I)          MU(I)          S(I)          NU(I)

        1           10.0         113.90         5.6460        4.7353
        2           20.0          56.70         5.6807        4.0378
        3           30.0          34.01         5.7143        3.5267
        4           40.0          21.30         5.7467        3.0587
        5           50.0          14.06         5.7781        2.6433
        6           60.0           9.87         5.8086        2.2895
        7           70.0           7.18         5.8382        1.9713
        8           80.0           5.42         5.8669        1.6901
        9           90.0           4.17         5.8948        1.4279


THE RESULTS GIVEN BY TLULIN FOR THE VARIOUS TEMPERATURES ARE

                   TSTAR         MUSTAR         MUPRIM

                    3.60         150.51         180.33      (EXTRAPOLATED)
                   10.00         113.90         113.90
                   15.00          85.30          80.12
                   35.00          27.65          26.86
                   37.00          25.11          24.47
                   41.37          20.31          20.11
                   45.00          17.68          17.28
                   50.00          14.06          14.06
                   75.00           6.30           6.23
                   80.00           5.42           5.42
                   85.00           4.79           4.75
                   90.00           4.17           4.17
                   97.80           3.20           3.42      (EXTRAPOLATED)
```

Discussion of Results

The function TLULIN works satisfactorily, indicating when extrapolation is needed and also reproducing the functional values exactly when TSTAR coincides with one of the base points. There is an appreciable difference between MUSTAR and MUPRIM, obtained by direct and logarithmic interpolation, respectively. A comparison with additional known values (not used in the program) shows the wisdom of transforming the functional values if their approximate behavior is known:

TSTAR	MUSTAR	MUPRIM	Exact μ[8]
15	85.30	80.12	74.9
35	27.65	26.86	26.62
45	17.68	17.28	17.24
75	6.30	6.23	6.22
85	4.79	4.75	4.75

6.4 Quadratic Interpolation

The linear type of interpolation, just discussed, will reproduce the base-point values $(x_i, f(x_i))$ and $(x_{i+1}, f(x_{i+1}))$ exactly. Between these base points, however, the predicted values of $f(x)$ are only approximate, except for the rather special case in which $f(x)$ is already a linear function of x (that is, $f(x)$ is a first-degree polynomial, $p_1(x)$). In order to try and reduce the resulting error, we can attempt to account for some of the curvature in $f(x)$ by approximating it with a quadratic or *second-degree* interpolating polynomial:

$$f(x) \doteq p_2(x) = a_0 + a_1 x + a_2 x^2. \tag{6.10}$$

The coefficients a_0, a_1, and a_2 can be determined by requiring that $p_2(x)$ pass through *three* base points, such as $(x_0, f(x_0))$, $(x_1, f(x_1))$, and $(x_2, f(x_2))$, shown in Fig. 6.8. It should be apparent that the

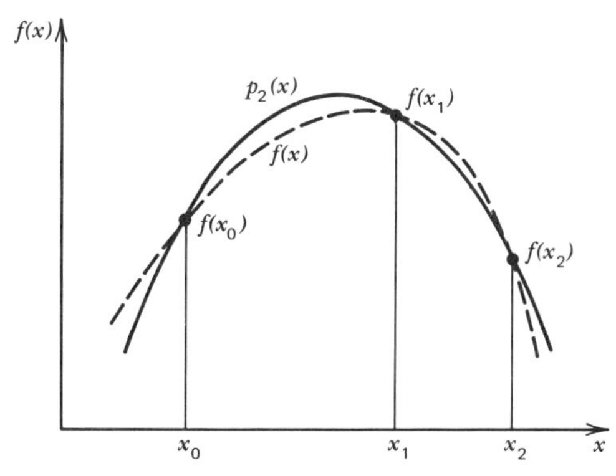

Figure 6.8 Quadratic interpolating polynomial.

base points can be assigned *any* convenient subscripts; for example, x_1, x_2, and x_3, or x_i, x_{i+1}, and x_{i+2} could be used instead. The present choice, which is the most conventional, has the slight advantage that the highest base-point subscript, 2, is identical with the degree of the interpolating polynomial.

We now have three simultaneous equations in a_0, a_1, and a_2:

$$\begin{aligned} f(x_0) &= a_0 + a_1 x_0 + a_2 x_0^2, \\ f(x_1) &= a_0 + a_1 x_1 + a_2 x_1^2, \\ f(x_2) &= a_0 + a_1 x_2 + a_2 x_2^2. \end{aligned} \tag{6.11}$$

The solution of (6.11) is rather tedious (see Problem 6.3), but eventually leads to

$$\begin{aligned} a_0 &= \frac{x_1 x_2 (x_2 - x_1) f(x_0) + x_2 x_0 (x_0 - x_2) f(x_1) + x_0 x_1 (x_1 - x_0) f(x_2)}{(x_0 - x_1)(x_1 - x_2)(x_2 - x_0)}, \\ a_1 &= \frac{(x_1^2 - x_2^2) f(x_0) + (x_2^2 - x_0^2) f(x_1) + (x_0^2 - x_1^2) f(x_2)}{(x_0 - x_1)(x_1 - x_2)(x_2 - x_0)}, \\ a_2 &= \frac{(x_2 - x_1) f(x_0) + (x_0 - x_2) f(x_1) + (x_1 - x_0) f(x_2)}{(x_0 - x_1)(x_1 - x_2)(x_2 - x_0)}. \end{aligned} \tag{6.12}$$

Substituting for a_0, a_1, and a_2 from (6.12) into (6.10) and simplifying leads to

$$\begin{aligned} f(x) \doteq p_2(x) = &\frac{(x - x_1)(x - x_2)}{(x_0 - x_1)(x_0 - x_2)} f(x_0) \\ &+ \frac{(x - x_0)(x - x_2)}{(x_1 - x_0)(x_1 - x_2)} f(x_1) \\ &+ \frac{(x - x_0)(x - x_1)}{(x_2 - x_0)(x_2 - x_1)} f(x_2). \end{aligned} \tag{6.13}$$

Hopefully, (6.13) will be a better approximation to $f(x)$ than was the first-degree interpolating polynomial of (6.7) and (6.8). (See Problem 6.4.)

6.5 The Lagrange Interpolating Polynomial

If we are still not satisfied with the accuracy afforded by (6.13), we could seek to construct a cubic or *third-degree* interpolating polynomial (involving four base points), and so on. However, the algebra involved in solving equations such as (6.5) and (6.11) for the polynomial coefficients becomes quite complicated. As a more systematic approach, we shall now investigate a general method for constructing an *nth-degree* interpolating polynomial $p_n(x)$. Note that the required number of base points is always one more than the degree of the interpolating polynomial; thus we shall generate $p_n(x)$ so that it passes exactly through the $n+1$ points $(x_i, f(x_i))$, $i = 0, 1, \ldots, n$.

To start, assume that the interpolating polynomial has the form

$$\begin{aligned} p_n(x) = {} & b_0(x-x_1)(x-x_2)\cdots(x-x_n) \\ & + b_1(x-x_0)(x-x_2)\cdots(x-x_n) \\ & + b_2(x-x_0)(x-x_1)(x-x_3)\cdots(x-x_n) \\ & + \cdots + b_i(x-x_0)(x-x_1)\cdots(x-x_{i-1})(x-x_{i+1})\cdots(x-x_n) \\ & + \cdots + b_{n-1}(x-x_0)(x-x_1)\cdots(x-x_{n-2})(x-x_n) \\ & + b_n(x-x_0)(x-x_1)\cdots(x-x_{n-2})(x-x_{n-1}). \end{aligned} \tag{6.14}$$

Although (6.14) may seem rather complicated, it is indeed a general nth-degree polynomial because there are $n+1$ arbitrary coefficients and all powers of x from x^0 through x^n are represented. The coefficients $b_0, b_1, \ldots, b_n$ are obtained, as before, by requiring that the polynomial must reproduce the base-point values exactly; that is, $p_n(x_i) = f(x_i)$, for $i = 0, 1, \ldots, n$. The technique is to equate x to $x_0, x_1, \ldots, x_n$ in turn, which causes all terms except one on the right-hand side of (6.14) to disappear. We obtain

$$b_0 = \frac{f(x_0)}{(x_0-x_1)(x_0-x_2)\cdots(x_0-x_n)},$$

$$b_1 = \frac{f(x_1)}{(x_1-x_0)(x_1-x_2)\cdots(x_1-x_n)},$$

and, in general,

$$b_i = \frac{f(x_i)}{(x_i-x_0)\cdots(x_i-x_{i-1})(x_i-x_{i+1})\cdots(x_i-x_n)} = f(x_i)\prod_{\substack{j=0 \\ j\neq i}}^{n} \frac{1}{(x_i-x_j)}. \tag{6.15}$$

(The symbol Π means a *repeated product*; for example, $\prod_{i=1}^{n} x_i$ or $\Pi_{i=1}^{n} x_i$ denotes $x_1 x_2 \ldots x_n$.)

Also, (6.14) can be rewritten as

$$p_n(x) = \sum_{i=0}^{n} b_i \prod_{\substack{j=0 \\ j\neq i}}^{n} (x-x_j). \tag{6.16}$$

Hence, from (6.15) and (6.16), the final form for $p_n(x)$, known as *Lagrange's interpolating polynomial*, is

$$p_n(x) = \sum_{i=0}^{n} L_i(x) f(x_i), \tag{6.17}$$

where

$$L_i(x) = \prod_{\substack{j=0 \\ j\neq i}}^{n} \left(\frac{x-x_j}{x_i-x_j}\right). \tag{6.18}$$

The advantage of the Lagrange form is that the notation is particularly compact, and that the necessary summation (Σ) and repeated multiplication (Π) can be achieved readily by using iterative loops in a computer program (see Problem 6.18). As mentioned previously, the choice of subscripts $i = 0, 1, \ldots, n$ for the base points is somewhat arbitrary. Suppose, for example, that we have a table containing a large number of base points whose subscripts range from 1 through 100. Theoretically, we could use all these points to construct a 99th-degree interpolating polynomial! The amount of computing involved would be out of all proportion to any gain in accuracy. Also, we would probably wish to interpolate in a small portion of the table only. Thus, it might be appropriate to use an nth-degree polynomial, where n has a much more modest value—between one and ten, for example. A total of $n+1$ base points would be involved, and for best accuracy these should be chosen to have subscripts $k, k+1, k+2, \ldots, k+n$, so that the value x for which interpolation is required is roughly centered between x_k and k_{k+n}. The situation is illustrated in Fig. 6.9.

To accommodate the renumbering of the base points, equations (6.17) and (6.18) would simply be modified so that the range on the summation became $\Sigma_{i=k}^{k+n}$, and likewise for the repeated product, $\Pi_{j=k, j\neq i}^{k+n}$.

Finally, note that the interpolating polynomial passing through a given set of base points is unique (assuming that none of the values x_i is repeated). Therefore, Lagrange's interpolating polynomial must be equivalent to (6.8) for $n=1$ (see the following example) and to (6.13) for $n=2$ (see Problem 6.5).

Approximate range for x

$$\begin{array}{cccccccc} & & \rightarrow| & & & |\leftarrow & & \\ x_1 & x_2 & x_k & x_{k+1} & x_{k+2} & x_{k+n} & x_{99} & x_{100} \\ f(x_1) & f(x_2)\cdots & f(x_k) & f(x_{k+1}) & f(x_{k+2})\cdots & f(x_{k+n})\cdots & f(x_{99}) & f(x_{100}) \end{array}$$

Points used in constructing the nth-degree interpolating polynomial.

Figure 6.9 Selection of base points.

EXAMPLE 6.3

For $n=1$ (linear interpolation), verify that Lagrange's form (6.17) and (6.18), is equivalent to equation (6.8). From (6.18), we have

$$L_0(x) = \frac{x-x_1}{x_0-x_1}, \qquad L_1(x) = \frac{x-x_0}{x_1-x_0}.$$

(Note that for the simple case of $n=1$, each "repeated" product consists of a single term only.) Substitution into (6.17) gives

$$p_1(x) = \frac{x-x_1}{x_0-x_1} f(x_0) + \frac{x-x_0}{x_1-x_0} f(x_1).$$

By calling the base points x_i and x_{i+1} instead of x_0 and x_1, (6.18) is obtained.

EXAMPLE 6.4

Obtain the second-degree interpolating polynomial that passes through the following base points and functional values:

i	x_i	$f(x_i)$
0	0	-5
1	1	1
2	3	25

Using Lagrange's formula, (6.18), we have:

$$L_0 = \frac{(x-x_1)(x-x_2)}{(x_0-x_1)(x_0-x_2)}$$

$$= \frac{(x-1)(x-3)}{(0-1)(0-3)} = \frac{1}{3}(x^2-4x+3),$$

$$L_1 = \frac{(x-x_0)(x-x_2)}{(x_1-x_0)(x_1-x_2)}$$

$$= \frac{(x-0)(x-3)}{(1-0)(1-3)} = \frac{1}{2}(-x^2+3x),$$

$$L_2 = \frac{(x-x_0)(x-x_1)}{(x_2-x_0)(x_2-x_1)}$$

$$= \frac{(x-0)(x-1)}{(3-0)(3-1)} = \frac{1}{6}(x^2-x).$$

Hence, from (6.17), the required polynomial is

$$p_2(x) = \sum_{i=0}^{2} L_i(x) f(x_i)$$

$$= L_0(-5) + L_1(1) + L_2(25)$$

$$= -\frac{5}{3}(x^2-4x+3) + \frac{1}{2}(-x^2+3x)$$

$$+ \frac{25}{6}(x^2-x)$$

$$= 2x^2+4x-5.$$

The reader should verify that $p_2(x)$ does indeed represent the three indicated functional values at the given base points.

6.6 Error Associated with the Interpolating Polynomial

We have already remarked that the nth-degree interpolating polynomial $p_n(x)$ will reproduce the $n+1$ base-point values exactly. However, for other values of x, $p_n(x)$ will, in general, only be an approximation to $f(x)$, a fact that can be reflected by writing

$$f(x) = p_n(x) + R_n(x). \qquad (6.19)$$

Here, $R_n(x)$ is the *remainder* or *error term* associated with the interpolating polynomial of degree n. In general, $R_n(x)$ may not be known precisely, but the following development will yield useful information concerning it.

We first suppose that the error term can more conveniently be expressed as

$$R_n(x) = \left[\prod_{i=0}^{n} (x - x_i)\right] G(x), \qquad (6.20)$$

where $G(x)$ is a function yet to be investigated. The particular choice of (6.20) is motivated partly by the fact that it obviously reduces to $R_n(x) = 0$, and hence conforms to the essential requirement $p_n(x_i) = f(x_i)$ at the base points x_i, $i = 0, 1, \ldots, n$.

For any particular value of x, next consider a new function $Q(t)$, of another variable t, defined by

$$Q(t) = f(t) - p_n(t) - \left[\prod_{i=0}^{n} (t - x_i)\right] G(x). \qquad (6.21)$$

Note that $Q(t) = 0$ when t assumes any of the base-point values x_i, $i = 0, 1, \ldots, n$; $Q(t)$ is also zero when $t = x$, by virtue of equations (6.19) and (6.20). Thus $Q(t)$ vanishes (equals zero) $n+2$ times on the smallest closed interval containing x *and* the $n+1$ base points. In other words, $Q(t) = 0$ possesses $n+2$ roots on the *closed* interval $[x, x_0, x_1, \ldots, x_n]$, which contains the smallest and largest of the specified values $x, x_0, x_1, \ldots, x_n$ as its end points.

If $f(t)$ is continuous and suitably differentiable, Rolle's theorem* requires that $Q'(t)$ (the first derivative, dQ/dt) vanish at least $n+1$ times on the interval. Application of Rolle's theorem repeatedly to successively higher derivatives shows that $Q''(t)$ must have at least n zeros, $Q^{(3)}(t)$ must have at least $n-1$ zeros, etc., and that $Q^{(n+1)}(t)$ must vanish at least once on the interval. Let such a point be at $t = \zeta$. That is, we can say that ζ is on the *open* interval $(x, x_0, x_1, \ldots, x_n)$ (the largest interval *not* including the smallest and largest of the values $x, x_0, x_1, \ldots, x_n$).

Upon differentiating (6.21) $n+1$ times with respect to t, we obtain:

$$Q^{(n+1)}(t) = f^{(n+1)}(t) - p_n^{(n+1)}(t) - (n+1)!G(x).$$

But $p_n^{(n+1)}(t) = 0$, since $p_n(t)$ is an nth-degree polynomial. At $t = \zeta$, since $Q^{(n+1)}(\zeta) = 0$, there results

$$G(x) = \frac{f^{(n+1)}(\zeta)}{(n+1)!}, \qquad \zeta \text{ in } (x, x_0, x_1, \ldots, x_n). \qquad (6.22)$$

The final expression for the error term is obtained by combining (6.20) and (6.22):

$$R_n(x) = \left[\prod_{i=0}^{n} (x - x_i)\right] \frac{f^{(n+1)}(\zeta)}{(n+1)!}, \qquad \zeta \text{ in } (x, x_0, x_1, \ldots, x_n). \qquad (6.23)$$

Note that the value of ζ is unknown, except that it lies *somewhere* on the interval between the smallest and largest of the values $x, x_0, x_1, \ldots, x_n$. Also, the $(n+1)$th derivative $f^{(n+1)}(x)$ can only be formulated if the function $f(x)$ is known completely—

**Rolle's Theorem:* Let $f(x)$ be continuous for $a \leqslant x \leqslant b$ and differentiable for $a < x < b$; if $f(a) = f(b)$, then $f'(\zeta) = 0$ for at least one ζ where $a < \zeta < b$. (This theorem has a simple geometrical interpretation, which the reader should investigate.)

an unlikely event if only tabulated values are supplied. In spite of these limitations, (6.23) *does* have some utility:

(a) It does enable an upper bound to be placed on the error if $f(x)$ *is* specified and can be differentiated $n+1$ times.

(b) It indicates that the use of $p_n(x)$ for *extrapolation* is likely to be inaccurate relative to its use for *interpolation* because the magnitude of the term in square brackets increases rapidly as x moves away from the interval containing the base points.

(c) It yields valuable information concerning the accuracy of integration formulas (see Chapter 7).

EXAMPLE 6.5

A second-degree interpolating polynomial, $p_2(x)$, is to be used to approximate $f(x) = \cos x$ on the interval $[0, 2]$. If the base points used are at $x = 0$, 1, and 2, obtain an upper bound for the error involved.

The error term (6.23) becomes

$$R_2(x) = x(x-1)(x-2)\frac{\sin\zeta}{6}.$$

Note first that the maximum magnitude of $x(x-1)(x-2)$ is $\sqrt{4/27} \doteq 0.385$, occurring when $x = 1 \pm \sqrt{1/3}$. Also, $\sin\zeta$ cannot exceed 1 on the interval under consideration. Hence the magnitude of the error cannot possibly exceed $0.385 \times 1/6 \doteq 0.064$, which is the required upper bound. Note that we have been forced to take a pessimistic viewpoint; in all probability, the actual error will be appreciably less than this upper bound.

6.7 Spline-Function Approximation

Spline-function approximation is essentially the numerical representation of an established drafting technique—that of using a flexible strip or *spline* for drawing a smooth curve through a set of points. The basic concept can be grasped readily if the following items are available: some bronze weather strip (or other thin flexible strip), a hammer, some nails or panel pins, and a wooden board. First, hammer the pins part way into the board, and pretend that they represent the n accurately known functional values $(x_i, f(x_i))$, $i = 1, 2, \ldots, n$, shown in Fig. 6.10. Then interweave the flexible strip, represented by the curve in Fig. 6.10, between the pins so that it presses against every pin. This practical demonstration serves to show that the strip is straight (that is, has zero curvature) at the two end pins, and that it has no sharp corners anywhere. We seek a mathematical representation of the shape of the strip, which is then taken as a spline-function approximation for $f(x)$ over the interval $[x_1, x_n]$.

An elementary knowledge of the mechanics of materials is helpful, but not essential, for the understanding of the following development. The strip in the above demonstration is essentially a simply supported beam, whose shape or deflection $y = y(x)$ is given approximately by the solution of the ordinary differential equation [27]:

$$EI\frac{d^2y}{dx^2} = M. \qquad (6.24)$$

Here, E is Young's modulus of elasticity and I is the cross-sectional moment of inertia; for our purposes, it suffices to know that they are both constants for a given beam. M is the bending moment at any point, and for a simply supported beam it is known to be continuous, to equal zero at the two end supports x_1 and x_n, and to vary linearly with x between any two supports such as x_i and x_{i+1}. Since the second derivative of y is therefore a linear function of x, it follows that the shape of the strip between any two pins is given by a third-degree or *cubic* polynomial, $p_3(x)$.

Thus, over successive intervals $[x_i, x_{i+1}]$, $i = 1, 2, \ldots, n-1$, we seek to approximate $f(x)$ by a series of cubic polynomials:

$$p_{3,i}(x), \qquad i = 1, 2, \ldots, n-1. \qquad (6.25)$$

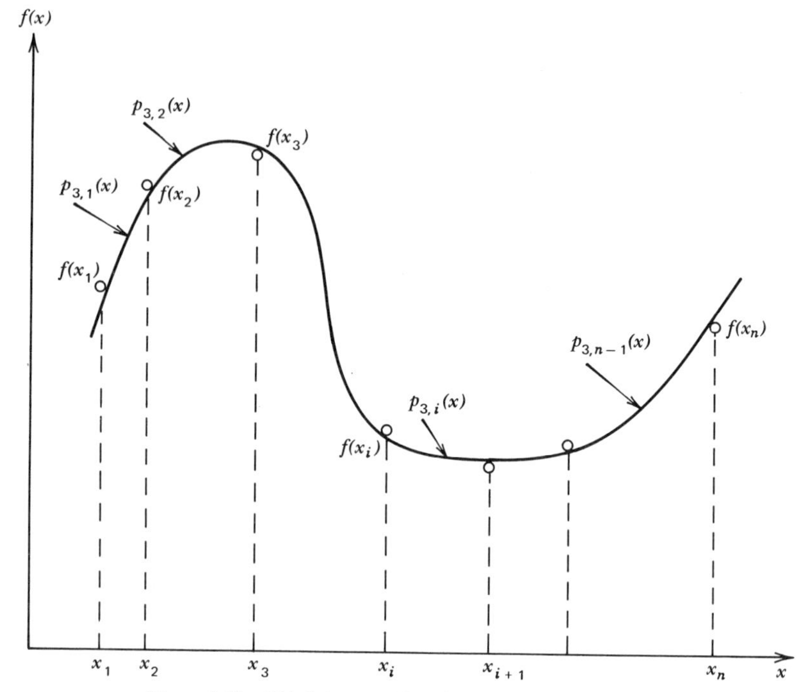

Figure 6.10 Third-degree spline-function approximation.

Since each cubic involves four arbitrary coefficients, there are altogether $4(n-1)$ coefficients whose values have to be determined. A knowledge of $p_{3,1}(x)$, $p_{3,2}(x), \ldots, p_{3,n-1}(x)$ then constitutes the required spline-function approximation for $f(x)$.

The requirement that each cubic polynomial $p_{3,i}(x)$ should match the known functional values at its two ends x_i and x_{i+1} affords $2(n-1)$ relations:

$$\left.\begin{aligned} p_{3,i}(x_i) &= f(x_i), \\ p_{3,i}(x_{i+1}) &= f(x_{i+1}), \end{aligned}\right\} i = 1, 2, \ldots, n-1. \tag{6.26}$$

A further $2(n-2)$ relations are obtained by requiring that the first and second derivatives of successive polynomials must match each other in value at the intermediate base points:

$$p'_{3,i}(x_i) = p'_{3,i-1}(x_i), \quad i = 2, 3, \ldots, n-1, \tag{6.27}$$

$$p''_{3,i}(x_i) = p''_{3,i-1}(x_i), \quad i = 2, 3, \ldots, n-1. \tag{6.28}$$

Conditions (6.27) and (6.28) correspond, respectively, to the known facts that the slope dy/dx and bending moment M are continuous along a simply supported beam. M is known to be zero at the end supports; from (6.24) this means that d^2y/dx^2 is also zero at these points, leading to

$$\begin{aligned} p''_{3,1}(x_1) &= 0, \\ p''_{3,n-1}(x_n) &= 0. \end{aligned} \tag{6.29}$$

Physically, conditions (6.29) correspond to the earlier observation that the strip has zero curvature at the two end pins.

Equations (6.26) through (6.29) amount to $4(n-1)$ relations, which, in principle, enable the $4(n-1)$ polynomial coefficients to be determined. The following approach enables the resulting algebra to be conducted systematically, although the details are still moderately involved. For ease of notation, define

$$f_i = f(x_i), \tag{6.30}$$

$$h_i = x_{i+1} - x_i, \tag{6.31}$$

$$\phi_i = p''_{3,i-1}(x_i) = p''_{3,i}(x_i). \tag{6.32}$$

Since each $p''_{3,i}(x)$ is, by definition, a linear function of x, it can be obtained on the interval $[x_i, x_{i+1}]$ by linear interpolation of the values ϕ_i and ϕ_{i+1} at each end of the interval:

$$p''_{3,i}(x) = \frac{x_{i+1}-x}{h_i}\phi_i + \frac{x-x_i}{h_i}\phi_{i+1}. \quad (6.33)$$

By integrating twice and imposing the conditions (6.26), it follows after some algebra that

$$p_{3,i}(x) = \frac{\phi_i}{6h_i}(x_{i+1}-x)^3 + \frac{\phi_{i+1}}{6h_i}(x-x_i)^3 + \left(\frac{f_{i+1}}{h_i} - \frac{h_i\phi_{i+1}}{6}\right)(x-x_i) + \left(\frac{f_i}{h_i} - \frac{h_i\phi_i}{6}\right)(x_{i+1}-x),$$
$$i = 1, 2, \ldots, n-1. \quad (6.34)$$

Once the ϕ_i, $i = 1, 2, \ldots, n$ have been determined as indicated below, substitution into (6.34) yields the individual cubic spline polynomials for use over successive intervals.

By differentiating (6.34) and imposing (6.27), we obtain the following system of linear equations in the ϕ_i:

$$\frac{h_{i-1}}{h_i}\phi_{i-1} + 2\left(1+\frac{h_{i-1}}{h_i}\right)\phi_i + \phi_{i+1} = \frac{6}{h_i}\left(\frac{f_{i+1}-f_i}{h_i} - \frac{f_i - f_{i-1}}{h_{i-1}}\right),$$
$$i = 2, 3, \ldots, n-1, \quad (6.35)$$

also with

$$\phi_1 = 0, \quad \phi_n = 0, \quad (6.36)$$

which follow from (6.29).

Note that (6.35) and (6.36) amount to n simultaneous linear equations in the n unknowns $\phi_1, \phi_2, \ldots, \phi_n$, whose values can then be determined. For small values of n (3, 4, or 5), the calculations can be done fairly easily by hand (see Example 6.6 below), but machine computation is advisable when larger numbers of points are involved. The simultaneous equations in the ϕ_i can be solved by the Gauss-Jordan method discussed in Section 8.4 and 8.6. However, (6.35) and (6.36) constitute a *tridiagonal* system, in which each equation involves three unknowns at most: ϕ_i and its two immediate neighbors ϕ_{i-1} and ϕ_{i+1}. An efficient algorithm is available for the solution of tridiagonal systems, and this is discussed further in Section 8.8.

The reader who wishes to learn more about spline-function approximation is referred to Ahlberg, Nilson, and Walsh [28]. A related technique, due to Akima [39], is also described in Problem 6.32.

EXAMPLE 6.6

Develop a cubic spline-function approximation for the following table of functional values, also illustrated in Fig. 6.11.

i	x_i	$f(x_i)$	h_i
1	1	1	1
2	2	3	2
3	4	4	1
4	5	2	–

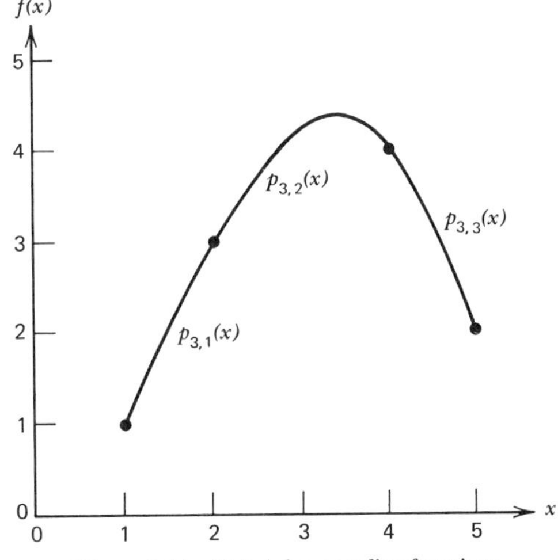

Figure 6.11 Third-degree spline functions.

Written out in full with numerical values inserted, equations (6.35) and (6.36) become:

$$i=1:\quad \phi_1 = 0;$$

$$i=2:\quad \frac{1}{2}\phi_1 + 2\left(1+\frac{1}{2}\right)\phi_2 + \phi_3 = \frac{6}{2}\left(\frac{4-3}{2} - \frac{3-1}{1}\right);$$

$$i=3:\quad 2\phi_2 + 2\left(1+\frac{2}{1}\right)\phi_3 + \phi_4 = \frac{6}{1}\left(\frac{2-4}{1} - \frac{4-3}{2}\right);$$

$$i=4:\quad \phi_4 = 0.$$

These simultaneous equations are readily solved, yielding:

$$\phi_1 = 0,\quad \phi_2 = -\frac{3}{4},\quad \phi_3 = -\frac{9}{4},\quad \phi_4 = 0. \tag{6.37}$$

Back-substitution into (6.34) then gives the individual cubic splines for use over each of the three successive intervals.

For example, the spline approximation over the second interval $[2,4]$ is, from (6.34), with $i=2$:

$$\begin{aligned} p_{3,2}(x) &= \frac{\phi_2}{6h_2}(x_3-x)^3 + \frac{\phi_3}{6h_2}(x-x_2)^3 \\ &\quad + \left(\frac{f_3}{h_2} - \frac{h_2\phi_3}{6}\right)(x-x_2) \\ &\quad + \left(\frac{f_2}{h_2} - \frac{h_2\phi_2}{6}\right)(x_3-x) \\ &= \frac{-3/4}{6\times 2}(4-x)^3 - \frac{9/4}{6\times 2}(x-2)^3 \\ &\quad + \left(\frac{4}{2} - \frac{2(-9/4)}{6}\right)(x-2) \\ &\quad + \left(\frac{3}{2} - \frac{2(-3/4)}{6}\right)(4-x) \\ &= -\frac{1}{8}x^3 + \frac{3}{8}x^2 + \frac{7}{4}x - 1. \end{aligned}$$

The other two approximations are obtained similarly. The complete cubic spline-function approximation to $f(x)$ is shown in Fig. 6.11 and is given by:

$$\left.\begin{aligned} &1 \leqslant x \leqslant 2: \\ &\qquad p_{3,1}(x) = -\frac{1}{8}x^3 + \frac{3}{8}x^2 + \frac{7}{4}x - 1; \\ &2 \leqslant x \leqslant 4: \\ &\qquad p_{3,2}(x) = -\frac{1}{8}x^3 + \frac{3}{8}x^2 + \frac{7}{4}x - 1; \\ &4 \leqslant x \leqslant 5: \\ &\qquad p_{3,3}(x) = \frac{3}{8}x^3 - \frac{45}{8}x^2 + \frac{103}{4}x - 33. \end{aligned}\right\} \tag{6.38}$$

Note that in this particular example, it so happens that $p_{3,1}(x)$ and $p_{3,2}(x)$ are identical.

6.8 Least-Squares Polynomial Approximation

We now reconsider the problem stated in the second part of Section 6.1—that of finding a suitable approximating function for the experimentally determined data points (x_i, y_i), $i = 1, 2, \ldots, m$. Recall that the individual x_i and y_i values are of questionable accuracy, and that it would be unrealistic to use an interpolating polynomial or spline function that passed through each point.

Assume that the points can be represented by the first-degree polynomial or straight line

$$y = a_0 + a_1 x, \tag{6.39}$$

as shown in Fig. 6.12. The intercept a_0 on the y axis at $x=0$ and the slope a_1 are, as yet, arbitrary; clearly, they can be adjusted so that the straight line (also called the *regression line*) follows the general trend of the points to a greater or lesser extent. Note from Fig. 6.12 that the vertical deviation δ_i of the ith point from the regression line is given by

$$\delta_i = y_i - a_0 - a_1 x_i. \tag{6.40}$$

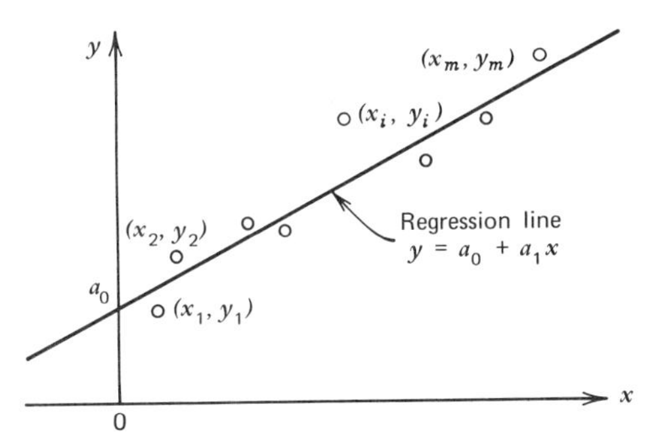

Figure 6.12 Linear regression.

Based on a certain assumed statistical model, it can be shown [1] that the regression line is considered the best representation of the data points if a_0 and a_1 are chosen so as to *minimize* a quantity S, the sum of the *squares* of the deviations δ_i for all the points:

$$S = \sum_{i=1}^{m} \delta_i^2 = \sum_{i=1}^{m} (y_i - a_0 - a_1x_i)^2. \quad (6.41)$$

Since the data points have already been determined, S is a function of the two parameters a_0 and a_1. Thus S can be minimized by equating the two partial derivatives $\partial S/\partial a_0$ and $\partial S/\partial a_1$ to zero, giving:

$$\sum_{i=1}^{m} (y_i - a_0 - a_1x_i) = 0,$$
$$\sum_{i=1}^{m} x_i(y_i - a_0 - a_1x_i) = 0. \quad (6.42)$$

Noting that $\Sigma_{i=1}^{m} a_0 = ma_0$ and $\Sigma_{i=1}^{m} a_1x_i = a_1\Sigma_{i=1}^{m} x_i$, rearrangement of equations (6.42) gives the following two simultaneous equations in a_0 and a_1:

$$a_0m \quad + a_1 \sum_{i=1}^{m} x_i = \sum_{i=1}^{m} y_i,$$
$$a_0 \sum_{i=1}^{m} x_i + a_1 \sum_{i=1}^{m} x_i^2 = \sum_{i=1}^{m} x_iy_i. \quad (6.43)$$

Equations (6.43), also known as the *normal* equations, have the solution

$$a_0 = \left(\sum_{i=1}^{m} y_i - a_1 \sum_{i=1}^{m} x_i\right) \Big/ m, \quad (6.44)$$

with

$$a_1 = \frac{m \sum_{i=1}^{m} x_iy_i - \sum_{i=1}^{m} x_i \sum_{i=1}^{m} y_i}{m \sum_{i=1}^{m} x_i^2 - \left(\sum_{i=1}^{m} x_i\right)^2}. \quad (6.45)$$

EXAMPLE 6.7

The following data have been published [11] for the tar (x) and nicotine (y) content (in milligrams) of several brands of king-size filter cigarettes:

Tar:	8.3	12.3	18.8	22.9	23.1	24.0	27.3	30.0	35.9	41.6
Nicotine:	0.32	0.46	1.10	1.32	1.26	1.44	1.42	1.96	2.23	2.20

These data are plotted in Fig. 6.13, and the general trend appears to be linear. Compute the least-squares first-degree polynomial that expresses y (the nicotine content) as a function of x (the tar content).

The main task is to compute the various sums needed in equations (6.44) and (6.45). The following values were obtained by hand.

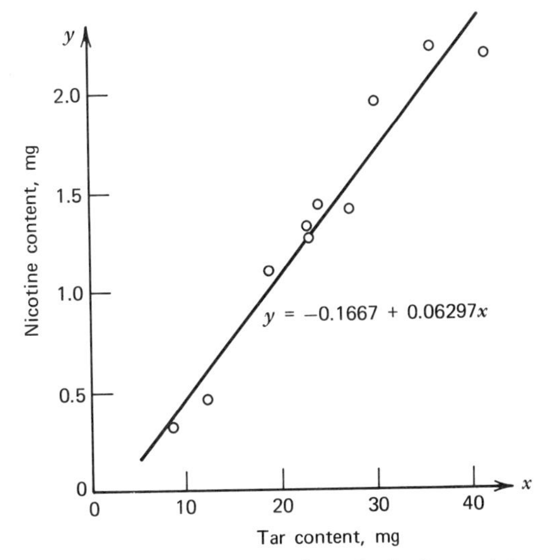

Figure 6.13 Linear regression of nicotine and tar content of cigarettes.

i	x_i	y_i	x_i^2	$x_i y_i$
1	8.3	0.32	68.89	2.656
2	12.3	0.46	151.29	5.658
3	18.8	1.10	353.44	20.680
4	22.9	1.32	524.41	30.228
5	23.1	1.26	533.61	29.106
6	24.0	1.44	576.00	34.560
7	27.3	1.42	745.29	38.766
8	30.0	1.96	900.00	58.800
9	35.9	2.23	1288.81	80.057
10	41.6	2.20	1730.56	91.520
Sum, $\Sigma_{i=1}^{10}$:	244.2	13.71	6872.30	392.031

From (6.45), the slope of the regression line is

$$a_1 = \frac{10\times 392.031 - 244.2\times 13.71}{10\times 6872.30 - (244.2)^2}$$
$$\doteq 0.06297.$$

From (6.44), the intercept of the regression line on the y axis at $x = 0$ is

$$a_0 \doteq (13.71 - 0.06297\times 244.2)/10$$
$$\doteq -0.1667.$$

Thus the required regression line has the equation

$$y = -0.1667 + 0.06297x,$$

which is drawn in Fig. 6.13.

The method of least squares can be extended to second-, third-, and higher-degree polynomials; the general approach is again to choose the coefficients of these polynomials so that S is minimized. The general case of an nth-degree regression polynomial $y = a_0 + a_1x + a_2x^2 + \cdots + a_nx^n$ leads to the following sum of squares of deviations:

$$S = \sum_{i=1}^{m} (y_i - a_0 - a_1x_i - a_2x_i^2 - \cdots - a_nx_i^n)^2. \tag{6.46}$$

By setting $\partial S/\partial a_0 = \partial S/\partial a_1 = \partial S/\partial a_2 = \cdots = \partial S/\partial a_n = 0$, we obtain the following system of $n+1$ simultaneous linear equations in the unknown coefficients a_0, $a_1, \ldots, a_n$:

$$\begin{aligned}
a_0m \quad &+ a_1\sum x_i \quad + a_2\sum x_i^2 \quad + \cdots + a_n\sum x_i^n \quad = \sum y_i,\\
a_0\sum x_i &+ a_1\sum x_i^2 \quad + a_2\sum x_i^3 \quad + \cdots + a_n\sum x_i^{n+1} = \sum x_i y_i,\\
&\vdots\\
a_0\sum x_i^n &+ a_1\sum x_i^{n+1} + a_2\sum x_i^{n+2} + \cdots + a_n\sum x_i^{2n} \quad = \sum x_i^n y_i.
\end{aligned} \tag{6.47}$$

Each summation in (6.47) is over the number m of data points; thus, for example, $\Sigma_{i=1}^{m} x_i y_i$ has been written more concisely as $\Sigma x_i y_i$. Once all the necessary summations have been performed, there remains the problem of solving (6.47) for the regression coefficients $a_0, a_1, \ldots, a_n$. Except for the simple case of $n = 1$ (giving equations (6.43), already studied), this is not a particularly easy task; suitable algorithms for solving a system of simultaneous linear equations, such as (6.47), are given in Sections 8.4 and 8.6.

6.9 Taylor's Expansion

So far, we have considered polynomial approximations only for situations involving a given number of tabulated values or experimentally determined points. However, as indicated in the third part of Section 6.1, we may sometimes be asked to provide a polynomial approximation,

$p_n(x)$, to a *specified function* $f(x)$, whose analytical form is given, such as $f(x) = \cos x$. This approximation can be achieved by performing *Taylor's expansion* of $f(x)$, provided that it is continuous and suitably differentiable; the outcome is a power series in x, which can be truncated at an appropriate stage to yield the desired polynomial approximation $p_n(x)$.

Taylor's theorem, stated here without proof, says that if a function $f(x)$, continuous in the closed interval $[x_0, x]$, possesses a continuous $(n+1)$th derivative everywhere on the open interval (x_0, x), then it can be represented by the finite power series:

$$f(x) = f(x_0) + (x-x_0)f'(x_0) + (x-x_0)^2 \times \frac{f''(x_0)}{2!} + (x-x_0)^3 \frac{f^{(3)}(x_0)}{3!} + \cdots + (x-x_0)^n \frac{f^{(n)}(x_0)}{n!} + R_n(x)$$

$$= p_n(x) + R_n(x), \qquad (6.48)$$

where $R_n(x)$, the *remainder*, is given by

$$R_n(x) = (x-x_0)^{n+1} \frac{f^{(n+1)}(\zeta)}{(n+1)!}, \qquad \zeta \text{ in } (x_0, x). \qquad (6.49)$$

The parameter ζ in (6.49) is an unknown function of x, so that it is usually impossible to evaluate the error or remainder term $R_n(x)$ exactly. Nevertheless, we shall see in the following example that (6.49) can prove useful in establishing an *upper bound* for the error incurred when $p_n(x)$ is used for approximating $f(x)$. An obvious consideration is that the error may be appreciable if there is a large deviation between x (the point of approximation) and x_0 (the point at which the function and its derivatives are evaluated).

EXAMPLE 6.8

Use Taylor's series to expand the function $f(x) = \cos x$ about the point $x_0 = 0$. Retain sufficient terms in the expansion so that a fifth-degree polynomial approximation, $p_5(x)$, is obtained. If the approximation is to be used for $0 \leqslant x \leqslant \pi/2$, obtain an upper bound for the maximum error involved. What are the actual errors involved at $x = \pi/4$ and at $x = \pi/2$?

The derivatives of $f(x) = \cos x$ and their numerical values at $x_0 = 0$ are:

i	$f^{(i)}(x)$	$f^{(i)}(0)$
0	$\cos x$	1
1	$-\sin x$	0
2	$-\cos x$	-1
3	$\sin x$	0
4	$\cos x$	1
5	$-\sin x$	0
6	$-\cos x$	-1

From (6.48) and (6.49), with $x_0 = 0$ and $n = 5$, we have:

$$p_5(x) = 1 + x \times 0 + x^2 \times \frac{-1}{2!} + x^3 \times \frac{0}{3!} + x^4 \times \frac{1}{4!} + x^5 \times \frac{0}{5!}$$

$$= 1 - \frac{x^2}{2} + \frac{x^4}{24},$$

$$R_5(x) = -\frac{x^6 \cos \zeta}{6!}.$$

Note that in this case it so happens that the coefficient of x^5 is zero; thus, we are fortunate in obtaining the accuracy afforded by a fifth-degree polynomial even though the highest power of x that actually appears is x^4.

The upper bound for the error involved is obtained by taking the most pessimistic view of $R_5(x)$: on the interval $[0, \pi/2]$ under consideration, x cannot exceed $\pi/2$, and $\cos \zeta$ is 1 at most. Thus,

$$R_5(x)_{max} = -\frac{(\pi/2)^6}{720} \doteq -0.02086.$$

For $x = \pi/4$, the approximation yields

$$p_5(\pi/4) = 1 - \frac{(\pi/4)^2}{2} + \frac{(\pi/4)^4}{24}$$

$$\doteq 1 - 0.30843 + 0.01585$$

$$\doteq 0.70743,$$

which, compared with the exact value of $\cos(\pi/4) = 1/\sqrt{2} \doteq 0.70710$, shows an error of -0.00033. The corresponding approximation at $x = \pi/2$ is $p_5(\pi/2) \doteq 0.01996$, which, compared with the exact value of $\cos(\pi/2) = 0$, shows an error of -0.01996. Note that the error in each case is smaller in absolute value than the upper bound of 0.02086. Also observe that the accuracy of the approximation is worst at $x = \pi/2$, the point most distant from $x_0 = 0$.

Problems

6.1 How many multiplication and addition steps are required for evaluating an nth-degree polynomial by Horner's rule? Write a function, named HORNER, that will use the rule to evaluate $p_n(x)$. Assume that a typical reference to the function will be

```
VALUE = HORNER (N, X, A)
```

Here, N and X stand for n and x, respectively, and A is the name of a vector in which the coefficients $a_0, a_1, \ldots, a_n$ of the polynomial are stored as A(1), A(2), . . . , A(N + 1).

Write a short main program that reads values for n, x, and the a_i for several different polynomials and then calls on HORNER to evaluate the polynomial.

6.2 Starting from equations (6.5), verify the truth of equation (6.8) for the first-degree interpolating polynomial.

6.3 Solve equations (6.11) for the coefficients a_0, a_1, and a_2 of the second-degree interpolating polynomial and check that equations (6.12) are indeed correct. Then verify equation (6.13) for $p_2(x)$.

6.4 Rewrite the second-degree interpolating polynomial, equation (6.13), for the special case of $x_1 - x_0 = x_2 - x_1 = h$. Then repeat Example 6.1 using second-degree interpolation; compare the results with those given in Example 6.2.

6.5 Write out Lagrange's interpolating polynomial, equations (6.17) and (6.18), for $n = 2$ and check that it agrees with equation (6.13).

Consider the following two sets of base points and functional values, which are the same except for a different order of presentation:

	Set 1		Set 2	
i	x_i	$f(x_i)$	x_i	$f(x_i)$
0	−6	17	2	15
1	2	15	7	4
2	7	4	−6	17

Would there be any difference between the second-degree Lagrange interpolating polynomials based on each of the two sets?

6.6 On page 304 the possibility was mentioned of constructing a 99th-degree interpolating polynomial that passed through 100 base points. Attempt to estimate how many arithmetic steps, such as multiplication, subtraction, etc., would be involved in such a venture.

6.7 Let M be the maximum magnitude of $f''(x)$ on the interval (x_0, x_1). By using (6.23), show that the error for linear interpolation for $f(x)$, using the functional values at x_0 and x_1, is bounded by

$$\frac{1}{8} M (x_1 - x_0)^2,$$

for $x_0 \leqslant x \leqslant x_1$. Does this same error bound apply for linear *extrapolation*, that is for $x < x_0$ or $x > x_1$?

6.8 Suppose we wish to prepare a table of functional values of $e^x \sin x$ for subsequent quadratic interpolation on the interval $0 \leqslant x \leqslant 2$. What equal base-point spacing should be used to insure that interpolation will be accurate to four decimal places for any argument in the indicated range?

6.9 The values shown in Table P6.9 are available [8] for the thermal conductivity, k (BTU/hr ft °F), of carbon dioxide gas, and for the viscosity, μ (lb/ft hr), of liquid ethylene glycol, at various temperatures T (°F).

Table P6.9

Carbon Dioxide		Ethylene Glycol	
T	k	T	μ
32	0.0085	0	242
212	0.0133	50	82.1
392	0.0181	100	30.5
572	0.0228	150	12.6
		200	5.57

In each case, determine the simplest interpolating polynomial that is likely to predict k and μ within 1% over the specified ranges of temperature. These polynomials will be needed in Problem 7.30.

Hint: ln μ is more nearly a simple function of T than μ itself.

6.10 When $f(x)$ is a single-valued function of x, and a value of x is required for which the dependent variable $f(x)$ assumes a specified value, the role of independent and dependent variable may be interchanged, and any of the appropriate interpolation formulas may be used. The process is usually termed *inverse interpolation*. Now suppose that $p_n(x)$ is the nth-degree interpolating polynomial passing through the points $(x_i, f(x_i))$, $i = k, k+1, \ldots, k+n$. For a given value $\bar{x}$, let the interpolated value be denoted as $\bar{f} = p_n(\bar{x})$. Suppose further that the same points are used to perform an inverse interpolation, starting with $\bar{f}$, and that the result yields x^*. Comment on the following statements:

(a) x^* will be identical to $\bar{x}$,

(b) the smaller the value of $|x^* - \bar{x}|$, the better the value of $\bar{f}$ will approximate $f(\bar{x})$.

6.11 Perform the necessary algebra to verify the basic equations, (6.34), (6.35), and (6.36), for cubic spline-function approximation.

6.12 The second-degree interpolating polynomial $p_2(x) = 2x^2 - 3x + 2$ passes through the three base points (0, 2), (1,1), and (3, 11). Derive the cubic spline-function approximation using the same base points, and compare it with $p_2(x)$.

6.13 Repeat Example 6.8, but expand instead about the point $x_0 = \pi/4$. Again obtain an upper bound for the error involved in using $p_5(x)$ on the interval $[0, \pi/2]$, and compare with the actual error at $x = \pi/2$.

6.14 Truncated after the first derivative, Taylor's expansion of (6.48) yields

$$f(x) \doteq f(x_0) + (x - x_0)f'(x_0).$$

Interpret this approximation graphically.

6.15 Truncated after the first derivative, Taylor's expansion of (6.48) yields

$$f(x) \doteq f(x_0) + (x - x_0)f'(x_0).$$

Investigate the relation, if any, between this formula and Newton's algorithm of equation (5.11).

6.16 For small values of x, the approximations

$$e^x \doteq 1 + x, \qquad \sin x \doteq x$$

are sometimes employed. In each case, use the error term from Taylor's expansion, (6.49), to estimate how large a value of x may be employed with the assurance that the error in the approximation is smaller than 0.01. Check your conclusions against tables of exponentials and sines.

6.17 Write a function, named QUAD (analogous to TLULIN in Example 6.2) that will perform a table look-up procedure followed by quadratic interpolation on a set of N base points and functional values stored in X(I), F(I), I = 1, 2, . . . , N. A typical reference to the function would be

```
FVAL = QUAD (X, F, XVAL, N, FLAG)
```

where XVAL is the value of the interpolant, FVAL is the corresponding interpolated value returned by QUAD, and the logical variable FLAG is set to .TRUE. in QUAD if extrapolation is required.

Test the function by writing a main program that handles input and output and calls on QUAD; for test data, use either the viscosities given in Example 6.2 or one of the tables in Problem 6.19.

6.18 Modify equations (6.17) and (6.18) so that they apply to the $n+1$ base points designated $x_k, x_{k+1}, \ldots, x_{k+n}$; that is, the subscripts start from k instead of zero.

Write a real function, named LAGR (analogous to TLULIN in Example 6.2) that will perform a table look-up procedure followed by Nth-degree Lagrangian interpolation on a set of M base points and functional values stored in X(I), F(I), I = 1, 2, . . . , M. Assume that the values X(I), I = 1, 2, . . . , M, are in ascending order. A typical reference to the function will be

```
FVAL = LAGR (X, F, XVAL, M, N, FLAG)
```

where XVAL is the value of the interpolant, FVAL is the corresponding interpolated value returned by LAGR, and the logical variable FLAG is set to .TRUE. in LAGR if extrapolation is required. As part of its table look-up procedure, LAGR should, if possible, automatically choose a sequence of base points X(K), X(K + 1), . . . , X(K + N) so that XVAL is centered fairly well between X(K) and X(K + N).

Test the function by writing a main program that handles input and output and calls on LAGR; for test data, use either the viscosities in Example 6.2 or one of the tables in Problem 6.19.

6.19 This is not a problem in itself, but presents tables of functional values that can be used as test data for various interpolating procedures.

Table P6.19.1 Densities of Water, ρ(g/ml) at Various Temperatures T *(°C) [8]*

T	ρ	T	ρ
0	0.9998679	30	0.9956756
5	0.9999919	35	0.9940594
10	0.9997277	40	0.9922455
15	0.9991265	45	0.99024
20	0.9982323	50	0.98807
25	0.9970739	55	0.98573
		60	0.98324

Table P6.19.2 Refractive Index, n, *of Aqueous Sucrose Solutions at 20°C;* P = *Percent Water [12]*

P	n	P	n
15	1.5033	60	1.3997
20	1.4901	65	1.3902
25	1.4774	70	1.3811
30	1.4651	75	1.3723
35	1.4532	80	1.3639
40	1.4418	85	1.3557
45	1.4307	90	1.3479
50	1.4200	95	1.3403
55	1.4096	100	1.3330

Table P6.19.3 Emf e *(microvolts) for the Pt-Pt/10% Rh Thermocouple at Various Temperatures* T *(°F) [13]*

e	T	e	T
0	32.0	3300	761.4
300	122.4	3500	799.0
500	176.0	4000	891.9
1000	296.4	4500	983.0
1500	405.7	5000	1072.6
1700	447.6	5300	1125.7
2000	509.0	5500	1160.8
2500	608.4	5900	1230.3
3000	704.7	6000	1247.5

6.20 This problem deals with two-dimensional interpolation, in which we consider the approximation of a function $f(x, y)$. Suppose that a total of mn functional values $f(x_i, y_j)$ are available, for all possible combinations of m levels of x_i ($i = 1, 2, \ldots, m$), and n levels of y_j ($j = 1, 2, \ldots, n$). For convenience, arrange these values $f_{ij} \equiv f(x_i, y_j)$ in a two-dimensional table, so that row i corresponds to $x = x_i$, and column j correspond to $y = y_j$. Then, given arbitrary x and y, the problem is to interpolate in the table to find an approximation to $f(x, y)$.

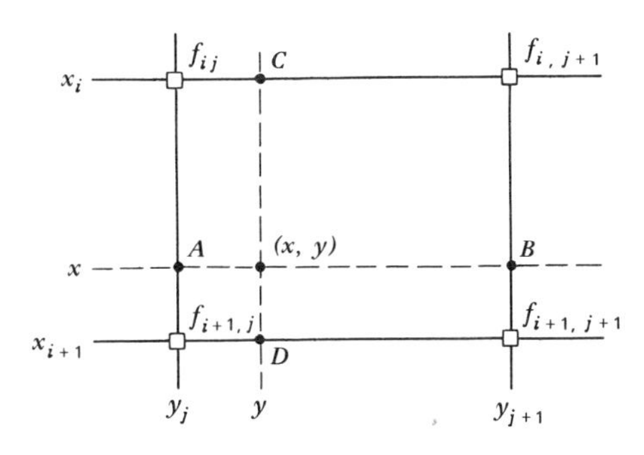

Figure P6.20.

Here, we consider linear interpolation. Let $x_i \leqslant x \leqslant x_{i+1}$, and $y_j \leqslant y \leqslant y_{j+1}$, as shown in Fig. P6.20. The symbol □ indicates points at which functional values are available. First, interpolate linearly through f_{ij} and $f_{i+1,j}$, and through $f_{i,j+1}$ and $f_{i+1,j+1}$, to obtain approximations f_A and f_B to $f(x, y)$ at the points A and B. Then interpolate linearly through f_A and f_B to obtain the final approximation to $f(x, y)$. If $\alpha = (x - x_i)/(x_{i+1} - x_i)$, and $\beta = (y - y_j)/(y_{j+1} - y_j)$, show that the result is

$$f(x, y) \doteq (1-\alpha)(1-\beta)f_{ij} + \beta(1-\alpha)f_{i,j+1} + \alpha(1-\beta)f_{i+1,j} + \alpha\beta f_{i+1,j+1}.$$

What would be the corresponding formula if the first interpolation were in the y direction, to give f_C and f_D, followed by interpolation in the x direction?

Give a simple graphical interpretation, based on relative areas, to the weight factors assigned in the above formula to f_{ij}, $f_{i,j+1}$, $f_{i+1,j}$, and $f_{i+1,j+1}$.

6.21 Write a real function, named LIN2D, that will perform the two-dimensional linear interpolation discussed in Problem 6.20. A typical call will be

FXY = LIN2D (F, X, Y, M, N, XVAL, YVAL)

Here, F, X, and Y are the arrays that have been preset with the tabulated values f_{ij}, x_i, and y_j, M and N have obvious meanings, and XVAL and YVAL correspond to the values x and y for which $f(x, y)$ is to be estimated. If x and/or y lie outside the range of the table, the function should print a message to this effect and return the value zero.

Also, write a main program that will read values for M, N, the matrix F, the vectors X and Y, and XVAL and YVAL. The main program should call on LIN2D as indicated above, then print the interpolated value FXY, and finally return to read additional pairs of values for XVAL and YVAL. Suggested test data are given in Problem 6.22.

6.22 This is not a problem in itself, but presents two tables that can be used as test data for various two-dimensional interpolation procedures.

(a) The values shown in Table P6.22.1 are given by Perry [8] for the specific volume, v (cu ft/lb), of superheated methane, at various temperatures and pressures. Estimate the specific volume of methane at (56.4°F, 12.7 psia), (56.4, 22.7), (56.4, 100), (411.2, 12.7), (411.2, 30.1), (−200, 10), and (0, 84.3). Also try a few temperatures and pressures beyond the scope of the table.

Table P6.22.1 Specific Volume of Superheated Methane, v*(cu ft/lb)*

Temp.	Pressure, psia						
°F	10	20	30	40	60	80	100
−200	17.15	8.47	5.57	4.12	2.678	1.954	1.518
−100	23.97	11.94	7.91	5.91	3.91	2.903	2.301
0	30.72	15.32	10.19	7.63	5.06	3.78	3.014
100	37.44	18.70	12.44	9.33	6.21	4.65	3.71
200	44.13	22.07	14.70	11.03	7.34	5.50	4.40
300	50.83	25.42	16.94	12.71	8.46	6.35	5.07
400	57.51	28.76	19.17	14.38	9.58	7.19	5.75
500	64.20	32.10	21.40	16.05	10.70	8.03	6.42

(b) The values shown in Table P6.22.2 are given by Perry [8] for the total vapor pressures, p (psia), of aqueous solutions of ammonia, at various temperatures and molal concentrations of ammonia.

Estimate the total vapor pressure at (126.5°F, 28.8 mole%), (126.5, 6.7), (126.5, 25.0), (60, 0), (237.5, 17.6), and (237.5, 35.0).

6.23 Suppose that $(m+1) \times (n+1)$ functional values $f(x_i, y_j)$ are available for all combinations of $m+1$ levels of x_i, $i = 0, 1, \ldots, m$, and $n+1$ levels of y_j, $j = 0, 1, \ldots, n$. Define Lagrangian interpolation coefficients comparable to equa-

Table P6.22.2 Vapor Pressure of Aqueous Ammonia, p(*psia*)

Temp. °F	Percentage Molal Concentration of Ammonia 0	10	20	25	30	35
60	0.26	1.42	3.51	5.55	8.65	13.22
80	0.51	2.43	5.85	9.06	13.86	20.61
100	0.95	4.05	9.34	14.22	21.32	31.16
140	2.89	9.98	21.49	31.54	45.73	64.78
180	7.51	21.65	44.02	62.68	88.17	121.68
220	17.19	42.47	81.91	113.81	156.41	211.24
250	29.83	66.67	124.08	169.48	229.62	305.60

tion (6.18) as follows:

$$X_{mi}(x) = \prod_{\substack{k=0 \\ k\neq i}}^{m} \frac{x - x_k}{x_i - x_k}, \qquad i = 0, 1, \ldots, m,$$

$$Y_{nj}(y) = \prod_{\substack{k=0 \\ k\neq j}}^{n} \frac{y - y_k}{y_j - y_k}, \qquad j = 0, 1, \ldots, n.$$

Show that

$$p_{mn}(x, y) = \sum_{i=0}^{m}\sum_{j=0}^{n} X_{mi}(x)Y_{nj}(y)f(x_i, y_j)$$

is a two-dimensional polynomial of degree m in x and degree n in y of the form

$$p_{mn}(x, y) = \sum_{i=0}^{m}\sum_{j=0}^{n} a_{ij}x^i y^j,$$

and satisfies the $(m+1)\times(n+1)$ conditions

$$p_{mn}(x_i, y_j) = f(x_i, y_j), \quad i = 0, 1, \ldots, m, \quad j = 0, 1, \ldots, n,$$

and therefore that $p_{mn}(x, y)$ may be viewed as a two-dimensional interpolating polynomial passing through the $(m+1)\times(n+1)$ points $(x_i, y_j, f(x_i, y_j))$, $i = 0, 1, \ldots, m, j = 0, 1, \ldots, n$.

6.24 Write a real function, named LAGR2D, that will perform a two-dimensional table look-up procedure followed by the two-dimensional Lagrangian interpolation of degree M in x and N in y discussed in Problem 6.23 on a set of P × Q base points stored in X(I), Y(J), F(I, J), I = 1, 2, . . . , P, J = 1, 2, . . . , Q. Assume that the X(I) and Y(J) are stored in ascending algebraic order. A typical reference to the function will be

```
FVAL = LAGR2D (X, Y, F, XVAL, YVAL,
               M, N, P, Q, FLAG)
```

where XVAL and YVAL are the specified values of X and Y, FVAL is the corresponding interpolated value returned by LAGR2D, and the logical variable FLAG is set to .TRUE. in LAGR2D if extrapolation is required. As part of its table look-up procedure, LAGR2D should, if possible, automatically choose sequences of base points within which XVAL and YVAL are fairly well centered.

Test the function by writing a main program that handles input and output and calls on LAGR2D; for test data, see Problem 6.22.

6.25 This problem should be attempted only in conjunction with Problem 8.17, which involves writing a subroutine for the solution of a tridiagonal system of simultaneous linear equations.

Write a function, named SPLINE, that will perform a cubic spline-function approximation on a given set of base points and functional values X(I), F(I), I = 1, 2, . . . , N. The function should: (a) compute and store the values ϕ_i, defined in equation (6.32), in the vector PHI, and (b) return as its value the approximation corresponding to a given argument XVAL (assumed to lie within the range of the base points). Anticipate that a typical reference to the function will be

```
FVAL = SPLINE (X, F, XVAL, PHI, N)
```

Test the function by writing a main program that handles input and output and calls on SPLINE, using test data selected from Examples 6.2 and 6.6, and Problem 6.19.

6.26 Write a program that will implement the linear regression scheme discussed in Section 6.8. The program should read values for m, the number of data points, and their coordinates (x_i, y_i), $i = 1, 2, \ldots, m$. Then compute and print a_0 and a_1 from the formulas (6.44) and (6.45), before returning to read another set of data. Test data may be taken from Example 6.7, and Problems 6.28, 6.29, 6.30, or 6.31.

6.27 Write a program that will implement the nth-degree polynomial regression discussed at the end of Section 6.8. The algorithm discussed in Section 8.4 or 8.6 (preferably the latter) should be used for solving the linear simultaneous equations (6.47) for the regression coefficients $a_0, a_1, \ldots, a_n$.

The program should accept values for the m points (x_i, y_i), $i = 1, 2, \ldots, m$, and n, the degree of the desired regression polynomial. The output from the program should consist of the coefficients in the regression polynomial $y = a_0 + a_1x + \cdots + a_nx^n$. Test the program for $n = 1, 2, 3, 4$, and 5 with data selected from Example 6.7 and Problems 6.28, 6.29, 6.30, or 6.31.

6.28 During the batch saponification reaction between equimolar amounts of sodium hydroxide and ethyl acetate, the concentration c (g moles/liter) of either reactant varies with time t (min) according to the equation

$$\frac{1}{c} = \frac{1}{c_0} + kt,$$

where c_0 is the initial concentration, and k (liter/g mole min) is the reaction rate constant. The following results were obtained in the laboratory, at a temperature of 77°F.

$1/c$:	24.7	32.4	38.4	45.0	52.3	65.6	87.6	102	135	154	192
t:	1	2	3	4	5	7	10	12	15	20	25

Obtain a least-squares estimate of: (a) the reaction rate constant, and (b) the initial concentration.

6.29 Nedderman [14] used stereoscopic photography of tiny air bubbles to determine the velocity profile close to the wall for flow of water in a 1 in. I.D. tube. At a Reynolds number of 1200, he obtained the following values:

y, distance from wall (cm)	u, velocity (cm/sec)	y	u
0.003	0.03	0.056	0.85
0.021	0.32	0.061	0.92
0.025	0.30	0.070	1.05
0.025	0.33	0.078	1.17
0.037	0.57	0.085	1.32
0.043	0.66	0.092	1.38
0.049	0.74	0.106	1.57
0.053	0.80	0.113	1.65
0.055	0.84		

Theoretically, the data should follow the law: $u = py + qy^2$, where p and q are constants. Use the method of least squares to estimate p and q.

6.30 Convection heat transfer data are often correlated using relations of the form

$$Nu = \alpha Re^{\beta} Pr^{1/3},$$

in which Nu, Re, and Pr are the Nusselt, Reynolds, and Prandtl numbers, and α and β are constants to be determined from experiment.

Use the method of least squares, based on a suitable rearrangement of the above equation, to obtain estimates of α and β for the following set of eight values, taken from experimental measurements on the air side of a finned-tube heat exchanger:

ln (Re):	8.5088	8.8133	9.0248	9.3009	9.4560	9.4772	9.6069	9.6932
ln ($Nu/Pr^{1/3}$):	3.5691	3.7803	4.0436	4.2076	4.3199	4.2994	4.4224	4.5405

6.31 A stockbroking company [23] has reported the following financial and operating figures for the decade 1960–1969, in terms of millions of dollars:

Year	Income from Operations	Operating Expenses	Income before Taxes	Estimated Income Taxes	Net Income
1960	130.50	102.25	28.25	13.94	14.31
1961	181.26	133.52	47.73	24.09	23.64
1962	147.10	122.30	24.80	12.49	12.31
1963	170.23	131.29	38.93	20.47	18.46
1964	180.58	140.59	40.00	18.98	21.01
1965	228.28	167.30	60.98	28.81	32.18
1966	288.69	201.97	86.72	41.82	44.90
1967	371.25	256.25	115.00	57.35	57.65
1968	425.98	313.15	112.83	58.34	54.49
1969	389.17	327.67	61.50	29.20	32.30

Use the program of Problem 6.27 to generate the nth-degree regression polynomials that represent each of the five sets of data given above; in each case, let $x = \text{year} - 1960$. Attempt to predict the corresponding figures for the years 1970, 1975, and 1980. Suggested values: $n = 1$, 2, 3, 4, and 5 in turn.

6.32 Akima [39] gives the following method for fitting a smooth curve through the set of n points (x_i, f_i), $i = 1, 2, \ldots, n$, where $f_i = f(x_i)$. The technique resembles the spline-function approximation of Section 6.7, except that each cubic polynomial segment is completely determined locally, from six neighboring points.

Consider a representative sequence of five points, A, B, C, D, and E, and let m_1, m_2, m_3, and m_4 denote the slopes of the four straight-line segments AB, BC, CD, and DE. Akima defines the slope t_C of his interpolating curve at point C as:

$$t_C = \frac{m_2|m_4 - m_3| + m_3|m_2 - m_1|}{|m_4 - m_3| + |m_2 - m_1|}.$$

Note that $t_C = m_2$ if $m_1 = m_2$, and $t_C = m_3$ if $m_3 = m_4$. In the event that $m_1 = m_2$ and $m_3 = m_4$, t_C is defined as $(m_2 + m_3)/2$.

The cubic interpolating polynomial for use on a typical interval $[x_i, x_{i+1}]$ is then given by:

$$\begin{aligned} p_{3,i}(x) &= f_i + t_i(x - x_i) \\ &\quad + \frac{1}{h_i}\left[\frac{3(f_{i+1} - f_i)}{h_i} - 2t_i - t_{i+1}\right] \\ &\quad \times (x - x_i)^2 + \frac{1}{h_i^2}\left[t_i + t_{i+1}\right. \\ &\quad \left. - \frac{2(f_{i+1} - f_i)}{h_i}\right](x - x_i)^3. \end{aligned}$$

Here, $h_i = x_{i+1} - x_i$, and t_i and t_{i+1} are slopes at the points (x_i, f_i) and (x_{i+1}, f_{i+1}), determined as indicated above.

The treatment for the two right-hand end intervals $[x_{n-2}, x_{n-1}]$ and $[x_{n-1}, x_n]$ is facilitated by introducing two hypothetical points (x_{n+1}, f_{n+1}) and (x_{n+2}, f_{n+2}). This is achieved by assuming that these points lie on the quadratic interpolating polynomial passing through (x_i, f_i), $i = n-2$, $n-1$, n, and letting $x_{n+2} - x_n = x_{n+1} - x_{n-1} = x_n - x_{n-2}$. Show from the Lagrange form of the interpolating polynomial that there results: $m_{n+1}-m_n=m_n-m_{n-1}=m_{n-1}-m_{n-2}$. A similar procedure is adopted for the extreme left-hand intervals.

Akima shows that the above technique can lead to even more natural-looking curves than those obtained from the spline polynomial.

Compare, as critically as possible, the above technique with the spline-function approximation of Section 6.7.

6.33 Write a function, named AKIMA, that will perform the interpolation discussed in Problem 6.32 on a given set of base points and functional values X(I), F(I), I = 1, 2, . . . , N. The function should return as its value the approximation corresponding to a given argument XVAL (assumed to lie within the range of the base points). Anticipate that a typical reference to the function will be

FVAL = AKIMA (X, F, XVAL, N)

Test the function by writing a main program that handles input and output and calls on AKIMA, using test data selected from Examples 6.2 and 6.6, and Problem 6.19.

CHAPTER 7
Numerical Integration

7.1 Introduction

We are concerned here with the evaluation of the definite integral

$$I = \int_a^b f(x)\,dx, \tag{7.1}$$

which is readily interpreted as the shaded area under the curve of $f(x)$ versus x, shown in Fig. 7.1. There are several

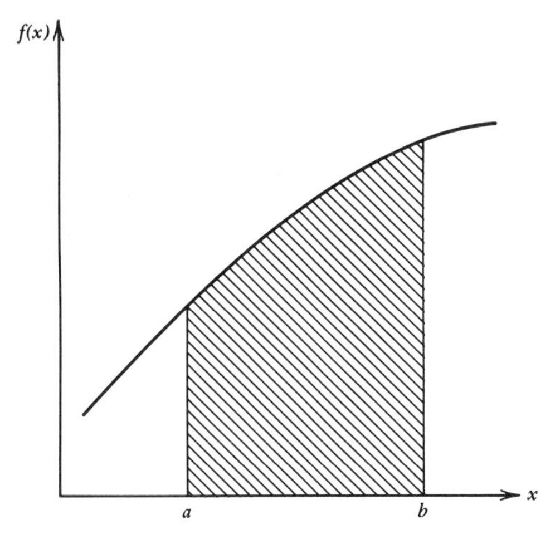

Figure 7.1 The integral $\int_b^a f(x)\,dx$.

situations that may arise, each depending on the exact nature of the integrand, $f(x)$:

(a) If $f(x)$ is a simple function of x, such as a polynomial, or an exponential, or a trigonometric function, or a simple combination of these, there is often little difficulty, and the integral can usually be obtained analytically.

(b) If $f(x)$ is a more complicated function, analytical integration may be very tedious, perhaps requiring a technique such as integration by parts, or a long search before a suitable change of variable is found. In many cases, analytical integration may be virtually impossible. Even the innocent-looking integral

$$I = \frac{2}{\sqrt{\pi}} \int_0^x e^{-\beta^2}\,d\beta$$

(which is, by definition, erf x, the *error function* of x) cannot be evaluated exactly, except in the limit $x \to \infty$, when it equals unity.

(c) In other cases, $f(x)$ may not even be a known analytical function of x, especially if it is derived from experimental data. Rather, numerical values will be available for the function corresponding to specified values of x. However, if $f(x)$ were subject to appreciable random experimental error, a statistical fit by the method of least squares (Section 6.8) might be appropriate before proceeding with the integration.

The numerical evaluation of integrals has useful applications in both (b) and (c) above. Thus, a typical situation to keep in mind is that $f(x)$ is a known but rather complicated function of x. Presumably, $f(x)$ will itself be a reasonably straightforward function to evaluate numerically. Since we are dealing with a function that defies simple analytical integration techniques, we seek to *approximate* it by another function that *can* be integrated easily; a polynomial in x, $p_n(x)$, for example, is chosen for this purpose. We therefore recognize at the outset that, in general, we can only hope to *approximate* the integral:

$$I = \int_a^b f(x)\,dx \doteq \int_a^b p_n(x)\,dx, \tag{7.2}$$

also illustrated in Fig. 7.2.

In this chapter, we shall most frequently take $p_n(x)$ to be an *interpolating* polynomial, but similar developments could

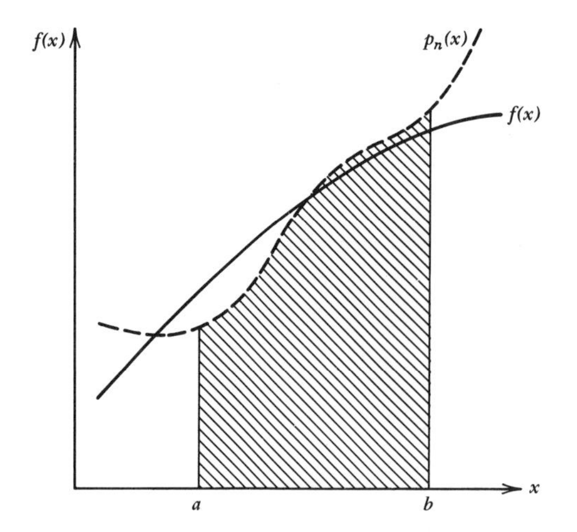

Figure 7.2 Integration of polynomial approximation.

also be made for the spline-function polynomial, the least-squares polynomial, and indeed for many other classes of polynomials not mentioned in Chapter 6.

7.2 The Trapezoidal Rule

The trapezoidal integration formula *could* be dismissed very quickly by saying that the area under the curve in Fig. 7.3 is roughly the area of the shaded trapezoid, which equals its base times its mean height, giving:

$$I = \int_a^b f(x)dx \doteq \frac{b-a}{2}[f(a)+f(b)]. \quad (7.3)$$

However, we shall go more deeply into

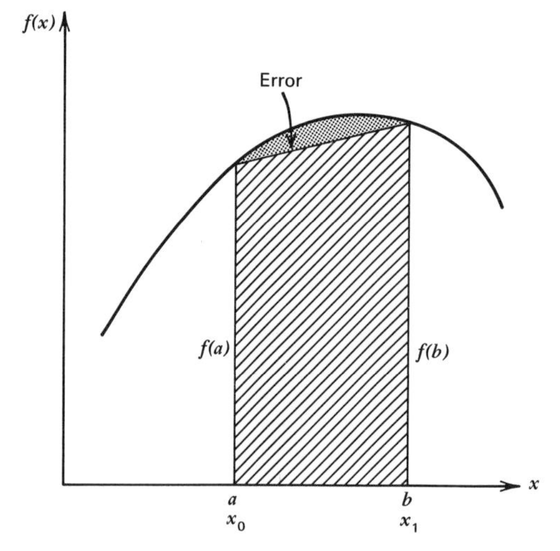

Figure 7.3 Trapezoidal rule.

an alternative derivation of (7.3), in order to emphasize the general basic procedure for generating more complicated formulas for approximating integrals. Referring to Fig. 7.4, draw the first-degree or linear

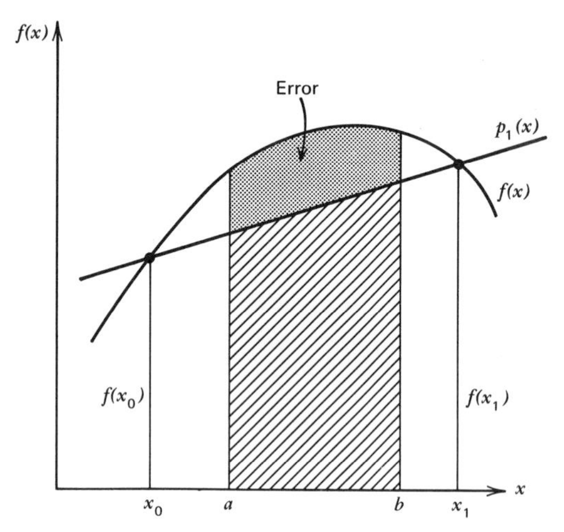

Figure 7.4 Integration of first-order interpolating polynomial.

interpolating polynomial $p_1(x)$ through the base points x_0 and x_1, *which need not necessarily coincide with the limits of integration a and b.* From equation (6.7), we have

$$p_1(x) = f(x_0) + (x-x_0)\frac{f(x_1)-f(x_0)}{x_1-x_0}.$$

The integral of (7.1) can then be approximated as

$$I = \int_a^b f(x)dx \doteq \int_a^b p_1(x)dx$$

$$\doteq \frac{b-a}{x_1-x_0}\left[x_1 f(x_0) - x_0 f(x_1) + \frac{b+a}{2}[f(x_1)-f(x_0)]\right]. \quad (7.4)$$

We can now specialize, if we wish, to the case in which x_0 and x_1 coincide with the limits of integration a and b, respectively. Letting $h = b-a = x_1-x_0$, (7.4) again reduces to the trapezoidal rule:

$$I = \int_a^b f(x)dx \doteq \frac{h}{2}[f(x_0)+f(x_1)]. \quad (7.5)$$

We now show that the trapezoidal rule overestimates the integral by the amount $h^3f''(\zeta)/12$, where $f''(\zeta)$ is the second derivative of $f(x)$ evaluated at some (unknown) point ζ in the open interval (x_0, x_1). From (6.23), the remainder term for $p_1(x)$, the first-degree interpolating polynomial passing through the base points $(x_0, f(x_0))$, $(x_1, f(x_1))$, is

$$R_1(x) = \frac{1}{2}(x-x_0)(x-x_1)f''(\zeta), \qquad \zeta \text{ in } (x, x_0, x_1).$$

Since $f(x) = p_1(x) + R_1(x)$, it follows that $\int_a^b f(x)dx = \int_a^b p_1(x)dx + \int_a^b R_1(x)dx$. The error involved when using the trapezoidal rule is clearly the missing term $\int_a^b R_1(x)dx$, that is:

$$\frac{1}{2}\int_{x_0}^{x_1} (x-x_0)(x-x_1)f''(\zeta)dx.$$

Here, ζ is itself an unknown function of x; since the integration is for $x_0 \leqslant x \leqslant x_1$, it follows that ζ lies in the open interval (x_0, x_1).

To proceed further, recall the *integral mean-value theorem* of the calculus, which states that if two functions, $g(x)$ and $q(x)$, are continuous for $a \leqslant x \leqslant b$, and if $g(x)$ is of constant sign for $a < x < b$, then

$$\int_a^b g(x)q(x)\,dx = q(\bar{\zeta})\int_a^b g(x)dx, \qquad \bar{\zeta} \text{ in } (a, b).$$

Also note that $(x-x_0)(x-x_1)$ is always negative for all x in the interval $a < x < b$.

It follows from the above that the error involved in using the trapezoidal rule is

$$-\frac{1}{2}f''(\bar{\zeta})\int_{x_0}^{x_1}(x-x_0)(x-x_1)dx = -\frac{1}{12}(x_1-x_0)^3f''(\bar{\zeta}).$$

Although ζ and $\bar{\zeta}$ are both unknown values somewhere on the open interval (x_0, x_1), they are not necessarily identical. However, there is little reason at this stage for preserving the distinction between the two and the superbar on $\bar{\zeta}$ can be omitted. Also, $x_1 - x_0$ is simply the *step-size* h. Thus the trapezoidal rule with its associated error term is

$$I = \int_a^b f(x)dx = \frac{h}{2}[f(x_0)+f(x_1)] - \frac{h^3}{12}f''(\zeta), \qquad \zeta \text{ in } (x_0, x_1). \tag{7.6}$$

Of course, if $f(x)$ is simply a straight line, $f''(\zeta)$ is zero for all ζ and the trapezoidal rule is exact. In general, the exact location of ζ will be unknown, so that the error is also unknown precisely. However, if $f(x)$ is a specified function, e^{-x^2} for example, the largest possible value of $f''(\zeta)$ for all ζ in (x_0, x_1) can be estimated, thus enabling an *upper bound* to be placed on the error.

7.3 Simpson's Rule

If we are unhappy with the relatively large error attending the trapezoidal rule (see the dotted area in Fig. 7.3), we have several alternatives available for improving the accuracy of the approximation:

(a) The interval $[a, b]$ can be *subdivided* and the trapezoidal rule can be applied repeatedly over each subdivision in turn (see Section 7.5).

(b) The two base points can be selected differently (no longer coinciding with a and b) according to a special formula given in Section 7.6 (Gauss-Legendre quadrature). A simple formula basically of the type (7.4) can then still be used.

(c) Higher-degree interpolating polynomials can be employed, involving three, four, or even more base points. One of the resulting integration formulas (Simpson's rule) is considered below. Others are mentioned in Section 7.4.

The situation involving three base points is shown in Fig. 7.5; we suppose, for the time being, that we are committed to equally spaced base points, the first

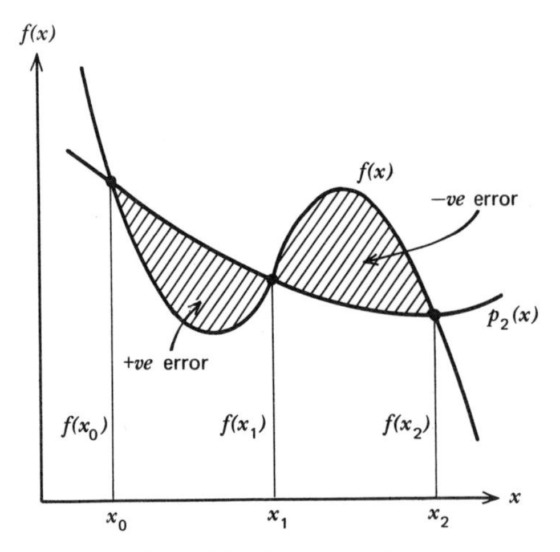

Figure 7.5 Simpson's rule.

and last of which (x_0 and x_2 in this case) coincide with the integration limits a and b. The second-degree interpolating polynomial, $p_2(x)$, is given by equation (6.13), which can be simplified somewhat for the present situation by putting $x_1-x_0=x_2-x_1=h$. We leave it as an exercise (Problem 7.4) to show that the resulting integration formula is

$$I=\int_{x_0}^{x_2} f(x)\,dx \doteq \int_{x_0}^{x_2} p_2(x)\,dx$$

$$=\frac{h}{3}[f(x_0)+4f(x_1)+f(x_2)]. \qquad (7.7)$$

Formula (7.7), which is called *Simpson's rule*, can also be written with its associated error term (derived in [1]):

$$I=\int_{x_0}^{x_2} f(x)dx=\frac{h}{3}[f(x_0)+4f(x_1)+f(x_2)]$$

$$-\frac{h^5}{90}f^{(4)}(\zeta), \qquad \zeta \text{ in } (x_0,x_2), \qquad (7.8)$$

where $f^{(4)}(\zeta)$ is the fourth derivative of $f(x)$ *somewhere* in the interval of integration, (x_0,x_2). Comparing (7.6) with (7.8), note that increasing the number of base points from two to three now allows a *third-degree* polynomial to be integrated *exactly*, since $f^{(4)}(\zeta)$ is zero for all cubic polynomials. As special cases of a cubic polynomial, Simpson's rule will of course also integrate quadratic and linear polynomials without error. Again comparing (7.6) and (7.8), note that the factor h^3 in the error term has now become h^5; for reasonably small values of h, this means that Simpson's rule is almost certain to be more accurate than the trapezoidal rule. This conclusion is not difficult to interpret graphically: the shaded areas in Fig. 7.5 *tend* to cancel each other, thus reducing the overall error.

7.4 Newton-Cotes Closed Integration Formulas

The trapezoidal rule and Simpson's rule are called *closed* integration formulas because none of the interval of integration lies outside the interval containing the base points. By taking four or more base points, constructing the appropriate interpolating polynomial, and integrating, an unlimited number of similar formulas can be developed. The first five of the resulting *Newton-Cotes* closed integration formulas are [1]:

$$\int_{x_0}^{x_1} f(x)dx=\frac{h}{2}[f(x_0)+f(x_1)]-\frac{h^3}{12}f''(\zeta), \qquad (7.9a)$$

$$\int_{x_0}^{x_2} f(x)dx=\frac{h}{3}[f(x_0)+4f(x_1)+f(x_2)]$$
$$-\frac{h^5}{90}f^{(4)}(\zeta), \qquad (7.9b)$$

$$\int_{x_0}^{x_3} f(x)dx=\frac{3h}{8}[f(x_0)+3f(x_1)+3f(x_2)$$
$$+f(x_3)]-\frac{3h^5}{80}f^{(4)}(\zeta), \qquad (7.9c)$$

$$\int_{x_0}^{x_4} f(x)dx=\frac{2h}{45}[7f(x_0)+32f(x_1)$$
$$+12f(x_2)+32f(x_3)+7f(x_4)]$$
$$-\frac{8h^7}{945}f^{(6)}(\zeta), \qquad (7.9d)$$

$$\int_{x_0}^{x_5} f(x)dx=\frac{5h}{288}[19f(x_0)+75f(x_1)$$
$$+50f(x_2)+50f(x_3)+75f(x_4)$$
$$+19f(x_5)]-\frac{275h^7}{12096}f^{(6)}(\zeta). \qquad (7.9e)$$

In each case, h is the (equal) spacing between successive base points, and ζ is somewhere on the open interval of integration. The formula involving four base points is called Simpson's *second* rule. Note that the formulas containing an *odd* number of base points are always advantageous from the standpoint of accuracy. In practice, the formulas involving many base points tend to be disregarded in favor of alternative techniques such as those discussed in Sections 7.5 and 7.6.

EXAMPLE 7.1

Use (a) the trapezoidal rule and (b) Simpson's rule to approximate numerically the integral

$$\int_1^3 f(x)\,dx = \int_1^3 (x^3 - 2x^2 + 7x - 5)\,dx$$

$$= \left[\frac{x^4}{4} - \frac{2x^3}{3} - \frac{7x^2}{2} - 5x\right]_1^3 = 20\tfrac{2}{3}. \quad (7.10)$$

The following functional values are available for $f(x) = x^3 - 2x^2 + 7x - 5$:

x	$f(x)$
1	1
$1\frac{1}{2}$	$4\frac{3}{8}$
2	9
$2\frac{1}{2}$	$15\frac{5}{8}$
3	25

(a) For the trapezoidal rule, shown in Fig. 7.6*a*, $h = 2$, $x_0 = 1$, $x_1 = 3$, $f(x_0) = 1$, and $f(x_1) = 25$. From (7.9a), we have

$$\int_1^3 f(x)\,dx = \frac{h}{2}\,[f(x_0) + f(x_1)] - \frac{h^3}{12} f''(\zeta)$$
$$= \frac{2}{2}(1 + 25) - \frac{8}{12} f''(\zeta)$$
$$= 26 - \frac{2}{3} f''(\zeta).$$

Since $f''(x) = 6x - 4$, the error is given by

$$-\frac{2}{3} f''(\zeta) = -\frac{2}{3}(6\zeta - 4), \qquad 1 < \zeta < 3,$$

which has an extreme value of $-9\frac{1}{3}$. The true error is $20\frac{2}{3} - 26 = -5\frac{1}{3}$, smaller than the bound as expected, but still quite large.

(b) For Simpson's rule, shown in Fig. 7.6*b*, $h = 1$, $x_0 = 1$, $x_1 = 2$, $x_2 = 3$, $f(x_0) = 1$, $f(x_1) = 9$, and $f(x_2) = 25$. From (7.9b), we have

$$\int_1^3 f(x)\,dx = \frac{h}{3}\,[f(x_0) + 4f(x_1) + f(x_2)] - \frac{h^5}{90} f^{(4)}(\zeta)$$
$$= \frac{1}{3}(1 + 36 + 25) - \frac{h^5}{90} f^{(4)}(\zeta)$$
$$= 20\tfrac{2}{3} - \frac{1}{90} f^{(4)}(\zeta).$$

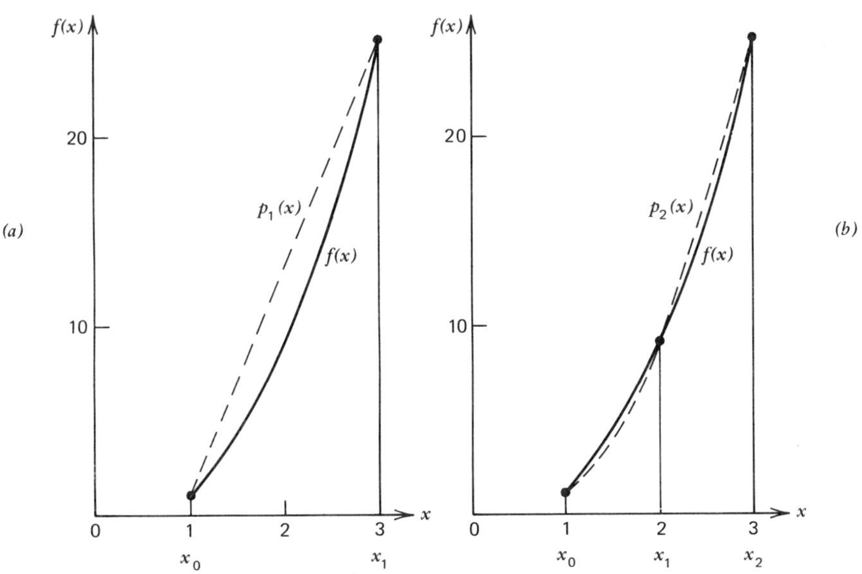

Figure 7.6 Numerical approximation of $\int_1^3 (x^3 - 2x^2 + 7x - 5)\,dx$. (a) Trapezoidal rule. (b) Simpson's rule.

Since the fourth derivative of $f(x)$ vanishes for all x, the error term is zero and the result is exact.

7.5 Composite Integration Formulas

One way for reducing the error associated with a low-order integration formula is to subdivide the interval of integration $[a, b]$ into smaller intervals and then to use the formula separately on each subinterval. Repeated application of a low-order formula is usually preferred to the single application of a high-order formula, partly because of the simplicity of the low-order formulas. Formulas of very high order have the added disadvantage that their weight factors (the coefficients of the $f(x_i)$) tend to be large with alternating signs, which can lead to serious round-off errors—that is, errors introduced because only a fixed, usually small, number of digits can be retained after each computer operation. Integration formulas resulting from interval subdivision and repeated application of low-order formulas are called *composite* integration formulas.

The simplest composite formula is generated by repeated application of the trapezoidal rule of (7.9a). Figure 7.7 illustrates one, two, and four such repeated

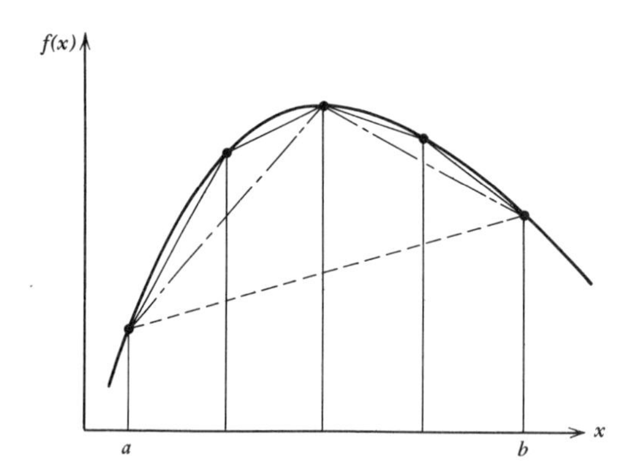

Figure 7.7 Repeated application of the trapezoidal rule; one (---), two (—-), and four (—) applications.

applications. The error involved, being the total area between the appropriate straight-line segments and the $f(x)$ curve, clearly falls rapidly as the number of applications increases. For generality, suppose that we have n repeated applications of the rule. Each subinterval is then of length $h = (b-a)/n$. Let $x_i = x_0 + ih$, $i = 1, 2, \ldots, n$. Then

$$\begin{aligned}\int_a^b f(x)\,dx &= \int_{x_0}^{x_n} f(x)\,dx = \int_{x_0}^{x_1} f(x)\,dx \\ &\quad + \int_{x_1}^{x_2} f(x)\,dx + \cdots \\ &\quad + \int_{x_{n-1}}^{x_n} f(x)\,dx \\ &= \frac{h}{2}[f(x_0) + f(x_1)] + \frac{h}{2}[f(x_1) \\ &\quad + f(x_2)] + \cdots + \frac{h}{2}[f(x_{n-1}) \\ &\quad + f(x_n)] - \frac{h^3}{12}\sum_{i=1}^{n} f''(\zeta_i), \\ &\qquad x_{i-1} < \zeta_i < x_i. \end{aligned} \tag{7.11}$$

Collecting common terms in (7.11), we obtain

$$\begin{aligned}\int_{x_0}^{x_n} f(x)\,dx &= \frac{h}{2}[f(x_0) + 2f(x_1) + 2f(x_2) \\ &\quad + \cdots + 2f(x_{n-1}) + f(x_n)] \\ &\quad - \frac{h^3}{12}\sum_{i=1}^{n} f''(\zeta_i) \\ &= \frac{h}{2}[f(x_0) + f(x_n)] \\ &\quad + h\sum_{i=1}^{n-1} f(x_i) - \frac{(b-a)^3}{12n^2} \\ &\quad \times f''(\zeta), \qquad a < \zeta < b. \end{aligned} \tag{7.12}$$

The error term of (7.11) simplifies to that in (7.12) because the continuous function $f''(x)$ assumes all values between its extreme values. Therefore, there must be some ζ in (a, b) for which $nf''(\zeta) = \Sigma_{i=1}^{n} f(\zeta_i)$, $x_{i-1} < \zeta_i < x_i$.

The error for the composite trapezoidal formula is proportional to $1/n^2$. Therefore, if we double the number of applications, the error will decrease *roughly* by a factor of four ($f''(\zeta)$ for two different values of n will not normally be identical). In fact, if we estimate the integral by using the composite formula twice, each with a different value of n, it is possible to estimate the error term by the following technique. Let I_n and E_n be, respectively, the estimate of the integral and the associated error for n applications of the composite trapezoidal formula. Then the true value of the integral is

$$I = I_{n_1} + E_{n_1} = I_{n_2} + E_{n_2}, \qquad (7.13)$$

where n_1 and n_2 are two different values of n. From (7.12),

$$\frac{E_{n_2}}{E_{n_1}} = \frac{-[(b-a)^3/12n_2^2]f''(\zeta_2)}{-[(b-a)^3/12n_1^2]f''(\zeta_1)}, \quad \zeta_1, \zeta_2 \text{ in } (a,b). \qquad (7.14)$$

Assuming that $f''(\zeta_1)$ and $f''(\zeta_2)$ are equal, (7.14) reduces to

$$E_{n_2} = \left(\frac{n_1}{n_2}\right)^2 E_{n_1}. \qquad (7.15)$$

Substitution of (7.15) into (7.13) leads to

$$I = I_{n_1} + \frac{I_{n_2} - I_{n_1}}{1 - (n_1/n_2)^2}. \qquad (7.16)$$

For the special case of $n_2 = 2n_1$, (7.16) becomes

$$I = \frac{4}{3}I_{n_2} - \frac{1}{3}I_{n_1}. \qquad (7.17)$$

This kind of approach, in which two approximations to an integral are employed to get a third (and hopefully better) approximation, is called *Richardson's deferred approach to the limit* or *Richardson's extrapolation*.

For n repeated applications of Simpson's rule, functional values are required at the $2n+1$ base points $x_0, x_1, \ldots, x_{2n}$. The composite Simpson's-rule formula is

$$\int_a^b f(x)\,dx = \int_{x_0}^{x_{2n}} f(x)\,dx = \frac{h}{3}[f(x_0) + 4f(x_1) + 2f(x_2) + 4f(x_3) + 2f(x_4) + \cdots + 2f(x_{2n-2}) + 4f(x_{2n-1}) + f(x_{2n})] - \frac{h^5}{90}\sum_{i=1}^{n} f^{(4)}(\zeta_i), \qquad (7.18)$$

where

$$x_{2i-2} < \zeta_i < x_{2i}, \qquad h = \frac{b-a}{2n},$$

and

$$x_i = x_0 + ih, \quad i = 0, 1, \ldots, 2n.$$

Equation (7.18) can also be rewritten in a form analogous to (7.12).

EXAMPLE 7.2

RADIANT HEAT FLUX USING THE REPEATED TRAPEZOIDAL RULE

Introduction

A *black-body radiator* is one that emits the maximum possible intensity of radiation for a given temperature. The rate $q\,d\lambda$ at which radiant energy leaves unit area of the surface of a black body, within the wavelength range λ to $\lambda + d\lambda$, is given by *Planck's law*:

$$q\,d\lambda = \frac{2\pi hc^2 d\lambda}{\lambda^5(e^{hc/k\lambda T} - 1)} \frac{\text{ergs}}{\text{cm}^2\,\text{sec}}, \quad (7.2.1)$$

in which c is the speed of light, 2.99793×10^{10} cm/sec, h is Planck's constant, 6.6256×10^{-27} erg sec, k is Boltzmann's constant, 1.38054×10^{-16} erg/°K, T is the absolute temperature of the surface, °K, and λ is the wavelength, cm.

Typical variations of q with wavelength are shown in Fig. 7.2.1. Note that as the temperature increases, the total amount of radiation increases and also the smaller wavelengths become the most intense. 1 Å (Angstrom unit) $= 10^{-8}$ cm, and 1 cal $= 4.184 \times 10^7$ erg.

The rate Q at which radiation is being emitted per unit time per unit area between wavelengths λ_1 and λ_2 is the corresponding area under the q curve:

$$Q = \int_{\lambda_1}^{\lambda_2} q\,d\lambda. \quad (7.2.2)$$

The integral of (7.2.2) does not, in general, reduce to a simple analytical expression. However, the integral *can* be evaluated exactly over the entire range of wavelengths $\lambda = 0$ to $\lambda = \infty$, giving the total rate of emission Q_{tot}, also known as the *emissive power* E, as

$$Q_{tot} = \int_0^\infty q\,d\lambda = \sigma T^4 = E, \quad (7.2.3)$$

where σ is the Stefan-Boltzmann constant, given by

$$\sigma = \frac{2\pi^5 k^4}{15c^2h^3}. \quad (7.2.4)$$

Problem Statement

Write a program whose data will include values for T, λ_1, and λ_2, and that will

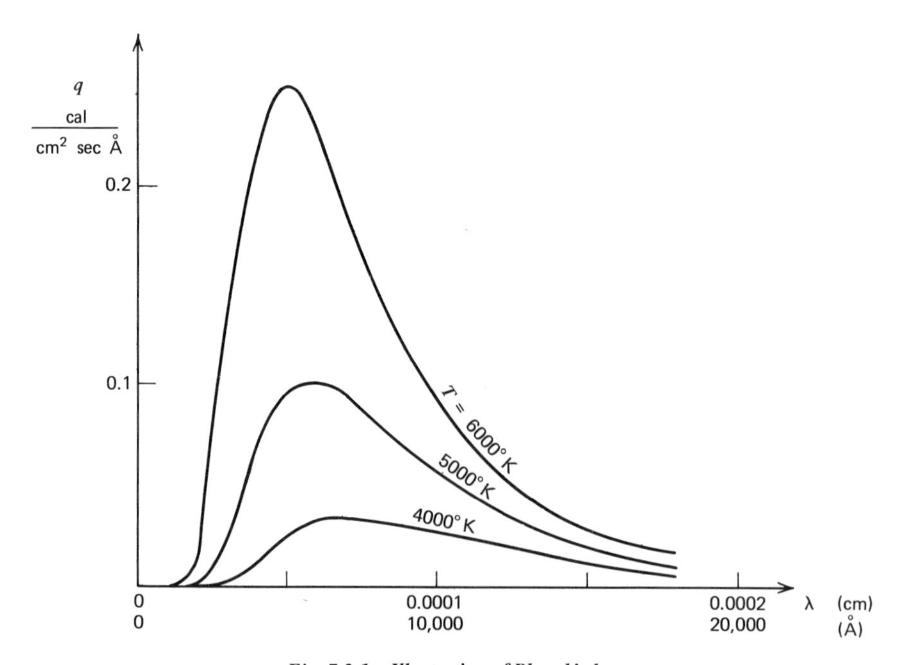

Fig. 7.2.1 Illustration of Planck's law.

compute and print the corresponding values for σ, E $(=\sigma T^4)$, Q (by numerical integration), Q_{tot} (by numerical integration), and $Q_{fract} = Q/E$ (the emission between λ_1 and λ_2 as a fraction of the total emission).

Suggested values are: $T = 2000$ and 6000°K, with $\lambda_1 = 3933.666$ Å (Ca^+ K line) and $\lambda_2 = 5895.923$ Å (Na D_1 line).

Method of Solution

The integrals (7.2.2) and (7.2.3) will be evaluated by n repeated applications of the trapezoidal rule according to equation (7.12), reproduced here in slightly different form and with the error term omitted:

$$I = \int_a^b f(x)\,dx \doteq h\left[\frac{f(a)+f(b)}{2} + \sum_{i=1}^{n-1} f(a+ih)\right]. \qquad (7.2.5)$$

A function TRAP is written for implementing (7.2.5). A general reference to TRAP would be $I \leftarrow$ TRAP (a, b, f, n), where the arguments have the same meanings as in (7.2.5).

A function named PLANCK is also written for returning q according to equation (7.2.1). In the flow diagram, note the precautions taken against: (a) dividing by $\lambda = 0$ (the corresponding q is actually zero) and (b) exceeding the allowable range in exponentiation (the corresponding q is virtually zero). Observe that $q = q(\lambda, T)$ really depends on both wavelength *and* temperature; however, the function TRAP can only integrate a function of a *single* variable, such as $f(x)$. Thus, although the function PLANCK has a single argument λ (over which the integration is being performed), the second variable T (which is actually constant for each complete integration) is assumed in PLANCK to have the same meaning as it does in the main flow diagram.

The integral of (7.2.3) is approximated as $\int_0^\infty q\,d\lambda \doteq \int_0^{\lambda_{max}} q\,d\lambda$, where λ_{max} is specified to be large enough so that its exact value is of no consequence; thus, we hope that $\int_{\lambda_{max}}^\infty q\,d\lambda$ will be negligibly small.

Flow Diagram

Main (c, h, and k are preset with the values following equation (7.2.1); T has a common meaning here and in the function PLANCK)

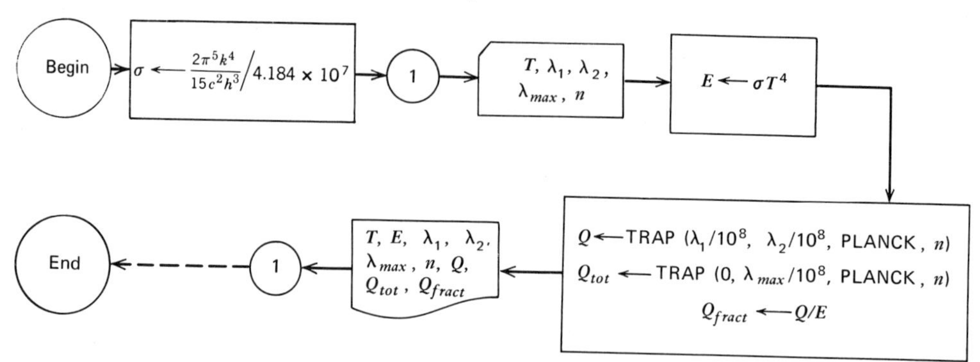

Function TRAP (a, b, f, n)

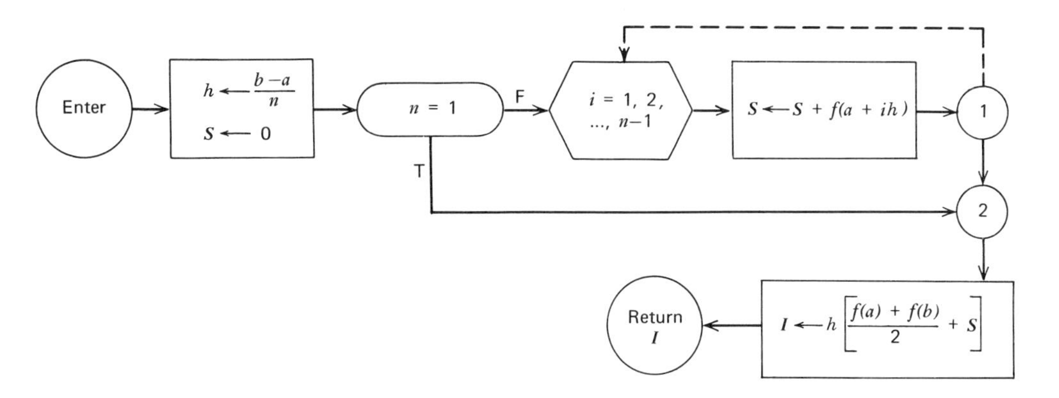

Function PLANCK (λ) (c, h, and k are preset with the values following equation (7.2.1); T has a common meaning here and in the main flow diagram)

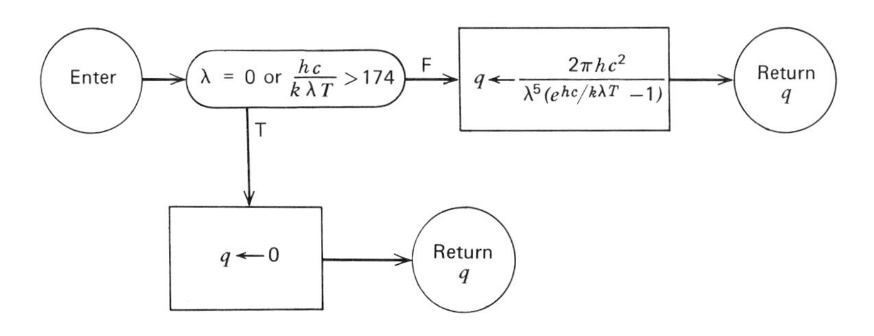

FORTRAN Implementation

List of Principal Variables

Program Symbol	Definition
(Main)	
C, H, K, PI	Physical constants c, h, k (see following equation (7.2.1)), and π.
E	Emissive power E, cal/sq cm sec.
LMAX	λ_{max}, upper limit that is effectively ∞ in the integral of (7.2.3).
LMBDA1, LMBDA2	λ_1 and λ_2, limits of integration for Q, Å.
N	Number of repeated applications, n, of the trapezoidal rule.
Q, QTOT, QFRACT	Q and Q_{tot}, cal/sq cm sec, and Q_{fract}, respectively.
SIGMA	Stefan-Boltzmann constant σ, cal/sq cm sec $(°K)^4$.
T	Absolute temperature T, °K.
(Function TRAP)	
A, B	Lower and upper limits of integration, a and b.
F	Name of the function being integrated, f.
H	Step size $h = (b-a)/n$.
I	Index used for counting the number of applications of the trapezoidal rule.
S	Storage for the summation S appearing in (7.2.5).
(Function PLANCK)	
LAMBDA	Wavelength λ, cm.

Program Listing

Main Program

```
C        DIGITAL COMPUTING                      EXAMPLE 7.2
C        RADIANT HEAT FLUX USING THE REPEATED TRAPEZOIDAL RULE.
C
      REAL  K, LMBDA1, LMBDA2, LMAX
      COMMON  T
      EXTERNAL  PLANCK
C
C     ..... PRESET LIGHT SPEED, PLANCK & BOLTZMANN CONSTANTS, & PI .....
      DATA  C, H, K, PI / 2.99793E10, 6.6256E-27, 1.38054E-16, 3.14159 /
C
C     ..... COMPUTE AND PRINT STEFAN-BOLTZMANN CONSTANT .....
      SIGMA = (2.*PI**5*K*(K/H)**3/(15.*C**2))/4.184E7
      WRITE (6,200)   SIGMA
C
C     ..... READ INPUT DATA, AND COMPUTE EMISSIVE POWER E .....
      READ (5,100,ERR=1)
    1 READ (5,101,END=999,ERR=1)   T, LMBDA1, LMBDA2, LMAX, N
      E = SIGMA*T**4
C
C     ..... USE FUNCTION TRAP TO DETERMINE RADIANT EMISSION Q BETWEEN
C           WAVELENGTHS LMBDA1 AND LMBDA2, AND OVER WHOLE SPECTRUM .....
      Q = TRAP (LMBDA1/1.E8, LMBDA2/1.E8, PLANCK, N)
      QTOT = TRAP (0., LMAX/1.E8, PLANCK, N)
C
C     ..... COMPUTE FRACTIONAL EMISSION, AND PRINT RESULTS .....
      QFRACT = Q/E
      WRITE (6,201)   T, E, LMBDA1, LMBDA2, LMAX, N, Q, QTOT, QFRACT
      GO TO 1
C
  999 CALL EXIT
C
C     ..... FORMATS FOR INPUT AND OUTPUT STATEMENTS .....
  100 FORMAT ( 1H  )
  101 FORMAT ( F10.1, 2F10.3, E10.1, I10 )
  200 FORMAT ( '1DIGITAL COMPUTING                       EXAMPLE 7.2'/
     1   ' RADIANT HEAT FLUX USING THE REPEATED TRAPEZOIDAL RULE'/
     2   '0THE STEFAN-BOLTZMANN CONSTANT IS     SIGMA =', 1PE12.5/
     3   '0' / '0  T', 10X, 'E', 8X, 'LMBDA1', 4X, 'LMBDA2', 4X, 'LMAX',
     4   7X, 'N', 7X, 'Q', 10X, 'QTOT', 7X, 'QFRACT' /' ' )
  201 FORMAT ( ' ', F6.1, E13.5, 2F10.3, E9.1, I6, 2E13.5, F10.6 )
C
      END
```

Function TRAP

```
      FUNCTION TRAP (A, B, F, N)
C
C       FUNCTION FOR INTEGRATING F(X) BETWEEN LOWER LIMIT A AND UPPER
C       LIMIT B, BY N REPEATED APPLICATIONS OF THE TRAPEZOIDAL RULE.
C
C     ..... COMPUTE STEP-SIZE H, AND SET SUM S TO ZERO .....
      H = (B - A)/N
      S = 0.
      IF (N .EQ. 1)  GO TO 2
C
C     ..... ADD UP FUNCTIONAL VALUES AT INTERMEDIATE POINTS .....
      NM1 = N - 1
      DO 1  I = 1, NM1
    1 S = S + F(A + I*H)
C
C     ..... ADD IN WEIGHTED END VALUES AND MULTIPLY BY STEP SIZE .....
    2 TRAP = H*((F(A) + F(B))/2. + S)
      RETURN
C
      END
```

Function PLANCK

```
      FUNCTION PLANCK (LAMBDA)
C
C       FUNCTION THAT COMPUTES Q, THE INTENSITY OF BLACK-BODY RADIATION
C       LEAVING UNIT AREA PER UNIT TIME PER UNIT WAVELENGTH
C
      REAL  K, LAMBDA
      COMMON  T
C
C     ..... PRESET LIGHT SPEED, PLANCK & BOLTZMANN CONSTANTS, & PI .....
      DATA  C, H, K, PI / 2.99793E10, 6.6256E-27, 1.38054E-16, 3.14159 /
C
C     ..... COMPUTE RADIANT EMISSION AT WAVELENGTH LAMBDA .....
      IF (LAMBDA .EQ. 0.)  GO TO 1
      EXPNT = H*C/(K*LAMBDA*T)
      IF (EXPNT .GT. 174.)  GO TO 1
      PLANCK = (2.*PI*H*C**2/(LAMBDA**5*(EXP(EXPNT) - 1.)))/4.1840E7
      RETURN
    1 PLANCK = 0.
      RETURN
C
      END
```

Data

```
   T       LMBDA1     LMBDA2       LMAX        N
2000.0   3933.666   5895.923     1.0E5         1
2000.0   3933.666   5895.923     1.0E6         1
2000.0   3933.666   5895.923     1.0E6         2
2000.0   3933.666   5895.923     1.0E6         5
2000.0   3933.666   5895.923     1.0E6        10
2000.0   3933.666   5895.923     1.0E6        25
2000.0   3933.666   5895.923     1.0E6        50
2000.0   3933.666   5895.923     1.0E6       100
2000.0   3933.666   5895.923     1.0E6       250
2000.0   3933.666   5895.923     1.0E6       500
2000.0   3933.666   5895.923     1.0E6      1000
2000.0   3933.666   5895.923     1.0E6      2500
2000.0   3933.666   5895.923     1.0E6      5000
2000.0   3933.666   5895.923     1.0E6     10000
2000.0   3933.666   5895.923     1.0E6     25000
6000.0   3933.666   5895.923     1.0E5         1
6000.0   3933.666   5895.923     1.0E6         1
6000.0   3933.666   5895.923     1.0E6         2
6000.0   3933.666   5895.923     1.0E6         5
6000.0   3933.666   5895.923     1.0E6        10
6000.0   3933.666   5895.923     1.0E6        25
6000.0   3933.666   5895.923     1.0E6        50
6000.0   3933.666   5895.923     1.0E6       100
6000.0   3933.666   5895.923     1.0E6       250
6000.0   3933.666   5895.923     1.0E6       500
6000.0   3933.666   5895.923     1.0E6      1000
6000.0   3933.666   5895.923     1.0E6      2500
6000.0   3933.666   5895.923     1.0E6      5000
6000.0   3933.666   5895.923     1.0E6     10000
6000.0   3933.666   5895.923     1.0E6     25000
```

Computer Output

```
DIGITAL COMPUTING                        EXAMPLE 7.2
RADIANT HEAT FLUX USING THE REPEATED TRAPEZOIDAL RULE

THE STEFAN-BOLTZMANN CONSTANT IS       SIGMA = 1.35510E-12
```

T	E	LMBDA1	LMBDA2	LMAX	N	Q	QTOT	QFRACT
2000.0	0.21682E 02	3933.666	5895.922	0.1E 06	1	0.62914E-01	0.42454E 00	0.002902
2000.0	0.21682E 02	3933.666	5895.922	0.1E 07	1	0.62914E-01	0.59943E-03	0.002902
2000.0	0.21682E 02	3933.666	5895.922	0.1E 07	2	0.44907E-01	0.95459E-02	0.002071
2000.0	0.21682E 02	3933.666	5895.922	0.1E 07	5	0.39945E-01	0.14048E 00	0.001842
2000.0	0.21682E 02	3933.666	5895.922	0.1E 07	10	0.39239E-01	0.93542E 00	0.001810
2000.0	0.21682E 02	3933.666	5895.922	0.1E 07	25	0.39041E-01	0.79733E 01	0.001801
2000.0	0.21682E 02	3933.666	5895.922	0.1E 07	50	0.39013E-01	0.20989E 02	0.001799
2000.0	0.21682E 02	3933.666	5895.922	0.1E 07	100	0.39006E-01	0.22360E 02	0.001799
2000.0	0.21682E 02	3933.666	5895.922	0.1E 07	250	0.39004E-01	0.21699E 02	0.001799
2000.0	0.21682E 02	3933.666	5895.922	0.1E 07	500	0.39003E-01	0.21677E 02	0.001799
2000.0	0.21682E 02	3933.666	5895.922	0.1E 07	1000	0.39001E-01	0.21677E 02	0.001799
2000.0	0.21682E 02	3933.666	5895.922	0.1E 07	2500	0.38999E-01	0.21677E 02	0.001799
2000.0	0.21682E 02	3933.666	5895.922	0.1E 07	5000	0.38998E-01	0.21677E 02	0.001799
2000.0	0.21682E 02	3933.666	5895.922	0.1E 07	10000	0.38989E-01	0.21646E 02	0.001798
2000.0	0.21682E 02	3933.666	5895.922	0.1E 07	25000	0.38951E-01	0.21645E 02	0.001797
6000.0	0.17562E 04	3933.666	5895.922	0.1E 06	1	0.42479E 03	0.16499E 01	0.241882
6000.0	0.17562E 04	3933.666	5895.922	0.1E 07	1	0.42479E 03	0.18423E-02	0.241882
6000.0	0.17562E 04	3933.666	5895.922	0.1E 07	2	0.44682E 03	0.30045E-01	0.254422
6000.0	0.17562E 04	3933.666	5895.922	0.1E 07	5	0.45305E 03	0.47483E 00	0.257971
6000.0	0.17562E 04	3933.666	5895.922	0.1E 07	10	0.45394E 03	0.35894E 01	0.258477
6000.0	0.17562E 04	3933.666	5895.922	0.1E 07	25	0.45419E 03	0.46717E 02	0.258618
6000.0	0.17562E 04	3933.666	5895.922	0.1E 07	50	0.45422E 03	0.27025E 03	0.258638
6000.0	0.17562E 04	3933.666	5895.922	0.1E 07	100	0.45423E 03	0.10663E 04	0.258643
6000.0	0.17562E 04	3933.666	5895.922	0.1E 07	250	0.45422E 03	0.18926E 04	0.258639
6000.0	0.17562E 04	3933.666	5895.922	0.1E 07	500	0.45421E 03	0.17451E 04	0.258632
6000.0	0.17562E 04	3933.666	5895.922	0.1E 07	1000	0.45420E 03	0.17560E 04	0.258628
6000.0	0.17562E 04	3933.666	5895.922	0.1E 07	2500	0.45420E 03	0.17558E 04	0.258625
6000.0	0.17562E 04	3933.666	5895.922	0.1E 07	5000	0.45395E 03	0.17558E 04	0.258484
6000.0	0.17562E 04	3933.666	5895.922	0.1E 07	10000	0.45375E 03	0.17558E 04	0.258372
6000.0	0.17562E 04	3933.666	5895.922	0.1E 07	25000	0.45365E 03	0.17536E 04	0.258310

Discussion of Results

The particular values for λ_1 and λ_2 were chosen because they lie within the visual range (roughly 4000 to 7000 Å). Also (see Jenkins and White [15]), $T = 2000°K$ and $T = 6000°K$ are representative of a hot tungsten filament and the surface of the sun, respectively.

Observe that the emissive power is far greater at the higher temperature; a threefold increase in T becomes an 81-fold increase in E. The fraction of the total radiation emitted between the specified λ_1 and λ_2 is also much larger at the higher temperature; this suggests that the visual range has adapted itself particularly well to those wavelengths that predominate in sunlight.

The computed QTOT should be very close to zero for N = 1, since the two values of the integrand used are $q_{\lambda=0}$ (exactly zero) and $q_{\lambda_{max}}$ (which lies on the tail of the curve and should be close to zero). But for N = 1 and LMAX = 10^5, QTOT is appreciable; however, QTOT *is* acceptably small for LMAX = 10^6, so this value for the upper limit is used in all further calculations. It takes many applications of the trapezoidal rule for QTOT to stabilize at a value close to E (N = 500 for T = 2000, and N = 1000 for T = 6000). The trouble is due partly to the fact that the tail of the q versus λ curve does not approach zero particularly rapidly for large λ.

In contrast, observe that Q (an integral over a much narrower range of λ) levels out at fairly constant values ($\doteq 0.03901$ for T = 2000, and 454.2 for T = 6000) at much smaller values of N (N = 50, for example). Thus, depending on the problem at hand, the trapezoidal rule, applied repeatedly, may or may not be satisfactory.

For very large values of N, the computed integrals start drifting away from their previously stable values. This phenomenon is due to round-off error; so many numbers are being added together when computing $\sum_{i=1}^{n-1} f(a+ih)$ that the least significant digits in the accumulated sum are being lost. Note that these very large values of N have been used only for the purposes of numerical experiment. In general, it would be far better practice to employ a much smaller number of applications of a formula that is basically more accurate than the trapezoidal rule.

7.6 Gauss-Legendre Quadrature

In previous sections, we have derived integration or *quadrature* formulas in which two of the base points (x_0 and x_n for an $(n+1)$-point formula) were always arranged to coincide with the limits of integration a and b. Also recall that the trapezoidal rule $(n=1)$ could integrate no higher than a first-degree polynomial exactly and that Simpson's rule $(n=2)$ could integrate no higher than a third-degree polynomial exactly.

If this restriction—of the coincidence of x_0 with a and x_n with b—were removed, and we were free to assign other strategic locations for the base points, we might hope to develop more accurate formulas for a given number of base points.

The general intent is shown in Fig. 7.8; for a certain purpose (mentioned below) the limits of integration have been standardized to $a=-1$ and $b=1$. Also, the variable of integration is renamed z and the integrand function is renamed $F(z)$. A linear interpolating polynomial $p_1(z)$ is drawn through the base points $(z_0, F(z_0))$ and $(z_1, F(z_1))$, and the integral is approximated as

$$\int_{-1}^{1} F(z)dz \doteq \int_{-1}^{1} p_1(z)dz. \quad (7.19)$$

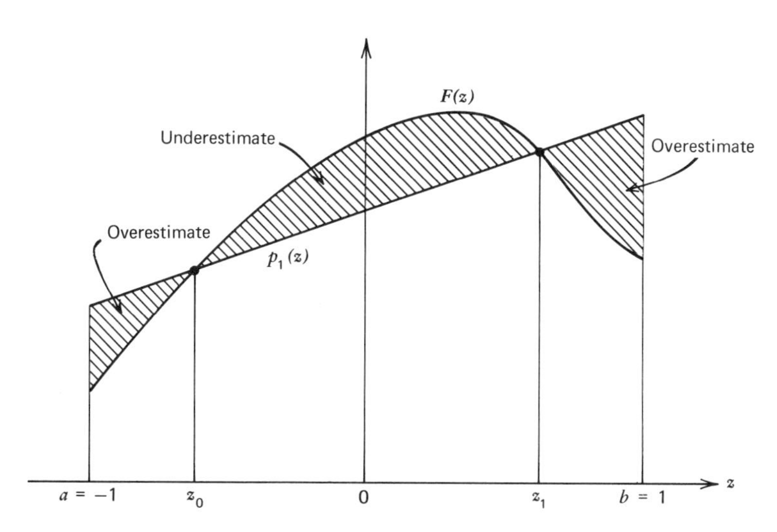

Figure 7.8 Base points not coinciding with limits of integration.

It is conceivably possible to locate z_0 and z_1 so that, even for a strongly varying function, such as that shown in Fig. 7.8, the shaded areas tend to cancel and the error in the approximation is small.

We now demonstrate that, by choosing $z_0 = -z_1$, it is possible to integrate *any* cubic polynomial $p_3(z) = a_0 + a_1z + a_2z^2 + a_3z^3$ *exactly* by such a technique. From (7.4), with $a=-1$ and $b=1$, the computed area simply becomes

$$\int_{-1}^{1} p_1(z)dz = f(-z_1) + f(z_1)$$
$$= 2(a_0 + a_2z_1^2). \quad (7.20)$$

However, the exact value of the integral is

$$\int_{-1}^{1} (a_0 + a_1z + a_2z^2 + a_3z^3)dz = 2\left(a_0 + \frac{a_2}{3}\right). \quad (7.21)$$

Clearly, if z_1 is chosen to be $1/\sqrt{3} \doteq 0.57735$, (7.20) agrees exactly with (7.21), and the case is proven.

The above is the simplest of the *Gauss-Legendre quadrature formulas*. By strategically locating the base points $z_0, z_1, \ldots, z_n$, a formula can be derived (see Carnahan, Luther, and Wilkes [1]) that will always integrate a $(2n+1)$th-degree polynomial exactly. Thus a three-point formula ($n=2$) would integrate $p_5(z)$ exactly. We do not pretend that we are always integrating just *polynomials*; however, since many functions can be *approximated* realistically by polynomials, there is a good chance that such quadrature formulas will have fairly successful applications.

All the Gauss-Legendre quadrature formulas are of the form

$$\int_{-1}^{1} F(z)dz \doteq \sum_{i=0}^{n} w_i F(z_i). \quad (7.22)$$

It can be shown [1] that the appropriate base points z_i are the zeros of the $(n+1)$th-degree Legendre polynomial $P_{n+1}(z)$, which can be generated using the following recursion relation

$$P_n(z) = \frac{2n-1}{n} zP_{n-1}(z) - \frac{n-1}{n} P_{n-2}(z), \quad (7.23)$$

with $P_0(z)=1$ and $P_1(z)=z$. Thus the first few Legendre polynomials are:

$$\begin{aligned} P_0(z) &= 1, \\ P_1(z) &= z, \\ P_2(z) &= \frac{1}{2}(3z^2-1), \\ P_3(z) &= \frac{1}{2}(5z^3-3z), \\ P_4(z) &= \frac{1}{8}(35z^4-30z^2+3). \end{aligned} \quad (7.24)$$

It can also be shown that the corresponding weight factors are given by

$$w_i = \int_{-1}^{1} \prod_{\substack{j=0 \\ j\neq i}}^{n} \left(\frac{z-z_j}{z_i-z_j}\right) dz; \quad (7.25)$$

thus once the base-point values z_i have been found, the weight factors can also be determined.

Table 7.1 shows the necessary base points and weight factors for the first few Gauss-Legendre quadrature formulas.

Normally, the limits of integration will

Table 7.1. Zeros of the Legendre Polynomials $P_{n+1}(z)$ and the Weight Factors for the Gauss-Legendre Quadrature [16]

$$\int_{-1}^{1} F(z)dz \doteq \sum_{i=0}^{n} w_i F(z_i)$$

Base Points, z_i		Weight Factors, w_i
	Two-Point Formula $n=1$	
$\pm$0.57735 02691 89626		1.00000 00000 00000
	Three-Point Formula $n=2$	
0.00000 00000 00000		0.88888 88888 88889
$\pm$0.77459 66692 41483		0.55555 55555 55556
	Four-Point Formula $n=3$	
$\pm$0.33998 10435 84856		0.65214 51548 62546
$\pm$0.86113 63115 94053		0.34785 48451 37454
	Five-Point Formula $n=4$	
0.00000 00000 00000		0.56888 88888 88889
$\pm$0.53846 93101 05683		0.47862 86704 99366
$\pm$0.90617 98459 38664		0.23692 68850 56189

not be -1 and 1, as required by the Gauss-Legendre quadrature formulas. The easiest approach to the evaluation of the general integral

$$\int_a^b f(x)\,dx, \tag{7.26}$$

where a and b are arbitrary but finite, is to transform the Gauss-Legendre formula from the standard interval $-1 \leqslant z \leqslant 1$ to the desired interval $a \leqslant x \leqslant b$, using

$$x = \frac{z(b-a)+a+b}{2}. \tag{7.27}$$

Then (7.26) becomes

$$\int_a^b f(x)\,dx = \frac{b-a}{2}\int_{-1}^{1} f\left(\frac{z(b-a)+a+b}{2}\right)dz \tag{7.28}$$

Finally, since the standard Gauss-Legendre quadrature is given by (7.22), the integral of (7.28) can be approximated by

$$\int_a^b f(x)\,dx \doteq \frac{b-a}{2}\sum_{i=0}^{n} w_i f\left(\frac{z_i(b-a)+a+b}{2}\right). \tag{7.29}$$

The general formulation of the Gauss-Legendre quadrature given by (7.29) is particularly suitable for machine computation because it does not require symbolic transformation of $f(x)$; instead, the base points z_i are transformed and the weight factors w_i are modified by the constant $(b-a)/2$.

Problems

7.1 Verify equation (7.4), which gives an approximation to an integral as the area under a first-degree interpolating polynomial for which the base points do not necessarily coincide with the limits of integration. Check that equation (7.4) yields the trapezoidal rule when $x_0 = a$ and $x_1 = b$.

7.2 It is stated on page 325 that the trapezoidal rule *overestimates* the true integral by a certain amount. Yet in Fig. 7.3 it apparently *underestimates* the integral. Can these seemingly conflicting statements be reconciled?

7.3 If the trapezoidal rule were used just once for estimating the integral $\int_5^6 \ln_e x\,dx$, what would be an upper bound for the error involved? What if Simpson's rule were used? In both cases, check to see if the actual error is less than the upper bound.

7.4 Starting with the appropriate second-degree interpolating polynomial, verify Simpson's rule, equation (7.7).

7.5 Let A, B, and C be points on the x axis such that $AB = BC$. Raise ordinates from these points to intersect the curve $y = y(x)$ at D, E, and F, respectively. Let P be the intersection of DF and BE, and let Q be the point between E and P such that $2EQ = PQ$.

Show that Simpson's rule amounts to saying that the area under the curve approximately equals $AC \times BQ$.

7.6 Comment on the following statement: "Simpson's (first) rule, (7.9b), is preferable to Simpson's second rule, (7.9c), because it has a smaller error term (cf. 1/90 with 3/80)."

7.7 The integral $I = \int_a^b f(x)\,dx$ is to be estimated by a quadratic-type rule, using three base points, x_1, x_2, and x_3, that are not necessarily equally spaced, and none of which necessarily coincides with a or b. Derive the appropriate integration formula and check that it reduces to Simpson's rule when $x_1 = a$, $x_3 = b$, and $(x_2 - x_1) = (x_3 - x_2) = h$.

7.8 Estimate the area of the sheet of metal shown in Fig. P7.8. The sides AB and CD are parallel and are separated by 8 inches.

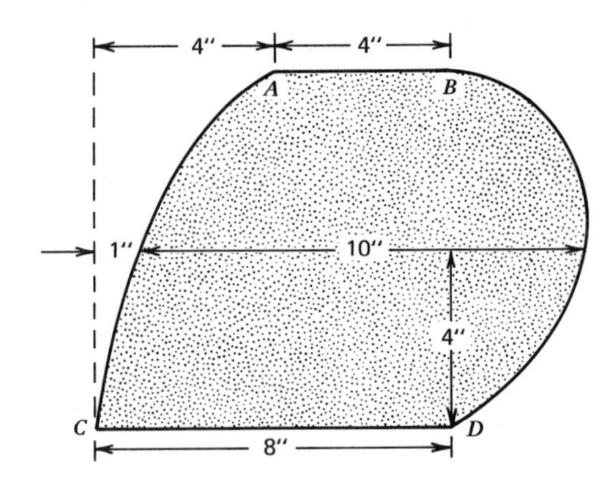

Figure P7.8.

7.9 Use $n = 1$, 2, 4, and 8 repeated applications of the trapezoidal rule to approximate the integral of equation (7.10). Plot the error involved in each case against n^2, and comment with reference to equation (7.12).

7.10 Verify that equation (7.16), Richardson's extrapolation based on the trapezoidal rule, follows from (7.13) and (7.15).

7.11 (a) Employ Richardson's extrapolation for the trapezoidal rule, equation (7.17), with $n_1 = 1$, to estimate the integral of equation (7.10). Comment on the result.

(b) Show that equation (7.17) is equivalent to n_1 repeated applications of Simpson's rule.

7.12 Rewrite equation (7.18), for the composite Simpson's rule, in a form analogous to equation (7.12).

7.13 Apply Richardson's extrapolation technique to Simpson's rule, and show that the resulting equation analogous to (7.17) for the trapezoidal rule is

$$I = \frac{16}{15} I_{n_2} - \frac{1}{15} I_{n_1}.$$

7.14 Approximate the integral of equation (7.10) by the two-point Gauss-Legendre quadrature.

7.15 As an engineer with a background

in numerical methods, you have been asked to obtain the best average value for the percentage of impurity in a product stream over a certain time period, during which this percentage is likely to fluctuate, although not very erratically. You may take a maximum of four samples and conduct at most two precise chemical analyses. If desired, samples may be mixed before analysis. What strategy do you recommend, and why?

7.16 Write a function named SIMPS, analogous to the function TRAP in Example 7.2, that will approximate the general integral of equation (7.1) by n repeated applications of Simpson's rule. Anticipate that a typical reference to the function will be

```
AREA = SIMPS (A, B, F, N)
```

in which the arguments have obvious meanings. To test the function, see Problem 7.17.

7.17 Write a main program that will call on the function SIMPS of Problem 7.16 to estimate the integrals of the following FORTRAN functions between the specified lower and upper limits:

Function	Lower Limit	Upper Limit
SIN	0.	3.141593
SIN	0.	6.283186
SQRT	1.	4.
SQRT	1.	10000.
EXP	0.	1.
ALOG	1.	10.
TAN	0.	1.55

For all cases, use successively 1, 2, 3, 5, 10, 20, 50, and 100 repeated applications of Simpson's rule, and compare the computed results with the known exact values for the integrals, if available.

7.18 The first executable statement on page 335 evaluates the Stefan-Boltzmann constant from the formula $\sigma = 2\pi^5k^4/15c^2h^3$, and also converts the result to units involving calories. Why was this statement *not* written more simply as:

```
SIGMA = (2.*PI**5*K**4/(15.*C**2*H**3))/4.184E7?
```

7.19 Write a function, named EXTRAP, that will evaluate a general integral of the form (7.1) by a technique based on Richardson's extrapolation of the trapezoidal rule, equation (7.17). Anticipate that a typical reference to the function will be

```
AREA = EXTRAP (A, B, F, N1)
```

where A and B are the limits of integration, F is the name of the integrand function, and N1 corresponds to n_1 in (7.17).

Write a main program that will call on EXTRAP to estimate the integrals of the FORTRAN functions given in Problem 7.17. Investigate the effect of using different values of N1.

7.20 Functional values $f(x_1), f(x_2), \ldots, f(x_n)$ are available at the base points $x_1, x_2, \ldots, x_n$ (arranged in ascending order, but not necessarily equally spaced). The integral $I = \int_a^b f(x)\,dx$ is to be estimated by repeated application of the trapezoidal-type formula given by equation (7.4). Write a function, named AREA1, to perform the required integration. A typical call will be

```
ANS = AREA1 (F, X, N, A, B)
```

in which F and X are the vectors used to store the functional values and base points, N is the number of points, and A and B are the integration limits. If possible, always arrange for the subintervals of integration to coincide with the known base points. In general, this will not be the case for the extreme subintervals involving the integration limits a and b. Avoid extrapolation if possible; that is, use a base point less than a and another greater than b, if available. To test the function, see Problem 7.22.

7.21 Write a function, named AREA2, similar to AREA1 of Problem 7.20, but now using repeated application of the quadratic-type integration formula developed in Problem 7.7. A typical call will be

```
ANS = AREA2 (F, X, N, A, B)
```

If possible, again completely embrace the limits of integration by known base points. To test the function, see Problem 7.22.

7.22 The *fugacity* f (atm) of a gas at a pressure P (atm) and a specified temperature T is given by Denbigh [17] as

$$\ln \frac{f}{P} = \int_0^P \frac{C-1}{p}\, dp,$$

where $C = Pv/RT$ is the experimentally determined compressibility factor at the same temperature T, R is the gas constant, and v is the molal volume. For a perfect gas, $C = 1$, and the fugacity is identical with the pressure. (Fugacity is important because the maximum shaft work that can be obtained during the steady isothermal flow of any gas between states 1 and 2 equals $RT \ln f_1/f_2$ per mole.)

Suppose that n values are available for C, namely, $C_1, C_2, \ldots, C_n$, for which the corresponding pressures are $p_1, p_2, \ldots, p_n$ (not necessarily equally spaced, but arranged in ascending order of magnitude). Assume that the value of p_1 (typically, 1 atm) is sufficiently low for the gas to be perfect in the range $0 \leqslant p \leqslant p_1$, so that

$$\ln \frac{f}{P} \doteq \int_{p_1}^P \frac{C-1}{p}\, dp.$$

Write a program that will read data values for n, C_i and p_i ($i = 1, 2, \ldots, n$), and P, and that will proceed to use one of the unequal-interval integration functions developed in Problems 7.20 and 7.21 to evaluate f. Suggested test data are: $P = 50$, 100, 150, 200, 250, 500, 750, 1000, 1500, and 2000 atm, with compressibility factors, derived from Perry [8], given in Table P7.22.

Table P7.22. Compressibility Factor, C

p (atm)	Methane (−70°C)	Ammonia (200°C)
1	0.9940	0.9975
10	0.9370	0.9805
20	0.8683	0.9611
30	0.7928	0.9418
40	0.7034	0.9219
50	0.5936	0.9020
60	0.4515	0.8821
80	0.3429	0.8411
100	0.3767	0.8008
120	0.4259	—
140	0.4753	—
160	0.5252	—
180	0.5752	—
200	0.6246	0.5505
250	0.7468	—
300	0.8663	0.4615
400	1.0980	0.4948
500	1.3236	0.5567
600	1.5409	0.6212
800	1.9626	0.7545
1000	2.3684	0.8914

7.23 In a study by Carnahan [18], the fraction f_{Th} of certain fission neutrons having energies above a threshold energy E_{Th} (Mev) was found to be

$$f_{Th} = 1 - 0.484 \int_0^{E_{Th}} \sinh(\sqrt{2E})e^{-E} dE.$$

Evaluate f_{Th} within ± 0.001 for $E_{Th} = 0.5$, 2.9, 5.3, and 8.6 Mev.

7.24 Consider the condensation of a pure vapor on the outside of a single cooled horizontal tube. According to the simple Nusselt theory of film condensation [2], it may be shown that the mean heat transfer coefficient is given by

$$h = \left(\frac{k^3\rho g\lambda}{\nu r\Delta T}\right)^{1/4}\left(\frac{2^{3/2}}{3\pi}\right)I^{3/4},$$

where

$$I = \int_0^\pi (\sin\beta)^{1/3} d\beta.$$

Here, k, ρ, and ν are the thermal conductivity, density, and kinematic viscosity of the condensed liquid film, r is the tube radius, λ is the latent heat of condensing vapor, g is the gravitational acceleration, ΔT is the difference between the vapor saturation temperature (T_v) and the tube wall temperature (T_w), and all these quantities are in consistent units.

For water, the group $\phi = (k^3\rho g\lambda/\nu)$, in $BTU^4/hr^4\,°F^3\,ft^7$, varies with temperature T (°F) as follows:

T:	100	110	120	130	140	150	160	170
$\phi \times 10^{-14}$:	0.481	0.536	0.606	0.670	0.748	0.820	0.892	0.976

T:	180	190	200	210	220	230	240	250
$\phi \times 10^{-14}$:	1.051	1.130	1.218	1.280	1.327	1.376	1.430	1.503

When used in the above formula for h, ϕ should be evaluated at the mean film temperature $\bar{T} = (T_v + T_w)/2$.

Write a program that uses the above equations to compute h. The input data should include values for T_v, T_w (both in °F), and d (tube diameter in inches); these values should also appear in the printed output. The program should then compute and print values of the integral I, the mean film temperature $\bar{T}$, the corresponding value of ϕ, and the resulting heat transfer coefficient h. The following four sets of test data are suggested:

T_v	T_w	d
212	208	0.75
212	208	2.00
212	210	0.75
120	116	0.75

7.25 Write a function, named GAUSS, that approximates $\int_a^b f(x)dx$ by a single application of an m-point Gauss-Legendre quadrature formula based on equation (7.29). Anticipate that a typical reference to the function will be

AREA = GAUSS (A, B, F, M)

in which the arguments have obvious meanings.

The function should incorporate, possibly by means of a DATA declaration, the necessary base points, z_i, and the corresponding weight factors, w_i, from Table 7.1, for the 2, 3, 4, and 5-point quadrature formulas.

Write a main program that will call on GAUSS to estimate the integrals of the FORTRAN functions given in Problem 7.17. In each case, use M = 2, 3, 4, and 5 in turn.

7.26 Write a function, named GAUSS2, that approximates $\int_a^b f(x)dx$ by n repeated applications of the two-point Gauss-Legendre quadrature, based on equation (7.29), over successive nonoverlapping subintervals each of length $(b-a)/n$. An anticipated function reference will be

AREA = GAUSS2 (A, B, F, N)

in which the arguments have obvious meanings.

Write a main program that will call on GAUSS2 to estimate the integrals of the FORTRAN functions given in Problem 7.17. In each case, use N = 1, 2, 3, 5, 10, and 20 in turn.

7.27 A simple pendulum of length L is shown in Fig. P9.17. If the pendulum is released from rest at $t=0$ with an initial angular displacement θ_0, the resulting period of oscillation can be shown [26] to equal

$$T = 4\sqrt{\frac{L}{g}}\int_0^{\pi/2} \frac{d\phi}{[1-\sin^2(\theta_0/2)\sin^2\phi]^{1/2}},$$

where g is the gravitational acceleration.

Use a numerical integration procedure to evaluate the dimensionless period of oscillation, $T\sqrt{g/L}$, for all values of θ_0 from $\pi/20$ to $\pi/2$ in increments of $\pi/20$.

7.28 A very light spring of length L has

Young's modulus E and a cross-sectional moment of inertia I; it is rigidly clamped at its lower end B and is initially vertical. A downwards force P at the free end A causes the spring to bend over. If θ is the angle of slope at any point and s is the

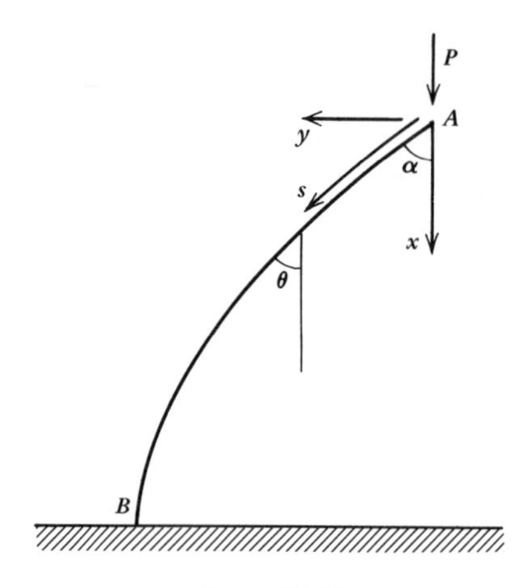

Figure P7.28.

distance along the spring measured from A, then integration of the exact governing equation $EI(d\theta/ds) = -Py$, noting that $dy/ds = \sin\theta$, leads to [7]:

$$ds = -\frac{d\theta}{[(2P/EI)(\cos\theta - \cos\alpha)]^{1/2}},$$

and

$$L = \sqrt{\frac{EI}{P}} \int_0^{\pi/2} [1 - \sin^2(\alpha/2)\sin^2\phi]^{-1/2} d\phi,$$

where α is the value of θ at A.

Show from the above that the Euler load P_e for which the spring just begins to bend is given by

$$P_e = \frac{\pi^2 EI}{4L^2}.$$

Let x_g and y_b denote the vertical distance of A above the datum plane and the horizontal distance of B from A, respectively. Compute the values of P/P_e for which $x_g/L = 0.99$, 0.95, 0.9, 0.5, and 0. What are the corresponding values of y_b/L and α?

7.29 The error function of a number $z\ (\geqslant 0)$ is defined as

$$\text{erf}\, z = \frac{2}{\sqrt{\pi}} \int_0^z e^{-\beta^2} d\beta.$$

Write a real function named MYERF that will compute the error function for an argument Z; anticipate that a typical call will be

```
X = MYERF (Z)
```

Write a short calling program that will call on MYERF to find the error function of the numbers from 0 to 4 in increments of 0.1, and compare the results with the values obtained by calling on the function ERF that is already in the computer library. Also see Problem 5.34 for an application.

7.30 The shell-and-tube heat exchanger shown in Fig. P7.30 is employed for

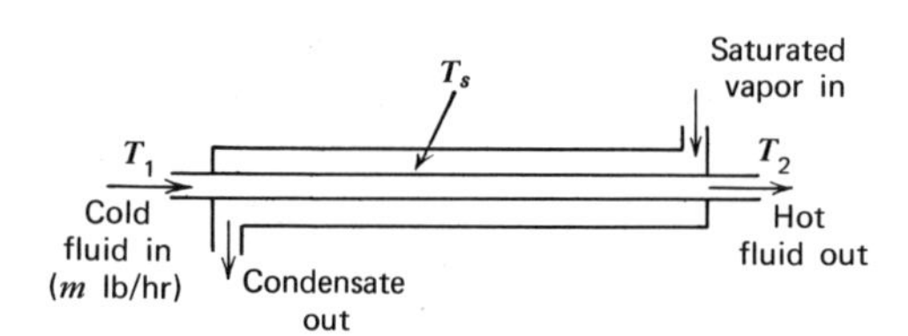

Figure P7.30.

heating a steady stream of m lb/hr of a fluid from an inlet temperature T_1 to an exit temperature T_2. This is achieved by continuously condensing a saturated vapor in the shell, thereby maintaining its temperature at T_s. By integrating a heat balance on a differential element of length, the required exchanger length is found to be

$$L = \frac{m}{\pi D} \int_{T_1}^{T_2} \frac{c_p dT}{h(T_s - T)},$$

where D is the tube diameter. The local heat transfer coefficient h is given by the correlation

$$h = \frac{0.023k}{D} \left(\frac{4m}{\pi D \mu}\right)^{0.8} \left(\frac{\mu c_p}{k}\right)^{0.4},$$

where c_p, μ, and k (the specific heat, viscosity, and thermal conductivity of the fluid, respectively) are functions of temperature T. All quantities in the above formulas must be in consistent units.

Write a program that will read values for m, T_1, T_2, T_s, D, and information concerning the temperature dependency of c_p, μ, and k, as data. The program should then compute the required exchanger length, L.

In making approximate heat-exchanger calculations, one often assumes mean values for c_p, μ, and k, evaluating them just once, at the mean fluid temperature $(T_1 + T_2)/2$. Let the program estimate the error involved in making this assumption.

Suggested Test Data

	Case A	*Case B*
Fluid	Carbon dioxide gas	Ethylene glycol liquid
m, lb/hr	22.5	45,000
T_1, °F	60	0
T_2, °F	280 and 500	90 and 180
T_s, °F	550	250
D, in.	0.495	1.032
c_p, BTU/lb °F	$0.251 + 3.46 \times 10^{-5}T - \dfrac{14{,}400}{(T+460)^2}$	$0.53 + 0.00065T$
k, BTU/hr ft °F	0.0085 (32°F) 0.0133 (212) 0.0181 (392) 0.0228 (572)	0.153 (constant)
μ, lb/ft hr	$0.0332\left(\dfrac{T+460}{460}\right)^{0.935}$	242 (0°F) 80.8 (50) 29.8 (100) 12.3 (150) 5.76 (200)

The above physical properties have been derived from Perry [8]. See Problem 6.9, concerning the interpolating polynomials resulting from the tabulated values of k for carbon dioxide and μ for ethylene glycol.

CHAPTER 8
Solution of Simultaneous Equations

8.1 Introduction

We have already encountered a few sets of simultaneous equations in Chapter 6. In particular: equations (6.5) and (6.11), giving the coefficients of the first- and second-degree interpolating polynomials; (6.35) and (6.36), giving the coefficients needed for the third-degree spline-function approximation; and (6.43) and (6.47), the normal equations resulting from the least-squares method. In these cases, when involving three equations at most, the solution can be obtained fairly simply. The equations are combined in such a way that the resulting equation contains a single variable; this variable can then be determined and back-substituted into the other equations to find the remaining variable or variables. Also note that all the simultaneous equations in Chapter 6 were *linear*: the unknowns, such as a_0 and a_1, always appeared individually and to the first power. Nonlinear terms such as a_0^2, a_1^3, $a_1a_2^2$, $\ln(a_1/a_2)$, etc., were completely absent.

To appreciate the complications that attend *nonlinearity*, the reader should spend a few minutes, first solving algebraically the equations

$$ax + by = c,$$
$$dx + ey = f,$$

for the unknowns x and y (a, b, c, d, e, and f are assumed known), then attempting to solve

$$ax + by = c,$$
$$dx^2 + ey^3 = f.$$

As the engineer, scientist, or mathematician progresses to more advanced problems, he will frequently encounter systems that not only contain more than two or three simultaneous equations, but that are sometimes nonlinear as well. The areas leading to simultaneous equations are diverse, involving, for example, the determination of: (a) stresses and deflections in a loaded structure, (b) currents and voltages in an electrical network, (c) flow rates and pressures in a network of pipes, (d) heat fluxes in furnaces, (e) chemical composition of a gas from mass-spectrometer measurements, (f) regression coefficients in the least-squares method, (g) optimum solutions in linear programming, and (h) numerical solutions to partial differential equations. In the last two cases, the number of simultaneous equations involved is typically several hundred.

There is, therefore, an obvious need for systematic methods of solving simultaneous linear and simultaneous nonlinear equations, and these will be discussed in Section 8.3 *et seq.* First, however, we present a short discussion concerning matrix and vector algebra. This material is not essential to an understanding of the solution of simultaneous equations. But if the reader knows what is meant by a matrix, a vector, the inverse and identity matrices, and matrix multiplication, he will then readily understand a few additional topics that are closely *related* to the solution of simultaneous equations.

8.2 Matrices and Vectors

An $m \times n$ matrix $\mathbf{A}$ is a rectangular array of elements having m rows and n columns. A typical element a_{ij}, lying in the ith row and jth column, may in general be a complex number, with real and imaginary parts; in our applications, however, all the elements will be real numbers. The notation $\mathbf{A} = (a_{ij})$ indicates the array of elements:

$$\mathbf{A} = (a_{ij}) = \begin{bmatrix} a_{11} & a_{12} & a_{13} & \cdots & a_{1n} \\ a_{21} & a_{22} & a_{23} & \cdots & a_{2n} \\ a_{31} & a_{32} & a_{33} & \cdots & a_{3n} \\ \vdots & \vdots & \vdots & & \vdots \\ a_{m1} & a_{m2} & a_{m3} & \cdots & a_{mn} \end{bmatrix}.$$

The *transpose* $\mathbf{A}^t$ of the $m \times n$ matrix $\mathbf{A}$ is the $n \times m$ matrix $\mathbf{B}$, with elements $b_{ij} = a_{ji}$; that is, the *rows* of $\mathbf{A}$ become the *columns* of $\mathbf{B}$. A one-column matrix is known as a *column vector* (or simply, a *vector*), denoted as

$$\mathbf{v} = \begin{bmatrix} v_1 \\ v_2 \\ \vdots \\ v_n \end{bmatrix} = [v_1\, v_2 \cdots v_n]^t.$$

A column vector is usually written as the transpose of a *row vector*, $[v_1\, v_2 \cdots v_n]^t$ for example, since this form economizes on space.

Consider two matrices $\mathbf{A} = (a_{ij})$ and $\mathbf{B} = (b_{ij})$. The matrix *sum* $\mathbf{S} = \mathbf{A} + \mathbf{B}$ exists if and only if both $\mathbf{A}$ and $\mathbf{B}$ have the same dimensions ($m \times n$, for example). $\mathbf{S}$ is then the $m \times n$ matrix with elements $s_{ij} = a_{ij} + b_{ij}$; that is, corresponding elements of $\mathbf{A}$ and $\mathbf{B}$ are simply added. The matrix difference $\mathbf{D} = \mathbf{A} - \mathbf{B}$ is similarly defined with $d_{ij} = a_{ij} - b_{ij}$. The matrix *product* $\mathbf{C} = \mathbf{AB}$ exists if and only if $\mathbf{A}$ and $\mathbf{B}$ have respective dimensions $m \times n$ and $n \times p$. $\mathbf{C}$ is then the $m \times p$ matrix with elements $c_{ij} = \sum_{k=1}^{n} a_{ik} b_{kj}$; that is, c_{ij} is formed by adding the products of elements in the ith row of $\mathbf{A}$ with corresponding elements in the jth column of $\mathbf{B}$. It is true that $\mathbf{A(BC)} = \mathbf{(AB)C}$ and that $(\mathbf{ABC})^t = \mathbf{C}^t\mathbf{B}^t\mathbf{A}^t$, but, in general, $\mathbf{AB} \neq \mathbf{BA}$. *Scalar multiplication* of a matrix $\mathbf{A}$ by a number c, written $c\mathbf{A}$, is achieved by multiplying every element of $\mathbf{A}$ by c.

EXAMPLE 8.1

Let

$$\mathbf{A} = \begin{bmatrix} 2 & 1 & 3 \\ -1 & 2 & 4 \end{bmatrix}, \qquad \mathbf{B} = \begin{bmatrix} 1 & 6 \\ 0 & 3 \\ 5 & -2 \end{bmatrix},$$

and

$$\mathbf{C} = \begin{bmatrix} -2 & 3 \\ 6 & 1 \\ 2 & 7 \end{bmatrix}.$$

Then

$$\mathbf{B} + \mathbf{C} = \begin{bmatrix} -1 & 9 \\ 6 & 4 \\ 7 & 5 \end{bmatrix}, \quad \mathbf{A}(\mathbf{B}+\mathbf{C}) = \begin{bmatrix} 25 & 37 \\ 41 & 19 \end{bmatrix},$$

$$(\mathbf{B}+\mathbf{C})\mathbf{A} = \begin{bmatrix} -11 & 17 & 33 \\ 8 & 14 & 34 \\ 9 & 17 & 41 \end{bmatrix},$$

and

$$\mathbf{A}^t = \begin{bmatrix} 2 & -1 \\ 1 & 2 \\ 3 & 4 \end{bmatrix}.$$

Observe that $\mathbf{A}(\mathbf{B}+\mathbf{C}) \neq (\mathbf{B}+\mathbf{C})\mathbf{A}$. Finally, if $\mathbf{x}$ is the vector $\mathbf{x} = [x_1\, x_2\, x_3]^t$, then

$$\mathbf{x}^t\mathbf{x} = [x_1 \quad x_2 \quad x_3] \begin{bmatrix} x_1 \\ x_2 \\ x_3 \end{bmatrix} = x_1^2 + x_2^2 + x_3^2.$$

Note that the system of n linear simultaneous equations in the unknowns x_1, $x_2, \ldots, x_n$,

$$\begin{aligned} a_{11}x_1 + a_{12}x_2 + \cdots + a_{1n}x_n &= b_1, \\ a_{21}x_1 + a_{22}x_2 + \cdots + a_{2n}x_n &= b_2, \\ \vdots \qquad \vdots \qquad\qquad \vdots \quad &\quad \vdots \\ a_{n1}x_1 + a_{n2}x_2 + \cdots + a_{nn}x_n &= b_n, \end{aligned} \tag{8.1}$$

can be represented concisely by the matrix equation

$$\mathbf{Ax} = \mathbf{b}. \tag{8.2}$$

Here, $\mathbf{A} = (a_{ij})$ is an $n \times n$ matrix containing the known coefficients on the left-hand side of (8.1), and $\mathbf{x}$ and $\mathbf{b}$ are the vectors $[x_1\, x_2 \cdots x_n]^t$ and $[b_1\, b_2 \cdots b_n]^t$, respectively. If two matrices are equal to each other, then every element in either must be identical to the corresponding element in the other. Note that both sides of (8.2) are n-element vectors; thus (8.2) implies n separate equalities which, when written in full, appear as equations (8.1).

The *identity* matrix $\mathbf{I}$ is an $n \times n$ matrix (that is, a *square* matrix) with every *diagonal* element $a_{ii} = 1$ and all *off-diagonal* elements $a_{ij} = 0$ for $i \neq j$. Thus, for $n = 3$,

$$\mathbf{I} = \begin{bmatrix} 1 & 0 & 0 \\ 0 & 1 & 0 \\ 0 & 0 & 1 \end{bmatrix}.$$

Note that $\mathbf{AI} = \mathbf{IA} = \mathbf{A}$, so that multiplication of a matrix by the identity matrix is analogous to multiplication of a scalar by unity. The *inverse* of a square matrix $\mathbf{A}$ is the square matrix denoted by $\mathbf{A}^{-1}$, such that $\mathbf{AA}^{-1} = \mathbf{I}$; it is also true that $\mathbf{A}^{-1}\mathbf{A} = \mathbf{I}$ and, if $\mathbf{A}$ and $\mathbf{B}$ are square, that $(\mathbf{AB})^{-1} = \mathbf{B}^{-1}\mathbf{A}^{-1}$. Multiplying equation (8.2) through by $\mathbf{A}^{-1}$, we obtain

$$\mathbf{A}^{-1}\mathbf{Ax} = \mathbf{Ix} = \mathbf{x} = \mathbf{A}^{-1}\mathbf{b}. \quad (8.3)$$

Thus, if the inverse of the coefficient matrix is known, the solution vector $\mathbf{x}$ of the system of equations (8.1) is readily obtained. In fact, once $\mathbf{A}^{-1}$ has been determined, equations (8.1) may be solved for as many different right-hand side vectors as needed, simply by forming the product $\mathbf{A}^{-1}\mathbf{b}$ each time. An efficient method for determining the inverse of a matrix is described in Section 8.5.

EXAMPLE 8.2

Consider the simultaneous equations

$$\begin{aligned} 3x_1 - x_2 - 2x_3 &= 9, \\ 2x_2 + x_3 &= 0, \\ x_1 + 5x_2 + 3x_3 &= 5. \end{aligned}$$

The coefficient matrix and its inverse (which can be checked by computing $\mathbf{AA}^{-1}$) are:

$$\mathbf{A} = \begin{bmatrix} 3 & -1 & -2 \\ 0 & 2 & 1 \\ 1 & 5 & 3 \end{bmatrix}, \mathbf{A}^{-1} = \begin{bmatrix} \frac{1}{6} & -\frac{7}{6} & \frac{3}{6} \\ \frac{1}{6} & \frac{11}{6} & -\frac{3}{6} \\ -\frac{2}{6} & -\frac{16}{6} & \frac{6}{6} \end{bmatrix}$$

Since the right-hand side vector is $\mathbf{b} = [9, 0, 5]^t$, the solution of the simultaneous equations is

$$\mathbf{x} = \mathbf{A}^{-1}\mathbf{b} = \begin{bmatrix} \frac{1}{6} & -\frac{7}{6} & \frac{3}{6} \\ \frac{1}{6} & \frac{11}{6} & -\frac{3}{6} \\ -\frac{2}{6} & -\frac{16}{6} & \frac{6}{6} \end{bmatrix} \begin{bmatrix} 9 \\ 0 \\ 5 \end{bmatrix} = \begin{bmatrix} 4 \\ -1 \\ 2 \end{bmatrix}.$$

That is, $x_1 = 4$, $x_2 = -1$, and $x_3 = 2$.

Two column vectors $\mathbf{v}_1$ and $\mathbf{v}_2$ are *orthogonal* if $\mathbf{v}_1^t\mathbf{v}_2 = 0$. The m vectors $\mathbf{v}_1, \mathbf{v}_2, \ldots, \mathbf{v}_m$ are *linearly dependent* if scalars $c_1, c_2, \ldots, c_m$, not all zero, can be found which give

$$c_1\mathbf{v}_1 + c_2\mathbf{v}_2 + \cdots + c_m\mathbf{v}_m = \mathbf{0}. \quad (8.4)$$

Here, $\mathbf{0}$ denotes the *null vector*, having all its elements zero. Orthogonal vectors can readily be shown to be linearly *independent*, that is, a relationship of the type (8.4) can not be found unless all the c_i are zero. A set of n linearly independent column vectors, each containing n elements, is said to constitute a *basis* for a vector space of dimension n. Any other vector in the space can then be expressed as a linear combination of such vectors. If, in addition, the vectors are orthogonal, they constitute an *initial basis*. Suppose an arbitrary vector $\mathbf{u}$ is to be expressed as the linear combination

$$\mathbf{u} = c_1\mathbf{v}_1 + c_2\mathbf{v}_2 + \cdots + c_n\mathbf{v}_n, \quad (8.5)$$

in which the $\mathbf{v}$'s are orthogonal. Multiplying equation (8.5) through by $\mathbf{v}_i^t$, and noting that all products on the right-hand side are zero except $\mathbf{v}_i^t c_i \mathbf{v}_i$, we have

$$c_i = \frac{\mathbf{v}_i^t\mathbf{u}}{\mathbf{v}_i^t\mathbf{v}_i}. \quad (8.6)$$

The *length* of a vector $\mathbf{u}$, denoted by $\|\mathbf{u}\|$, is $(\mathbf{u}^t\mathbf{u})^{1/2}$. A *normalized* vector is one whose length is unity. Note that $\mathbf{u}/\|\mathbf{u}\|$ is a normalized vector. A set of normalized orthogonal vectors is an *orthonormal* set.

EXAMPLE 8.3.

The vectors

$$\mathbf{v}_1=\begin{bmatrix}1\\2\\3\end{bmatrix},\quad \mathbf{v}_2=\begin{bmatrix}0\\3\\-2\end{bmatrix},\quad \text{and } \mathbf{v}_3=\begin{bmatrix}-13\\2\\3\end{bmatrix},$$

are orthogonal and linearly independent. They form an initial basis for the set of all vectors $\mathbf{u}=[x\,y\,x]^t$. Suppose we wish to express the vector $\mathbf{u}=[4,1,2]^t$ as the linear combination

$$\mathbf{u}=c_1\mathbf{v}_1+c_2\mathbf{v}_2+c_3\mathbf{v}_3.$$

From equation (8.6),

$$c_1=\frac{\mathbf{v}_1{}^t\mathbf{u}}{\mathbf{v}_1{}^t\mathbf{v}_1}=\frac{1\times 4+2\times 1+3\times 2}{1^2+2^2+3^2}=\frac{6}{7}.$$

Similarly, $c_2=-1/13$, and $c_3=-22/91$. If the elements of the vectors $\mathbf{v}_1$, $\mathbf{v}_2$, and $\mathbf{v}_3$ are divided by their lengths $\sqrt{14}$, $\sqrt{13}$, and $\sqrt{182}$, respectively, the resulting vectors will form an orthonormal set. Their elements would then correspond to the direction cosines of three mutually perpendicular axes.

The *determinant* $\det(\mathbf{A})$ or $|\mathbf{A}|$, of an $n\times n$ matrix $\mathbf{A}$ is defined by

$$\det(\mathbf{A})=\sum \pm a_{1p}a_{2q}\cdots a_{nt}, \qquad (8.7)$$

where the summation of $n!$ terms extends over all permutations $p, q, \ldots, t$ of the column subscripts $1, 2, \ldots, n$. The plus or minus sign depends on whether this permutation is even (plus sign) or odd (minus sign), that is, whether an even or odd number of paired interchanges is needed to restore $p, q, \ldots, t$ to the natural order $1, 2, \ldots, n$. For example,

$$\begin{vmatrix}a_{11}&a_{12}&a_{13}\\a_{21}&a_{22}&a_{23}\\a_{31}&a_{32}&a_{33}\end{vmatrix}=\begin{array}{l}a_{11}a_{22}a_{33}-a_{11}a_{23}a_{32}\\-a_{12}a_{21}a_{33}+a_{12}a_{23}a_{31}\\+a_{13}a_{21}a_{32}-a_{13}a_{22}a_{31}.\end{array}$$

Note that in this case, with $n=3$, $n!=6$ terms appear in the determinant.

The addition of a constant multiple of the elements in one row (or column) of a matrix $\mathbf{A}$ to the corresponding elements in another row (or column) does not alter $\det(\mathbf{A})$. If c is a scalar, then $\det(c\mathbf{A})=c\det(\mathbf{A})$. If $\det(\mathbf{A})\neq 0$, then the system of equations $\mathbf{Ax}=\mathbf{b}$ has a unique solution for $\mathbf{b}\neq\mathbf{0}$. Also, the homogeneous equations $\mathbf{Ax}=\mathbf{0}$ will only have a nontrivial (that is, nonzero) solution if $\det(\mathbf{A})=0$.

8.3 Solution of Two and Three Simultaneous Equations

Consider the following two simultaneous equations:

$$3x_1+4x_2=29, \qquad (8.8)$$
$$6x_1+10x_2=68. \qquad (8.9)$$

To solve for x_1 and x_2, we could multiply (8.8) by two and subtract (8.9) from the result:

$$\begin{array}{lrl}(8.8)\times 2 & 6x_1+8x_2= & 58,\\ (8.9) & 6x_1+10x_2= & 68,\\ \hline \text{Subtract:} & -2x_2= & -10,\\ & x_2= & 5.\end{array}$$

Back-substitution into either (8.8) or (8.9) then gives, for example,

$$3x_1+4\times 5=29,$$

that is,

$$x_1=3.$$

The complete solution is: $x_1=3$ and $x_2=5$. Clearly, starting by multiplying (8.9) by 0.4 and subtracting from (8.8) would have led to the same result. What we next seek is a completely systematic method of solution that can then automatically be extended to the solution of three or more simultaneous equations.

Starting again with the original two equations, we could alternatively have proceeded as follows:

Step 1. Divide equation (8.8) through by the coefficient of x_1:

$$x_1 + \frac{4}{3}x_2 = \frac{29}{3}, \qquad (8.10)$$

$$6x_1 + 10x_2 = 68. \qquad (8.11)$$

Step 2. Subtract a suitable multiple (6, in this case) of equation (8.10) from equation (8.11), so that x_1 is eliminated from (8.11):

$$x_1 + \frac{4}{3}x_2 = \frac{29}{3}, \qquad (8.12)$$

$$2x_2 = 10. \qquad (8.13)$$

Step 3. Divide (8.13) through by the coefficient of x_2:

$$x_1 + \frac{4}{3}x_2 = \frac{29}{3}, \qquad (8.14)$$

$$x_2 = 5. \qquad (8.15)$$

Step 4. Subtract a suitable multiple, namely 4/3, of (8.15) from (8.14), so that x_2 is eliminated from (8.14):

$$x_1 \qquad = \frac{9}{3} \; (= 3), \qquad (8.16)$$

$$x_2 = 5.$$

The solution is again complete, with $x_1 = 3$ and $x_2 = 5$. Note the following:

(a) Steps 1 and 3 are called *normalization* steps because they normalize to unity the coefficients of x_1 and x_2. This was achieved by dividing by the coefficients 3 (in Step 1) and 2 (in Step 3); we shall call these coefficients, which appear on the "diagonal" of the left-hand side, the *pivot coefficients* or *pivot elements*.

(b) Steps 2 and 4 are called *reduction* or *elimination* steps because they completely remove, for example, x_1 from equation (8.11) and x_2 from equation (8.14).

(c) As far as the arithmetic is concerned, we can actually dispense with x_1, x_2, and the equals signs. Thus the original equations can be written symbolically as

$$\begin{bmatrix} 3 & 4 & 29 \\ 6 & 10 & 68 \end{bmatrix}.$$

Here, we are giving the *matrix* or two-dimensional array of the relevant coefficients. Steps 1 to 4 can then be rephrased symbolically in terms of the following matrices:

Step 1.

$$\begin{bmatrix} 1 & \frac{4}{3} & \frac{29}{3} \\ 6 & 10 & 68 \end{bmatrix}$$

Step 2.

$$\begin{bmatrix} 1 & \frac{4}{3} & \frac{29}{3} \\ 0 & 2 & 10 \end{bmatrix}$$

Step 3.

$$\begin{bmatrix} 1 & \frac{4}{3} & \frac{29}{3} \\ 0 & 1 & 5 \end{bmatrix}$$

Step 4.

$$\begin{bmatrix} 1 & 0 & 3 \\ 0 & 1 & 5 \end{bmatrix}$$

At this stage, we can reinterpret this matrix by reinstating x_1, x_2, and the equals signs, giving:

$$\begin{aligned} x_1 \qquad &= 3, \\ x_2 &= 5. \end{aligned}$$

To advance one step further, next consider the solution of the *three* simultaneous equations

$$\begin{aligned} 2x_1 - 7x_2 + 4x_3 &= 9, \\ x_1 + 9x_2 - 6x_3 &= 1, \\ -3x_1 + 8x_2 + 5x_3 &= 6, \end{aligned}$$

which can again be represented symbolically in the matrix form:

$$\begin{bmatrix} 2 & -7 & 4 & 9 \\ 1 & 9 & -6 & 1 \\ -3 & 8 & 5 & 6 \end{bmatrix}.$$

To solve the equations, we proceed in the same way as before:

Step 1. Normalize row 1, corresponding to the first equation, by dividing through by the pivot element, 2:

$$\begin{bmatrix} 1 & -\frac{7}{2} & 2 & \frac{9}{2} \\ 1 & 9 & -6 & 1 \\ -3 & 8 & 5 & 6 \end{bmatrix}.$$

Steps 2 and 3. Reduce to zero the elements in the second and third rows of the first column. That is, remove or eliminate x_1 from the second and third equations. This is achieved by subtracting 1 times the first row from the second row, and subtracting -3 times (that is, adding 3 times) the first row from (to) the third row:

$$\begin{bmatrix} 1 & -\frac{7}{2} & 2 & \frac{9}{2} \\ 0 & \frac{25}{2} & -8 & -\frac{7}{2} \\ 0 & -\frac{5}{2} & 11 & \frac{39}{2} \end{bmatrix}.$$

Step 4. Normalize the second row by dividing through by its pivot, 25/2:

$$\begin{bmatrix} 1 & -\frac{7}{2} & 2 & \frac{9}{2} \\ 0 & 1 & -\frac{16}{25} & -\frac{7}{25} \\ 0 & -\frac{5}{2} & 11 & \frac{39}{2} \end{bmatrix}.$$

Steps 5 and 6. Subtract suitable multiples ($-7/2$ and $-5/2$) of the second row from the first and third rows, in order to eliminate x_2 from the first and third equations:

$$\begin{bmatrix} 1 & 0 & -\frac{6}{25} & \frac{88}{25} \\ 0 & 1 & -\frac{16}{25} & -\frac{7}{25} \\ 0 & 0 & \frac{47}{5} & \frac{94}{5} \end{bmatrix}.$$

Step 7. Normalize the third row:

$$\begin{bmatrix} 1 & 0 & -\frac{6}{25} & \frac{88}{25} \\ 0 & 1 & -\frac{16}{25} & -\frac{7}{25} \\ 0 & 0 & 1 & 2 \end{bmatrix}.$$

Steps 8 and 9. Eliminate x_3 from the first and second equations:

$$\begin{bmatrix} 1 & 0 & 0 & 4 \\ 0 & 1 & 0 & 1 \\ 0 & 0 & 1 & 2 \end{bmatrix}.$$

The process is now complete, and we can reinstate the x's and equals signs:

$$\begin{aligned} x_1 \qquad\qquad &= 4, \\ x_2 \qquad &= 1, \\ x_3 &= 2. \end{aligned}$$

The above procedure is called *Gauss-Jordan reduction.* We shall next seek a way of implementing it on the computer, in order to solve a general system of n equations. The matrix notation of equation (8.1) will be used for the coefficients, modified so that the right-hand side elements $b_1, b_2, \ldots, b_n$ are now included as the last or $(n+1)$th column of matrix $\mathbf{A}$. Thus, for $n = 3$:

$$\begin{aligned} a_{11}x_1 + a_{12}x_2 + a_{13}x_3 &= a_{14}, \\ a_{21}x_1 + a_{22}x_2 + a_{23}x_3 &= a_{24}, \\ a_{31}x_1 + a_{32}x_2 + a_{33}x_3 &= a_{34}. \end{aligned}$$

8.4 The Gauss-Jordan Method for Solving n Simultaneous Equations

We have just seen that n simultaneous equations can be represented by the following matrix of coefficients:

$$\begin{bmatrix} a_{11} & a_{12} & \cdots & a_{1n} & a_{1,n+1} \\ a_{21} & a_{22} & \cdots & a_{2n} & a_{2,n+1} \\ \vdots & \vdots & & \vdots & \vdots \\ a_{n1} & a_{n2} & \cdots & a_{nn} & a_{n,n+1} \end{bmatrix}. \quad (8.17)$$

Matrix (8.17) is known as the *augmented matrix* because it also contains, in its $(n+1)$th column, the numbers that appear to the right-hand side of the equals signs.

Recall that the solution was effected in the 2×2 and 3×3 cases by a series of: (1) *normalization* and (2) *reduction* steps. We now extend this procedure to the system of n equations.

The key to the solution involves the selection of successive *pivot* elements, which follow down the leading diagonal, a_{11}, a_{22}, etc. If we establish a *pivot counter*, k, then the pivot elements will be in the a_{kk} *position* at the time of the normalization steps:

$$a_{kk}, \qquad k=1,2,\ldots,n.$$

The algebraic forms of the normalization and reduction steps now follow.

1. Normalization. The kth pivot appears in row k. Everything in row k to the left of the pivot has already been reduced to zero. Therefore, we wish to divide all elements by a_{kk}, from the pivot itself (column k), to the extreme right (column $n+1$). Using j as a column counter,

$$a_{kj} \leftarrow a_{kj}/a_{kk}, \quad j=k+1,k+2,\ldots,n+1,$$
$$a_{kk} \leftarrow 1. \qquad (8.18)$$

Note that the pivot element itself must be the *last* to be divided by a_{kk}, otherwise all the other divisions would merely be by unity.

2. Reduction. Because of earlier steps, a typical situation (shown here for $k=3$) will be:

$$\begin{array}{c} \text{Pivot} \\ \text{column} \\ \downarrow \end{array}$$
$$\begin{bmatrix} 1 & 0 & a_{1k} & \cdots\cdots & a_{1,n+1} \\ 0 & 1 & a_{2k} & \cdots\cdots & a_{2,n+1} \\ 0 & 0 & a_{kk}\ (=1) & \cdots & a_{k,n+1} \\ \cdots & \cdots & \cdots & \cdots & \cdots \\ 0 & 0 & a_{nk} & \cdots\cdots & a_{n,n+1} \end{bmatrix}.$$

We wish to reduce to zero all elements in the pivot column k, *except* the pivot element itself (that is, a_{kk}, which is already 1). Consider the relation of row i $(i \neq k)$ to row k:

$$\begin{array}{lccccccc} \text{Row } i\text{:} & 0 & 0 & \cdots & \boxed{a_{ik}} & \cdots & a_{ij} & \cdots & a_{i,n+1} \\ \text{Row } k\text{:} & 0 & 0 & \cdots & a_{kk}\ (=1) & \cdot & a_{kj} & \cdots & a_{k,n+1} \\ & & & & \uparrow & & \uparrow & & \uparrow \\ & & & & \text{Pivot column, } k & & \text{General column, } j & & \text{Last column, } n+1 \end{array}$$

The encircled element a_{ik} is to be reduced to zero. Clearly, the multiple of row k to be subtracted from row i is a_{ik} because this will then give

$$a_{ik} \leftarrow a_{ik} - a_{ik}a_{kk} = 0.$$

An element such as a_{ij} then becomes modified to

$$a_{ij} \leftarrow a_{ij} - a_{ik}a_{kj}.$$

The complete algorithm for the reduction step then becomes:

$$\left.\begin{array}{l} a_{ij} \leftarrow a_{ij} - a_{ik}a_{kj}, \\ j=k+1,k+2,\ldots,n+1, \\ a_{ik} \leftarrow 0. \end{array}\right\} \begin{array}{l} i=1,2,\ldots,n, \\ i\neq k. \end{array} \qquad (8.19)$$

Complete Algorithm for Gauss-Jordan Reduction

$$\left.\begin{array}{l}
a_{kj} \leftarrow a_{kj}/a_{kk}, \\
\qquad j = k+1, k+2, \ldots, n+1, \\
a_{kk} \leftarrow 1. \\
\\
\left.\begin{array}{l}
a_{ij} \leftarrow a_{ij} - a_{ik}a_{kj}, \\
\qquad j = k+1, k+2, \ldots, n+1, \\
a_{ik} \leftarrow 0.
\end{array}\right\} \begin{array}{l} i = 1, 2, \ldots, n, \\ i \neq k, \end{array}
\end{array}\right\} k = 1, 2, \ldots, n. \tag{8.20}$$

If **A** is the $n \times n$ matrix that comprises the left-hand side coefficients of the n simultaneous equations, it can be shown [1] that the determinant of **A** equals the product of the successive pivot elements that appear just before each normalization step. That is, $\det(\mathbf{A}) = \Pi_{k=1}^{n} a_{kk}$.

EXAMPLE 8.4

POTENTIALS AND CURRENT IN A BRIDGE NETWORK

Problem Statement

Figure 8.4.1 shows a resistance bridge that is connected through a resistance R_s to a source of constant emf E. Write a program that will read values for E (volts) and R_s, R_g (the galvanometer resistance), R_a, R_b, R_c, and R_d (ohms). The program should compute and print values for the potentials V_1, V_2, and V_3 (volts) at the nodes numbered 1, 2, and 3, and the current I_g (amps) through the galvanometer. The data should include the case of a balanced bridge ($R_a/R_b = R_c/R_d$), for which I_g should be zero.

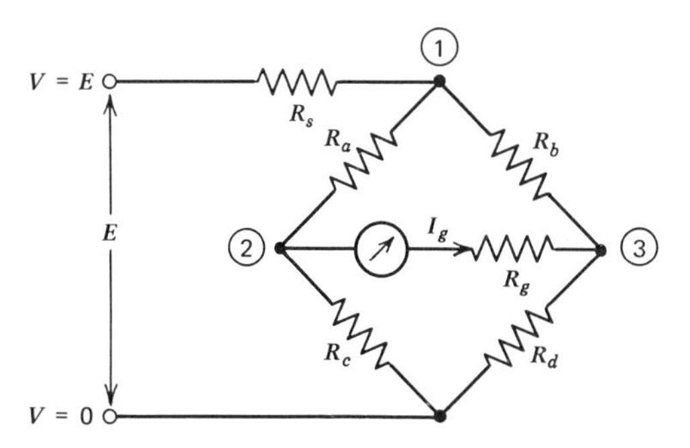

Figure 8.4.1 Resistance bridge network.

Method of Solution

We apply Kirchoff's current law, which states that the instantaneous sum of the currents into any node is zero. Also, from Ohm's law, the current I resulting from a potential difference ΔV across a resistance R is $I = \Delta V/R$, which can also be written as $I = \Delta VG$, where $G = 1/R$ is the *conductance.*

Thus, for nodes 1, 2, and 3 in turn:

$$
\begin{aligned}
(E-V_1)G_s + (V_2-V_1)G_a + (V_3-V_1)G_b &= 0,\\
(V_1-V_2)G_a + (V_3-V_2)G_g \qquad - V_2G_c &= 0, \\
(V_1-V_3)G_b + (V_2-V_3)G_g \qquad - V_3G_d &= 0.
\end{aligned}
\tag{8.4.1}
$$

Collecting coefficients of V_1, V_2, and V_3 in (8.4.1) gives:

$$
\begin{aligned}
(G_s+G_a+G_b)V_1 \quad - G_aV_2 \quad - G_bV_3 &= EG_s,\\
-G_aV_1 \quad + (G_a+G_g+G_c)V_2 \quad - G_gV_3 &= 0,\\
-G_bV_1 \quad - G_gV_2 \quad + (G_b+G_g+G_d)V_3 &= 0.
\end{aligned}
\tag{8.4.2}
$$

The simultaneous equations (8.4.2) are solved by the Gauss-Jordan algorithm, (8.20), for the special case of $n = 3$. The augmented matrix must first be established, and this is achieved by setting $a_{11} = G_s + G_a + G_b$, $a_{12} = -G_a$, $a_{13} = -G_b$, etc. The conductances G have been used above for compactness, but the program works directly in terms of the reciprocals of the resistances; for example, $a_{11} = 1/R_s + 1/R_a + 1/R_b$, $a_{12} = -1/R_a$, $a_{13} = -1/R_b$, etc.

By way of variation, the algorithm of (8.20) is modified slightly. By counting backwards from the fourth or $(n+1)$th column and ending at the pivot element, the normalization step can be expressed as the single operation

$$a_{kj} \leftarrow a_{kj}/a_{kk}, \qquad j = 4, 3, \ldots, k,$$

which is equivalent to the following two operations in (8.20):

$$
\begin{aligned}
&a_{kj} \leftarrow a_{kj}/a_{kk}, \quad j = k+1, k+2, \ldots, n+1,\\
&a_{kk} \leftarrow 1.
\end{aligned}
$$

A similar variation is introduced into the reduction steps.

After the Gauss-Jordan algorithm has been implemented, the required potentials V_1, V_2, and V_3 will be found in a_{14}, a_{24}, and a_{34}, respectively. The galvanometer current is then computed as $I_g = (V_2 - V_3)/R_g$.

Flow Diagram

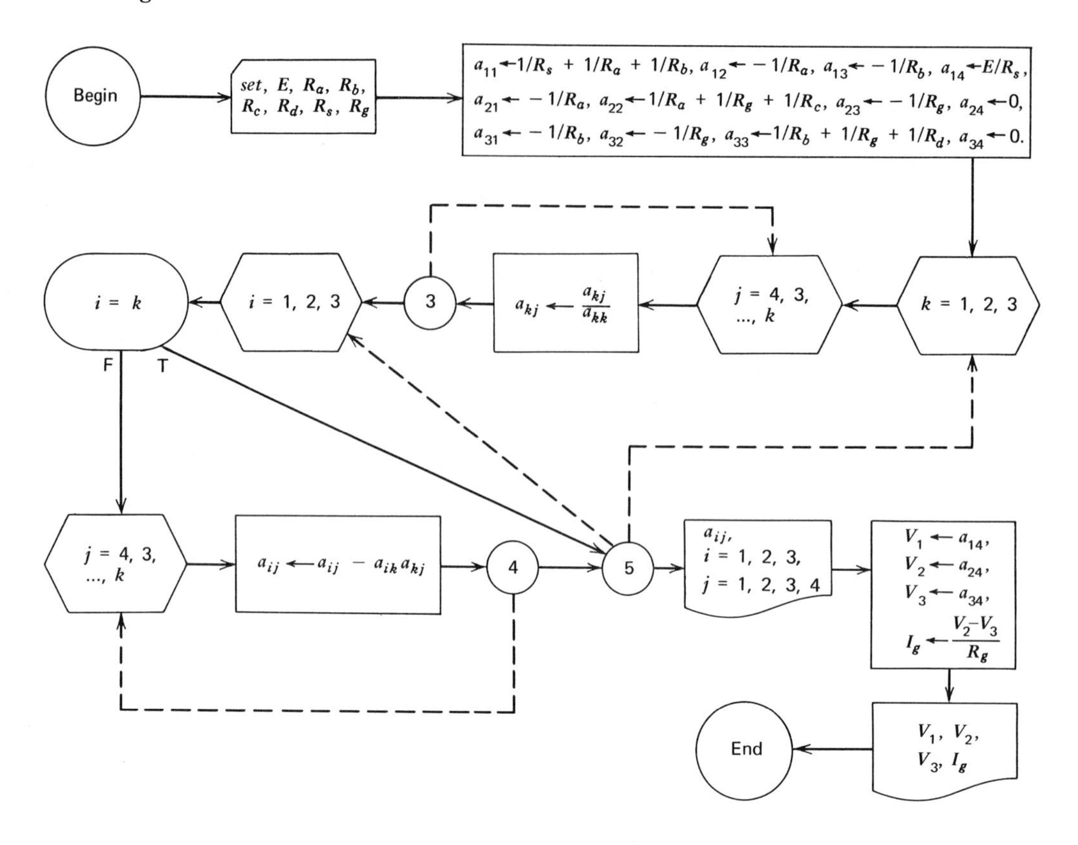

FORTRAN Implementation

List of Principal Variables

Program Symbol	Definition
A	3 × 4 augmented matrix **A** for Gauss-Jordan reduction.
E	Constant emf E of source, volts.
I	Row index, i.
IG	Galvanometer current I_g, amps (positive from node 2 *to* node 3).
J	Column index, j. Because of FORTRAN limitations (the increment in a DO statement must be positive), the steps $j = 4, 3, \ldots, k$ in the flow diagram cannot be programmed directly. The reader should convince himself that the construction used in the following program, involving two auxiliary variables JB and JBMAX, is both legal FORTRAN and also accomplishes the intent of the flow diagram. We shall welcome the day when this restriction is removed.
K	Pivot counter, k.
RA, RB, RC, RD, RS, RG	Resistances R_a, R_b, R_c, R_d, R_s, and R_g, ohms, as shown in Fig. 8.4.1.
SET	Integer identification number for each set of data.
V1, V2, V3	Potentials at nodes 1, 2, and 3, volts.

The following program achieves format-free input by using a NAMELIST declaration in conjunction with an appropriate input statement. The abbreviated data sets after the first emphasize that the entire list need not be read; those variables not reassigned values will automatically retain their previous values, as can readily be checked by examining the computer output.

Program Listing

```
C        DIGITAL COMPUTING                       EXAMPLE 8.4
C        POTENTIALS AND CURRENT IN A BRIDGE NETWORK.
C
      INTEGER SET
      REAL IG
      DIMENSION A(3,4)
      NAMELIST /DATA/ SET, E, RA, RB, RC, RD, RS, RG
C
C     ..... PRINT HEADING, READ AND CHECK DATA .....
      WRITE (6,200)
    1 READ (5,DATA,END=999,ERR=1)
      WRITE (6,201) SET, E, RA, RB, RC, RD, RS, RG
C
C     ..... SET UP THE AUGMENTED MATRIX & PRINT IT OUT .....
      A(1,1) = 1./RS + 1./RA + 1./RB
      A(1,2) = - 1./RA
      A(1,3) = - 1./RB
      A(1,4) = E/RS
      A(2,1) = - 1./RA
      A(2,2) = 1./RA + 1./RG + 1./RC
      A(2,3) = - 1./RG
      A(2,4) = 0.
      A(3,1) = - 1./RB
      A(3,2) = - 1./RG
      A(3,3) = 1./RB + 1./RG + 1./RD
      A(3,4) = 0.
      WRITE (6,202)  ((A(I,J), J = 1, 4), I = 1, 3)
C
C     ..... INCREMENT THE PIVOT COUNTER K FROM 1 THROUGH 3 .....
      DO 5  K = 1, 3
C
C     ..... NORMALIZE THE PIVOT ROW, COUNTING BACK FROM 4TH COLUMN .....
      JBMAX = 5 - K
      DO 3  JB = 1, JBMAX
      J = 5 - JB
    3 A(K,J) = A(K,J)/A(K,K)
C
C     ..... PERFORM REDUCTION STEPS FOR ROWS 1, 2, AND 3 .....
      DO 5  I = 1, 3
      IF (I .EQ. K)  GO TO 5
C
C     ..... AGAIN COUNTING BACK FROM 4TH COLUMN TO PIVOT COLUMN .....
      DO 4  JB = 1, JBMAX
      J = 5 - JB
    4 A(I,J) = A(I,J) - A(I,K)*A(K,J)
    5 CONTINUE
C
C     ..... PRINT OUT THE TRANSFORMED AUGMENTED MATRIX .....
      WRITE (6,203)  ((A(I,J), J = 1, 4), I = 1, 3)

C
C     ..... COMPUTE AND PRINT VOLTAGES AND GALVANOMETER CURRENT .....
      V1 = A(1,4)
      V2 = A(2,4)
      V3 = A(3,4)
      IG = (V2 - V3)/RG
      WRITE (6,204) V1, V2, V3, IG
      GO TO 1
C
  999 CALL EXIT
C
C     ..... FORMATS FOR OUTPUT STATEMENTS .....
  200 FORMAT ('1DIGITAL COMPUTING                       EXAMPLE 8.4'/
     1 ' POTENTIALS AND CURRENT IN A BRIDGE NETWORK.')
  201 FORMAT ('0'/ '0THE INPUT DATA FOR SET NO.', I3, ' ARE'/
     1 '0     E  =', F5.1, 8X, 'RA =', F5.1, 8X, 'RB =', F5.1/
     2 '      RC =', F5.1, 8X, 'RD =', F5.1, 8X, 'RS =', F5.1/
     3 '      RG =', F5.1)
  202 FORMAT ('0THE INITIAL AUGMENTED MATRIX A IS'/ ' '/ (3X, 4F11.5))
  203 FORMAT ('0THE FINAL TRANSFORMED MATRIX A IS'/ ' '/ (3X, 4F11.5))
  204 FORMAT ('0VOLTAGES AND GALVANOMETER CURRENT ARE'/
     1         '0     V1    =', F9.5/ '       V2    =', F9.5/
     2         '      V3    =', F9.5/ '       IG    =', F9.5)
C
      END
```

Data

```
&DATA SET = 1, E = 10., RA = 2., RB = 1., RC = 1., RD = 2.,
RS = 0.1, RG = 5. &END
&DATA SET = 2, RA = 10., RB = 10., RC = 10., RD = 10., RS = 0.1,
RG = 100. &END
&DATA SET = 3, RC = 10.5 &END
&DATA SET = 4, RC = 11.0 &END
&DATA SET = 5, RC = 12.0 &END
&DATA SET = 6, RC = 15.0 &END
&DATA SET = 7, RC = 20.0 &END
&DATA SET = 8, RC = 25.0 &END
&DATA SET = 9, RC = 30.0 &END
```

Computer Output (For data sets 1, 2, 3, and 4)

```
DIGITAL COMPUTING                          EXAMPLE 8.4
POTENTIALS AND CURRENT IN A BRIDGE NETWORK.


THE INPUT DATA FOR SET NO.  1 ARE

   E  = 10.0          RA =  2.0          RB =  1.0
   RC =  1.0          RD =  2.0          RS =  0.1
   RG =  5.0

THE INITIAL AUGMENTED MATRIX A IS

     11.50000   -0.50000   -1.00000  100.00003
     -0.50000    1.70000   -0.20000    0.0
     -1.00000   -0.20000    1.70000    0.0

THE FINAL TRANSFORMED MATRIX A IS

      1.00000   -0.0       -0.0        9.35960
     -0.0        1.00000   -0.0        3.44828
     -0.0       -0.0        1.00000    5.91133

VOLTAGES AND GALVANOMETER CURRENT ARE

    V1    =  9.35960
    V2    =  3.44828
    V3    =  5.91133
    IG    = -0.49261


THE INPUT DATA FOR SET NO.  2 ARE

   E  = 10.0          RA = 10.0          RB = 10.0
   RC = 10.0          RD = 10.0          RS =  0.1
   RG =100.0

THE INITIAL AUGMENTED MATRIX A IS

     10.20000   -0.10000   -0.10000  100.00003
     -0.10000    0.21000   -0.01000    0.0
     -0.10000   -0.01000    0.21000    0.0

THE FINAL TRANSFORMED MATRIX A IS

      1.00000   -0.0       -0.0        9.90099
     -0.0        1.00000   -0.0        4.95050
     -0.0       -0.0        1.00000    4.95049

VOLTAGES AND GALVANOMETER CURRENT ARE

    V1    =  9.90099
    V2    =  4.95050
    V3    =  4.95049
    IG    =  0.00000
```

```
THE INPUT DATA FOR SET NO.  3 ARE

     E  = 10.0          RA  = 10.0          RB  = 10.0
     RC = 10.5          RD  = 10.0          RS  =  0.1
     RG =100.0

THE INITIAL AUGMENTED MATRIX A IS

      10.20000    -0.10000    -0.10000   100.00003
      -0.10000     0.20524    -0.01000     0.0
      -0.10000    -0.01000     0.21000     0.0

THE FINAL TRANSFORMED MATRIX A IS

       1.00000    -0.0        -0.0         9.90218
      -0.0         1.00000    -0.0         5.06623
      -0.0        -0.0         1.00000     4.95657

VOLTAGES AND GALVANOMETER CURRENT ARE

     V1    =  9.90218
     V2    =  5.06623
     V3    =  4.95657
     IG    =  0.00110


THE INPUT DATA FOR SET NO.  4 ARE

     E  = 10.0          RA  = 10.0          RB  = 10.0
     RC = 11.0          RD  = 10.0          RS  =  0.1
     RG =100.0

THE INITIAL AUGMENTED MATRIX A IS

      10.20000    -0.10000    -0.10000   100.00003
      -0.10000     0.20091    -0.01000     0.0
      -0.10000    -0.01000     0.21000     0.0

THE FINAL TRANSFORMED MATRIX A IS

       1.00000    -0.0        -0.0         9.90332
      -0.0         1.00000    -0.0         5.17625
      -0.0        -0.0         1.00000     4.96235

VOLTAGES AND GALVANOMETER CURRENT ARE

     V1    =  9.90332
     V2    =  5.17625
     V3    =  4.96235
     IG    =  0.00214
```

Discussion of Results

The computed results are displayed above for data sets 1, 2, 3, and 4. The value in the a_{14} position of all the initial augmented matrices should equal $E/R_s = 10.0/0.1 = 100.0$. However, the printed value is 100.00003, which reflects that: (a) the decimal fraction 0.1 cannot be stored with complete accuracy in a binary machine, and/or (b) the format has requested more significant figures than are held in a standard IBM 360 word. Observe in each case that the final transformed matrix contains the identity matrix in its first three columns, indicating that the scheme has worked successfully. The source resistance has the fairly low value of $R_s = 0.1$ throughout, resulting in a minor potential drop across R_s, so that V_1 is fairly close to the source emf E in most cases.

The first data set is for a highly unbalanced situation, and gives $I_g = -0.49261$; the minus sign indicates that the current flows from node 3 to node 2. The remaining data sets (2 through 9) are for a bridge that is initially balanced

Table 8.4.1 Variation of Galvanometer Current I_g with Unbalanced Resistance R_c.

R_c (ohms)	I_g (amps)
10.0	0
10.5	0.00110
11.0	0.00214
12.0	0.00408
15.0	0.00893
20.0	0.01480
25.0	0.01896
30.0	0.02206

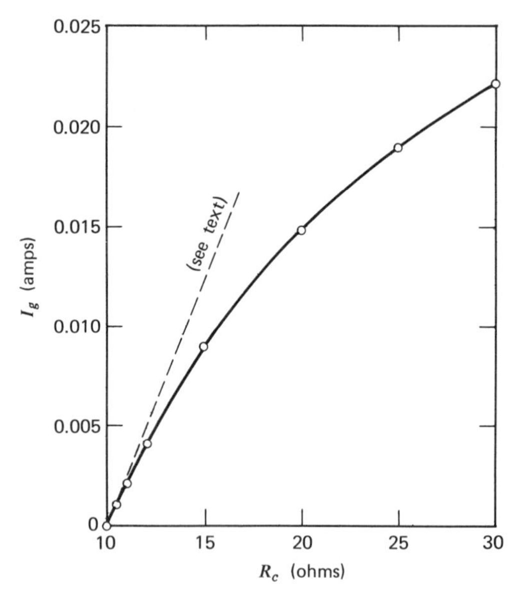

Figure 8.4.2 Variation of galvanometer current I_g with unbalanced resistance R_c.

($R_a = R_b = R_c = R_d = 10$ ohms); one arm (R_c) is then progressively increased. The resulting galvanometer current I_g is shown as a function of R_c in Table 8.4.1 and Fig. 8.4.2.

It may be shown [25] that for an initially balanced bridge with an infinite galvanometer resistance and a constant applied emf E, the potential difference $V_2 - V_3$ is given approximately by

$$V_2 - V_3 \doteq \frac{E\Delta R_c}{4R_c}, \qquad (8.4.3)$$

provided that the fractional increase $\Delta R_c/R_c$ is small. The broken line shown in Fig. 8.4.2 (obtained by setting $E \doteq 9.9$ and dividing $V_2 - V_3$ by R_g) represents equation (8.4.2), and appears to be valid provided that $\Delta R_c/R_c$ is small. In general, I_g is clearly a nonlinear function of R_c.

8.5 Computing the Inverse Matrix

As explained previously, the augmented matrix, **C** for example, consists of the left-hand side coefficient matrix **A**, with the right-hand side vector **b** appended as an extra column:

$$\mathbf{C} = [\mathbf{A} \,\vdots\, \mathbf{b}].$$

The vertical dotted line emphasizes that **b** is merely appended, and that the product **Ab** is not being formed. At the conclusion of the manipulations dictated by the Gauss-Jordan algorithm we have a modified augmented matrix, **D** for example:

$$\mathbf{D} = [\mathbf{I} \,\vdots\, \mathbf{x}],$$

which contains the identity matrix **I** with the solution vector **x** appended. The sequence of Gauss-Jordan steps is actually equivalent to forming the matrix product $\mathbf{D} = \mathbf{A}^{-1}\mathbf{C}$, that is, to multiplying **C** by $\mathbf{A}^{-1}$, the inverse of the coefficient matrix **A**; this must be so because we have already seen in Section 8.2 that $\mathbf{A}^{-1}\mathbf{A} = \mathbf{I}$ and $\mathbf{A}^{-1}\mathbf{b} = \mathbf{x}$. The implication of this is that if we augment **C** still further by including the identity matrix **I** in columns $n+2, n+3, \ldots, 2n+1$, then these columns will ultimately contain $\mathbf{A}^{-1}$:

$$\mathbf{C} = [\mathbf{A} \,\vdots\, \mathbf{b} \,\vdots\, \mathbf{I}],$$
$$\mathbf{D} = \mathbf{A}^{-1}\mathbf{C} = [\,\mathbf{I} \,\vdots\, \mathbf{x} \,\vdots\, \mathbf{A}^{-1}].$$

It is understood that the algorithm of (8.20) would be modified so that the column counter j would have an upper limit of $2n+1$ instead of $n+1$. That is, the usual reduction and normalization steps would automatically include the additional columns appended.

In fact, *several* right-hand side vectors $\mathbf{b}_1, \mathbf{b}_2, \ldots$ could also be included in the augmented matrix, in which case we would obtain several solution vectors $\mathbf{x}_1, \mathbf{x}_2, \ldots$ at once. For example:

$$\mathbf{C} = [\mathbf{A} \,\vdots\, \mathbf{b}_1 \,\vdots\, \mathbf{b}_2 \,\vdots\, \mathbf{b}_3 \,\vdots\, \mathbf{I}]$$
$$\mathbf{D} = \mathbf{A}^{-1}\mathbf{C} = [\,\mathbf{I} \,\vdots\, \mathbf{x}_1 \,\vdots\, \mathbf{x}_2 \,\vdots\, \mathbf{x}_3 \,\vdots\, \mathbf{A}^{-1}].$$

In writing the corresponding algorithm, we conventionally assume that the augmented matrix has $n+m$ columns—that is, m extra columns are appended to the left-hand side coefficient matrix **A**. In (8.20), the only modification needed is to the upper limit of the column counter, which now becomes $j = k+1, k+2, \ldots, n+m$.

EXAMPLE 8.5.

Starting with

$$\begin{bmatrix} 2 & -7 & 4 & 9 & -46 & 0 & 1 & 0 & 0 \\ 1 & 9 & -6 & 1 & 52 & 1 & 0 & 1 & 0 \\ -3 & 8 & 5 & 6 & 54 & 28 & 0 & 0 & 1 \end{bmatrix}, \tag{8.21}$$

the reader should show that the Gauss-Jordan algorithm produces

$$\begin{bmatrix} 1 & 0 & 0 & 4 & -2 & 1 & \frac{93}{235} & \frac{67}{235} & \frac{6}{235} \\ 0 & 1 & 0 & 1 & 6 & 2 & \frac{13}{235} & \frac{22}{235} & \frac{16}{235} \\ 0 & 0 & 1 & 2 & 0 & 3 & \frac{7}{47} & \frac{1}{47} & \frac{5}{47} \end{bmatrix}. \tag{8.22}$$

What is the interpretation of these two augmented matrices?

8.6 Maximum-Pivot Strategy

The reader will doubtless have asked by now: "What happens if a zero pivot element is encountered?" This can easily happen, for example with

$$\begin{aligned} 3x_2 &= 6, \\ 2x_1 + 5x_2 &= 12, \end{aligned}$$

which have the solution $x_1 = 1$ and $x_2 = 2$. Yet even the first normalization step will fail because of division by zero. A simple remedy, if it can be recognized at the outset, is to rearrange the equations; the solution of

$$\begin{aligned} 2x_1 + 5x_2 &= 12, \\ 3x_2 &= 6, \end{aligned}$$

for example, presents no difficulty. However, a zero pivot may be generated during the calculations, and, if there are several equations, it may not be obvious as to how they should be rearranged.

A related problem arises when the absolute value of a pivot is small in relation to the other coefficients. Consider, for instance, the system

$$\begin{aligned} 0.0003x_1 + 3.0000x_2 &= 2.0001, \\ 1.0000x_1 + 1.0000x_2 &= 1.0000, \end{aligned} \tag{8.23}$$

which has the exact solution $x_1 = 1/3$ and $x_2 = 2/3$. If the equations are solved using pivots on the matrix diagonal, as indicated in the previous examples, there results:

$$\begin{aligned} 1.0000x_1 + 10000x_2 &= 6667, \\ x_2 &= 6666/9999. \end{aligned} \tag{8.24}$$

If x_2 from the second equation is taken to be 0.6667, then, from the first equation, $x_1 = 0.0000$; for $x_2 = 0.66667$, $x_1 = 0.30000$; for $x_2 = 0.666667$, $x_1 = 0.330000$, etc. The solution depends highly on the number of figures retained. If the equations are solved in reverse order, that is, by interchanging the two rows and proceeding as before, then x_2 is found to be $1.9998/2.9997 \doteq 0.66667$, with $x_1 = 0.33333$. This example indicates the advisability of choosing as the pivot the coefficient of largest absolute value in a column, rather than merely the diagonal element.

These difficulties can normally be overcome by using a technique known as Gauss-Jordan reduction with *maximum-pivot strategy*. This strategy requires that the kth pivot $(k = 1, 2, \ldots, n)$ be the element of greatest magnitude that is still available for pivoting, that is, the largest element not in a row or column already containing a pivot. In order to carry out the reduction without interchanging rows (or columns), it is necessary to keep a list of the subscripts of the pivot elements. Let r_k and c_k be, respectively, the row and column subscripts of the kth pivot element, p_k.

The method is probably best illustrated with a numerical example, after which the algorithm will be stated. Consider the set of equations used in Section 8.4, but taken in a different order to illustrate the procedure better. The solutions, as before, are $\mathbf{x} = [4,\ 1,\ 2]^t$:

$$\begin{aligned} -3x_1 + 8x_2 + 5x_3 &= 6, \\ 2x_1 - 7x_2 + 4x_3 &= 9, \\ x_1 + 9x_2 - 6x_3 &= 1. \end{aligned}$$

The augmented coefficient matrix for this system is

$$\begin{bmatrix} -3 & 8 & 5 & 6 \\ 2 & -7 & 4 & 9 \\ 1 & 9 & -6 & 1 \end{bmatrix}.$$

Then, for $k = 1$, that is, for the first pass, element a_{32} has the greatest absolute value of the nine possible coefficients, so that $p_1 = 9$, $r_1 = 3$, and $c_1 = 2$. After the normalization steps for the first cycle, the transformed matrix is

$$\begin{bmatrix} -3 & 8 & 5 & 6 \\ 2 & -7 & 4 & 9 \\ \frac{1}{9} & 1 & -\frac{2}{3} & \frac{1}{9} \end{bmatrix}.$$

The reduction steps lead to

$$\begin{bmatrix} -\frac{35}{9} & 0 & \frac{31}{3} & \frac{46}{9} \\ \frac{25}{9} & 0 & -\frac{2}{3} & \frac{88}{9} \\ \frac{1}{9} & 1 & -\frac{2}{3} & \frac{1}{9} \end{bmatrix}.$$

For the second cycle, only a_{11}, a_{13}, a_{21}, and a_{23} are available as potential pivots. Noting that a_{13} has the greatest magnitude of these four candidates, we have $p_2 = 31/3$, $r_2 = 1$, and $c_2 = 3$. After the normalization and reduction steps, the matrix becomes

$$\begin{bmatrix} -\frac{35}{93} & 0 & 1 & \frac{46}{93} \\ \frac{235}{93} & 0 & 0 & \frac{940}{93} \\ -\frac{13}{93} & 1 & 0 & \frac{41}{93} \end{bmatrix}.$$

For the third and last cycle, only a_{21} is available for pivoting. Then $p_3 = 235/93$, $r_3 = 2$, and $c_3 = 1$. After normalization and reduction, the matrix becomes

$$\begin{bmatrix} 0 & 0 & 1 & 2 \\ 1 & 0 & 0 & 4 \\ 0 & 1 & 0 & 1 \end{bmatrix}.$$

Note that the first three columns contain a permuted identity matrix, and that the fourth contains the solutions, although not in their natural order. By reinstating the x's and equals signs, we readily see that $x_3 = 2$, $x_1 = 4$, and $x_2 = 1$. However, for automatic implementation on the digital computer, the key to unravelling the solutions is to examine the r and c vectors, which contain the values:

k	r_k	c_k
1	3	2
2	1	3
3	2	1

From this example, the reader should convince himself that the value of the c_kth element in the solution vector $\mathbf{x}$ is to be found in the r_kth row of the last or $(n+1)$th column of the matrix. Thus, in general,

$$x_{c_k} = a_{r_k,n+1}, \qquad k = 1, 2, \ldots, n.$$

The complete algorithm for Gauss-Jordan reduction with the maximum pivot strategy is:

Choosing the pivot

$$\left.\begin{array}{ll}
p_k = a_{r_k c_k}, & |a_{r_k c_k}| = \max\limits_{i,j} |a_{ij}|, \\
& i = 1, 2, \ldots, n, \\
& i \neq r_1, r_2, \ldots, r_{k-1}, \\
& j = 1, 2, \ldots, n, \\
& j \neq c_1, c_2, \ldots, c_{k-1}, \\
\textit{Normalization} & \\
a_{r_k j} \leftarrow a_{r_k j}/p_k, & j = 1, 2, \ldots, n+1, \\
\textit{Reduction} & \\
a_{ij} \leftarrow a_{ij} - a_{ic_k} a_{r_k j}, & \\
& j = 1, 2, \ldots, n+1, \\
& i = 1, 2, \ldots, n, \\
& i \neq r_k,
\end{array}\right\} k = 1, 2, \ldots, n. \qquad (8.25)$$

8.7 Solution of Simultaneous Nonlinear Equations

The previous sections have primarily been concerned with the solution of simultaneous *linear* equations. The more general case of a system of n simultaneous *nonlinear* equations is:

$$\begin{array}{r}
f_1(x_1, x_2, \ldots, x_n) = 0, \\
f_2(x_1, x_2, \ldots, x_n) = 0, \\
\vdots \qquad\qquad \vdots \\
f_n(x_1, x_2, \ldots, x_n) = 0.
\end{array} \qquad (8.26)$$

That is, we have n functions involving the variables $x_1, x_2, \ldots, x_n$; the problem is, of course, to find the values of these variables for which each of the functions becomes zero. Just as it is sometimes possible for the single equation $f(x) = 0$

to have more than one root, it is also sometimes possible for (8.26) to have multiple solutions—in contrast to the usual situation with simultaneous *linear* equations.

By defining the vector

$$\mathbf{x} = [x_1 \; x_2 \ldots x_n]^t, \qquad (8.27)$$

equations (8.26) can be written more concisely as

$$f_i(\mathbf{x}) = 0, \qquad i = 1, 2, \ldots, n, \quad (8.28)$$

analogous to (5.4), $f(x) = 0$, for the case of a single variable.

In general, the solution of (8.28) is a challenging problem, particularly if the functions f_i are of any complexity, and a direct solution in a finite number of operations is usually impossible. The best we can hope for is to develop an iterative type of solution, in which an initial estimate for the solution vector $\mathbf{x}$ is improved at each iteration. There are two main possibilities, analogous to the *successive substitution* and *Newton* methods for solving the single equation $f(x) = 0$.

The first possibility is to rearrange (8.28) in the form

$$x_i = F_i(\mathbf{x}), \qquad i = 1, 2, \ldots, n, \quad (8.29)$$

and also to denote the kth estimate of the solution vector as

$$\mathbf{x}_k = [x_{1k} \; x_{2k} \cdots x_{nk}]^t, \qquad (8.30)$$

where $x_{1k}, x_{2k}, \ldots, x_{nk}$ are the kth estimates of the individual variables $x_1, x_2, \ldots, x_n$. A *successive-substitution* scheme is then given by the algorithm

$$x_{ik} = F_i(x_{1,k-1}, x_{2,k-1}, \ldots, x_{n,k-1})$$

or

$$x_{ik} = F_i(\mathbf{x}_{k-1}), \qquad i = 1, 2, \ldots, n, \quad k = 2, 3, \ldots \qquad (8.31)$$

in which the initial approximations are assumed to be contained in the vector $\mathbf{x}_1$. A sufficient condition for convergence of this scheme can be shown [1] to be

$$\left|\frac{\partial F_i}{\partial x_1}\right| + \left|\frac{\partial F_i}{\partial x_2}\right| + \cdots + \left|\frac{\partial F_i}{\partial x_n}\right| < 1, \qquad i = 1, 2, \ldots, n, \qquad (8.32)$$

for all values of the x_i between their initial approximations and the final solutions. If there is an obvious rearrangement of (8.28) to the form (8.29), such that (8.32) is satisfied, then this successive-substitution scheme has the merit of simplicity, although convergence may not be especially rapid.

The second possibility is the *Newton-Raphson* method, analogous to *Newton's* method for solving the single equation $f(x) = 0$. Accordingly, it is instructive to reconsider Newton's method for a moment (see Section 5.4): given a kth approximation x_k, such that $f(x_k)$ is not zero, the goal is to change x_k slightly and obtain a further approximation x_{k+1}, such that $f(x_{k+1})$ is as close as possible to zero. It follows from (5.11) that the appropriate change δx_k, such that $x_{k+1} = x_k + \delta x_k$, is given by

$$f'(x_k)\delta x_k = -f(x_k). \qquad (8.33)$$

The Newton-Raphson method for solving (8.28) follows the same general philosophy. Suppose that at the kth stage of iteration we have an approximation $\mathbf{x}_k$ that does not quite satisfy (8.28), that is,

$$f_i(\mathbf{x}_k) \neq 0, \qquad i = 1, 2, \ldots, n.$$

What we would like to do is to alter $\mathbf{x}_k$ slightly, to give $\mathbf{x}_{k+1}$, so that each f_i is changed by an amount $-f_i(\mathbf{x}_k)$, that is, by minus its current value, thereby reducing it to zero.

First note that a differential change in f_i is

$$df_i = \frac{\partial f_i}{\partial x_1}dx_1 + \frac{\partial f_i}{\partial x_2}dx_2 + \cdots + \frac{\partial f_i}{\partial x_n}dx_n.$$

Thus, any *finite* change in f_i is given approximately by

$$\delta f_i \doteq \frac{\partial f_i}{\partial x_1}\delta x_1 + \frac{\partial f_i}{\partial x_2}\delta x_2 + \cdots + \frac{\partial f_i}{\partial x_n}\delta x_n.$$

Hence the appropriate changes to be made in the x_i at the kth iteration are given by the solution, $\delta x_{1k}, \delta x_{2k}, \ldots, \delta x_{nk}$, of the simultaneous *linear* equations:

$$\begin{aligned}
\frac{\partial f_1}{\partial x_1}\delta x_{1k} + \frac{\partial f_1}{\partial x_2}\delta x_{2k} + \cdots + \frac{\partial f_1}{\partial x_n}\delta x_{nk} &= -f_1(\mathbf{x}_k),\\
\frac{\partial f_2}{\partial x_1}\delta x_{1k} + \frac{\partial f_2}{\partial x_2}\delta x_{2k} + \cdots + \frac{\partial f_2}{\partial x_n}\delta x_{nk} &= -f_2(\mathbf{x}_k),\\
\vdots \qquad\qquad \vdots \qquad\qquad\qquad \vdots \qquad\quad &\quad \vdots\\
\frac{\partial f_n}{\partial x_1}\delta x_{1k} + \frac{\partial f_n}{\partial x_2}\delta x_{2k} + \cdots + \frac{\partial f_n}{\partial x_n}\delta x_{nk} &= -f_n(\mathbf{x}_k).
\end{aligned} \tag{8.34}$$

In (8.34), it is understood that each partial derivative is evaluated using the current approximation $\mathbf{x}_k$. Equations (8.34) may be expressed more concisely by defining a coefficient matrix $\mathbf{\Phi}(\mathbf{x})$ as

$$\mathbf{\Phi}(\mathbf{x}) = (f_{ij}) = \left(\frac{\partial f_i(\mathbf{x})}{\partial x_j}\right), \quad 1 \leqslant i, j \leqslant n, \tag{8.35}$$

a solution vector $\boldsymbol{\delta}\mathbf{x}_k$ as

$$\boldsymbol{\delta}\mathbf{x}_k = [\delta x_{1k}\, \delta x_{2k} \cdots \delta x_{nk}]^t, \tag{8.36}$$

and a right-hand side vector $\mathbf{f}(\mathbf{x}_k)$ as

$$\mathbf{f}(\mathbf{x}_k) = [f_1(\mathbf{x}_k)\ f_2(\mathbf{x}_k) \cdots f_n(\mathbf{x}_k)]^t. \tag{8.37}$$

Equations (8.34) can then be reexpressed as

$$\mathbf{\Phi}(\mathbf{x}_k)\boldsymbol{\delta}\mathbf{x}_k = -\mathbf{f}(\mathbf{x}_k), \tag{8.38}$$

which should be compared with (8.33), the corresponding equation for Newton's solution of the single equation $f(x) = 0$. Once (8.38) has been solved for $\boldsymbol{\delta}\mathbf{x}_k$ (the Gauss-Jordan algorithm with the maximum-pivot criterion would be particularly appropriate), the next estimate of the solution of (8.28) is given by

$$\mathbf{x}_{k+1} = \mathbf{x}_k + \boldsymbol{\delta}\mathbf{x}_k. \tag{8.39}$$

Starting from an initial estimate $\mathbf{x}_1$, the above procedure can clearly be repeated for successive iterations, $k = 1, 2, \ldots$ until convergence according to some predetermined criterion has been satisfied, or a specified number of iterations has been performed. In effect, the Newton-Raphson method has reduced the solution of the simultaneous *nonlinear* equations to the *repeated* solution of a system of simultaneous *linear* equations.

Convergence of the Newton-Raphson method to a solution of (8.28) is guaranteed [1] if the starting vector is "close enough" to the solution vector and the determinant of $\mathbf{\Phi}(\mathbf{x}_k)$ is never zero. If convergence does occur, it usually occurs fairly rapidly. A notable disadvantage of the method is that algebraic expressions must usually be derived by hand for each of the partial derivatives $f_{ij} = \partial f_i/\partial x_j$, which total n^2 in number. If there are several equations to be solved, and if the f_i are moderately involved in form, this is no trivial task. However, the derivatives may also be approximated numerically.

8.8 Tridiagonal Systems

Consider the following system of simultaneous linear equations in the unknowns $v_1, v_2, \ldots, v_N$:

$$\begin{aligned}
b_1v_1 + c_1v_2 \qquad\qquad\qquad\qquad &= d_1\\
a_2v_1 + b_2v_2 + c_2v_3 \qquad\qquad &= d_2\\
a_3v_2 + b_3v_3 + c_3v_4 \qquad &= d_3\\
\cdots\cdots\cdots\cdots\cdots\cdots &\\
a_iv_{i-1} + b_iv_i + c_iv_{i+1} &= d_i\\
\cdots\cdots\cdots\cdots\cdots\cdots &\\
a_{N-1}v_{N-2} + b_{N-1}v_{N-1} + c_{N-1}v_N &= d_{N-1}\\
a_Nv_{N-1} + b_Nv_N &= d_N.
\end{aligned} \tag{8.40}$$

Note that the first and last equations contain just two unknowns, and that the remainder contain only three unknowns, such as v_i and its two immediately adjacent neighbors v_{i-1} and v_{i+1}. The matrix of coefficients a_i, b_i, and c_i alone is called a *tridiagonal* matrix. Such systems of tridiagonal equations occur occasionally in engineering, notably in the finite-difference solution of ordinary and partial differential equations, not considered in this text. However, the spline-function approximation of Section 6.7 also leads to equations of the type (8.40), so it is appropriate to discuss the solution of tridiagonal systems here.

Equations (8.40) could be solved by the Gauss-Jordan method, but the following technique is much faster, since it takes into account the tridiagonal nature of the system.

We first demonstrate the validity of a *recursion solution* of the form

$$v_i = \gamma_i - \frac{c_i}{\beta_i} v_{i+1},$$

in which the constants β_i and γ_i are to be determined. Substitution into the ith equation of (8.40) gives

$$a_i\left(\gamma_{i-1} - \frac{c_{i-1}}{\beta_{i-1}} v_i\right) + b_i v_i + c_i v_{i+1} = d_i.$$

That is,

$$v_i = \frac{d_i - a_i\gamma_{i-1}}{b_i - \dfrac{a_i c_{i-1}}{\beta_{i-1}}} - \frac{c_i v_{i+1}}{b_i - \dfrac{a_i c_{i-1}}{\beta_{i-1}}},$$

which verifies the above form, subject to the following recursion relations:

$$\beta_i = b_i - \frac{a_i c_{i-1}}{\beta_{i-1}}, \qquad \gamma_i = \frac{d_i - a_i \gamma_{i-1}}{\beta_i}.$$

Also, from the first equation of (8.40),

$$v_1 = \frac{d_1}{b_1} - \frac{c_1}{b_1} v_2,$$

whence $\beta_1 = b_1$ and $\gamma_1 = d_1/\beta_1$. Finally, substitution of the recursion solution into the last equation of (8.40) yields

$$v_N = \frac{d_N - a_N v_{N-1}}{b_N} = \frac{d_N - a_N\left(\gamma_{N-1} - \dfrac{c_{N-1}}{\beta_{N-1}} v_N\right)}{b_N},$$

whence

$$v_N = \frac{d_N - a_N \gamma_{N-1}}{b_N - \dfrac{a_N c_{N-1}}{\beta_{N-1}}} = \gamma_N.$$

To summarize, the complete algorithm for the solution of the tridiagonal system is

$$v_N = \gamma_N,$$

$$v_i = \gamma_i - \frac{c_i v_{i+1}}{\beta_i}, \quad i = N-1, N-2, \ldots, 1, \tag{8.41}$$

where the β's and γ's are determined from the recursion formulas

$$\beta_1 = b_1, \quad \gamma_1 = d_1/\beta_1,$$

$$\beta_i = b_i - \frac{a_i c_{i-1}}{\beta_{i-1}}, \qquad i = 2, 3, \ldots, N,$$

$$\gamma_i = \frac{d_i - a_i \gamma_{i-1}}{\beta_i}, \qquad i = 2, 3, \ldots, N. \tag{8.42}$$

Problems

8.1 If **A** and **B** are two matrices, prove that $(\mathbf{AB})^t = \mathbf{B}^t\mathbf{A}^t$.

8.2 If **A**, **B**, and **C** are nonsingular square matrices, prove that $(\mathbf{ABC})^{-1} = \mathbf{C}^{-1}\mathbf{B}^{-1}\mathbf{A}^{-1}$.

8.3 Show that if $\mathbf{B}_1$ is a close approximation to $\mathbf{A}^{-1}$, the inverse of a square matrix **A**, then an even better approximation $\mathbf{B}_2$ is given by

$$\mathbf{B}_2 = \mathbf{B}_1(2\mathbf{I} - \mathbf{A}\mathbf{B}_1).$$

8.4 A point P has position vector **x** relative to an origin O in rectangular coordinates. If the axes are rotated, but still remain orthogonal, the new position vector of P may be expressed by the linear transformation **Ax**. By noting that the distance OP is unchanged, show that **A** is an *orthogonal* matrix, that is, one for which $\mathbf{A}^t = \mathbf{A}^{-1}$.

8.5 What does the following represent?

$$\sum_{j=1}^{n} a_{ij}x_j = b_i, \qquad i = 1, 2, \ldots, n.$$

8.6 In Example 8.2 on page 351, verify that the matrix labeled $\mathbf{A}^{-1}$ is indeed the inverse of matrix **A**.

8.7 Write the following simultaneous equations in matrix form:

$$\begin{aligned} x_1 + 3x_2 &= 5, \\ 4x_1 - x_2 &= 12. \end{aligned}$$

By means of a hand calculation, implement the Gauss-Jordan reduction scheme with the usual diagonal pivots to compute: (a) x_1 and x_2, (b) the inverse of the coefficient matrix.

8.8 By means of hand calculations, solve the following three equations using the Gauss-Jordan reduction scheme with the maximum pivot criterion:

$$\begin{aligned} x_1 + 2x_2 + 8x_3 &= -3, \\ 2x_1 - 4x_2 - 6x_3 &= 0, \\ -2x_1 + 4x_2 + 2x_3 &= 4. \end{aligned}$$

Do not interchange rows or columns. Show the status of the augmented matrix after each complete pass of the elimination procedure.

8.9 Draw a flow diagram that represents the complete Gauss-Jordan reduction algorithm of equation (8.20).

8.10 Starting with the augmented matrix (8.21) in Example 8.5 on page 365, verify that the Gauss-Jordan algorithm does indeed generate the matrix shown in (8.22). What is the interpretation of these two augmented matrices—that is, what do their individual columns represent?

8.11 Horizontal and vertical loads X and Y, and a moment M are applied to the cantilever of length L shown in Fig. P8.11.

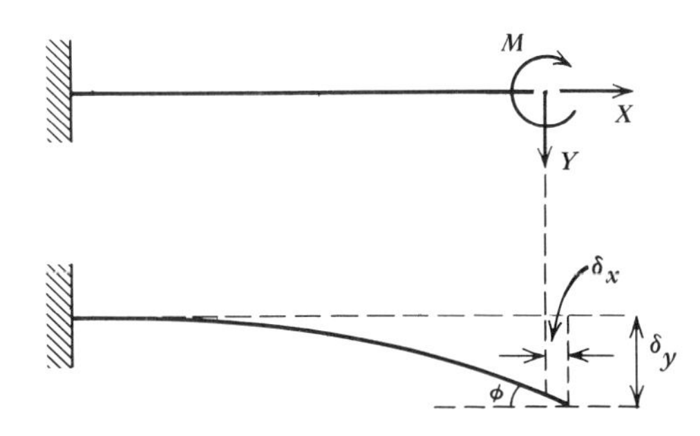

Figure P8.11.

At the free end, the horizontal extension δ_x, the downwards deflection δ_y, and the slope ϕ (approximately equal to the angle with the horizontal for small ϕ) are related to the loads as follows:

$$\begin{bmatrix} \delta_x \\ \delta_y \\ \phi \end{bmatrix} = \begin{bmatrix} \dfrac{L}{EA} & 0 & 0 \\ 0 & \dfrac{L^3}{3EI} & \dfrac{L^2}{2EI} \\ 0 & \dfrac{L^2}{2EI} & \dfrac{L}{EI} \end{bmatrix} \begin{bmatrix} X \\ Y \\ M \end{bmatrix}.$$

Here, E = Young's modulus, A = cross-sectional area, and I = second moment of area for the cantilever. The above 3×3 matrix is the *flexibility matrix* $\mathbf{\Gamma}$ for the cantilever.

Find $\mathbf{K} = \mathbf{\Gamma}^{-1}$, the *stiffness matrix* for the cantilever.

8.12 Jenkins and White [15] give precisely determined values, shown here in

Table P8.12, for the refractive index, n, at various wavelengths, λ (angstrom units), for borosilicate crown glass.

Table P8.12

λ	n
6563	1.50883
6439	1.50917
5890	1.51124
5338	1.51386
5086	1.51534
4861	1.51690
4340	1.52136
3988	1.52546

Use the second, fourth, and seventh entries in Table P8.12 to determine the constants A, B, and C in Cauchy's equation for refractive index:

$$n = A + \frac{B}{\lambda^2} + \frac{C}{\lambda^4}.$$

Investigate the merits of Cauchy's equation for predicting the remaining entries in the table.

8.13 This problem should be preceded by Problem 8.9. Write a logical function, named GJ, that implements the Gauss-Jordan reduction procedure on the N × (N + 1) augmented matrix A whose elements are stored in A(1, 1) through A(N, N + 1). Use the diagonal-pivot algorithm of equation (8.20). Normally, on exit from GJ, the array A will contain the transformed matrix and the value of the function itself will be returned as .FALSE. However, if at any stage a pivot smaller in absolute value than EPS is encountered, GJ should immediately return the value .TRUE. to the calling program. Thus, typical references to the function might be

```
IF (GJ(A, N, EPS))  GO TO 10
```

or

```
SWITCH = GJ(ARRAY, 10, 1.0E-20)
```

where SWITCH is a logical variable. Test the function by writing a main program that handles input and output and calls on GJ. For test data, use the three simultaneous equations on page 353, equations (8.23), and any of the simultaneous equations discussed in Problems (8.12), (8.18), (8.19), (8.22), or (8.23), or in Example 8.4.

8.14 Write a logical function, named GJNNPM, that implements the Gauss-Jordan reduction procedure on the N × (N + M) augmented matrix A whose elements are stored in A(1, 1) through A(N, N + M). Use the diagonal-pivot algorithm of equation (8.20), modified according to the comment at the top of page 365. Normally, on exit from GJNNPM, the array A will contain the transformed matrix and the value of the function itself will be returned as .FALSE. However, if at any stage a pivot smaller in absolute value than EPS is encountered, GJNNPM should immediately return the value .TRUE. to the calling program. Thus, a typical reference to the function might be

```
CHECK = GJNNPM (A, M, N, EPS)
```

where CHECK is a logical variable. Test the function by writing a main program that handles input and output and calls on GJNNPM. For test data, use equations (8.21) and any of the test data mentioned in Problem 8.13.

8.15 Write a logical function, named MAXPIV, that implements the Gauss-Jordan reduction procedure on the N × (N + 1) augmented matrix A whose elements are stored in A(1, 1) through A(N, N + 1). Use the maximum-pivot strategy of equations (8.25). Normally, on exit from MAXPIV, the array A will contain the transformed matrix and the value of the function itself will be returned as .FALSE. However, if at any stage a pivot smaller in absolute value than EPS is encountered, MAXPIV should immediately return the value .TRUE. to the calling program. Thus, a typical reference to the function might be

SWITCH = MAXPIV (A, N, EPS)

where SWITCH is a logical variable. Test the function by writing a main program that handles input and output and calls on MAXPIV. For test data, use equations (8.23) and the three simultaneous equations on page 353, and any of the test data mentioned in Problem 8.13.

8.16 Is the algorithm of (8.41) and (8.42) for the solution of a tridiagonal system of equations still valid for: (a) a single equation ($n = 1$), and (b) two equations ($n = 2$)?

8.17 Write a subroutine, named TRIDAG, that will solve a tridiagonal system of simultaneous linear equations (see Section 8.8). Assume that before entry to TRIDAG, the coefficients a_i, b_i, and c_i, and the right-hand side values d_i have already been stored in the vectors A, B, C, and D. Upon exit from TRIDAG, the solution should be stored in the vector V.

The algorithm should be based on equations (8.41) and (8.42), but should be modified so that the unknowns have first and last subscripts f and l, respectively; that is, the unknowns are assumed to be $v_f, v_{f+1}, \ldots, v_{l-1}, v_l$. Assume that a typical call on the subroutine will be

CALL TRIDAG (F, L, A, B, C, D, V)

Here, F and L are the program symbols for the integers f and l.

The subroutine TRIDAG may be used in conjunction with Problem 6.25, which deals with spline-function approximation.

8.18 Consider the electrical network shown in Fig. P8.18; the values of the resistances are in ohms (Ω), and the potential applied between the terminals A and B is 100 volts (V). We wish to find the potentials at the nodes or junctions numbered 1 through 6 and the current in amperes through each resistance.

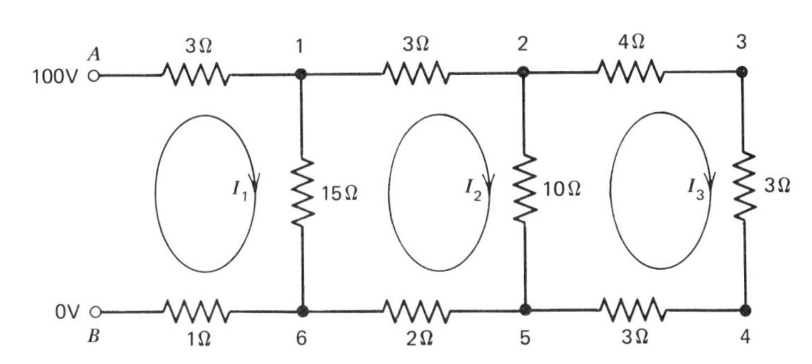

Figure P8.18.

The problem may be solved by the *loop* method, in which currents I_1, I_2, and I_3 are supposed to flow round the individual loops as shown in Fig. P8.18. The potential drop across any resistance is given by Ohm's law as the product of the resistance and the current flowing through it; for example, $V_1 - V_2 = 3I_2$, $V_2 - V_3 = 10(I_2 - I_3)$. By applying Kirchoff's potential law (the algebraic sum of the potential drops round a closed loop is zero), show that the loop currents obey the following simultaneous linear equations:

$$\begin{aligned} 19I_1 - 15I_2 \qquad\qquad &= 100, \\ -15I_1 + 30I_2 - 10I_3 &= 0, \\ -10I_2 + 20I_3 &= 0. \end{aligned}$$

Use the Gauss-Jordan method by hand to solve for I_1, I_2, and I_3, and hence predict the node potentials $V_1, V_2, \ldots, V_6$.

8.19 (a) The electrical network of Problem 8.18 may also be solved by the *node* method, which is essentially the application of Ohm's law in conjunction with Kirchoff's current law, already discussed in Example 8.4. Show that the node method leads to the following system of simultaneous linear equations in the node potentials:

$$
\begin{array}{rrrrrrl}
11V_1 & -\;5V_2 & & & & -\;V_6 & = 500, \\
-20V_1 & +41V_2 & -15V_3 & & -\;6V_5 & & = 0, \\
 & -\;3V_2 & +\;7V_3 & -\;4V_4 & & & = 0, \\
 & & -\;V_3 & +\;2V_4 & -\;V_5 & & = 0, \\
 & -\;3V_2 & & -10V_4 & +28V_5 & -15V_6 & = 0, \\
-\;2V_1 & & & & -15V_5 & +47V_6 & = 0.
\end{array}
$$

(b) Use the node method to derive the corresponding set of equations for the network shown in Fig. P8.19.

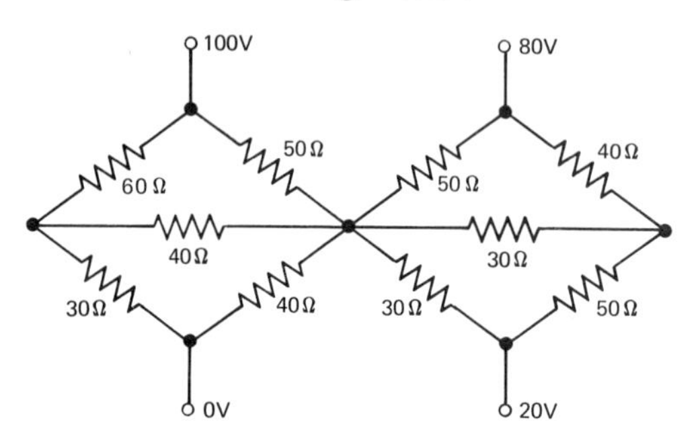

Figure P8.19.

(c) Write a main program that will accept in turn the augmented matrices for the node equations in (a) and (b) above, call on one of the Gauss-Jordan functions written in Problem 8.13, 8.14, or 8.15, and print out the final transformed matrix together with the potential at each node.

8.20 Explain why the program of Example 8.4 would fail as written if any of the resistances were zero. Could the program be modified so that if one or more of the resistances were zero, the node potentials would still be computed and printed? If so, how?

8.21 A particular electrical network consists of several resistors that are joined at $n+m$ nodes, numbered $i=1, 2, \ldots, n, n+1, \ldots, n+m$. The potential V_i volts is specified at m of these nodes, numbered $i=n+1,\ n+2, \ldots,\ n+m$. There is at most a single resistance R_{ij} $(=R_{ji})$ ohms connected between any two nodes i and j.

Show that the unknown potentials $V_1, V_2, \ldots, V_n$ obey the simultaneous linear equations

$$\sum_{j=1}^{n} a_{ij}V_j = b_i, \qquad i = 1, 2, \ldots, n,$$

in which

$a_{ij} = 0$ if nodes i and j are unconnected,

$a_{ij} = 1/R_{ij}$ if nodes i and j are connected,

$a_{ii} = -\Sigma_k\, 1/R_{ik}$, where the summation is for $1 \leqslant k \leqslant n$, but only including nodes k that are connected to node i,

$b_i = -\Sigma_k\, V_k/R_{ik}$, where the summation is for $n+1 \leqslant k \leqslant n+m$, but only including nodes k that are connected to node i.

Write a program that will accept information describing the network and that will proceed to compute and print values for: (a) the unknown potentials $V_1, V_2, \ldots, V_n$, and (b) an $(n+m)\times(n+m)$ matrix for which a typical element such as I_{ij} gives the current in amperes flowing from node i to node j.

Hints for solution. The description of the network is facilitated by reading in an $(n+m)\times(n+m)$ connection or *incidence* matrix C of logical values, whose general element C(I, J) is T (true) if nodes I and J are connected by a resistor and F (false) if they are not. A second $(n+m)\times(n+m)$ matrix, R, should also be read as data; a typical element R(I, J) will contain the value of the resistance joining nodes I and J. Note that if nodes I and J are unconnected, this information will be registered in C(I, J) (= F), so there will be no need to store an infinite value in R(I, J)! Assume for simplicity that zero resistances will not be encountered. The program may be tested with the networks of Example 8.4, and Problems 8.18 and 8.19.

8.22 When a pure sample gas is bombarded by low-energy electrons in a mass spectrometer, the galvanometer shows peak heights that correspond to individual m/e (mass-to-charge) ratios for the resulting mixture of ions. For the ith peak produced by a pure sample j, one can then assign a sensitivity s_{ij} (peak height per micron of Hg sample pressure). These sensitivity coefficients are unique for each type of gas.

A distribution of peak heights may also be obtained for an n-component gas *mixture* that is to be analyzed for the partial pressures $p_1, p_2, \ldots, p_n$ of each of its constituents. The height h_i of a certain peak is a linear combination of the products of the individual sensitivities and partial pressures:

$$\sum_{j=1}^{n} s_{ij} p_j = h_i.$$

In general, more than n peaks may be available. However, if exactly n peaks (usually, the most distinct) are chosen, we have $i = 1, 2, \ldots, n$, so that the individual partial pressures are given by the solution of n simultaneous linear equations.

Write a program that will read values for the number of components N, the sensitivities S(1, 1) . . . S(N, N), and the peak heights H(1) . . . H(N). The program should then call on one of the functions mentioned in Problems 8.13, 8.14, or 8.15 for solving the resulting N simultaneous equations, and then print values for the individual partial pressures P(1) . . . P(N).

Suggested Test Data

The sensitivities given in Table P8.22 on page 376 were reported by Carnahan [18], in connection with the analysis of a hydrocarbon gas mixture.

A particular gas mixture (sample #39G) produced the following peak heights: $h_1 = 17.1$, $h_2 = 65.1$, $h_3 = 186.0$, $h_4 = 82.7$, $h_5 = 84.2$, $h_6 = 63.7$, and $h_7 = 119.7$. The measured total pressure of the mixture was 39.9 microns of Hg, which can be compared with the sum of the computed partial pressures.

8.23 Consider the seven-member plane truss shown in Fig. P8.23.1. Each *member* can be considered to be a rigid rod. The members are pin-jointed at the nodes or *joints* of the truss. Joint 1 is fastened to the foundation, while joint 5 is supported

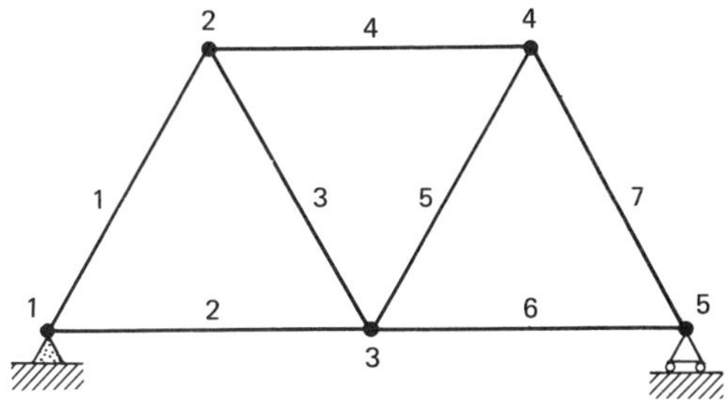

Figure P8.23.1.

on rollers, and hence free to move in the horizontal direction to compensate for structural expansion from heat. Since the truss is stationary, the net forces in the horizontal and vertical directions at each of the joints must be zero. Two kinds of forces act at the joints: *externally applied* forces, usually called *loadings*, and *member forces*. The member forces may be *tensile* (the member tends to pull the joint in the direction of the member) or *compressive* forces. Consider joint 2, shown in Fig. P8.23.2. Assuming that all member

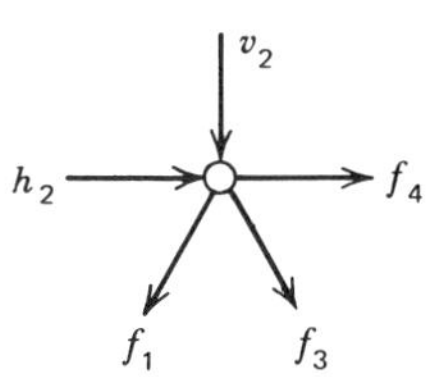

Figure P8.23.2.

forces f are tensile, a horizontal force balance to the right gives

$$h_2 + f_4 + sf_3 - sf_1 = 0,$$

in which $s = \sin 30°$. A similar balance vertically downwards gives

$$v_2 + cf_1 + cf_3 = 0,$$

in which $c = \cos 30°$; h_2 and v_2 denote the loadings at joint 2 (positive to the right and down). The triangles in Fig. P8.23.1 are equilateral.

Table P8.22

Peak		Component Index, j						
		1	2	3	4	5	6	7
	m/e	Hydrogen	Methane	Ethylene	Ethane	Propylene	Propane	n-Pentane
1	2	16.87	0.1650	0.2019	0.3170	0.2340	0.1820	0.1100
2	16	0.0	27.70	0.8620	0.0620	0.0730	0.1310	0.1200
3	26	0.0	0.0	22.35	13.05	4.4200	6.001	3.043
4	30	0.0	0.0	0.0	11.28	0.0	1.110	0.3710
5	40	0.0	0.0	0.0	0.0	9.8500	1.684	2.108
6	44	0.0	0.0	0.0	0.0	0.2990	15.98	2.107
7	72	0.0	0.0	0.0	0.0	0.0	0.0	4.670

By considering each of the joints in turn, show that there are seven simultaneous linear equations in the unknowns $f_1, f_2, f_3, f_4, f_5, f_6$, and f_7, and that the corresponding augmented coefficient matrix is

$$\begin{bmatrix} s & 0 & -s & -1 & 0 & 0 & 0 & h_2 \\ -c & 0 & -c & 0 & 0 & 0 & 0 & v_2 \\ 0 & 1 & s & 0 & -s & -1 & 0 & h_3 \\ 0 & 0 & c & 0 & c & 0 & 0 & v_3 \\ 0 & 0 & 0 & 1 & s & 0 & -s & h_4 \\ 0 & 0 & 0 & 0 & -c & 0 & -c & v_4 \\ 0 & 0 & 0 & 0 & 0 & 1 & s & h_5 \end{bmatrix}$$

Write a main program that will read this augmented matrix, call on one of the functions written in Problem 8.13, 8.14, or 8.15 for implementing the Gauss-Jordan algorithm, and print out the final transformed matrix. *Suggested test data*: $h_2 = -5$T (tons), $v_2 = 3$T, $h_3 = 0$, $v_3 = 10$T, $h_4 = 4$T, $v_4 = 0$, and $h_5 = 3$T.

8.24 Consider Q gpm of water flowing from point 1 to point 2 in an inclined pipe of length L ft and diameter D in. Starting from equation (5.4.2), including an extra term to account for elevation changes, and assuming a constant value of the Moody friction factor f_M, the following equation can be shown to hold approximately for turbulent flow in pipes of average roughness:

$$(z_2 - z_1) + 2.31\,(p_2 - p_1) + 8.69 \times 10^{-4} Q^2 L/D^5 = 0.$$

Here, p is the pressure in psig and z is the elevation in feet. Also assume that a typical head-discharge curve for a centrifugal pump can be represented by $\Delta p = \alpha - \beta Q^2$, in which Δp is the pressure increase in psig across the pump, Q is the flow rate in gpm, and α and β are constants depending on the particular pump.

Consider the piping system shown in Fig. P8.24. The pressures p_1 and p_5 are both atmospheric (0 psig); there is an increase in elevation between points 4 and 5, but pipes C and D are horizontal. Based on the above, the governing equations are:

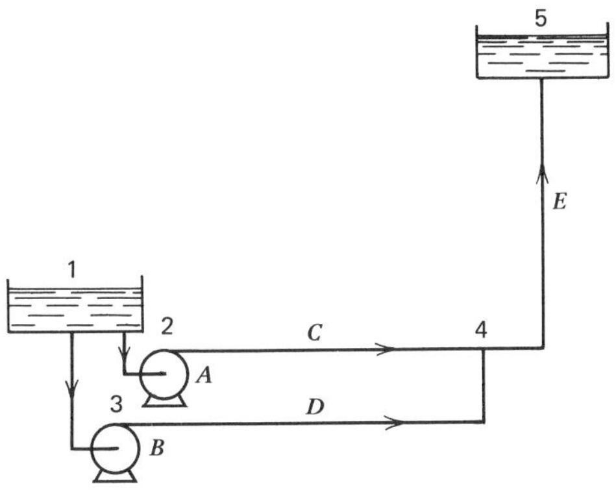

Figure P8.24.

$$Q_E = Q_C + Q_D, \qquad 2.31\,(p_4 - p_2) + 8.69 \times 10^{-4} Q_C^2 L_C / D_C^5 = 0,$$
$$p_2 = \alpha_A - \beta_A Q_C^2, \qquad 2.31\,(p_4 - p_3) + 8.69 \times 10^{-4} Q_D^2 L_D / D_D^5 = 0,$$
$$p_3 = \alpha_B - \beta_B Q_D^2, \qquad z_5 - z_4 + 2.31\,(0 - p_4) + 8.69 \times 10^{-4} Q_E^2 L_E / D_E^5 = 0.$$

Write a program that will accept values for α_A, β_A, α_B, β_B, $z_5 - z_4$, D_C, L_C, D_D, L_D, D_E, and L_E, and that will solve the above equations for the unknowns Q_C, Q_D, Q_E, p_2, p_3, and p_4. One suggested set of test data is $z_5 - z_4 = 70$ ft, with:

Pump	α, psi	β, psi/(gpm)^2
A	156.6	0.00752
B	117.1	0.00427

Pipe	*D*, in.	*L*, ft
C	1.278	125
D	2.067	125
E	2.469	145

Assume that the above pipe lengths have already included the equivalent lengths of all fittings and valves.

8.25 The isothermal irreversible second-order constant-volume reaction $A + B \rightarrow C + D$ has a velocity constant k liter/g mole min. A volumetric flow rate v liter/min of a solution containing equal inlet concentrations a_0 g mole/liter each of A and B is fed to n CSTRs (continuous stirred tank reactors) in series, each of volume V liter. If a_i denotes the exit concentration of A from the ith reactor, a material balance on the ith CSTR gives:

$$Vka_i^2 = v\,(a_{i-1} - a_i).$$

Letting $\beta = kV/v$, we have the following n simultaneous nonlinear equations, analogous to (8.28):

$$f_i = \beta a_i^2 + a_i - a_{i-1} = 0, \quad i = 1, 2, \ldots, n.$$

Assume that values for k, v, n, a_0, and a_n have been specified. The problem here is to solve the above simultaneous equations by the Newton-Raphson method in order to find the necessary reactor volume V and the intermediate concentrations a_1, $a_2, \ldots, a_{n-1}$.

By defining $x_i = a_i$, $i = 1, 2, \ldots, n-1$, and $x_n = \beta$, show that the only nonzero elements of the matrix $\mathbf{\Phi}$ of equation (8.35) are:

$$f_{11} = 1 + 2x_1x_n, \qquad f_{1n} = x_1^2;$$
$$f_{i,i-1} = -1, \qquad f_{ii} = 1 + 2x_ix_n,$$
$$f_{in} = x_i^2, \qquad i = 2, 3, \ldots, n-1;$$
$$f_{n,n-1} = -1, \qquad f_{nn} = a_n^2.$$

Write a program that will accept values for a_0, a_n, k, v, n, a convergence tolerance, an upper limit for the number of iterations, and a vector of initial estimates for the unknowns. The program should then implement the Newton-Raphson method to compute and print the values of a_1, $a_2, \ldots, a_{n-1}$, β (and V). One of the Gauss-Jordan functions developed in Problems 8.13, 8.14, and 8.15 should be used for solving the simultaneous linear equations (8.34) that arise at each stage of the iteration.

Suggested Test Data

$a_0 = 2.0$, $a_n = 1.0$, 0.5, and 0.1 g mole/liter, $k = 0.075$ liter/g mole min, $v = 30$ liter/min, with $n = 2$, 3, 4, 5, and 10 for each value of a_n.

CHAPTER 9

Solution of Ordinary Differential Equations

9.1 Introduction

There is scarcely a branch of engineering or science in which ordinary differential equations (ODEs) are not involved, particularly if transient or time-dependent processes are being studied. Typical situations that readily come to mind involve: electrical oscillations, bending and vibration of structures, motion of bodies in electrical and gravitational fields, growth and decay of bacteriological populations, rates of chemical reactions, and so on. Thus, it is clearly very important to be able to solve ODEs. Although many such equations can be solved by standard analytical techniques, an even greater number of practically important ODEs cannot be so solved, particularly if nonlinear elements are present. In such cases, numerical methods can usually be employed to give approximate solutions of acceptable accuracy. This chapter serves to introduce the reader to such numerical techniques.

A very simple form of a first-order ODE is the following:

$$\frac{dy}{dx} = f(x), \tag{9.1}$$

where $f(x)$ is a known function, and x and y are the *independent* and *dependent* variables, respectively. If we are also supplied with the *initial condition* that $y = y_0$ at $x = x_0$, the solution to equation (9.1) is simply obtained by separating the variables and integrating:

$$\int_{y_0}^{y} dy = \int_{x_0}^{x} f(x)\,dx,$$

that is,

$$y = y_0 + \int_{x_0}^{x} f(x)\,dx. \tag{9.2}$$

Should the integral of equation (9.2) be analytically intractable, then a standard numerical integration routine may be used to give a good approximation to the solution.

In this chapter, however, we shall be concerned with developing approximate numerical solutions to the more general form of a first-order ODE, namely

$$\frac{dy}{dx} = f(x, y), \tag{9.3}$$

which cannot be integrated quite so easily.

Note that any nth-order differential equation (one in which the highest derivative is of order n) can usually be reduced to an equivalent system of n first-order equations by introducing $n-1$ new variables. For example, consider the following second-order equation:

$$\frac{d^2y}{dt^2} = -k^2y, \tag{9.4}$$

in which k is a known constant. By defining a new variable $z = dy/dt$, (9.4) can be rewritten as the following pair of first-order equations:

$$\begin{aligned} \frac{dz}{dt} &= -k^2y, \\ \frac{dy}{dt} &= z. \end{aligned} \tag{9.5}$$

Since any higher-order equation can be decomposed similarly, it will suffice to consider only first-order ODEs in the subsequent discussions of the various methods of solution.

9.2 Euler's Method

In order to solve equation (9.3), we attempt a procedure, shown in Figs. 9.1

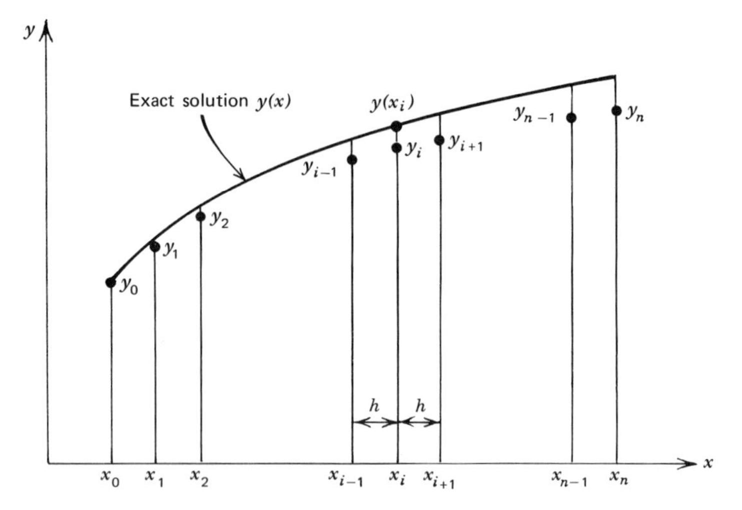

Figure 9.1 Numerical approximations y_i to exact solution y(x).

and 9.2, which consists of a sequence of stepwise integrations, each over a small increment h in the x direction. Successive values of x are denoted by x_0 (the initial value), $x_1, x_2, \ldots, x_i, \ldots, x_n$, where

$$x_i = x_0 + ih,$$

$$h = \frac{(x_n - x_0)}{n},$$

and x_n is the upper limit at which the solution is required. Denote the exact solution to (9.3) as $y = y(x_i)$ at •ny point x_i, and let the corresponding numerical approximation, to be determined by the method given below, be designated by y_i.

Suppose temporarily that we know $y(x_i)$ at some point x_i, as shown in Fig. 9.2.

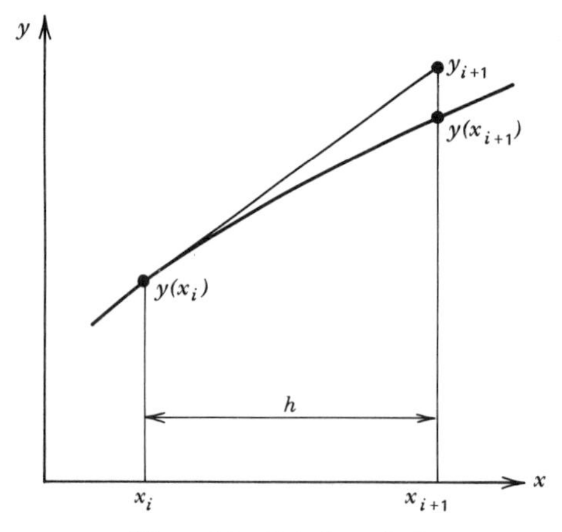

Figure 9.2 Euler's method.

From equation (9.3), integrating across the interval,

$$\int_{y(x_i)}^{y(x_{i+1})} dy = \int_{x_i}^{x_{i+1}} f(x, y(x))dx,$$

or,

$$y(x_{i+1}) = y(x_i) + \int_{x_i}^{x_{i+1}} f(x, y(x))dx. \quad (9.6)$$

Since $y(x)$ is presumably unknown for $x > x_i$, we shall have to approximate the integral of equation (9.6). One way of doing this is to treat $f(x, y)$ as though it were constant across the interval $[x_i, x_{i+1}]$, being equal to its value at the beginning of the step. Thus,

$$\begin{aligned} y(x_{i+1}) &\doteq y(x_i) + f(x_i, y(x_i)) \int_{x_i}^{x_{i+1}} dx \\ &\doteq y(x_i) + hf(x_i, y(x_i)). \quad (9.7) \end{aligned}$$

The geometrical interpretation of equation (9.7) is to draw a tangent of slope $f(x_i, y(x_i))$ at the initial point $(x_i, y(x_i))$ and to project it to $x = x_{i+1}$, giving the approximation y_{i+1} at the end of the step. The deviation $y_{i+1} - y(x_{i+1})$ between the computed and exact solutions at any stage is called the *truncation error.*

In practice, of course, the exact solution $y(x_i)$ at the beginning of the step will be unknown. Instead, we shall have to use its computed approximation y_i, available from a similar calculation across the preceding step. For purposes of computation, (9.7) then becomes

$$y_{i+1} = y_i + hf(x_i, y_i). \tag{9.8}$$

Starting from the initial condition

$$y = y_0 \quad \text{at} \quad x = x_0, \tag{9.9}$$

equation (9.8) can be used repeatedly over successive steps $i = 0, 1, 2, \ldots, n-1$, until the approximation is completed over the range $x = x_0$ to $x = x_n$. This procedure for approximating the solution of an ODE is called *Euler's method.*

Better accuracy would be expected for smaller values of h, and this is generally observed in practice. It can be shown, on the basis of Taylor's expansion (see [1], for example), that as h becomes very small $(h \rightarrow 0)$, the truncation error at any point becomes directly proportional to the step-size h; in other words, the truncation error is said to be $O(h)$ (of *order* h). Euler's method is therefore called a *first-order* procedure. It is also a *one-step* method, since we can advance from x_i to x_{i+1} by knowing y_i at the beginning of the step only, and not necessarily at any preceding step such as y_{i-1}, y_{i-2}, etc.

In implementing Euler's method on the computer, it is seldom necessary or even desirable to retain the subscripts on either x or y. As an illustration, Fig. 9.3 represents the essential steps in Euler's method for solving, for example, the differential equation $dy/dx = x^2 + y^2$. The program variables N, H, XZERO, and YZERO are assumed to have been preset with appropriate values for n, h, x_0, and y_0, respectively.

Flow Diagram

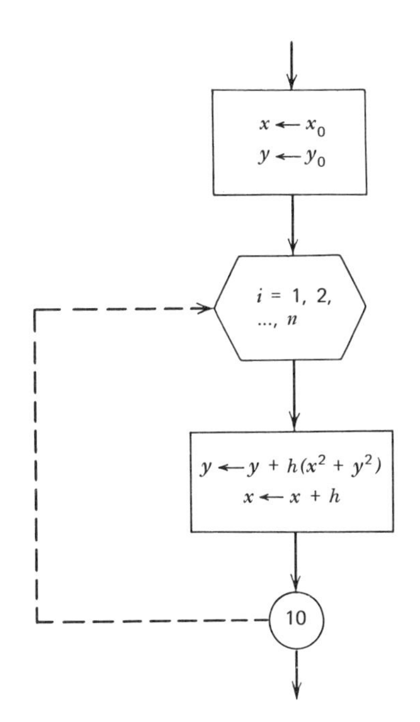

Program Segment

```
      X = XZERO
      Y = YZERO
      DO  10  I = 1, N
      Y = Y + H*(X**2 + Y**2)
   10 X = X + H
```

Figure 9.3 Illustration of Euler's method.

Note that as soon as a new X is computed, it simply replaces the previous value, and the same is true for Y. In most programs, provision would also be made for printing out some or all of these intermediate values of X and Y, which would otherwise be lost permanently.

9.3 Taylor's Expansion Approach

This section presents an alternative development of Euler's method and also indicates possible further steps leading to higher-order methods. Since y is a function of x, Taylor's expansion (page 313) about a base point x_i yields:

$$y(x_{i+1}) = y(x_i+h) = y(x_i) + h\frac{dy}{dx} + \frac{h^2}{2!}\frac{d^2y}{dx^2} + \frac{h^3}{3!}\frac{d^3y}{dx^3} + \cdots, \quad (9.10)$$

in which h is the usual step size and all derivatives on the right-hand side are evaluated at x_i. But, from (9.3), the first derivative dy/dx is the known function $f(x, y)$, and if the series is truncated at this point, we simply obtain Euler's algorithm:

$$y(x_{i+1}) \doteq y(x_i) + hf(x_i, y(x_i)). \quad (9.7)$$

For small values of h, the error involved in using (9.7) to proceed across a single step is approximately the next term in the series, namely $(h^2/2)d^2y/dx^2$; that is, the *local* truncation error in advancing y across a *single* step would be approximately proportional to h^2, or $O(h^2)$. But in using the method to proceed all the way from x_0 to x_n, the number of repeated applications of Euler's method is n, which is *inversely* proportional to the step size h; that is, n is $O(1/h)$. Thus we might suspect for small step sizes that the overall truncation error, compounded over n steps, would be proportional to h^2/h, or simply $O(h)$. As mentioned in Section 9.2, this somewhat tenuous conclusion can actually also be proved by a much more rigorous (and lengthy) argument.

In an attempt to obtain better accuracy—to develop a method in which the overall truncation error is $O(h^2)$, for example—it seems natural to retain the next higher order derivative in (9.10). This second derivative, d^2y/dx^2, may be evaluated in terms of the known function $f(x, y)$ and its partial derivatives by using the *chain rule*:

$$\frac{d^2y}{dx^2} = \frac{d}{dx}\left(\frac{dy}{dx}\right) = \frac{df}{dx} = \frac{\partial f}{\partial x} + \frac{\partial f}{\partial y}\frac{dy}{dx} = \frac{\partial f}{\partial x} + f\frac{\partial f}{\partial y}. \quad (9.11)$$

In principle, further algorithms, in which the truncation error is $O(h^3)$, or $O(h^4)$, etc., can be developed by retaining sufficient higher-order derivatives in (9.10), each of which is then expressed in terms of $f(x, y)$ and its partial derivatives by repeated application of the chain rule. In practice, however, such approaches are extremely tedious unless $f(x, y)$ is a particularly simple function. (For example, let the reader attempt to develop d^2y/dx^2, d^3y/dx^3, and d^4y/dx^4, for the case in which $f(x, y) = \tan^{-1}(x^2/\sqrt{1+y^4}) \ln x^3$.)

Fortunately, techniques do exist—in particular, the *Runge-Kutta* and *multistep* methods, to be discussed in Sections 9.4 and 9.6—which afford the additional accuracy corresponding to retention of the higher-order derivatives in the Taylor's expansion of (9.10), yet which do not require the accompanying tedious algebraic manipulations.

EXAMPLE 9.1

TRANSIENT CURRENT IN INDUCTANCE AND RESISTANCE BY EULER'S METHOD

Problem Statement

At time $t = 0$, the constant voltage source E shown in Fig. 9.1.1 is suddenly connected to the resistance R and inductance L in series. Write a program that will accept values for E, R, and L as data, and that will use Euler's method to compute the current i as a function of time until 95% of the maximum current $i_{max} = E/R$ is attained. Compare the computed solution with the known analytical solution, and, by repeating the calculations with different time steps Δt, investigate to see if Euler's method is indeed first order.

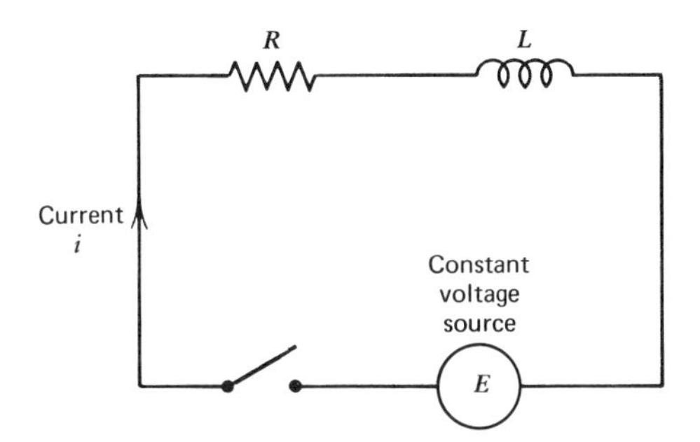

Figure 9.1.1 Resistance and inductance in series.

Method of Solution

The differential equation for the current is

$$L\frac{di}{dt} + Ri = E, \qquad (9.1.1)$$

subject to the initial condition

$$i = 0 \quad \text{at} \quad t = 0. \qquad (9.1.2)$$

The analytical solution for the current at any time t is

$$i = \frac{E}{R}(1 - e^{-Rt/L}), \qquad (9.1.3)$$

which can readily be verified by noting that it satisfies both the ODE (9.1.1) and the initial condition (9.1.2). That is, the current rises from zero and asymptotically reaches its limiting value E/R for large values of t.

The procedure employs Euler's method expressed in equations (9.8) and (9.9). Equation (9.1.1) is first rearranged to give

$$\frac{di}{dt} = \frac{E - Ri}{L}. \qquad (9.1.4)$$

Across a step of duration Δt between time t_j and t_{j+1} (the subscript j is used here to avoid confusion with the current i), Euler's representation of (9.1.4) is

$$i_{j+1} = i_j + \Delta t\frac{E - Ri_j}{L}.$$

In the program, there is no need to retain the subscript on i and the key computational step is basically

$$i \leftarrow i + \Delta t\frac{E - Ri}{L}, \qquad j = 1, 2, \ldots, jmax. \qquad (9.1.5)$$

The upper limit $jmax$ on the number of time steps can be precalculated from (9.1.3) by substituting $i = 0.95E/R$, solving for t, and dividing by the time step Δt, giving

$$jmax = -\frac{L \ln 0.05}{R\Delta t},$$

which should be rounded up to the next higher integer.

Note that for very small time steps Δt, printout of the results at every step of the calculations might not only be wasteful but might also generate so much computer output as to obscure the meaning of the results. Therefore, printout of the time t, the computed approximation i, the exact solution i_{true}, and the truncation error $\delta = i - i_{true}$ is performed only when the counter j is an integral multiple of a preassigned integral frequency $freq$.

Flow Diagram

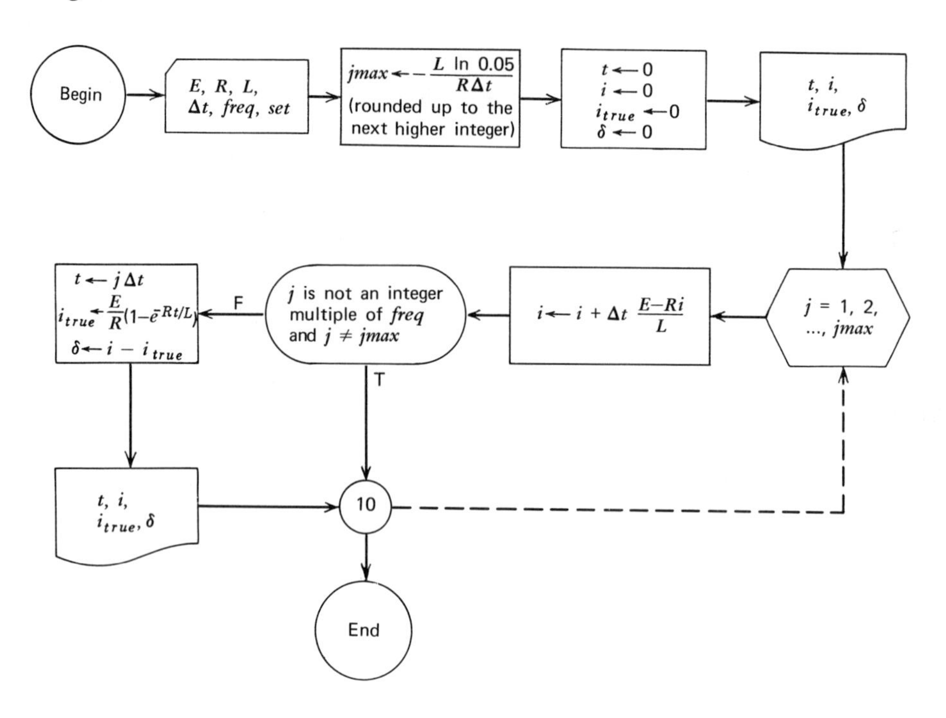

FORTRAN Implementation

List of Principal Variables

Program Symbol	Definition
DELTA	Truncation error, $\delta = i - i_{true}$.
DT	Time step, Δt, sec.
E	Source emf, E, volts.
FREQ	Integer frequency of printout, *freq*.
I	Computed current, i, amps.
ITRUE	Exact value of current, i_{true}, from equation (9.1.3), amps.
J	Counter on number of time steps, j.
JMAX	Upper limit on number of time steps, *jmax*.
L	Inductance, L, henries.
R	Resistance, R, ohms.
SET	Data set identification number, *set*.
T	Time, t, sec.

Program Listing

```
C         DIGITAL COMPUTING                        EXAMPLE 9.1
C         TRANSIENT CURRENT IN INDUCTANCE AND RESISTANCE, EULER'S METHOD.
C
      INTEGER FREQ, SET
      REAL I, ITRUE, L
      WRITE (6,200)
C
C     ..... READ & CHECK INPUT DATA, AND SET INITIAL VALUES .....
    1 READ (5,100,END=999,ERR=1)  E, R, L, DT, FREQ, SET
      JMAX = - L/R*ALOG(0.05)/DT + 1.
      WRITE (6,201) SET, E, R, L, DT, FREQ, JMAX
      T = 0.
      I = 0.
      ITRUE = 0.
      DELTA = 0.
      WRITE (6,202) T, I, ITRUE, DELTA
C
C     ..... COMPUTE CURRENT I OVER SUCCESSIVE TIME STEPS
C           DT UNTIL 95% OF MAXIMUM CURRENT IS ATTAINED .....
      DO 10  J = 1, JMAX
      I = I + DT*(E - R*I)/L
C
C     ..... PRINT VALUES WHEN J IS AN INTEGER MULTIPLE
C           OF FREQ, AND COMPARE WITH EXACT CURRENT .....
      IF (J/FREQ*FREQ.NE.J .AND. J.NE.JMAX)  GO TO 10
         T = J*DT
         ITRUE = (1. - EXP(-R*T/L))*E/R
         DELTA = I - ITRUE
         WRITE (6,202) T, I, ITRUE, DELTA
   10 CONTINUE
      GO TO 1
C
  999 CALL EXIT
C
C     ..... FORMATS FOR INPUT AND OUTPUT STATEMENTS .....
  100 FORMAT (3X, F5.1, 2(7X, F4.1), 8X, F6.3, 10X, I4, 9X, I3)
  200 FORMAT ('1DIGITAL COMPUTING                        EXAMPLE 9.1'/ 'TRANSIEN
     1T CURRENT IN INDUCTANCE AND RESISTANCE BY EULER''S METHOD.')
  201 FORMAT ('1THE INPUT DATA AND THE VALUE OF JMAX ARE'/ '0    SET    =
     1', I5/  '      E      =', F7.1/ '      R      =', F7.1/
     2        '      L      =', F7.1/ '      DT     =', F9.3/
     3        '      FREQ   =',   I5/ '      JMAX   =',   I5/
     4 '0THE COMPUTED SOLUTION NOW FOLLOWS'/ '0        T          I
     5ITRUE       DELTA'/ ' ')
  202 FORMAT (F10.3, 3F11.5)
C
      END
```

Data

```
E = 10.0,    R = 2.0,    L = 5.0,    DT = 1.000,    FREQ =   1,    SET =  1
E = 10.0,    R = 2.0,    L = 5.0,    DT = 0.500,    FREQ =   1,    SET =  2
E = 10.0,    R = 2.0,    L = 5.0,    DT = 0.100,    FREQ =   5,    SET =  3
E = 10.0,    R = 2.0,    L = 5.0,    DT = 0.050,    FREQ =  10,    SET =  4
E = 10.0,    R = 2.0,    L = 5.0,    DT = 0.010,    FREQ =  50,    SET =  5
E = 10.0,    R = 2.0,    L = 5.0,    DT = 0.005,    FREQ = 100,    SET =  6
E = 10.0,    R = 2.0,    L = 5.0,    DT = 0.001,    FREQ = 500,    SET =  7
```

Computer Output

```
DIGITAL COMPUTING                    EXAMPLE 9.1
TRANSIENT CURRENT IN INDUCTANCE AND RESISTANCE BY EULER'S METHOD.
```

```
THE INPUT DATA AND THE VALUE OF JMAX ARE

    SET   =    1
    E     =   10.0
    R     =    2.0
    L     =    5.0
    DT    =    1.000
    FREQ  =    1
    JMAX  =    8

THE COMPUTED SOLUTION NOW FOLLOWS
```

T	I	ITRUE	DELTA
0.0	0.0	0.0	0.0
1.000	2.00000	1.64840	0.35160
2.000	3.20000	2.75336	0.44664
3.000	3.92000	3.49403	0.42597
4.000	4.35200	3.99052	0.36148
5.000	4.61120	4.32332	0.28788
6.000	4.76672	4.54641	0.22031
7.000	4.86003	4.69595	0.16408
8.000	4.91602	4.79619	0.11983

```
THE INPUT DATA AND THE VALUE OF JMAX ARE

    SET   =    3
    E     =   10.0
    R     =    2.0
    L     =    5.0
    DT    =    0.100
    FREQ  =    5
    JMAX  =   75

THE COMPUTED SOLUTION NOW FOLLOWS
```

T	I	ITRUE	DELTA
0.0	0.0	0.0	0.0
0.500	0.92314	0.90635	0.01679
1.000	1.67583	1.64840	0.02744
1.500	2.28956	2.25594	0.03362
2.000	2.78998	2.75335	0.03663
2.500	3.19801	3.16060	0.03741
3.000	3.53070	3.49403	0.03668
3.500	3.80197	3.76701	0.03496
4.000	4.02316	3.99052	0.03264
4.500	4.20351	4.17350	0.03000
5.000	4.35056	4.32332	0.02724
5.500	4.47046	4.44598	0.02448
6.000	4.56823	4.54641	0.02182
6.500	4.64794	4.62863	0.01931
7.000	4.71294	4.69595	0.01699
7.500	4.76594	4.75106	0.01487

```
THE INPUT DATA AND THE VALUE OF JMAX ARE

    SET   =    5
    E     =   10.0
    R     =    2.0
    L     =    5.0
    DT    =    0.010
    FREQ  =   50
    JMAX  =  749

THE COMPUTED SOLUTION NOW FOLLOWS
```

T	I	ITRUE	DELTA
0.0	0.0	0.0	0.0
0.500	0.90799	0.90635	0.00164
1.000	1.65106	1.64840	0.00267
1.500	2.25920	2.25594	0.00326
2.000	2.75690	2.75335	0.00355
2.500	3.16422	3.16060	0.00362
3.000	3.49757	3.49403	0.00354
3.500	3.77038	3.76701	0.00337
4.000	3.99365	3.99052	0.00314
4.500	4.17638	4.17350	0.00288
5.000	4.32593	4.32332	0.00260
5.500	4.44832	4.44598	0.00233
6.000	4.54848	4.54641	0.00207
6.500	4.63046	4.62863	0.00183
7.000	4.69754	4.69595	0.00159
7.490	4.75145	4.75007	0.00139

```
THE INPUT DATA AND THE VALUE OF JMAX ARE

    SET   =    7
    E     =   10.0
    R     =    2.0
    L     =    5.0
    DT    =    0.001
    FREQ  =  500
    JMAX  = 7490

THE COMPUTED SOLUTION NOW FOLLOWS
```

T	I	ITRUE	DELTA
0.0	0.0	0.0	0.0
0.500	0.90650	0.90635	0.00015
1.000	1.64846	1.64840	0.00006
1.500	2.25588	2.25594	-0.00006
2.000	2.75318	2.75335	-0.00017
2.500	3.16031	3.16060	-0.00029
3.000	3.49364	3.49403	-0.00039
3.500	3.76653	3.76701	-0.00048
4.000	3.98994	3.99052	-0.00057
4.500	4.17285	4.17350	-0.00065
5.000	4.32260	4.32332	-0.00073
5.500	4.44519	4.44598	-0.00079
6.000	4.54557	4.54641	-0.00084
6.500	4.62774	4.62863	-0.00089
7.000	4.69502	4.69595	-0.00093
7.490	4.74910	4.75007	-0.00097

Discussion of Results

Throughout the calculations, the circuit parameters E, R, and L are held constant, only the time-step Δt being varied in each of the seven data sets. The results are displayed for data sets 1, 3, 5, and 7, for which $\Delta t = 1$, 0.1, 0.01, and 0.001 sec, respectively.

Observe, as expected, that the accuracy of the computed solution improves as Δt decreases. Using logarithmic coordinates, the truncation error $\delta = i - i_{true}$ at $t = 3$ sec is plotted against Δt for data sets 1 through 6 in Fig. 9.1.2. The straight line drawn on the figure has a slope of unity, verifying that Euler's method is indeed first order—that is, for sufficiently small Δt, the truncation error δ is proportional to Δt. This last statement is strictly true only if the computations are being performed to an adequate number of significant figures. Clearly, with the limited standard word size of the IBM 360/67 computer (equivalent to about 6 or 7 significant decimal digits), round-off errors will probably accumulate and overwhelm any potential gain in accuracy for very small values of Δt. Note in particular the value of *jmax* for data set 7: *thousands* of time steps are involved.

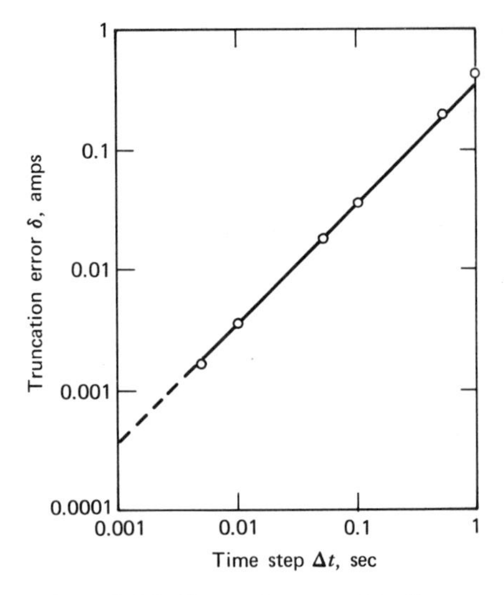

Figure 9.1.2 Truncation error at $t = 3$ sec.

Also observe for data set 7 that most of the truncation errors are of opposite sign (negative in this case) compared to those for the other data sets; this accounts for the absence of a point on Fig. 9.1.2 at $\Delta t = 0.001$. The broken line emphasizes the region in which round-off errors are probably starting to manifest themselves.

9.4 Runge-Kutta Methods

In Euler's method, the derivative $dy/dx = f(x, y)$ is treated as through it were constant across a single step h and equal to its value at the beginning of that step. We now ask if there is a more suitable *average* derivative available for use between x_i and x_{i+1}. If, in a hypothetical ideal situation, we knew $y(x_i)$ and $y(x_{i+1})$, we could estimate an average derivative as the mean of the values at the beginning and end of the step:

$$\bar{f} = \frac{1}{2}[f(x_i, y(x_i)) + f(x_{i+1}, y(x_{i+1}))]. \tag{9.12}$$

This is wishful thinking, of course, since the exact solutions $y(x_i)$ and $y(x_{i+1})$ are unknown. However, we can at least approximate equation (9.12) with the formula

$$\bar{f} = \frac{1}{2}[f(x_i, y_i) + f(x_{i+1}, y^*_{i+1})], \tag{9.13}$$

where y_i is the usual computed approximation to the solution at the beginning of the step, and y^*_{i+1} is a *preliminary estimate* for the approximation at the end of the step. But one way of estimating y^*_{i+1} is to employ Euler's method:

$$y^*_{i+1} = y_i + hf(x_i, y_i). \tag{9.14}$$

The final approximation y_{i+1} at the end of the step is given essentially by using Euler's method again, but now with the average derivative $\bar{f}$:

$$y_{i+1} = y_i + h\bar{f}. \tag{9.15}$$

Restating equations (9.13), (9.14), and (9.15), the complete algorithm for advancing the computed approximation between x_i and x_{i+1} is:

Compute First Approximation

$$y^*_{i+1} = y_i + hf(x_i, y_i),$$

Obtain Average Derivative

$$\bar{f} = \frac{1}{2}[f(x_i, y_i) + f(x_{i+1}, y^*_{i+1})], \tag{9.16}$$

Proceed Across Step with Average Derivative

$$y_{i+1} = y_i + h\bar{f}.$$

This procedure, known as the *improved Euler method*, and also as *Heun's method*, is illustrated in Fig. 9.4. The algorithm of (9.16) is a simple example of a *predictor-corrector method*, since the first *predicted* approximation y^*_{i+1} is subsequently refined to give the final *corrected* approximation y_{i+1}.

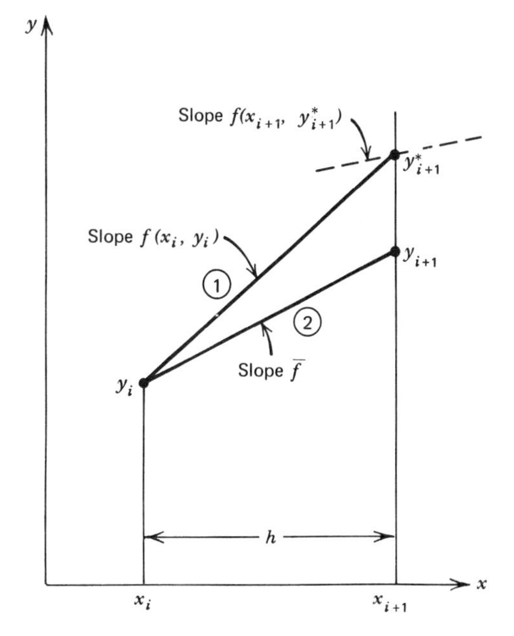

Figure 9.4 *Improved Euler method. (A second-order Runge-Kutta procedure.)*

Another similar procedure is to advance to the halfway point $(x_{i+1/2}, y^*_{i+1/2})$ using Euler's method, and then to take the value for $\bar{f}$ entirely at this point, as follows:

Compute Approximation at Half-Way Point

$$y^*_{i+1/2} = y_i + \frac{h}{2}f(x_i, y_i),$$

Obtain Average Derivative

$$\bar{f} = f(x_{i+1/2}, y^*_{i+1/2}), \tag{9.17}$$

Proceed Across Step with Average Derivative

$$y_{i+1} = y_i + h\bar{f}.$$

This procedure, known as the *modified Euler method*, is illustrated in Fig. 9.5.

Both the above methods are special cases of a general class usually known as *Runge-Kutta* procedures, all of which go to various lengths to evaluate a suitable average derivative $\bar{f}$ over the interval h. The above methods are also *second-order* procedures, since it can be shown that their associated overall truncation error is $O(h^2)$ (that is, proportional to h^2 as $h \to 0$), representing a distinct advantage over Euler's method, even taking into account the relative amounts of computation needed by each. The mathematics of the higher-order Runge-Kutta methods, that is, ones with even smaller truncation errors, is fairly involved, and the reader is referred to Henrici [19]. Generally, however, $\bar{f}$ involves more and more intermediate steps in its computation as the order of accuracy of the method improves.

The following is typical of the many *fourth-order* Runge-Kutta algorithms that have been developed. Observe that four derivative evaluations are needed.

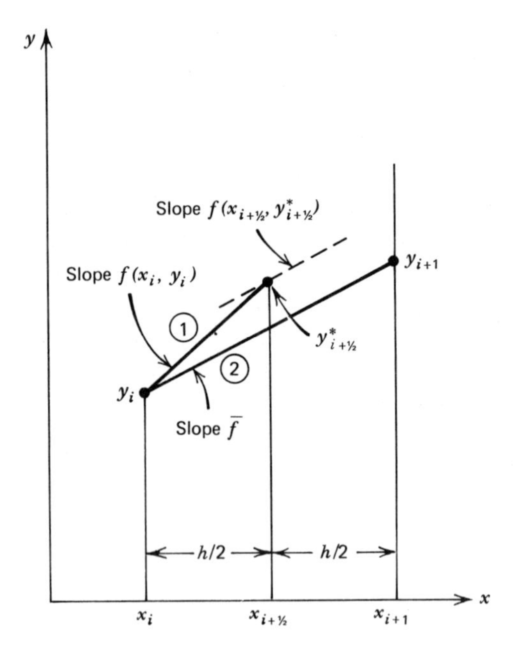

***Figure 9.5** Modified Euler method. (A second-order Runge-Kutta procedure.)*

$$y_{i+1} = y_i + \frac{h}{6}(k_1 + 2k_2 + 2k_3 + k_4),$$

where

$$\left.\begin{aligned} k_1 &= f(x_i, y_i), \\ k_2 &= f\left(x_i + \frac{1}{2}h, y_i + \frac{1}{2}hk_1\right), \\ k_3 &= f\left(x_i + \frac{1}{2}h, y_i + \frac{1}{2}hk_2\right), \\ k_4 &= f(x_i + h, y_i + hk_3). \end{aligned}\right\} \quad (9.18)$$

Finally, note that the Runge-Kutta procedures are also one-step methods, since we need only a knowledge of y_i (and not y_{i-1}, y_{i-2}, etc.) to be able to proceed to y_{i+1}.

9.5 Simultaneous Differential Equations

In a more general situation, we could have n dependent variables $y_1, y_2, \ldots, y_n$, each related to the single independent variable x by the following system of n simultaneous first-order ODEs:

$$\begin{aligned} \frac{dy_1}{dx} &= f_1(x, y_1, y_2, \ldots, y_n), \\ \frac{dy_2}{dx} &= f_2(x, y_1, y_2, \ldots, y_n), \\ &\cdots\cdots\cdots\cdots\cdots\cdots, \\ \frac{dy_n}{dx} &= f_n(x, y_1, y_2, \ldots, y_n). \end{aligned} \quad (9.19)$$

Note that the subscripts on the y's now refer to the individual variables and no longer to the value at a particular step x_i. There will also be n associated initial conditions for each of the y's at $x = x_0$. Both the Euler and Runge-Kutta methods can readily be applied to the solution of such a system. The procedure then essentially amounts to applying the usual algorithms to each equation in turn. Fairly obviously, every y must be advanced across a given step h before proceeding to the next step. Also, when using the Runge-Kutta method, each dependent variable is advanced to the intermediate stage (or stages) before proceeding across the whole step. In this way, the proper values of the y's will always be available for use in the functions on the right-hand sides of equations (9.19).

While the Euler and Runge-Kutta methods apply to first-order equations, it is important to reemphasize that most higher-order linear ODEs can easily be reduced to a system of first-order equations. For example, consider the third-order equation

$$\frac{d^3y}{dx^3}+a\frac{d^2y}{dx^2}+b\frac{dy}{dx}+c=F(x,y). \quad (9.20)$$

By introducing the new variables $y_1=y$, $y_2=dy/dx$, and $y_3=d^2y/dx^2$, (9.20) can be rephrased as the system:

$$\begin{aligned}\frac{dy_1}{dx}&=y_2 \qquad (=f_1(x,y_1,y_2,y_3)),\\ \frac{dy_2}{dx}&=y_3 \qquad (=f_2(x,y_1,y_2,y_3)), \quad (9.21)\\ \frac{dy_3}{dx}&=F(x,y_1)-ay_3-by_2-c\\ &\qquad (=f_3(x,y_1,y_2,y_3)).\end{aligned}$$

EXAMPLE 9.2

TRANSIENT CURRENT IN INDUCTANCE AND RESISTANCE BY RUNGE-KUTTA METHOD

Problem Statement

Repeat Example 9.1, now using the second-order Runge-Kutta algorithm of (9.16), also known as Heun's method.

Method of Solution

The whole development, including the program and data, is almost identical with that of Example 9.1, with the following exception. Recall that the key computational step of Euler's method, starting with the DO statement on page 386, was

```
DO 10 J = 1, JMAX
I = I + DT*(E - R*I)/L
(etc.)
```

This segment must now be replaced with the following sequence, corresponding to the Runge-Kutta algorithm of (9.16). The comment declarations have been added for clarity.

```
      DO 10 J = 1, JMAX
C
C     .....COMPUTE DERIVATIVE F AT BEGINNING OF STEP.....
      F = (E - R*I)/L
C
C     .....COMPUTE FIRST APPROXIMATION, ISTAR, AT END OF STEP.....
      ISTAR = I + DT*F
C
C     .....COMPUTE AVERAGE DERIVATIVE FBAR.....
      FBAR = 0.5*(F + (E - R*ISTAR)/L)
C
C     .....PROCEED ACROSS STEP WITH AVERAGE DERIVATIVE.....
      I = I + DT*FBAR
      (etc.)
```

With this modification (and with ISTAR declared to be of real mode), the program has been run with the same data sets as in Example 9.1.

Computer Output

```
DIGITAL COMPUTING                    EXAMPLE 9.2
TRANSIENT CURRENT IN INDUCTANCE AND RESISTANCE BY HEUN'S METHOD.
```

```
THE INPUT DATA AND THE VALUE OF JMAX ARE

    SET    =    1
    E      =   10.0
    R      =    2.0
    L      =    5.0
    DT     =    1.000
    FREQ   =    1
    JMAX   =    8

THE COMPUTED SOLUTION NOW FOLLOWS

      T          I          ITRUE       DELTA

    0.0        0.0        0.0         0.0
    1.000      1.60000    1.64840    -0.04840
    2.000      2.68800    2.75336    -0.06536
    3.000      3.42784    3.49403    -0.06619
    4.000      3.93093    3.99052    -0.05959
    5.000      4.27303    4.32332    -0.05029
    6.000      4.50566    4.54641    -0.04075
    7.000      4.66385    4.69595    -0.03210
    8.000      4.77142    4.79619    -0.02477
```

```
THE INPUT DATA AND THE VALUE OF JMAX ARE

    SET    =    3
    E      =   10.0
    R      =    2.0
    L      =    5.0
    DT     =    0.100
    FREQ   =    5
    JMAX   =   75

THE COMPUTED SOLUTION NOW FOLLOWS

      T          I          ITRUE       DELTA

    0.0        0.0        0.0         0.0
    0.500      0.90612    0.90635    -0.00023
    1.000      1.64803    1.64840    -0.00037
    1.500      2.25548    2.25594    -0.00046
    2.000      2.75285    2.75335    -0.00050
    2.500      3.16009    3.16060    -0.00051
    3.000      3.49352    3.49403    -0.00050
    3.500      3.76653    3.76701    -0.00048
    4.000      3.99006    3.99052    -0.00045
    4.500      4.17309    4.17350    -0.00042
    5.000      4.32294    4.32332    -0.00038
    5.500      4.44564    4.44598    -0.00034
    6.000      4.54610    4.54641    -0.00031
    6.500      4.62836    4.62863    -0.00027
    7.000      4.69571    4.69595    -0.00024
    7.500      4.75085    4.75106    -0.00021
```

```
THE INPUT DATA AND THE VALUE OF JMAX ARE

    SET    =    5
    E      =   10.0
    R      =    2.0
    L      =    5.0
    DT     =    0.010
    FREQ   =   50
    JMAX   =  749

THE COMPUTED SOLUTION NOW FOLLOWS

      T          I          ITRUE       DELTA

    0.0        0.0        0.0         0.0
    0.500      0.90634    0.90635    -0.00000
    1.000      1.64838    1.64840    -0.00002
    1.500      2.25590    2.25594    -0.00004
    2.000      2.75330    2.75335    -0.00005
    2.500      3.16053    3.16060    -0.00007
    3.000      3.49395    3.49403    -0.00008
    3.500      3.76693    3.76701    -0.00009
    4.000      3.99042    3.99052    -0.00009
    4.500      4.17341    4.17350    -0.00010
    5.000      4.32322    4.32332    -0.00010
    5.500      4.44588    4.44598    -0.00011
    6.000      4.54630    4.54641    -0.00011
    6.500      4.62852    4.62863    -0.00011
    7.000      4.69584    4.69595    -0.00011
    7.490      4.74996    4.75007    -0.00011
```

```
THE INPUT DATA AND THE VALUE OF JMAX ARE

    SET    =    7
    E      =   10.0
    R      =    2.0
    L      =    5.0
    DT     =    0.001
    FREQ   =  500
    JMAX   = 7490

THE COMPUTED SOLUTION NOW FOLLOWS

      T          I          ITRUE       DELTA

    0.0        0.0        0.0         0.0
    0.500      0.90633    0.90635    -0.00001
    1.000      1.64819    1.64840    -0.00021
    1.500      2.25555    2.25594    -0.00039
    2.000      2.75282    2.75335    -0.00054
    2.500      3.15994    3.16060    -0.00066
    3.000      3.49327    3.49403    -0.00075
    3.500      3.76618    3.76701    -0.00083
    4.000      3.98962    3.99052    -0.00090
    4.500      4.17255    4.17350    -0.00096
    5.000      4.32233    4.32332    -0.00100
    5.500      4.44495    4.44598    -0.00103
    6.000      4.54535    4.54641    -0.00106
    6.500      4.62755    4.62863    -0.00108
    7.000      4.69485    4.69595    -0.00110
    7.490      4.74895    4.75007    -0.00112
```

Discussion of Results

The computed results are displayed for $\Delta t = 1$, 0.1, 0.01, and 0.001 sec, and the absolute values of the truncation error at $t = 3$ sec are plotted in Fig. 9.2.1. Even though it requires twice as many derivative evaluations per time step, the Runge-Kutta procedure is clearly superior to Euler's method (also shown in Fig. 9.2.1 for comparison). The steeper straight line in Fig. 9.2.1 has a slope of two, indicating that the truncation error is $O((\Delta t)^2)$. Inevitably, however, round-off error predominates for small Δt (broken line), and, unless double-precision arithmetic is used, there is no point in employing a Δt of appreciably less than 0.05. The full black dot indicates the point, missing from Fig. 9.1.2, in which the truncation error of Euler's method actually changed sign; it is purely fortuitous that it appears to lie on the broken straight line.

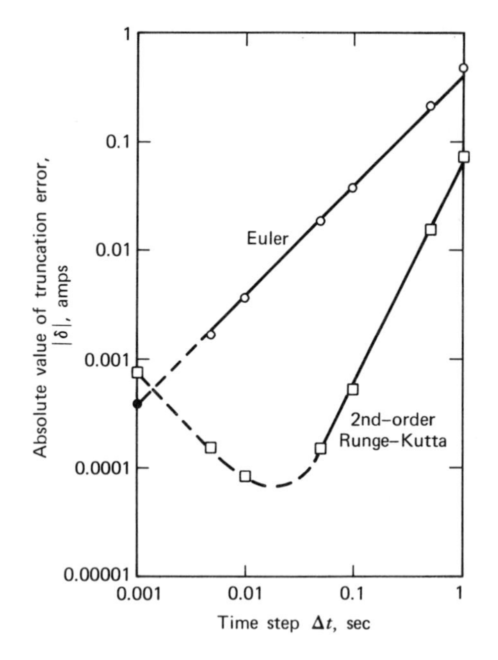

Figure 9.2.1 Truncation errors at $t = 3$ sec for Euler and second-order Runge-Kutta methods.

9.6 Multistep Methods

Consider again the integration of the single ODE

$$\frac{dy}{dx} = f(x, y). \qquad (9.3)$$

Suppose that we have already proceeded to a point x_i, at which the approximate solution y_i has just been computed, and that the next value, y_{i+1}, is required. For this purpose, recall that the one-step methods of Sections 9.2 and 9.4 used algorithms of the basic form

$$y_{i+1} = y_i + \int_{x_i}^{x_{i+1}} f(x, y(x))\,dx, \qquad (9.22)$$

in which the integral across the single interval $[x_i, x_{i+1}]$ was approximated in some particular way.

In contrast, the *multistep methods* use algorithms of the basic form

$$y_{i+1} = y_{i-k} + \int_{x_{i-k}}^{x_{i+1}} f(x, y(x))\,dx, \qquad (9.23)$$

in which the integral extends over $k+1$ intervals, each of width h. Depending on the particular choice of k and the method used for approximating the integral of (9.23), a virtually endless assortment of multistep formulas may be developed. Here, as an introduction, we consider just a couple of possibilities.

Suppose that a value of $k = 3$ has been selected. Evaluation of the integral of (9.23) then amounts to estimating the shaded area in Fig. 9.6. One possibility might be to use the five-point Newton-Cotes closed integration formula of equation (7.9d). However, this would require a knowledge of $f_{i+1} = f(x_{i+1}, y_{i+1})$, which is unfortunately unavailable, since y_{i+1} has not yet been obtained. A realistic alternative is to use an *open* integration formula, in which the range of integration is not completely spanned by the base points.

Although open integration formulas are not discussed in detail in this text, the basic plan is similar to that used in developing the closed formulas of Chapter 7: integrate the interpolating polynomial that passes through suitably selected base points. One possibility is to derive the second-degree interpolating polynomial, $p_2(x)$, that passes through the three base points shown in Fig. 9.6, namely (x_{i-2}, f_{i-2}), (x_{i-1}, f_{i-1}), and (x_i, f_i), and to integrate this polynomial between the limits

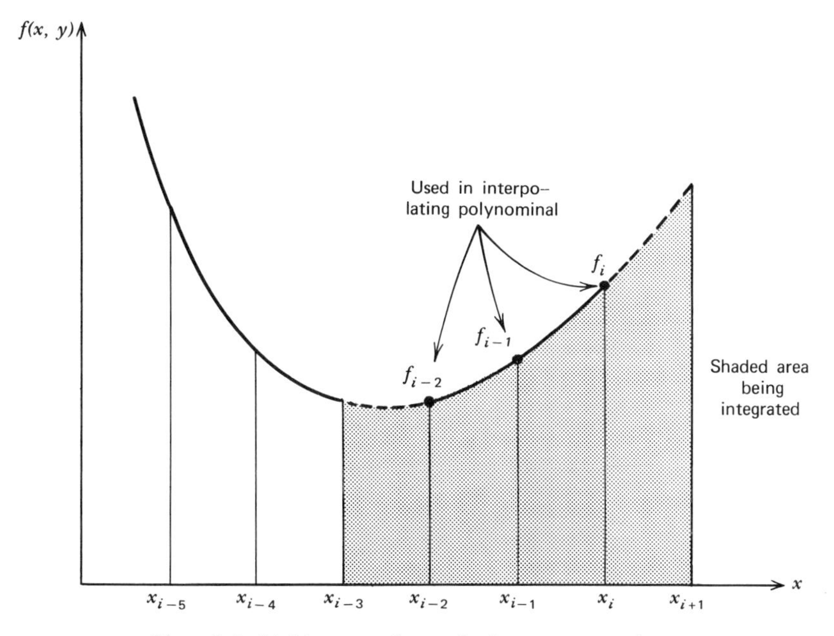

Figure 9.6 Multistep open integration between x_{i-3} and x_{i+1}.

x_{i-3} and x_{i+1}. Note that $f_{i-2} = f(x_{i-2}, y_{i-2})$, $f_{i-1} = f(x_{i-1}, y_{i-1})$, and $f_i = f(x_i, y_i)$ will already be available if the precaution has been taken of saving the necessary values from previous calculations. The result of the integration (details of the algebra, together with many other open integration formulas, can be found in [1]) is:

$$\int_{x_{i-3}}^{x_{i+1}} p_2(x)\,dx = \frac{4h}{3}(2f_{i-2} - f_{i-1} + 2f_i) + \frac{14}{45}h^5 f^{(4)}(\zeta), \quad \zeta \text{ in } (x_{i-3}, x_{i+1}). \tag{9.24}$$

Thus the algorithm corresponding to (9.23) and (9.24) is

$$y_{i+1} = y_{i-3} + \frac{4h}{3}(2f_{i-2} - f_{i-1} + 2f_i). \tag{9.25}$$

Equation (9.25) may be used repeatedly in order to complete the solution of the ODE over successive steps. However, the accuracy can be improved by noting that the preliminary estimate of y_{i+1} from (9.25) now enables $f_{i+1} = f(x_{i+1}, y_{i+1})$ to be computed. A *closed* integration formula can then be employed to yield a refined estimate for y_{i+1}. Referring to equation (9.23) and Fig. 9.7, it is appropriate to select $k = 1$, and to use f_{i-1}, f_i, and the newly computed f_{i+1} as the base points for another second-degree interpolating polynomial which is then integrated between x_{i-1} and x_{i+1}. The result is, of course, Simpson's rule, (7.9b):

$$\int_{x_{i-1}}^{x_{i+1}} p_2(x)\,dx = \frac{h}{3}(f_{i-1} + 4f_i + f_{i+1}) - \frac{h^5}{90}f^{(4)}(\zeta), \quad \zeta \text{ in } (x_{i-1}, x_{i+1}). \tag{9.26}$$

Thus the algorithm corresponding to (9.23) and (9.26) is

$$y_{i+1} = y_{i-1} + \frac{h}{3}(f_{i-1} + 4f_i + f_{i+1}). \tag{9.27}$$

Note that the two integration formulas just used, (9.24) and (9.26), both involve errors of $O(h^5)$, but that the closed formula of (9.26) has a much lower associated numerical factor than the open formula of (9.26) (namely, $-1/90$, in contrast to 14/45). The obvious procedure is to use (9.25) to *predict* a preliminary estimate of y_{i+1}, and to employ (9.27) to *correct* this estimate. The resulting combination is

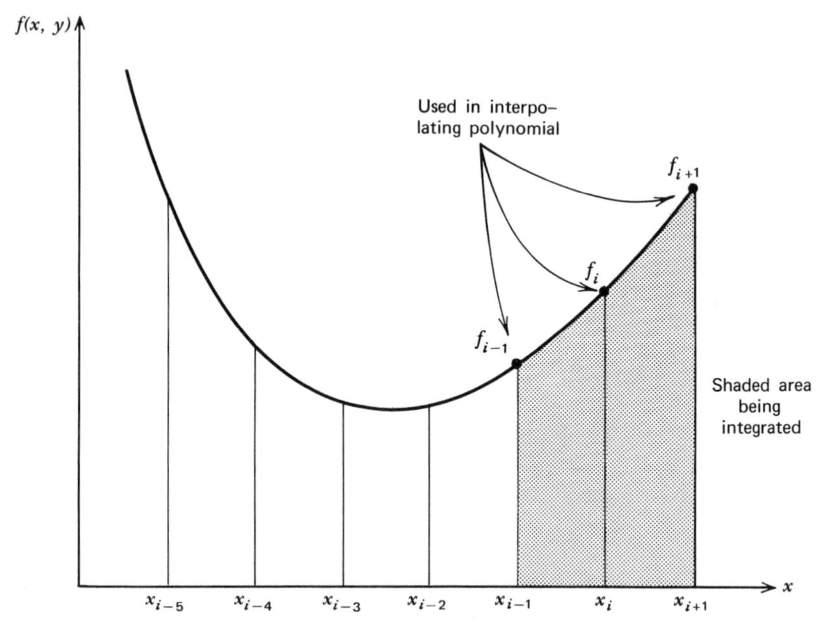

Figure 9.7 Multistep closed integration between x_{i-1} and x_{i+1}.

called *Milne's fourth-order predictor-corrector method* (it can be shown that the overall truncation error is $O(h^4)$), summarized in Table 9.1.

Table 9.1 Fourth-Order Milne Method

Predictor:

$$y_{i+1} = y_{i-3} + \frac{4h}{3}(2f_{i-2} - f_{i-1} + 2f_i),$$

Corrector:

$$y_{i+1} = y_{i-1} + \frac{h}{3}(f_{i-1} + 4f_i + f_{i+1}).$$

Observe that the corrector equation can be applied iteratively, each time using the most recently available value for f_{i+1}, until y_{i+1} shows little further change. In practice, however, it is generally thought preferable to apply it just once, and, if necessary, to spend the effort on reducing the step size instead. The improved Euler method, in which a first approximation y^*_{i+1} was computed (see equation (9.16)), is another example of a predictor-corrector technique (but not of a multistep method).

The principal advantage of multistep methods, of which we have considered only one example, is that they require fewer derivative evaluations per step than the corresponding Runge-Kutta methods of comparable accuracy. They have three main disadvantages, however. First, they are not "self-starting"; usually, beginning with $y = y_0$ at $x = x_0$, a fourth-order Runge-Kutta method (or even Euler's method with a very tiny step size) must be employed to compute y_1, y_2, and y_3 (and hence f_1, f_2, and f_3) before (9.25) can be used at all. Second, it is rather complicated to change the step size h once the calculations are under way. Third, catastrophic numerical instabilities may occur for certain types of ODEs, rendering the computed results meaningless; this point is discussed much more fully in [1]. Broadly, however, the multistep methods should be regarded favorably, since they frequently afford considerable economies of computing time when handling large problems involving ODEs.

Problems

9.1 What is the solution of the ODE

$$\frac{dc}{dt} = -kc^2,$$

subject to the initial condition that $c = c_0$ at $t = 0$?

9.2 Express the second-order ODE

$$x^2\frac{d^2y}{dx^2} + x\frac{dy}{dx} + (x^2 - p^2)y = 0$$

as a pair of first-order ODEs.

9.3 Modify the flow diagram and program segment of Fig. 9.3 so that the values of x and y (X and Y) are displayed or printed: (a) before the iterations start, and (b) at the end of every kth iteration, where k is a known integer.

9.4 Use the Taylor's expansion approach of Section 9.3 to obtain a solution to the ODE

$$\frac{dy}{dx} = x^2,$$

subject to the initial condition $y = y_0$ at $x = x_0$. Compare your result with the analytical solution.

9.5 Use the Taylor's expansion approach of Section 9.3 to obtain a solution to the ODE

$$\frac{dy}{dx} = 2y,$$

subject to the initial condition $y = y_0$ at $x = x_0$. Compare your result with the analytical solution.

9.6 Verify that (9.1.3) satisfies both the ODE (9.1.1) and the initial condition (9.1.2).

9.7 If y is plotted against x on logarithmic paper, in the manner of Fig. 9.2.1, what is the significance if the result is a straight line of slope m?

9.8 Comment on the use of the fourth-order Runge-Kutta algorithm of (9.18) for solving the equation $dy/dx = f(x)$, in which the derivative is a function of x only.

9.9 If the *corrector* equation for the fourth-order Milne method of Table 9.1 on page 397 were to be used *iteratively* for computing y_{i+1}, under what conditions for the step size would you expect convergence to occur?

9.10 Discounting the error term, verify the open integration formula of equation (9.24).

9.11 Repeat Example 9.1, now using the modified Euler method of equation (9.17). Investigate the order of the method.

9.12 Repeat Example 9.1, now using Milne's predictor-corrector method of Table 9.1 on page 397. Investigate the order of the procedure.

9.13 Repeat Problem 7.30, with the following variation. Assume that the exchanger length L is a known quantity, to be included in the data, and that the resulting exit temperature T_2 is to be computed by the program. First note that the differential form of the heat balance given in Problem 7.30 is

$$\frac{dT}{dx} = \frac{\pi Dh(T_s - T)}{mc_p},$$

where x is the distance from the heat-exchanger inlet. The problem then amounts to a step-by-step solution of a first-order ODE, with the initial condition $T = T_1$ at $x = 0$; the solution is terminated when the specified length L is reached, at which stage T is the required temperature T_2.

9.14 A periodic voltage $V = V_m \sin \omega t$ is applied to terminal A of the rectifying and ripple-filter circuit shown in Fig. P9.14.

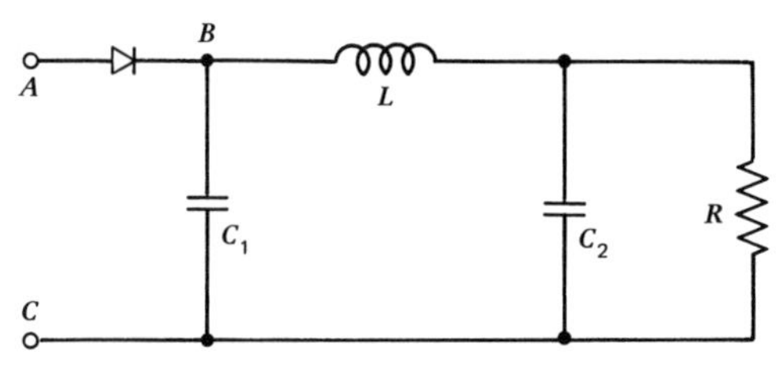

Figure P9.14.

The particular diode rectifier used has essentially negligible resistance for current flowing from A to B, but has infinite resistance in the reverse direction. Terminal C is grounded.

After many cycles of continuous operation, determine the variation with time of: (a) the voltage across capacitor C_1, (b) the voltage across the load resistor R, and (c) the current through the inductance. For the load, also determine the ripple factor (root-mean-square deviation of the voltage from its mean value, divided by that mean value).

Suggested Test Data

$V_m = 160$ volts, $\omega = 120\pi$ sec^{-1}, $C_1 = 4$, $C_2 = 8$, then $C_1 = 8$, $C_2 = 4$ microfarads, $L = 5$ henries, with $R = 220$ and 2200 ohms for each combination of capacitances.

9.15

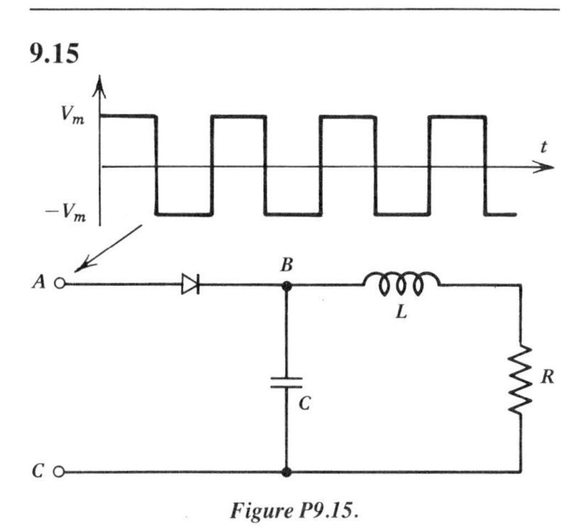

Figure P9.15.

A periodic square-wave voltage of amplitude V_m and period T is applied to terminal A in Fig. P9.15; terminal C is grounded. The rectifier is such that for $v_A \geqslant v_B$, the current, i_{AB} flowing from A to B obeys the relation

$$i_{AB} = k(v_A - v_B)^{3/2},$$

in which k is a constant; however, when $v_A < v_B$, no current flows between A and B in either direction.

Write a program that will compute the variation with time t of the voltages across: (a) the capacitor and (b) the rectifier. Discontinue the calculations when either t exceeds an upper limit t_{max} or the fractional voltage change from one cycle to the next (compared at corresponding times) falls below a small value ε.

Suggested Test Data

$V_m = 100$ volts, $T = 0.1, 0.01$, and 0.001 sec, $C = 10$ microfarads, $L = 2.5$ henries, $R = 500$ ohms, with $k = 0.003$ and 0.0003 amps/(volts)$^{3/2}$ for each value of T; $\varepsilon = 0.001$ and $t_{max} = 100T$.

9.16

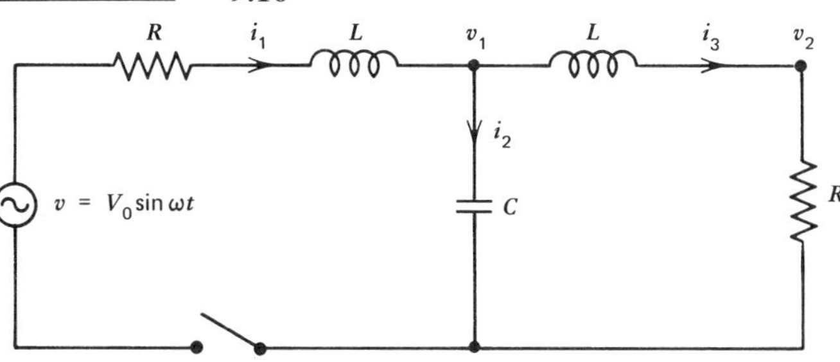

Figure P9.16.

Figure P9.16 shows an ac voltage source $v = V_0 \sin \omega t$, in series with a resistance R ohms, that is connected to a load resistance of equal magnitude. The intervening T section is a rudimentary low-pass filter that is intended to attenuate frequencies above a cutoff value $f_c = \omega_c/2\pi$ cps. Standard design techniques (see Skilling [31], for example) show that the proper choices of inductance and capacitance are: $L = R/\omega_c$ henries, and $C = 2/\omega_c R$ farads.

At time $t = 0$, with no current flowing and the capacitor uncharged, the switch is closed. Derive the differential equations and initial conditions that govern the subsequent variations with time of the potentials v_1 and v_2, and the currents i_1, i_2, and i_3. Write a computer program that uses a standard numerical technique for computing and printing v, v_1, v_2, i_1, i_2, and i_3 as functions of time.

Suggested Test Data

$R = 500$ ohms, $\omega_c = 20{,}000$ sec^{-1}, $V_0 = 10$ volts, with ω from 5000 to 40,000 sec^{-1} in steps of 5000 sec^{-1}. Also include appropriate values for Δt (the time step used in the solution), *tmax* (the maximum time for which the solution is to be computed), and *freq* (an integer frequency of printout, as in Example 9.1); *tmax* should be sufficiently large so that the transient effect caused by closing the switch has essentially died away.

Examine v_2 in particular and comment on the effectiveness of the filter. Note that considerably more selective filters can be designed by incorporating special additional sections, as indicated by Skilling [31].

9.17

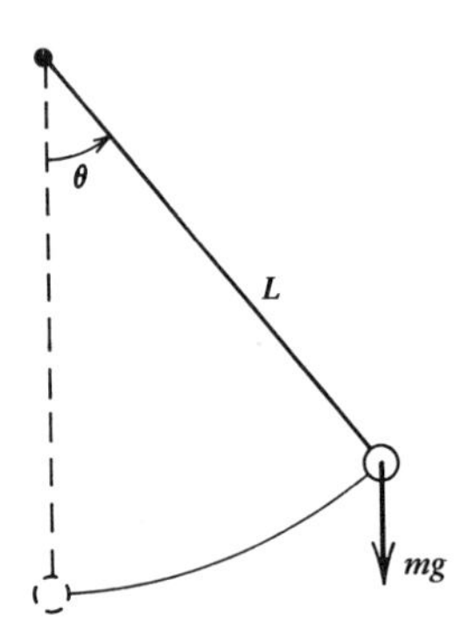

Figure P9.17.

The angular displacement θ of the simple pendulum of length L shown in Fig. P9.17 obeys the differential equation

$$\frac{d^2\theta}{dt^2} = -\frac{g}{L}\sin\theta,$$

where t is time and g is the gravitational acceleration.

By defining a dimensionless time $\tau = t\sqrt{g/L}$, show that the above equation of motion can be rewritten as

$$\frac{d^2\theta}{d\tau^2} = -\sin\theta.$$

The pendulum is released from rest with $\theta = \theta_0$ at $\tau = 0$. Write a program that will use a numerical method to compute and tabulate θ and $d\theta/d\tau$ as functions of τ. The input data should include values for θ_0, $\Delta\tau$ (the time step used in the solution), τ_{max} (the maximum time for which the solution is to be computed), and *freq* (an integer frequency of printout, as in Example 9.1). Examine the output to see if the motion is periodic; if so, compare the period with the known value of 2π for small values of θ_0.

Suggested Test Data

θ_0 from $\pi/20$ to $\pi/2$ in increments of $\pi/20$; $\tau_{max} = 20$; $\Delta\tau$ and *freq* will depend on the particular solution technique used, but $\Delta\tau$ should be taken sufficiently small so that its exact value does not appreciably influence the computed solution.

9.18

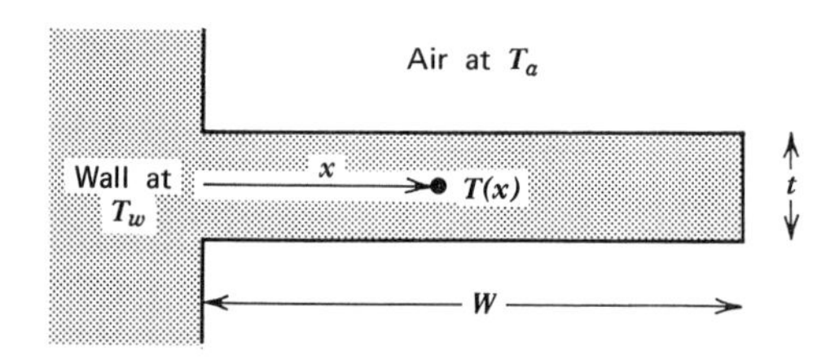

Figure P9.18.

Figure P9.18 shows a cross section of a long cooling fin of width W, thickness t, and thermal conductivity k, that is bonded to a hot wall, maintaining its base (at $x = 0$) at a temperature T_w. Heat is conducted steadily through the fin and is lost from the sides of the fin by convection to the surrounding air with a heat transfer coefficient h. (For simplicity, radiation is neglected at sufficiently low temperatures.)

Show that the fin temperature T, assumed here to depend only on the distance *along* the fin, obeys the differential equation

$$kt\frac{d^2T}{dx^2} = 2h(T - T_a),$$

where T_a is the mean temperature of the surrounding air. (Fourier's law gives the heat flux density as $q = -k\,dT/dx$ for steady conduction in one dimension.) If the surface of the fin is vertical, h obeys the dimensional correlation $h = 0.21(T - T_a)^{1/3}$ BTU/hr sq ft °F, with both temperatures in °F (Rohsenow and Choi [6]).

What are the boundary conditions on T? Write a program that will compute: (a) the temperature distribution along the fin, and (b) the total rate of heat loss from the fin to the air per foot length of fin.

Suggested Test Data

$T_w = 200$°F, $T_a = 70$°F, $t = 0.25$ in.; investigate all eight combinations of $k = 25.9$ (steel) and 220 (copper) BTU/hr ft °F, $W = 0.5, 1, 2,$ and 5 in.

9.19

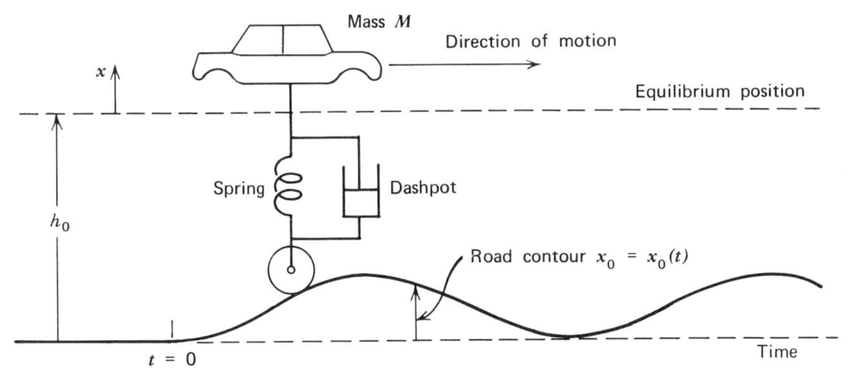

Figure P9.19.

Let the suspension of a vehicle of mass M be represented by the lumped spring/dashpot system shown in Fig. P9.19. At time $t = 0$ the vehicle is travelling steadily with its center of gravity at an equilibrium distance h_0 above the ground, and it has no vertical velocity. For subsequent times ($t > 0$) the contour of the road (the displacement of the road surface from the reference level at time zero) is given by an arbitrary function $x_0 = x_0(t)$.

If the stiffness (restoring force/extension) of the spring is k, and the damping coefficient of the dashpot is r, then show that $x(t)$, the displacement of the center of gravity of the vehicle as a function of time, is the solution of the second-order differential equation

$$M\frac{d^2x}{dt^2} = -k(x - x_0) - r\left(\frac{dx}{dt} - \frac{dx_0}{dt}\right),$$

subject to the initial conditions $x = dx/dt = 0$ at $t = 0$.

Let the road contour function be given by $x_0(t) = A(1 - \cos\omega t)$, where $2A$ is the full displacement of the road surface from the reference level. Note that the under-, critically-, and over-damped cases correspond to $r/2\sqrt{kM}$ being less than, equal to, and greater than unity, respectively.

Write a program that uses the model outlined above to simulate the behavior of a specified suspension system, over the time period $0 \leqslant t \leqslant t_{max}$. The program should read values for M, R, K, A, W, TMAX, DT, and FREQ, where W corresponds to ω, DT is the time-step size to be used in integrating the differential equation, and FREQ is an integer used for controlling the printing frequency, such that the values of t, x, dx/dt, d^2x/dt^2, and $x_0(t)$ are printed initially and every FREQ time steps thereafter. The remaining symbols have obvious interpretations. A one-step method is suggested for solving the equation, and the program should allow for the processing of several sets of data.

Suggested Test Data

$M = 3680\ \text{lb}_m$, $K = 640\ \text{lb}_f/\text{in.}$, $A = 2$ in., $W = 7$ rad/sec, and TMAX $= 10$ sec. Investigate $R = 80$, 160, and 240 lb_f sec/in. in turn. The values of DT and FREQ will depend on the solution technique chosen, but in any event the effect of varying DT should be investigated. For each situation, plot x, x_0, dx/dt, and d^2x/dt^2 as functions of time.

9.20 Let the damping coefficient, r, of Problem 9.19 be a nonlinear function of the relative velocities of the two parts of the dashpot:

$$r = r_0\left(1 + c\left|\frac{dx}{dt} - \frac{dx_0}{dt}\right|^{1/2}\right).$$

Rerun the simulation of Problem 9.19 with this definition of the damping coefficient. Investigate all nine combinations of $r_0 =$ 80, 160, and 240 lb_f sec/in., with $c = 0.1$, 1, and 10 $(\text{sec/in.})^{1/2}$.

9.21

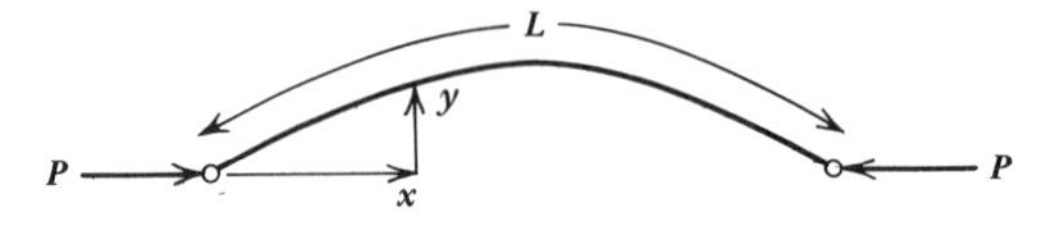

Figure P9.21.

The pin-ended strut shown in Fig. P9.21 is subjected to an axial load P. The lateral deflection y is governed by [20]

$$\frac{d^2y}{dx^2} + \frac{P}{EI}\left[1 + \left(\frac{dy}{dx}\right)^2\right]^{3/2} y = 0,$$

where $E =$ Young's modulus, and I is the appropriate cross-sectional moment of inertia of the strut. (For small deflections—not the situation here—the slope dy/dx is frequently neglected.)

If the length of the strut is L, assumed constant, at what value of PL^2/EI will the central deflection y_c be such that $y_c/L = 0.3$?

9.22 The installation shown in Fig. P9.22 delivers water from a reservoir of constant elevation H to a turbine. The surge tank

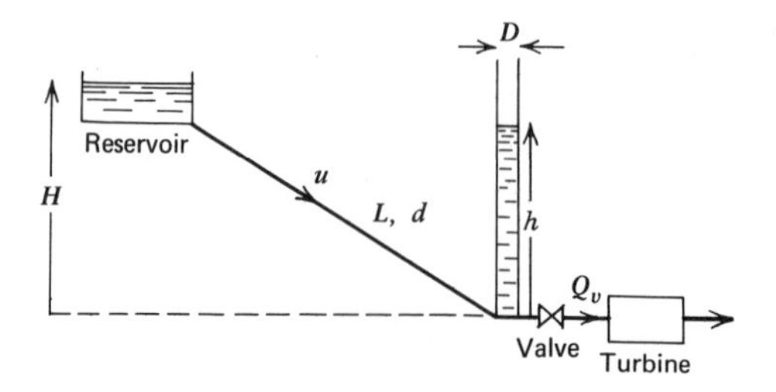

Figure P9.22.

of diameter D is intended to prevent excessive pressure rises whenever the valve is closed quickly during an emergency.

Assuming constant density, and neglecting the effect of acceleration of water in the surge tank, a momentum balance on the water in the pipe leads to

$$g(H-h) - \frac{1}{2} f_M \frac{L}{d} u|u| = L\frac{du}{dt}.$$

Here, $h =$ height of water in the surge tank (h_0 under steady conditions), $f_M =$ Moody friction factor, $u =$ mean velocity in the pipe, $g =$ gravitational acceleration, and $t =$ time. Also, continuity requires that

$$\frac{\pi d^2}{4} u = \frac{\pi D^2}{4}\frac{dh}{dt} + Q_v.$$

The flow rate Q_v through the valve in the period $0 \leqslant t \leqslant t_c$ during which it is being closed is approximated by

$$Q_v = k\left(1 - \frac{t}{t_c}\right)(h - h^*)^{1/2},$$

where the constant k depends on the particular valve, and the downstream head h^* depends on the particular emergency at the turbine.

During and after an emergency shutdown of the valve, compute the variation with time of the level in the surge tank.

Suggested Test Data

$g = 32.2$ ft/sec^2, $H = 100$ ft, $h_0 = 88$ ft, $f_M = 0.024$, $L = 2000$ ft, $d = 2$ ft, $t_c = 6$ sec, $k = 21.4$ ft$^{2.5}$/sec, and $D = 4$, 6, 10, and 15 ft. Take two extreme values for h^*: (a) its original steady value, $h_0 - Q_{v0}^2/k^2$, where Q_{v0} is the original steady flow rate, and (b) zero.

9.23 A projectile of mass m and maximum diameter d is fired with muzzle velocity V_0 at an angle of fire θ_0 above the horizontal. The subsequent velocity components u and v in the horizontal and vertical directions x and y, respectively, obey the ODEs

$$m\frac{du}{dt} = -F_D \cos\theta,$$

$$m\frac{dv}{dt} = -mg - F_D \sin\theta.$$

Here, θ = angle of projectile path above the horizontal, and g = gravitational acceleration. The drag force F_D is given by $F_D = C_D A \rho V^2/2$, where V = speed of projectile, $A = \pi d^2/4$ = projected area, and ρ = density of air. Consistent units are assumed in the above equations.

The drag coefficient C_D varies with the Mach number $M = V/c$, where c = velocity of sound in air. Representative values (see Streeter [21], for example) are:

M:	0	0.5	0.75	1	1.25	1.5	2	2.5	3
C_D:	1.6	1.8	2.1	4.8	5.8	5.4	4.8	4.2	3.9

The range on horizontal terrain is to be R. Write a program that will determine if this range can be achieved and, if so, the two possible angles of fire, the total time of flight, and the velocities on impact. In each case, also determine the error in the range that would result from a ± 10 minute of arc error in the firing angle.

Suggested Test Data

$m = 100$ lb$_m$, $g = 32.2$ ft/sec^2, $d = 6$ in., $V_0 = 2154$ ft/sec, $\rho = 0.0742$ lb$_m$/cu ft, $c = 1137$ ft/sec, with $R = 0.5, 1, 2, 4, 6, 8$, and 10 thousand yards.

9.24

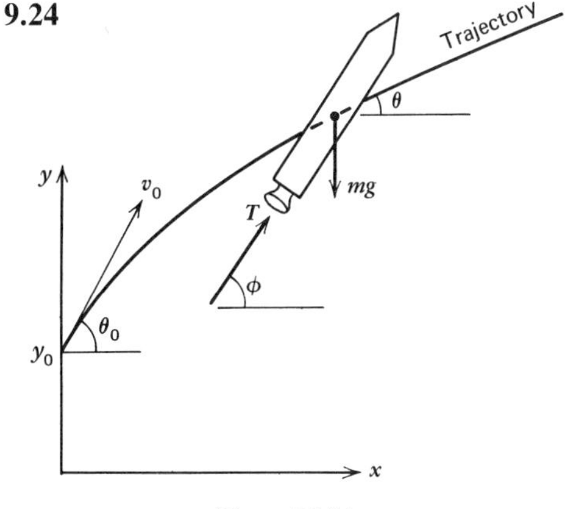

Figure P9.24.

This problem introduces the task of predicting the trajectory of a rocket, but is open-ended in that the programmer is left to investigate special cases (ballistics or satellite placement, for example) as he sees fit.

Neglecting aerodynamic drag, the equations of motion for the rocket shown in Fig. P9.24 are

$$\frac{du}{dt} = \frac{T}{m}\cos\phi,$$

$$\frac{dv}{dt} = \frac{T}{m}\sin\phi - g.$$

Here, u and v (ft/sec) are the x and y components of the velocity, g (ft/sec^2) is the gravitational acceleration, and m (lb$_m$) is the instantaneous mass of the rocket. The thrust T is given by

$$T = -g_c I \frac{dm}{dt},$$

in which g_c is the conversion factor 32.2 lb$_m$ ft/lb$_f$ sec^2 and I (lb$_f$ sec/lb$_m$) is the specific impulse of the propellant (typically in the range 200 to 400 for chemical fuels). The thrust attitude angle ϕ may or may not coincide with the trajectory angle θ at any instant.

The program should accept initial values (at $t = 0$) for θ_0, y_0 (vertical height; $x_0 = 0$ for simplicity), v_0 (speed), and m_0 (mass), and then use a numerical technique to compute and print values for u, v, x, y, θ, T, ϕ, and m as functions of time. The programmer should also specify how the fuel consumption ($-dm/dt$) and thrust attitude angle ϕ are to be scheduled during the flight; multiple stages may also be considered. For further information, the reader is referred to Chapter 8 of Thomson [22].

9.25 Consider an ecological study that involves two wildlife animal species: A, which has an abundant supply of natural food, and B, which depends entirely for its food supply by preying on A. At a given time t, let N_A and N_B be the populations of A and B, respectively, in a given natural area.

By making certain simple assumptions, show that the following differential equations could describe the growth or decay of the two populations:

$$\frac{dN_A}{dt} = \alpha N_A - \beta N_A N_B,$$

$$\frac{dN_B}{dt} = -\gamma N_B + \delta N_A N_B.$$

What significance do you attach to the positive constants α, β, γ, and δ?

Write a program that will use a standard numerical technique to solve the above equations for a wide variety of initial conditions and values for α, β, γ, and δ. If possible, arrange for the program to plot N_A and N_B as functions of time.

CHAPTER 10

An Introduction to Optimization Methods

10.1 Introduction

The optimization problem has been described succinctly by Aris [33] as "getting the best you can out of a given situation." Problems amenable to solution by mathematical optimization techniques have the general characteristics that: (a) there are one or more independent variables whose values must be chosen to yield a viable solution, and (b) there is some measure of *goodness* available to distinguish between the many viable solutions generated by different choices of these variables. Mathematical optimization techniques are used for guiding the problem solver to that choice of variables that *maximizes* the *goodness measure* (profit, for example) or that *minimizes* some *badness measure* (cost, for example).

One of the most important areas for the application of mathematical optimization techniques is in engineering design. Consequently, this chapter will be written from the viewpoint of a designer who wishes to establish optimum characteristics for his engineering design. However, the methods have wide applicability for large classes of problems involving the search for extreme functional values. Applications vary from the generation of "best" functional representations (curve fitting, for example) to the design of optimal control systems.

10.2 Engineering Design and the Optimization Problem

Virtually every engineering design evolves in an iterative or trial-and-error fashion. The design problem is often stated in an ambiguous and open-ended way. The designer's first task is to decide what the problem is and to identify the requirements or specifications for any proposed solution. The designer then generates a design concept, usually in the form of a rough configuration, for the object, system, or process being designed. These phases of the design activity are the most difficult and require the greatest ingenuity and engineering judgment.

The designer next attempts to *model* the object, system, or process. This model may assume a wide variety of forms, but the usually preferred ones are mathematical in character. The engineer may use many weapons from his arsenal of mathematical and scientific tools, from the most fundamental of nature's laws to the most empirical of data correlations, from the most rigorous analytical mathematical tools to the least formal cut-and-try methods, for assisting in the analysis of the proposed design.

The mathematical model normally contains several (n, for example) parameters known as the *design variables*, $x_1, x_2, \ldots, x_n$. To simplify the notation, we will use the notion of a *design vector*, $\mathbf{x} = [x_1, x_2, \ldots, x_n]^t$. We may view each design variable as one dimension in the *space* of the design variables, and any particular set of values for the design variables (that is, any particular design vector) as a *point* in this space. It may happen that certain equality constraints

$$\begin{aligned} g_1(\mathbf{x}) &= 0, \\ &\vdots \\ g_r(\mathbf{x}) &= 0, \end{aligned} \tag{10.1}$$

and/or inequality constraints

$$\begin{aligned} g_{r+1}(\mathbf{x}) &< 0, \\ &\vdots \\ g_m(\mathbf{x}) &< 0, \end{aligned} \tag{10.2}$$

are imposed on the space so that only points in a portion of the total space may even be considered as acceptable candidates for solution of the design problem. Such a region is termed the *feasible space*; points in this region are termed *feasible solutions* or feasible design alternatives.

The equality constraints come from functional relationships that must be strictly satisfied among the n variables; these constraints often follow from elementary conservation principles. If there is to be a design problem, that is, if there is to be a choice in the values which at least some of the design variables may assume, then r must be smaller than n. The inequality constraints usually arise from some specified design limitations (maximum permissible stress, minimum allowable temperature, etc.). There is no upper limit on the value of m.

Figure 10.1 illustrates a simple situation involving the design of a chemical reactor. For simplicity, let us assume that to solve the problem the designer must choose just two independent variables, temperature T and pressure P; the design vector in this case is $\mathbf{x} = [T, P]^t$. The design space is two-dimensional, with temperature and pressure as the coordinates. In practice, there will be certain inequality constraints placed upon the range of these variables. As shown in Fig. 10.1 there might be constant lower limits for the temperature and pressure, and an upper limit for each that depends on the magnitude of the other. The latter might come from limitations due to the strength of the reactor vessel material; qualitatively, this constraint might be viewed as follows: "the higher the temperature in the reactor, the lower the strength of the steel walls of the reactor vessel; hence the lower the maximum allowable pressure in the reactor." The region of feasible solutions to the chemical problem is the region enclosed by the constraints.

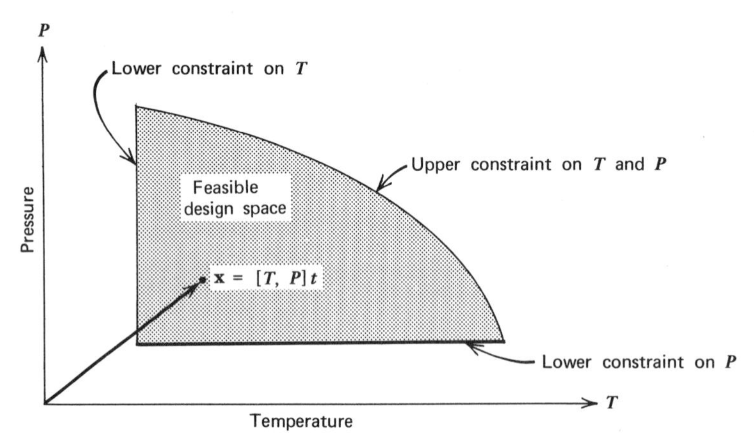

Figure 10.1 Feasible design space for operation of a chemical reactor.

Briefly stated, the optimization problem is to find the one design vector or point $\mathbf{x}^*$ in the feasible solution space that corresponds to the best design. In order to locate the best feasible solution of the usually infinite number of feasible solutions using a mathematical optimization technique, it is essential that a meaningful, computable, quantitative, single-valued function $f(\mathbf{x})$ of the design variables be formulated to serve as a measuring stick for comparison of feasible design alternatives. This function is usually called the *merit*, *effectiveness*, *criterion*, *utility*, or *objective* function. Throughout this chapter we shall label $f(\mathbf{x})$ the *objective function.*

Thus, the optimization problem is to find that design vector $\mathbf{x}^*$ that *maximizes* the objective function $f(\mathbf{x})$, that is,

$$f(\mathbf{x}^*) = \max_{\mathbf{x}_{feasible}} f(\mathbf{x}), \tag{10.3}$$

subject to constraints (10.1) and (10.2). In other situations, or by rephrasing the problem, the optimization problem may be to find the $\mathbf{x}^*$ that *minimizes* an objective function $f(\mathbf{x})$, that is,

$$f(\mathbf{x}^*) = \min_{\mathbf{x}_{feasible}} f(\mathbf{x}), \qquad (10.4)$$

subject to constraints (10.1) and (10.2).

One of the designer's principal responsibilities is to construct a suitable objective function to be maximized. In the simple chemical problem illustrated in Fig. 10.1, the objective function might be the yield from the reaction, that is, the optimization problem is to find the operating temperature and pressure that will produce the maximum amount of product from a given amount of raw materials. Or, we might rephrase the problem as a minimization problem, to find that combination of temperature and pressure that will minimize the amount of raw material required to yield a fixed amount of product.

The objective function may be quite simple and relatively easy to calculate (the total weight of a storage tank to hold a specified amount of some material), or quite complicated and difficult to calculate (the rate of return over a 20-year period for a proposed oil refinery, taking into account likely variations in markets, raw materials, taxes, etc.). The objective function may be very illusive due to the presence of conflicting or dimensionally incompatible subobjectives; for example, one might be asked to design a plant of minimum cost that also minimizes air and water pollution and that is aesthetically appealing. In such cases, it may not be possible to find a quantitative objective function and hence to automate the selection of better design alternatives using one of the mathematical optimization procedures. In what follows, we assume that a suitable objective function can be formulated by the designer.

10.3 Classification of Optimization Methods

The mathematical optimization methods can be categorized in a variety of ways; one such classification is shown in Table 10.1.

Table 10.1 Classification of Optimization Methods

1. Optimization methods for functions of one variable.
2. Optimization methods for unconstrained functions of n variables ($n > 1$).
3. Optimization methods for constrained objective functions.
4. Optimization methods that exploit system or information flow structure.

In this introductory text, we shall only examine methods that fall into the first two classes of Table 10.1. The term *unconstrained* in the second class does not imply the absence of constraints *per se*, but only the absence of constraints that place the *global* extreme value for the objective function (that is, the extreme value of the objective function with no imposed constraints) outside the feasible region. The third class of methods in Table 10.1 involves constrained objective functions, meaning that the global extreme value does indeed lie outside the feasible region. This additional feature considerably complicates the optimization problem. The term *mathematical programming* has been applied to the methods of classes 2 and 3 in Table 10.1. A special technique, called *linear programming*, applies when the constraints (10.1) and (10.2) and objective function (10.3) are linear in the design variables. Methods in the fourth class are based on a technique called *dynamic programming*, that makes use of special features of the structural organization of the proposed process or design. Unfortunately, these latter topics cannot be introduced without the assumption of considerable mathematical sophistication. The reader is referred to Gass [37] and

Nemhauser [38], respectively, for further discussions of linear programming and of dynamic programming.

10.4 Optimization Methods for Functions of One Variable

Some of the optimization methods for functions of one variable are listed in Table 10.2. The traditional tool for solution of such problems is the differential calculus. Necessary and sufficient conditions for locating a *local* extreme (maximum or minimum) point $(x^*, f(x^*))$ of an unconstrained continuous and differentiable objective function $f(x)$ with continuous higher-order derivatives in the neighborhood of the extreme point are:

(1) $f'(x^*) = 0$.
(2) If k denotes the order of the first non-vanishing derivative, $f^{(k)}(x^*)$, then k must be *even*.

$$(10.5)$$

If (1) and (2) are satisfied, then $(x^*, f(x^*))$ is a local minimum point if $f^{(k)}(x^*) > 0$ and a local maximum point if $f^{(k)}(x^*) < 0$.

Table 10.2 Tools for Optimizing Objective Functions of One Variable, f(x)

1. The calculus
2. Numerical mathematics
   Root-finding procedures for $f'(x) = 0$
   1. Newton's method
   2. Successive substitution methods
   3. False-position methods
   4. Half-interval method
   5. Special methods for polynomial functions
3. Searching schemes
   (a) Simultaneous methods
      1. Exhaustive search
   (b) Sequential methods
      1. Dichotomous search
      2. Equal-interval multipoint searches
      3. Golden-section search
      4. Fibonacci search

Unfortunately, the applicability of the calculus to design problems is severely limited by the complex, nonlinear, and frequently discontinuous and nondifferentiable nature of many objective functions.

However, some of the numerical root-finding procedures listed in Table 10.2 for solving $f'(x) = 0$ have already been discussed in Chapter 5. Newton's method requires the computation of the second derivative $f''(x)$ as well as the first derivative $f'(x)$, and requires continuity and differentiability of $f(x)$ and $f'(x)$. The false position and half-interval methods require only that $f'(x)$ be continuous and computable. Should the objective function be a polynomial, then one of several root-finding procedures for polynomials might also prove useful.

10.5 Searching Schemes

In contrast with the numerical methods, the searching schemes listed under (3a) and (3b) of Table 10.1 require only that the objective function $f(x)$ be computable; for most of these procedures, neither differentiability nor continuity of the objective function is essential.

In each case, the optimization problem is to locate the global extreme value of an objective function $f(x)$ on some arbitrary starting interval $a \leqslant x \leqslant b$, such that the final optimal x value found by the algorithm and the true optimal value, x^*, lie within a specified *interval of uncertainty*,† that is, an interval of specified length *known* to contain the extreme value. To simplify comparison of the various methods, we will assume that an arbitrary starting interval $[a, b]$ has been mapped onto the unit interval $[0, 1]$ by a suitable linear transformation. Hence, the initial interval of uncertainty for each method will have length 1. Other definitions used

†Note that by selecting a particular interval for examination, we may have created a *constrained* one-dimensional problem.

in common throughout Sections 10.6 to 10.12 are:

α_i = length of the interval of uncertainity after i iterations for the sequential searching strategies (see below),
α_n = length of *final* interval of uncertainty,
N = total number of evaluations of $f(x)$ required to reduce the length of the interval of uncertainty from 1 to α_n,
n = the number of iterations for the sequential searching strategies (see below).

10.6 Exhaustive Search

The one-dimensional searching schemes may be classified into two categories: *simultaneous* searches and *sequential* searches. In the simultaneous schemes, all points at which the function is to be evaluated are selected *a priori*. One such approach, termed the *exhaustive search*, is illustrated in Fig. 10.2. Here the function, which need not be differentiable or even continuous in some cases, is evaluated at intervals equally spaced by $\alpha_n/2$, assumed to be small enough to catch all pertinent features of the function. The total number of functional evaluations required is

$$N = \frac{2}{\alpha_n} + 1. \qquad (10.6)$$

Clearly, this is not a very effective approach for small α_n, particularly when the function is difficult to evaluate. For example, to reduce the starting interval of length 1 to a final interval of uncertainty of length 0.001, so that the root would be known within ± 0.0005, requires $N =$ 2,001 functional evaluations.

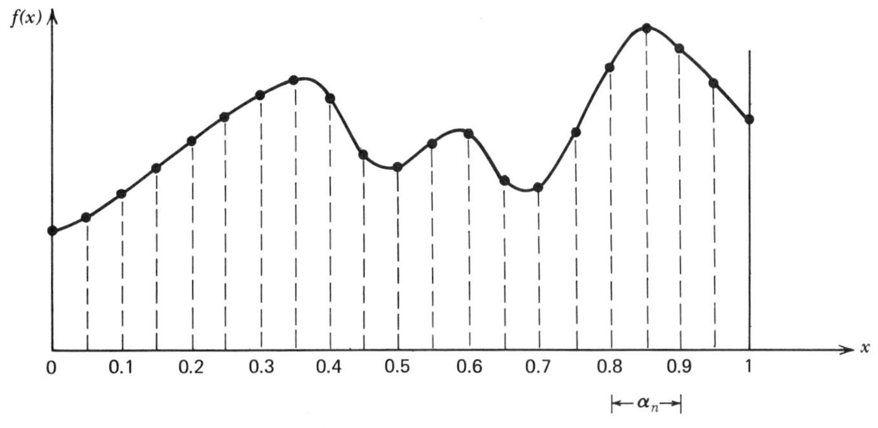

Figure 10.2 Exhaustive search.

10.7 Sequential Search Methods

When the sequential methods listed in Table 10.2 are used, the arguments for which the object function will be evaluated cannot be known *a priori*; instead, the sequence of argument values depends on the values of $f(x)$ already observed. All these methods require that $f(x)$ exhibit *unimodal* character, that is, exhibit just *one local extreme point* on the interval of interest, as shown in Fig. 10.3. More formally, unimodality implies that $f(x) < f(x-h)$ for $a \leqslant x < x^*$ and $f(x) < f(x+h)$ for $x^* < x \leqslant b$ (for a minimum) or $f(x) > f(x-h)$ for $a \leqslant x < x^*$ and $f(x) > f(x+h)$ for $x^* < x \leqslant b$ (for a maximum). The function need be neither differentiable nor continuous.

The sequential methods all involve the computation of the function at one or more arguments which, because of the unimodality assumption, permit reduction of the length of the interval of uncertainty. A second set of functional values is then established for the reduced interval and a yet smaller interval of uncertainty is

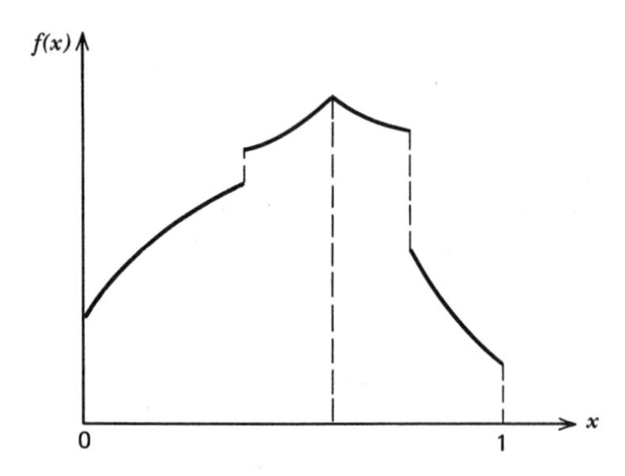

Figure 10.3 A function unimodal on the interval [0, 1].

found. The process is repeated until the interval length is reduced to the specified value α_n. All these schemes have the property that the *number* of functional evaluations required to achieve a specified terminal interval of uncertainty is known *a priori*, although the particular *arguments* used cannot be.

10.8 The Dichotomous Search

Clearly, the evaluation of the function at just *one* point in the interval yields insufficient information for reducing the interval of uncertainty. At least *two* functional evaluations must be made at *different* arguments in the interval. The locations of these arguments are arbitrary, but some choices are better than others. Supposing that a maximum is required, one of these one-dimensional searching schemes is illustrated in Fig. 10.4.

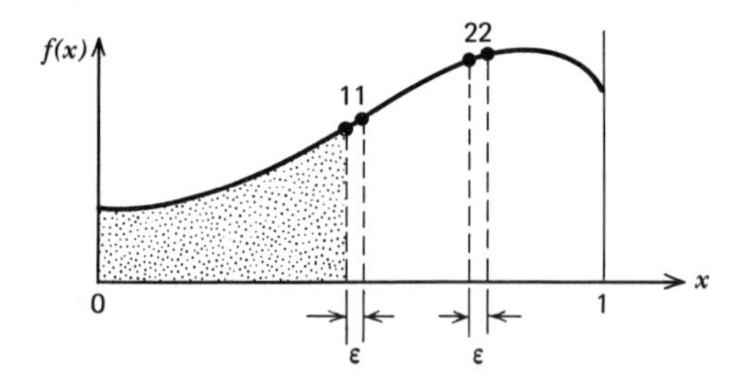

Figure 10.4 Dichotomous search.

If, on the first pass of the sequential interval-slicing process, we evaluate the function at two arguments centered about the midpoint and separated by a small distance ε, that is, if we compute $f(1/2+\varepsilon/2)$ and $f(1/2-\varepsilon/2)$ as shown in Fig. 10.4, then the *smaller* subinterval of length $(1-\varepsilon)/2$ can be rejected on the basis of the relative functional values at the two arguments (the points labeled 1 in Fig. 10.4).

In this case, the interval $[(1-\varepsilon)/2, 1]$ becomes the new interval of uncertainty since $f((1-\varepsilon)/2) < f((1+\varepsilon)/2)$. Next, two new functional evaluations are made at arguments centered about the midpoint of the new subinterval (points 2) and the subinterval subsequently is reduced again, based on these new functional values. This process, which is termed the *dichotomous search*, reduces the interval of uncertainty *approximately* by half for each iteration of the procedure. N and α_n are related to the number of iterations n as follows:

$$\alpha_n = \frac{1}{2^n} + \left(1 - \frac{1}{2^n}\right)\varepsilon, \tag{10.7}$$

$$N = 2n, \tag{10.8}$$

$$N = 2\frac{\ln\left(\dfrac{1-\varepsilon}{\alpha_n - \varepsilon}\right)}{\ln 2} \doteq 2.89 \ln\left(\frac{1-\varepsilon}{\alpha_n - \varepsilon}\right). \tag{10.9}$$

10.9 Equal-Interval Multipoint Search Methods

Other choices of argument locations for the evaluation of $f(x)$ on each iteration are illustrated in Figs. 10.5 and 10.6. In both cases, the selected points are *equally spaced* in the interval. Fig. 10.5 shows the two-point equal-interval method applied to a unimodal function,

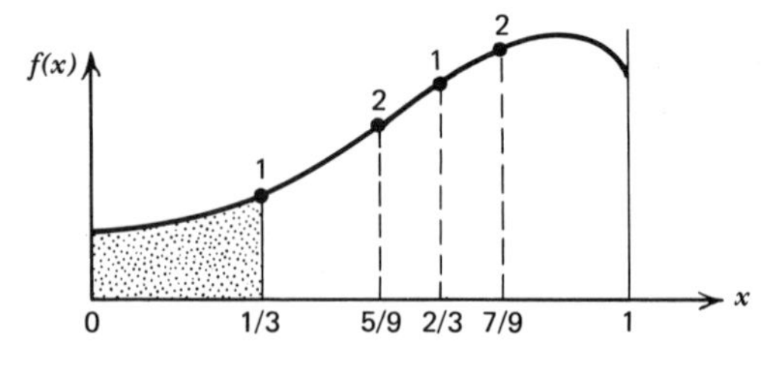

Figure 10.5 Two-point equal-interval search.

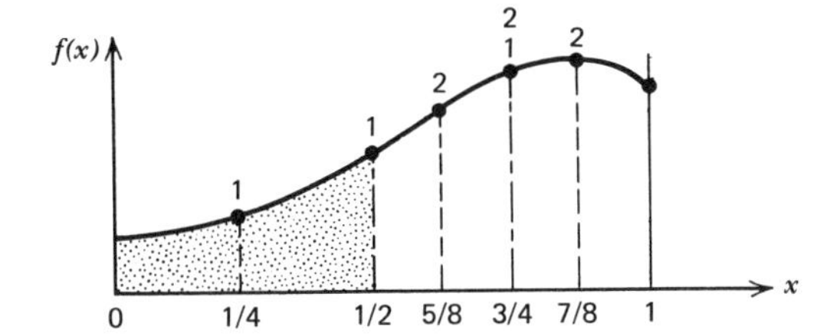

Figure 10.6 Three-point equal-interval search.

and we again suppose that a maximum is required. On the first pass, the function is evaluated at $x = 1/3$ and $x = 2/3$ (points 1).

The two-thirds of the original interval that must contain the extreme functional value is retained as the new interval of uncertainty for the second pass; for the function shown, this would be the interval $[1/3, 1]$. On the next pass, the function is again evaluated at 1/3 and 2/3 of the span of the current interval, in this case at $x = 5/9$ and $x = 7/9$ (points 2). Again, the two-thirds of the interval known to contain the extreme functional value is saved, in this case, the interval $[5/9, 1]$.

The procedure is repeated until the interval of uncertainty is reduced to the specified α_n. The parameters α_n, n, and N are related by:

$$\alpha_n = \left(\frac{2}{3}\right)^n, \qquad (10.10)$$

$$N = 2n, \qquad (10.11)$$

$$N \doteq -4.95 \ln \alpha_n. \qquad (10.12)$$

A similar three-point equal-interval search is shown in Fig. 10.6. In this case, the function is first evaluated at $x = 1/4$, $x = 1/2$, and $x = 3/4$ (points 1). That half of the original interval centered about the argument of largest functional value ($x = 3/4$ in this case) is retained. In the second pass of the sequential scheme, the function is evaluated at $x = 5/8$, $x = 3/4$, and $x = 7/8$. In this case, and in fact for all the odd-point equal-interval methods, there will be one functional value from the preceding iteration that will appear again and need not be recomputed (such as $f(3/4)$ here). The parameters α_n, n, and N are related by:

$$\alpha_n = \left(\frac{1}{2}\right)^n, \qquad (10.13)$$

$$N = 2n + 1, \qquad (10.14)$$

$$N \doteq 1 - 2.89 \ln \alpha_n. \qquad (10.15)$$

It can be shown that the three-point equal-interval search illustrated in Fig. 10.6 is the most efficient (that is, requires the fewest functional evaluations N, for a given α_n) of all possible equal-interval searches.

10.10 The Golden-Section Method

Intuitively, we might guess that rejection of one-half of the uncertainty interval per iteration of a sequential interval-slicing procedure would be the best we could achieve on the average, given no advance information about $f(x)$ except for its unimodal character. And from the viewpoint of the *minimax strategy*, the strategy that minimizes the likelihood of the worst possible outcome, this is the case. However, to reduce the interval of uncertainty by one-half using either the dichotomous or three-point equal-interval search requires *two* functional evaluations on each iteration. Are there other yet more efficient interval-slicing methods?

Yes! One such method is called the *golden-section search* and is illustrated in Fig. 10.7. In this strategy, two selected

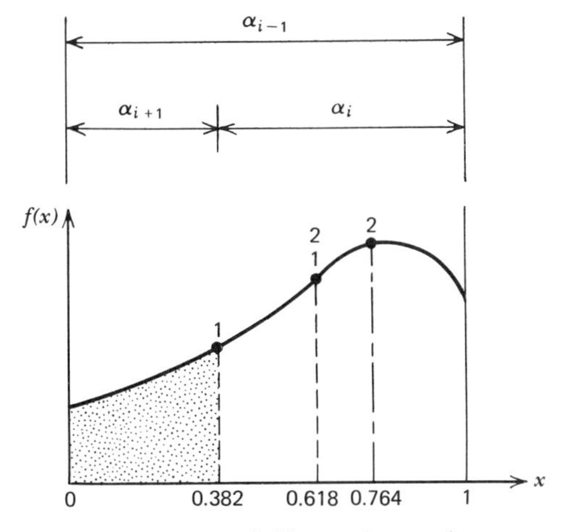

Figure 10.7 Golden-section search.

arguments are positioned symmetrically about the center of each interval of uncertainty in such a way that τ, the ratio of the lengths of successive intervals of uncertainty, satisfies the relationships

$$\tau = \frac{\alpha_{i+1}}{\alpha_i} = \frac{\alpha_i}{\alpha_{i-1}}, \tag{10.16}$$

where

$$\alpha_{i-1} = \alpha_i + \alpha_{i+1}. \tag{10.17}$$

That is, τ remains constant, is approximately equal to 0.618 (see Problem 10.6), and is called the *golden-section number.* Hence, each iteration of this sequential technique reduces the interval to approximately 61.8% of its previous value. This method would not appear to be as effective as the interval-halving achieved by one of the methods of Section 10.9. However, closer examination shows that, after the first pass, one of the two arguments chosen for the ith iteration will coincide with one of the arguments from the $(i-1)$th iteration (for example, $x = 0.618$ in Fig. 10.7). Thus, only *one* new functional evaluation is required for all iterations after the first. The parameters α_n, n, and N are related by:

$$\alpha_n \doteq (0.618)^n, \tag{10.18}$$

$$N = n + 1, \tag{10.19}$$

$$N \doteq 1 - 2.08 \ln \alpha_n. \tag{10.20}$$

10.11 The Fibonacci Search

It has been shown that the very best strategy for optimizing a unimodal function is the *Fibonacci search,* based on the Fibonacci number sequence F_j, where

$$\left.\begin{aligned} F_0 &= 1, \\ F_1 &= 1, \\ F_j &= F_{j-1} + F_{j-2}, \quad j \geqslant 2. \end{aligned}\right\} \tag{10.21}$$

The first few numbers of the Fibonacci sequence generated by the recursion formula of (10.21) are shown in Table 10.3.

Table 10.3 Fibonacci Numbers

j	F_j	j	F_j
0	1	5	8
1	1	6	13
2	2	7	21
3	3	8	34
4	5	9	55

The arguments at which the objective function $f(x)$ is to be evaluated at each iteration of this sequential strategy are determined by *ratios* of appropriate Fibonacci numbers. Given a starting interval of length F_N, the Fibonacci search requires at most N functional evaluations to reduce the interval of uncertainty to length 1.

To simplify the explanation, consider the special case of maximizing an objective function $f(x)$ on the starting interval $[0, 13]$, shown in Fig. 10.8. Suppose it is desired to reduce the length of the interval of uncertainty to 1. We then examine the Fibonacci sequence, and find the F_N that corresponds to (or is next greater than) the length of the initial interval of uncertainty α_0, in this case 13. Since $F_6 = 13$, no more than $N = 6$ functional evaluations should be required.

The first two arguments are positioned by the ratios F_4/F_6 and F_5/F_6, namely, at 5/13 and 8/13 of the span of the interval, or at $x = 5$ and $x = 8$ (the points marked 1 in Fig. 10.8). Since $f(5) > f(8)$, the new interval of uncertainty for the second pass of the iterative scheme is $[0, 8]$, so that $\alpha_1 = 8$. The arguments in the new interval are positioned by the ratios F_3/F_5 and F_4/F_5, that is, at 3/8 and 5/8 of the span of the interval, or at $x = 3$ and $x = 5$ (points 2). In general, the ratios involved on the ith iteration are given by F_{N-i-1}/F_{N-i+1} and F_{N-i}/F_{N-i+1}. One of the arguments for the ith pass will always coincide with one of the arguments for the $(i-1)$th pass. Hence, just one new functional evaluation need be made for each iteration after the first. On the last or $n =$ fifth pass, one of the arguments is located within a small distance ε of the other, thereby halving the interval of uncertainty after the $N-2$ or fourth pass.

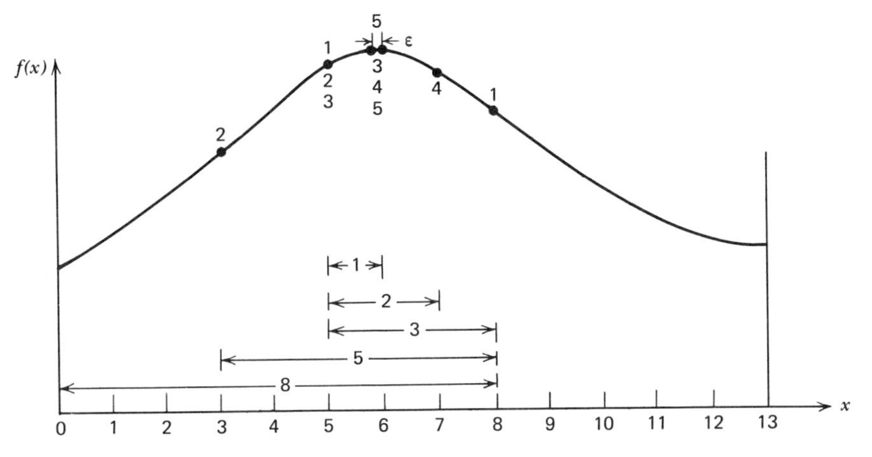

Figure 10.8 Fibonacci search.

If the problem is reformulated in terms of the starting interval of length 1 used for illustrating the other searching strategies, the parameters α_n, n, and N are related by:

$$\alpha_n = 1/F_N, \tag{10.22}$$

$$N = n + 1. \tag{10.23}$$

For $j > 5$, the ratio $F_{j-1}/F_j \doteq 0.618$; thus, the method is closely related to the golden-section search, and usually no more than one functional evaluation can be saved using the Fibonacci search rather than the golden-section search.

10.12 A Comparison of the One-Dimensional Searching Methods

A comparison of the number of functional evaluations required to reduce an initial unit interval to a specified value of α_n is shown in Table 10.4. The sequential strategies are clearly superior to the simultaneous exhaustive search.

In passing, note that all of the above one-dimensional searching strategies assume that a starting interval has been selected; that is, upper and lower limits have been established for the single independent variable. Thus, these methods are applicable to the solution of *constrained* one-dimensional problems as well as certain *unconstrained* one-dimensional problems (those whose global extreme value lies in the *open* starting interval).

One might reasonably ask why so much emphasis has been placed on one-dimensional methods, when most optimization problems involve several variables. The principal reason is that

Table 10.4 A Comparison of the One-Dimensional Searching Methods

α_n	Exhaustive	3-Point Even-Interval	Dichotomous[a]	Golden Section	Fibonacci
0.1	21	8	7	6	6
0.01	201	15	14	11	11
0.001	2001	21	20	16	16
0.0001	20001	28	27	21	20

[a]Depends on ε.

many of the multidimensional searching techniques, such as those discussed in Section 10.15, involve *sequences* of *unidirectional* searches (searches along successive straight-line paths), and the one-dimensional methods can readily be adapted to handle these suboptimization problems.

EXAMPLE 10.1

OPTIMIZATION OF A CONDENSER USING THE THREE-POINT EQUAL-INTERVAL SEARCH

Introduction

The heat exchanger shown in Fig. 10.1.1 is used for removing Q BTU/hr from a hydrocarbon vapor that condenses at a uniform temperature T_v °F on the outside of a bundle of tubes of total surface area A sq ft. A flow of m lb/hr of cooling water enters the tubes at T_1 °F and leaves at T_2 °F. The overall heat-transfer coefficient between the cooling water and condensing vapor is assumed to be constant at U BTU/hr sq ft °F. If the specific heat of the water is c_p BTU/lb °F, it may be shown [2] that:

$$Q = mc_p(T_2 - T_1) = \frac{UA(T_2 - T_1)}{\ln[(T_v - T_1)/(T_v - T_2)]}. \quad (10.1.1)$$

Thus, if Q, c_p, T_1, T_v, and U are specified, the required values for m and A corresponding to an arbitrary T_2 are:

$$m = \frac{Q}{c_p(T_2 - T_1)}, \quad (10.1.2)$$

$$A = \frac{Q \ln[(T_v - T_1)/(T_v - T_2)]}{U(T_2 - T_1)}. \quad (10.1.3)$$

The combined cost of the cooling water and condenser, C \$/hr, may be expressed as:

$$C = m[c_1 + c_2(T_2 - T_1)^{1.5}] + c_3A^{0.725}. \quad (10.1.4)$$

Here, c_1 \$/lb is the base cost of the cooling water; the term involving c_2 represents a premium that increases substantially when the water is returned for reprocessing at a temperature T_2 that is significantly above its supply temperature T_1. The term $c_3A^{0.725}$ represents the hourly cost of the condenser, obtained by amortizing its installed cost over a given number of years.

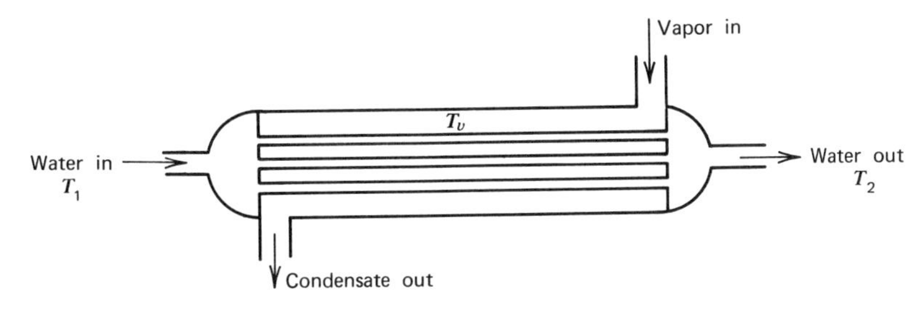

Figure 10.1.1 Condenser.

Problem Statement

Write a program that reads values for c_1, c_2, c_3, c_p, ΔT, Q, T_1, T_v, and U as data, and that proceeds to find the exit temperature T_2 that minimizes the overall cost C. The three-point equal-interval search should be used for locating T_2 within a final interval of uncertainty ΔT.

Method of Solution

The algorithm is expressed below in terms of a main flow diagram and flow diagrams for two functions, CNDNSR and EXTREM. The function EXTREM implements the three-point equal-interval search, described earlier, and will locate either the maximum or minimum of a function f according to whether the logical variable *max* is given as *true* or *false*, respectively. A general reference to EXTREM would be:

$$fval \leftarrow \text{EXTREM}\ (max, x_L, x_R, f, \alpha, x)$$

Here, x_L and x_R are the lower and upper

limits of the starting interval and α is the final interval of uncertainty. EXTREM returns two values: *fval*, the extreme value of the function on the interval, and x, the corresponding value of the independent variable.

The function CNDNSR returns the total hourly cost according to equation (10.1.4), for a specified value of T_2. The other variables (c_1, c_2, etc.), which are constant for each set of data, are assumed to have the same values in CNDNSR as they do in the main flow diagram.

The main flow diagram handles input and output, and uses EXTREM to find the minimum overall cost C and the corresponding water exit temperature T_2. Note that T_2 must lie on the open interval (T_1, T_v); thus the starting interval is arbitrarily taken to be the closed interval $[T_1+0.01, T_v-0.01]$.

Three-Point Equal-Interval Search

The three-point search is fairly easily implemented. First, the base-point spacing δ is computed as $(x_R-x_L)/4$, and the base points for the first interval are calculated as:

$$x_1 = x_L + \delta,$$

$$x_2 = x_L + 2\delta,$$

$$x_3 = x_L + 3\delta.$$

The function f is then called on to give:

$$y_1 = f(x_1),$$

$$y_2 = f(x_2),$$

$$y_3 = f(x_3).$$

The *extreme* functional value then becomes the *central* functional value y_2 for the next pass of the interval-slicing procedure; the independent variable value that corresponds to this extreme functional value is then assigned as the new value of x_2 for the next pass. After halving δ, the new values for the base points x_1, x_3, and the corresponding functional values y_1, y_3, are computed from:

$$x_1 = x_2 - \delta,$$

$$y_1 = f(x_1),$$

$$x_3 = x_2 + \delta,$$

$$y_3 = f(x_3).$$

The extreme of the three new functional values y_1, y_2, and y_3 then serves to determine the new x_2 and y_2 for the next pass of the algorithm. This iterative process continues until

$$2\delta \leqslant \alpha,$$

that is, until the interval of uncertainty is no larger than the specified value α.

The final extreme value is assigned as the value of the function EXTREM, and the final value of the independent variable is assigned to x upon return to the calling program.

Flow Diagram

Main (c_1, c_2, c_3, c_p, Q, T_1, T_v, and U have a common meaning here and in the function CNDNSR.)

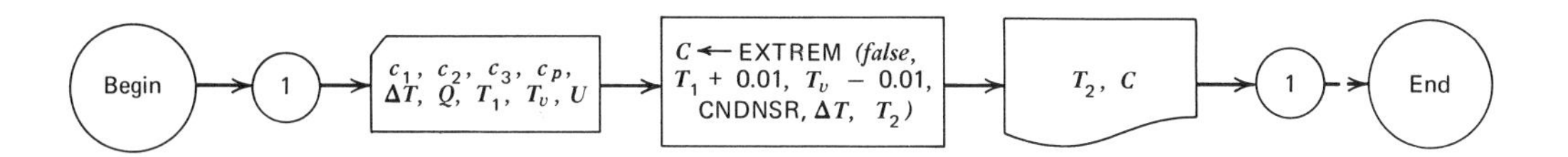

Function CNDNSR (T_2): (c_1, c_2, c_3, c_p, Q, T_1, T_v, and U have a common meaning here and in the main flow diagram.)

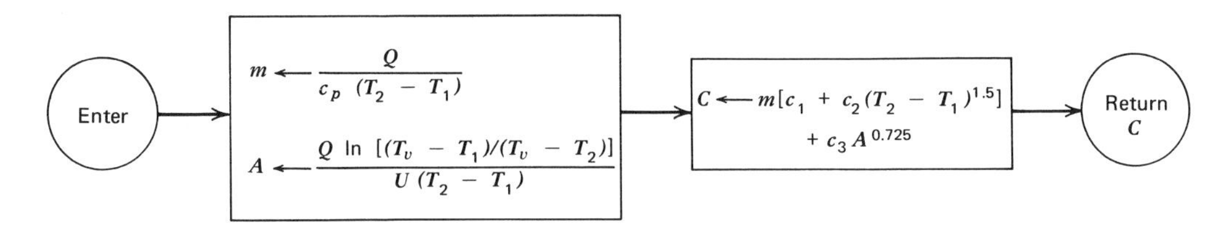

Function EXTREM $(max, x_L, x_R, f, \alpha, x)$

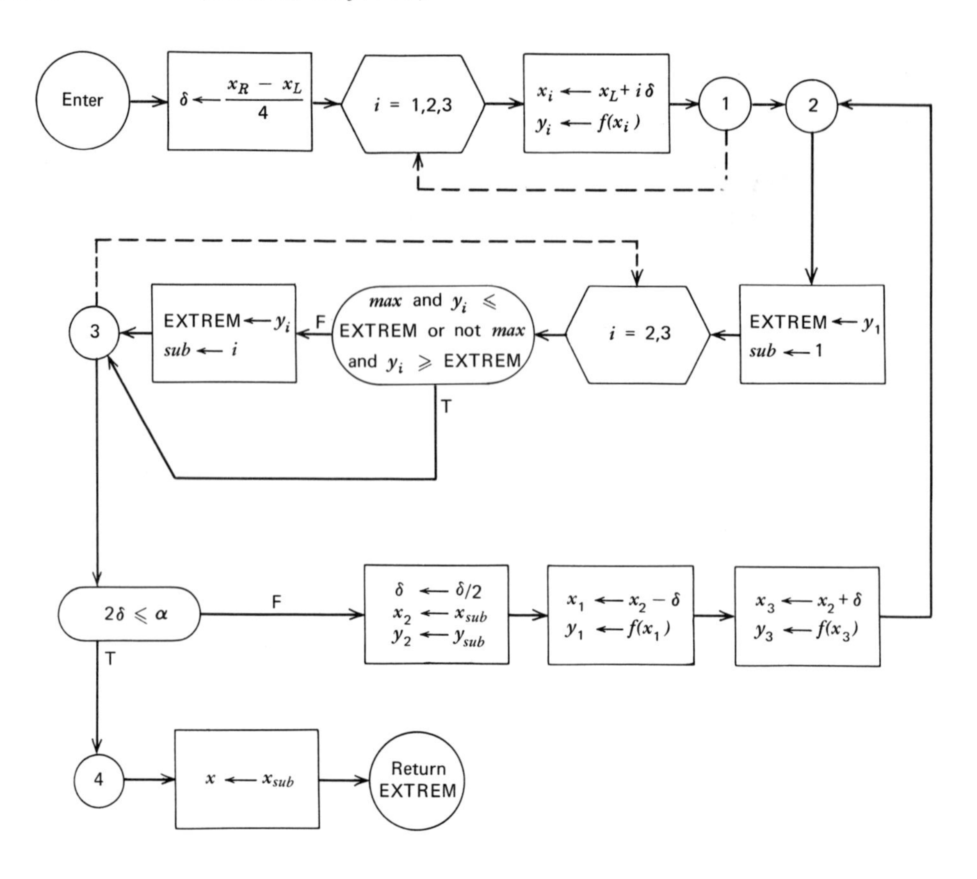

FORTRAN Implementation

List of Principal Variables

Program Symbol	Definition
(Main)	
A	Heat-exchange surface area of the condenser, A, sq ft.
C1, C2	Constants c_1 and c_2 for determining cost of cooling water in dollars; as read, they refer to 1000 gallons, but are converted internally to a 1-lb basis.
C3	Constant c_3 for determining cost of condenser in dollars; as read, c_3 is for a 1-year period, but is converted internally to a 1-hr basis.
CP	Specific heat c_p of cooling water, BTU/lb °F.
DT	Final interval of uncertainty, ΔT °F, for T_2.
GPM	Flow rate of cooling water, gpm.
HRINYR	Number of hours in a year. (The condenser is assumed to operate full time.)
M	Mass flow rate of cooling water, m lb/hr.
Q	Rate of heat removal from condensing vapor, Q BTU/hr.
SET	Integer data-set identification number.
T1, T2	Cooling water inlet and exit temperatures, T_1 and T_2 °F, respectively.
TV	Condensing temperature of vapor, T_v °F.
U	Overall heat-transfer coefficient between cooling water and condensing vapor, U BTU/hr sq ft °F.
$CNDNR	Annual cost of condenser, $/yr.
$TOTAL	Total annual cost, $/yr.
$WATER	Annual cooling-water cost, $/yr.
(Function EXTREM)	
ALPHA	Final interval of uncertainty, α.
DELTA	Spacing between base points, δ.
F	Name of the function, f, whose extreme value is required.
MAX	Logical variable, *max*; according to whether MAX is *true* or *false*, EXTREM searches for a maximum or a minimum functional value, respectively.
SUB	Subscript of the elements in the X and Y vectors corresponding to the extreme point for each pass of the three-point search.
X	Vector of base points x_1, x_2, and x_3.
XEXT	Value of x corresponding to the extreme functional value.
XL, XR	Lower and upper limits of the starting interval, x_L and x_R, respectively.
Y	Vector of functional values $y_i = f(x_i)$, $i = 1, 2, 3$.

Note that in addition to finding the optimum water exit temperature, the following program also computes and prints the corresponding water flow rate in gpm, the condenser area A sq ft, and the condenser, water, and total costs in dollars per annum. The program achieves format-free input by using a NAMELIST declaration in conjunction with an appropriate input statement. The abbreviated data sets after the first emphasize that the entire list need not be read; those variables not reassigned values will automatically retain their previous values, as can readily be checked by examining the computer output.

Program Listing

Main Program

```
C        DIGITAL COMPUTING                        EXAMPLE 10.1
C        OPTIMIZATION OF CONDENSER BY 3-PT EQUAL-INTERVAL SEARCH.
C
      COMMON C1, C2, C3, CP, Q, T1, TV, U
      EXTERNAL CNDNSR
      INTEGER SET
      REAL M
      NAMELIST /DATA/ SET, C1, C2, C3, CP, DT, Q, T1, TV, U
C
C     ..... PRINT HEADING, READ & CHECK INPUT DATA .....
      WRITE (6,200)
    1 READ (5,DATA,END=999)
      WRITE (6,201) SET, C1, C2, C3, CP, DT, Q, T1, TV, U

C
C     ..... CONVERT C1, C2, & C3 TO PROPER UNITS .....
      HRINYR = 24.*365.
      C1 = C1/8.33E3
      C2 = C2/8.33E3
      C3 = C3/HRINYR
C
C     ..... USE EXTREM TO FIND OPTIMUM EXIT TEMPERATURE T2
C           AND THE CORRESPONDING MINIMUM COSTS PER ANNUM .....
      $TOTAL = HRINYR*EXTREM(.FALSE., T1+0.01, TV-0.01, CNDNSR, DT, T2)
      M = Q/(CP*(T2 - T1))
      GPM = M/(60.*8.33)
      A = Q*ALOG((TV - T1)/(TV - T2))/(U*(T2 - T1))
      $WATER = M*(C1 + C2*(T2 - T1)**1.5)*HRINYR
      $CNDNR = A**0.725*C3*HRINYR
      WRITE (6,202) T2, $TOTAL, $WATER, $CNDNR, GPM, A
      GO TO 1
C
  999 CALL EXIT
C
C     ..... FORMATS FOR OUTPUT STATEMENTS .....
  200 FORMAT ('1DIGITAL COMPUTING                     EXAMPLE 10.1'/ ' OPTIMI
     1ZATION OF CONDENSER BY 3-PT EQUAL-INTERVAL SEARCH.')
  201 FORMAT ('0'/ '0THE INPUT VALUES FOR DATA SET NO.', I3, ' ARE'/
     1   '0     C1      =',  F9.3/ '      C2      =', F11.5/
     2   '      C3      =',  F7.1/ '      CP      =',   F9.3/
     3   '      DT      =',  F9.3/ '      Q       =', E13.3/
     4   '      T1      =',  F7.1/ '      TV      =',   F7.1/
     5   '      U       =',  F7.1)
  202 FORMAT ('0OPTIMUM CONDITIONS ARE:'/
     1   '0     T2     =', F10.1/ '      $TOTAL =', F10.1/
     2   '      $WATER =', F10.1/ '      $CNDNR =', F10.1/
     3   '      GPM    =', F10.1/ '      A      =', F10.1)
C
      END
```

Function CNDNSR

```
      FUNCTION CNDNSR (T2)
C
C        COMPUTES THE CONDENSER & COOLING WATER COST, $/HR.
C
      COMMON C1, C2, C3, CP, Q, T1, TV, U
      REAL M
C
      M = Q/(CP*(T2 - T1))
      A = Q*ALOG((TV - T1)/(TV - T2))/(U*(T2 - T1))
      CNDNSR = M*(C1 + C2*(T2 - T1)**1.5) + A**0.725*C3
      RETURN
C
      END
```

Function EXTREM

```
      FUNCTION EXTREM (MAX, XL, XR, F, ALPHA, XEXT)
C
C        FUNCTION TO FIND EXTREME VALUE (MAXIMUM OR MINIMUM) OF A
C        UNIMODAL FUNCTION F ON THE INTERVAL (XL,XR), WITHIN A FINAL
C        INTERVAL OF UNCERTAINTY ALPHA.
C
      LOGICAL MAX
      INTEGER SUB
      DIMENSION X(3), Y(3)
C
C     ..... SET INITIAL BASE POINTS AND FUNCTIONAL VALUES .....
      DELTA = (XR - XL)/4.
      DO 1  I = 1, 3
      X(I) = XL + I*DELTA
    1 Y(I) = F(X(I))
C
C     ..... FIND THE EXTREME OF THE THREE FUNCTIONAL VALUES .....
    2 EXTREM = Y(1)
      SUB = 1
      DO 3  I = 2, 3
      IF (MAX .AND. Y(I).LE.EXTREM .OR. .NOT.MAX .AND. Y(I).GE.EXTREM)
     1                                                          GO TO 3
      EXTREM = Y(I)
      SUB = I
    3 CONTINUE
C
C     ..... IS INTERVAL OF UNCERTAINTY SMALL ENOUGH? .....
      IF (2.*DELTA .LE. ALPHA)  GO TO 4
C
C     ..... SET UP BASE POINTS AND FUNCTIONAL VALUES FOR NEXT INTERVAL .....
      DELTA = DELTA/2.
      X(2) = X(SUB)
      Y(2) = Y(SUB)
      X(1) = X(2) - DELTA
      Y(1) = F(X(1))
      X(3) = X(2) + DELTA
      Y(3) = F(X(3))
      GO TO 2
C
C     ..... ASSIGN X ITS VALUE AT THE EXTREME POINT & RETURN .....
    4 XEXT = X(SUB)
      RETURN
C
      END
```

Data

```
 &DATA SET = 1, C1 = 0.05, C2 = 0.00032, C3 = 10.7, CP = 1.000,
 DT = 0.1, Q = 1.E6, T1 = 75.0, TV = 175.0, U = 100. &END
 &DATA SET = 2, C1 = 0.10, C2 = 0.00032, C3 = 10.7 &END
 &DATA SET = 3, C1 = 0.05, C2 = 0.00064, C3 = 10.7 &END
 &DATA SET = 4, C1 = 0.05, C2 = 0.00032, C3 = 21.4 &END
```

Computer Output

```
DIGITAL COMPUTING                   EXAMPLE 10.1
OPTIMIZATION OF CONDENSER BY 3-PT EQUAL-INTERVAL SEARCH.


THE INPUT VALUES FOR DATA SET NO.  1 ARE

    C1      =     0.050
    C2      =     0.00032
    C3      =    10.7
    CP      =     1.000
    DT      =     0.100
    Q       =     0.100E 07
    T1      =    75.0
    TV      =   175.0
    U       =   100.0

OPTIMUM CONDITIONS ARE:

    T2      =       118.7
    $TOTAL  =      3795.5
    $WATER  =      3427.9
    $CNDNR  =       367.6
    GPM     =        45.8
    A       =       131.4


THE INPUT VALUES FOR DATA SET NO.  2 ARE

    C1      =     0.100
    C2      =     0.00032
    C3      =    10.7
    CP      =     1.000
    DT      =     0.100
    Q       =     0.100E 07
    T1      =    75.0
    TV      =   175.0
    U       =   100.0

OPTIMUM CONDITIONS ARE:

    T2      =       140.5
    $TOTAL  =      4757.8
    $WATER  =      4329.1
    $CNDNR  =       428.7
    GPM     =        30.6
    A       =       162.4


THE INPUT VALUES FOR DATA SET NO.  3 ARE

    C1      =     0.050
    C2      =     0.00064
    C3      =    10.7
    CP      =     1.000
    DT      =     0.100
    Q       =     0.100E 07
    T1      =    75.0
    TV      =   175.0
    U       =   100.0

OPTIMUM CONDITIONS ARE:

    T2      =       103.5
    $TOTAL  =      5777.4
    $WATER  =      5437.9
    $CNDNR  =       339.4
    GPM     =        70.2
    A       =       117.7


THE INPUT VALUES FOR DATA SET NO.  4 ARE

    C1      =     0.050
    C2      =     0.00032
    C3      =    21.4
    CP      =     1.000
    DT      =     0.100
    Q       =     0.100E 07
    T1      =    75.0
    TV      =   175.0
    U       =   100.0

OPTIMUM CONDITIONS ARE:

    T2      =       116.8
    $TOTAL  =      4161.0
    $WATER  =      3433.6
    $CNDNR  =       727.4
    GPM     =        47.9
    A       =       129.5
```

Discussion of Results

Four sets of data have been processed: a base case (Set 1), and three other cases in which c_1, c_2, and c_3 have been arbitrarily doubled (Sets 2, 3, and 4, respectively).

In all instances, the water cost dominates that of the condenser. In fact, if c_3 (relating to the condenser) is doubled, the optimum total cost rises by only 9.5%. However, a doubling of c_1 or c_2 (relating to the water cost) results in an optimum total cost increase of 25.2% or 52.1%, respectively. There is clearly a price to be paid for reducing "thermal pollution."

The above treatment is only a first approximation to reality. In practice, the optimum design of a condenser would include more variables. For example, since the heat-transfer coefficient U depends on the water velocity and tube diameter, consideration would also have to be given to the proper selection of the number of tubes, their length, and their diameter.

10.13 Functions of Two or More Variables

In Sections 10.14 to 10.16 we introduce a few methods that can be used for optimizing functions such as $f(\mathbf{x}) = f(x_1, x_2, \ldots, x_n)$, of n independent variables. However, since these methods have simple geometrical interpretations for the special case of $n = 2$, we shall frequently concentrate on the optimization of the function $f(x_1, x_2)$. The resulting techniques can then be extended fairly readily to the n-dimensional case.

A typical such function $f(\mathbf{x}) = f(x_1, x_2)$ is shown in Fig. 10.9. The locus of functional values in the direction normal to

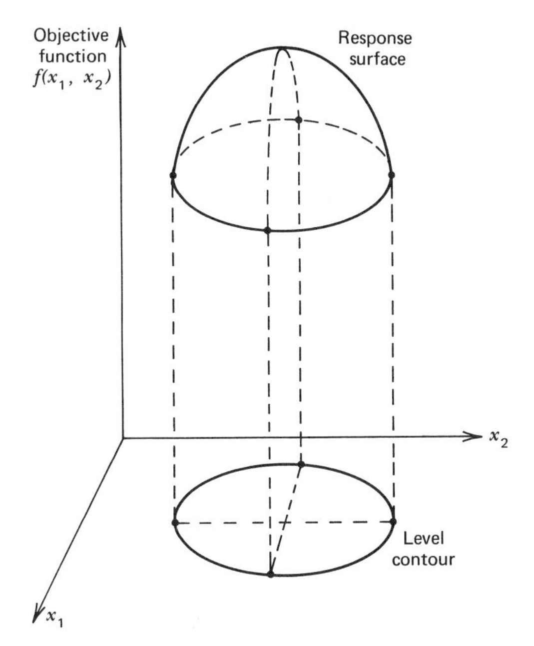

Figure 10.9 Response surface for a function of two variables.

the x_1-x_2 plane is sometimes called the *response surface*. The intersection of the response surface with a plane parallel to the x_1-x_2 plane generates a curve along which the functional value is constant; the projection of this curve onto the x_1-x_2 plane is called a *level contour*. Assuming that $f(\mathbf{x})$ is everywhere continuous, the level contour will be a connected curve; if $f(\mathbf{x})$ is also everywhere differentiable, the level contour will be a smooth curve. Several such level contours, each corresponding to a certain functional value, are shown in Fig. 10.10, which is essentially a two-dimensional "topographical map" of the function.

If $f(\mathbf{x})$ is continuous and differentiable, it can be expanded about a point $\mathbf{a}$ (see

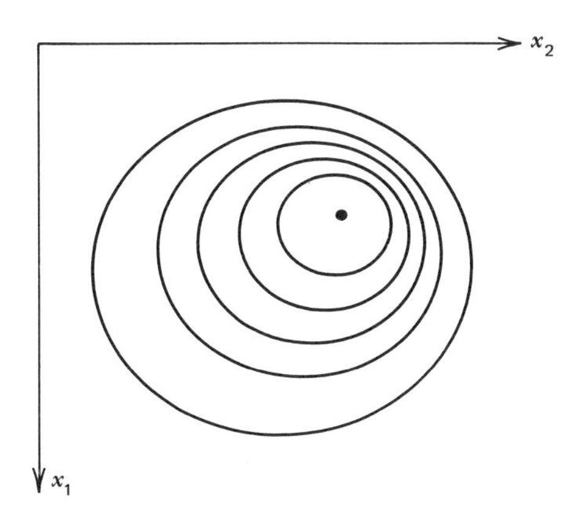

Figure 10.10 Level contours for a function of two variables.

Fig. 10.11, for example) in a Taylor's series. Here, $\mathbf{a}$ has coordinates (a_1, a_2), so that it can also be represented as the column vector $\mathbf{a} = [a_1\ a_2]^t$. If only the linear terms in the series are retained, then the equation of the *tangent plane* to the response surface at $\mathbf{a}$ is given by

$$y(\mathbf{x}) = f(\mathbf{a}) + (x_1 - a_1)f_{x_1}(\mathbf{a}) + (x_2 - a_2)f_{x_2}(\mathbf{a}), \qquad (10.24)$$

in which f_{x_1} and f_{x_2} are the partial derivatives $\partial f/\partial x_1$ and $\partial f/\partial x_2$, respectively.

If the intersection of the contour plane (the plane passing through $\mathbf{a}$ that is parallel to the x_1-x_2 plane) and the tangent plane at $\mathbf{a}$ is projected onto the x_1-x_2 plane, a line tangent to the level contour at $\mathbf{a}$ is generated. This line, shown in Fig. 10.12, is called the *contour tangent*, and has the equation

$$(x_1 - a_1)f_{x_1}(\mathbf{a}) + (x_2 - a_2)f_{x_2}(\mathbf{a}) = 0. \qquad (10.25)$$

The projection onto the x_1-x_2 plane of that line in the tangent plane along which the function $y(\mathbf{x})$ increases most rapidly per unit distance is called the *direction of steepest ascent* or simply the *gradient*.

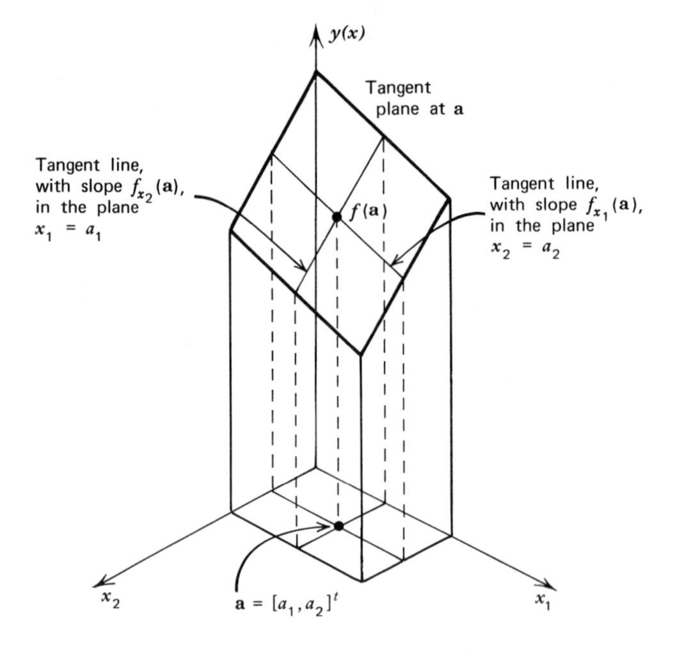

Figure 10.11 Tangent plane for a function of two variables.

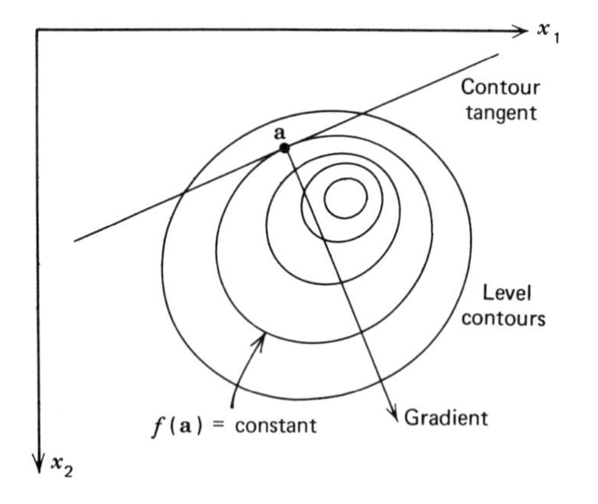

Figure 10.12 Contour tangent and gradient for a function of two variables.

The gradient at **a** is orthogonal to the contour tangent at **a** and its direction cosines d_1 and d_2 are given by:

$$d_1 = f_{x_1}(\mathbf{a})/S, \qquad d_2 = f_{x_2}(\mathbf{a})/S,$$

where

$$S = \sqrt{(f_{x_1}(\mathbf{a}))^2 + (f_{x_2}(\mathbf{a}))^2}. \tag{10.26}$$

The n-dimensional generalizations of the tangent plane and contour tangent are *hyperplanes*, the n-dimensional analogue of a plane in three-dimensional space. Although impossible to picture graphically, their mathematical formulations follow similarly as:

$$y(\mathbf{x}) = f(\mathbf{a}) + \sum_{i=1}^{n} (x_i - a_i) f_{x_i}(\mathbf{a}), \tag{10.27}$$

and

$$\sum_{i=1}^{n} (x_i - a_i) f_{x_i}(\mathbf{a}) = 0. \tag{10.28}$$

The direction cosines of the gradient, which remains a line even in n-dimensional space, are:

$$d_i = f_{x_i}(\mathbf{a})/S, \quad i = 1, 2, \ldots, n.$$

where

$$S = \sqrt{\sum_{i=1}^{n} (f_{x_i}(\mathbf{a}))^2}. \tag{10.29}$$

A point **a** is a *stationary point* when all the first partial derivatives vanish, that is, when $f_{x_i}(\mathbf{a}) = 0, \ i = 1, 2, \ldots, n$. An *extreme point* is a stationary point **a** at which the scalar

$$(\mathbf{x} - \mathbf{a})^t \mathbf{A} (\mathbf{x} - \mathbf{a}), \tag{10.30}$$

known as the *quadratic form*, maintains

the same sign for all $\mathbf{x} \neq \mathbf{0}$. Here, **A** is the matrix of second-order partial derivatives:

$$\mathbf{A} = \begin{bmatrix} f_{x_1x_1} & f_{x_1x_2} & \cdots & f_{x_1x_n} \\ f_{x_2x_1} & f_{x_2x_2} & \cdots & f_{x_2x_n} \\ \vdots & \vdots & & \vdots \\ f_{x_nx_1} & f_{x_nx_2} & \cdots & f_{x_nx_n} \end{bmatrix}, \qquad (10.31)$$

evaluated at the point **a**. If **A** is *positive definite* (that is, if (10.30) is positive for all $\mathbf{x} \neq \mathbf{0}$), it can be shown that the extreme point is a *minimum* point; if **A** is *negative definite*, the extreme point is a *maximum* point.

In this connection, the determinants D_i of the various submatrices formed from **A** are also of interest:

$$D_i = \begin{vmatrix} f_{x_1x_1} & \cdots & f_{x_1x_i} \\ \vdots & & \vdots \\ f_{x_ix_1} & \cdots & f_{x_ix_i} \end{vmatrix}, \qquad i = 1, 2, \ldots, n. \qquad (10.32)$$

It can be shown (a) that **A** is positive definite if $D_i > 0$, $i = 1, 2, \ldots, n$, and (b) that **A** is negative definite if $D_i > 0$, $i = 2, 4, 6, \ldots$ and $D_i < 0$, $i = 1, 3, 5, \ldots$.

The concept of unimodality carries over from one- to n-dimensional space naturally; unimodality implies that the response surface has just one extreme point in the region under consideration. Three types of unimodality are illustrated in Fig. 10.13. A function is unimodal if, given an extreme point **a**, there is *some* path, from every other point **x** in the space, along which the objective function values rise or fall monotonically, as in Fig. 10.13*a*. Fig. 10.13*b* shows a function that is not unimodal: given an extreme point **a**, it is impossible to find a strictly rising or falling path from point **b** to point **a**. A somewhat more stringent form of unimodality, called *strong unimodality* (Fig. 10.13*c*), requires that the *straight line* path from every point **x** to the extreme point **a** be strictly rising

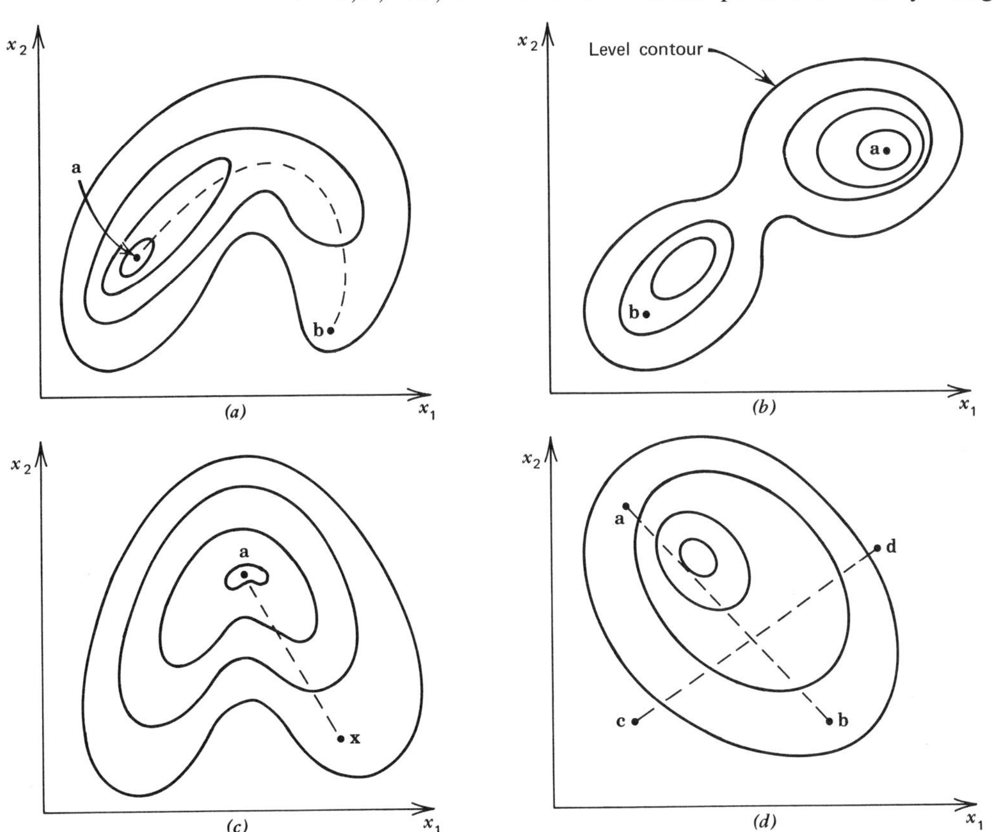

Figure 10.13 Unimodal behavior for a function of two variables. (a) Unimodality. (b) Not unimodal. (c) Strong unimodality. (d) Linear unimodality.

or falling. A yet more restrictive form, called *linear unimodality*, is illustrated in Fig. 10.13*d*: *all* straight line paths between *any* two points in the space (for example, between **a** and **b**, or between **c** and **d**) must yield objective functional values that exhibit unimodal behavior.

While a strictly unimodal function would have just one *local* extreme point equivalent to the one and only *global* extreme point, it is often possible to confine observation to a small region of interest in which the only concern is that the *local* behavior of the function be unimodal. The optimization procedures described in Sections 10.14 and 10.15 require that the objective function be unimodal, at least locally. Generally, these procedures are more efficient when the function is strongly or linearly unimodal.

10.14 Optimization Methods for Unconstrained Objective Functions of Two or More Variables

The optimization methods available for finding extreme values for functions $f(\mathbf{x}) = f(x_1, x_2, \ldots, x_n)$ of n variables can be classed into categories similar to those given in Table 10.2 for the single-variable case, namely:

(1) the calculus,
(2) numerical root-finding procedures for $f_{x_i}(\mathbf{x}) = 0,\ i = 1, 2, \ldots, n$,
(3a) simultaneous searching schemes, and
(3b) sequential searching schemes.

The calculus is the classical tool for solving extrema problems in n dimensions, requiring the solution of the n equations

$$f_{x_i}(\mathbf{x}) = 0,\ i = 1, 2, \ldots, n. \qquad (10.33)$$

Unfortunately, its direct utility in engineering design is quite limited because of the difficulties of: (a) obtaining explicit expressions for the required derivatives, and (b) solving algebraically the resulting system of usually quite nonlinear simultaneous equations.

However, it may be possible to use one of the numerical root-finding procedures for solving the generated system of nonlinear equations $f_{x_i}(\mathbf{x}) = 0,\ i = 1, 2, \ldots, n$. In particular, the Newton-Raphson method has already been discussed in Section 8.7; if it converges, it usually does so rapidly, and is probably the best method when applicable. Note that the method requires that all second partial derivatives $f_{x_i x_j}(\mathbf{x})$ be computable, preferably from analytical expressions, although numerical approximations to these derivatives can be devised.

Fortunately, there are several searching schemes that only require the evaluation of $f(\mathbf{x})$, and not its derivatives. Hence, these methods are particularly useful when such derivatives are difficult to evaluate, frequently the case in engineering design work.

The n-dimensional analogue of the one-dimensional simultaneous-type exhaustive search, already discussed, requires the evaluation of the objective function at all nodes of a grid structure overlaid in the region of interest. Consider, for example, the n-dimensional *hypercube* (the n-dimensional equivalent of a rectangle in two dimensions or of a cube in three dimensions) bounded by $a_i \leqslant x_i \leqslant b_i,\ i = 1, 2, \ldots, n$. If α_i is the desired final interval of uncertainty for the variable x_i, the total number of functional evaluations required is

$$N = \prod_{i=1}^{n} \left[\frac{2(b_i - a_i)}{\alpha_i} + 1 \right]. \qquad (10.34)$$

Note that even for a three-dimensional problem with a starting hypercube for which $b_i - a_i = 1$ $(i = 1, 2, 3)$, and with a modest final interval of uncertainty such as $\alpha_i = 0.001$ $(i = 1, 2, 3)$, the total number of functional evaluations required is about 8×10^9; thus, from a practical standpoint, this approach is virtually useless.

10.15 Sequential Searching Methods Requiring Evaluation of $f(x)$ Only

In this section, we describe the lattice, univariate, rotating-coordinate, and direct-search methods, none of which require the computation of derivatives of the objective function.

Lattice Method. As illustrated in Fig. 10.14 for two dimensions, a grid of selected coarseness is overlaid in the region of interest. A node of the grid is chosen as a central starting point. The function is evaluated there and at the $3^n - 1$ adjacent node points, where n is the number of dimensions. The adjacent point with the greatest (or least) functional value is then selected as the next starting point and the procedure repeated, except that the functional values already computed need not be recomputed. The process is continued until the greatest (or least) functional value occurs at the starting point. The grid size is then halved in each dimension and the search is begun afresh with the reduced grid mesh. The grid-size reduction is repeated, when necessary, until the interpoint distance in each dimension is smaller than some specified tolerance.

Univariate Search. In this search, only one variable is changed at a time, typically in the order $x_1, x_2, \ldots, x_n$; thus, the function is minimized (or maximized) in one coordinate direction at a time. Starting from an arbitrary point $\mathbf{x}_0$, successive intermediate points $\mathbf{x}_1, \mathbf{x}_2, \ldots, \mathbf{x}_n$ are found by performing linear minimizations in the directions

$$\mathbf{p}_0 = \begin{bmatrix} 1 \\ 0 \\ \vdots \\ 0 \end{bmatrix}, \mathbf{p}_1 = \begin{bmatrix} 0 \\ 1 \\ \vdots \\ 0 \end{bmatrix}, \ldots, \mathbf{p}_{n-1} = \begin{bmatrix} 0 \\ 0 \\ \vdots \\ 1 \end{bmatrix},$$

such that $\mathbf{x}_{i+1}$ is the point of minimum functional value along the line of direction $\mathbf{p}_i$ passing through the point $\mathbf{x}_i$. Thus,

$$\mathbf{x}_{i+1} = \mathbf{x}_i + \beta_i \mathbf{p}_i, \qquad i = 0, 1, \ldots, n-1,$$

where $|\beta_i|$ is the distance between $\mathbf{x}_i$ and $\mathbf{x}_{i+1}$.

Although it need not be computed, note that the gradient $\mathbf{g}(\mathbf{x}_{i+1}) = \mathbf{g}_{i+1}$ at the new point $\mathbf{x}_{i+1}$ is such that

$$\mathbf{p}_i^t \mathbf{g}_{i+1} = 0. \qquad (10.35)$$

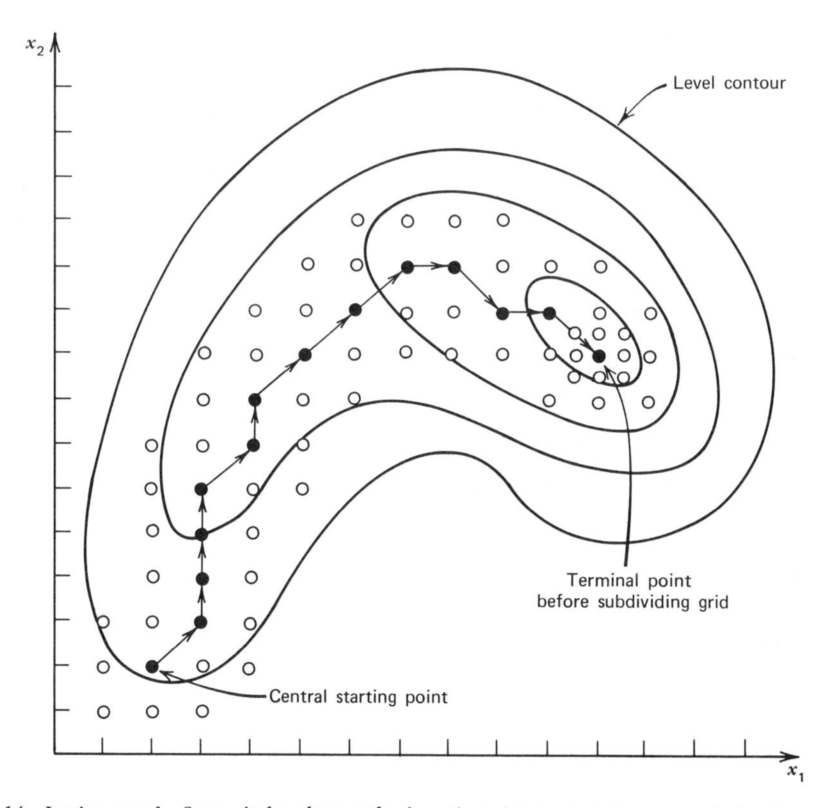

***Figure 10.14** Lattice search. Open circles show nodes investigated but rejected; opaque circles show sequence of starting nodes.*

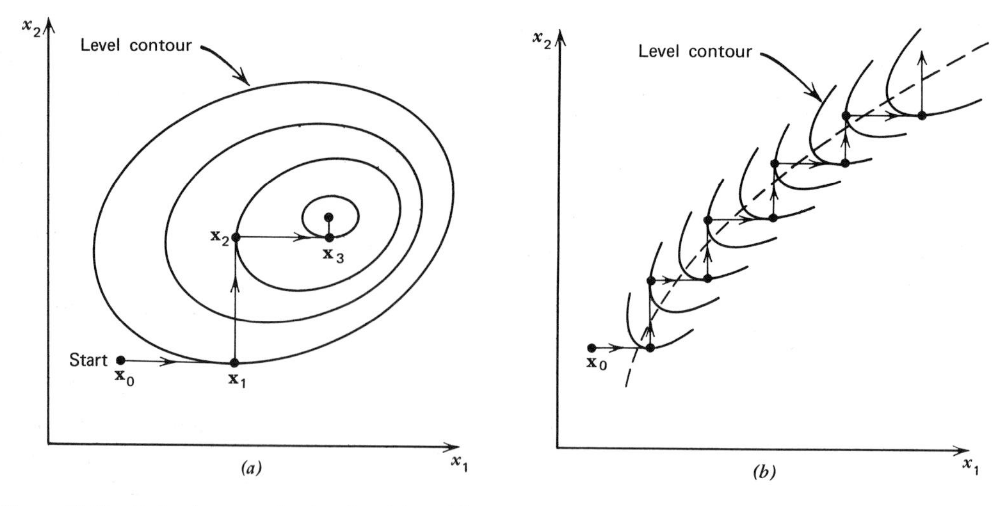

Figure 10.15 Univariate search, showing (a) rapid progress, and (b) slow progress along a ridge with strong interaction.

That is, $\mathbf{p}_i$ is tangent to the contour at $\mathbf{x}_{i+1}$, as illustrated in Fig. 10.15*a*.

After completing one round, a second round is performed, and so on, until each $|\beta_i|$ is smaller than some corresponding tolerance α_i.

The univariate search can often be very slow if the variables $x_1, x_2, \ldots, x_n$ interact strongly, as with the ridgelike structure shown in Fig. 10.15*b*. Here, the normalized gradient vector on the ridge has all direction cosines approximately equal in magnitude. If there is no interaction at all, that is, if the principal axes of the surface are in the coordinate directions, then the method will find the extreme value quickly.

Any of the one-dimensional searching algorithms already discussed can be used for finding the point of minimum functional value in each coordinate axial direction, once a unimodal segment has been isolated.

Rotating Coordinate Method. In an attempt to overcome the ridge-orientation problem of the univariate search, Rosenbrock [35] has developed a variation of the univariate method that involves rotation of the axes of the search.

The procedure begins with a standard one-round univariate search, starting at an arbitrary point $\mathbf{x}_0$; that is, n successive linear minimizations (or maximizations) are performed in the coordinate directions $\mathbf{p}_i$, giving

$$\mathbf{x}_{i+1} = \mathbf{x}_i + \beta_i \mathbf{p}_i, \qquad i = 0, 1, \ldots, n-1$$

such that

$$\mathbf{p}_i^t \mathbf{g}_{i+1} = 0.$$

Here, as before, $\mathbf{p}_i = [0\,0 \ldots 1 \ldots 0\,0]^t$, in which the $(i+1)$th element is unity.

Next, a new set of searching directions $\mathbf{p}_n, \mathbf{p}_{n+1}, \ldots, \mathbf{p}_{2n-1}$ is generated, such that

$$\mathbf{p}_n = \frac{\mathbf{x}_n - \mathbf{x}_0}{\|\mathbf{x}_n - \mathbf{x}_0\|}, \tag{10.36}$$

and $\mathbf{p}_{n+1}, \ldots, \mathbf{p}_{2n-1}$ satisfy the orthogonality conditions

$$\mathbf{p}_i^t \mathbf{p}_j = 0, \qquad i, j = n, n+1, \ldots, 2n-1, \quad (i \neq j). \tag{10.37}$$

As before,

$$\mathbf{x}_{i+1} = \mathbf{x}_i + \beta_i \mathbf{p}_i, \quad i = n, n+1, \ldots, 2n-1,$$

such that

$$\mathbf{p}_i^t \mathbf{g}_{i+1} = 0.$$

In general, each round of the method consists of n one-dimensional searches in n orthogonal directions. In the first round,

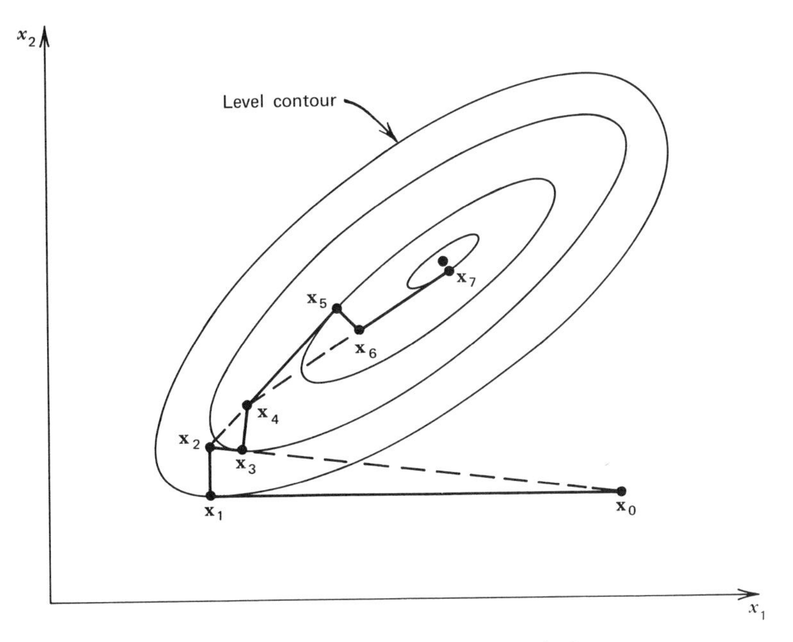

Figure 10.16 Rotating coordinate method.

the directions used are those of the n coordinates. In the second round, the first search, giving $\mathbf{x}_{n+1}$, is conducted in the direction given by a line passing through the points $\mathbf{x}_0$ and $\mathbf{x}_n$; the following $n-1$ searches are in the mutually orthogonal directions $\mathbf{p}_{n+1}, \ldots, \mathbf{p}_{2n-1}$, which are also orthogonal to the first direction $\mathbf{p}_n$. In successive rounds, new sets of orthogonal vectors are generated, so that for the $(k+1)$th round, the first search is in the direction

$$\mathbf{p}_{nk} = \frac{\mathbf{x}_{nk} - \mathbf{x}_{n(k-1)}}{\|\mathbf{x}_{nk} - \mathbf{x}_{n(k-1)}\|}, \qquad k = 1, 2, \ldots \tag{10.38}$$

A method for generating the necessary orthogonal vectors is discussed in Problem 10.17.

Direct Search. The direct search of Hooke and Jeeves [34] is a trial-and-error technique that has been used successfully by many investigators. There appears to be no one set of rules that constitutes *the* direct search, but all the approaches consist of a local *exploratory search* followed by a global *pattern move*. The exploratory search consists of a restricted univariate search, such that at most a single step of length δ_j, usually small, is taken in the jth coordinate direction. Based on information found in the exploratory search, the pattern move then consists of a relatively large step in a certain direction.

For the minimization problem, the steps in the complete algorithm, also illustrated in Fig. 10.17, are:

(1) Start with an arbitrary *base point* $\mathbf{b}_1$.

(2) Let $\boldsymbol{\delta}_1 = [\delta_1\, 0\, 0 \ldots 0]^t$, and evaluate the function at $\mathbf{b}_1 + \boldsymbol{\delta}_1$, at $\mathbf{b}_1$, and (if necessary) at $\mathbf{b}_1 - \boldsymbol{\delta}_1$, keeping the best point, also called the first new *temporary point* $\mathbf{t}_{11}$. That is:

$$\mathbf{t}_{11} = \begin{cases} \mathbf{b}_1 + \boldsymbol{\delta}_1 & \text{if } f(\mathbf{b}_1 + \boldsymbol{\delta}_1) < f(\mathbf{b}_1), \\ \mathbf{b}_1 - \boldsymbol{\delta}_1 & \text{if } f(\mathbf{b}_1 - \boldsymbol{\delta}_1) < f(\mathbf{b}_1), \\ \mathbf{b}_1 & \text{if } f(\mathbf{b}_1) \leq \min\,[f(\mathbf{b}_1 + \boldsymbol{\delta}_1), f(\mathbf{b}_1 - \boldsymbol{\delta}_1)]. \end{cases} \tag{10.39}$$

(3) Next, let $\boldsymbol{\delta}_2 = [0\, \delta_2\, 0 \ldots 0]^t$, and obtain a second temporary point $\mathbf{t}_{12}$

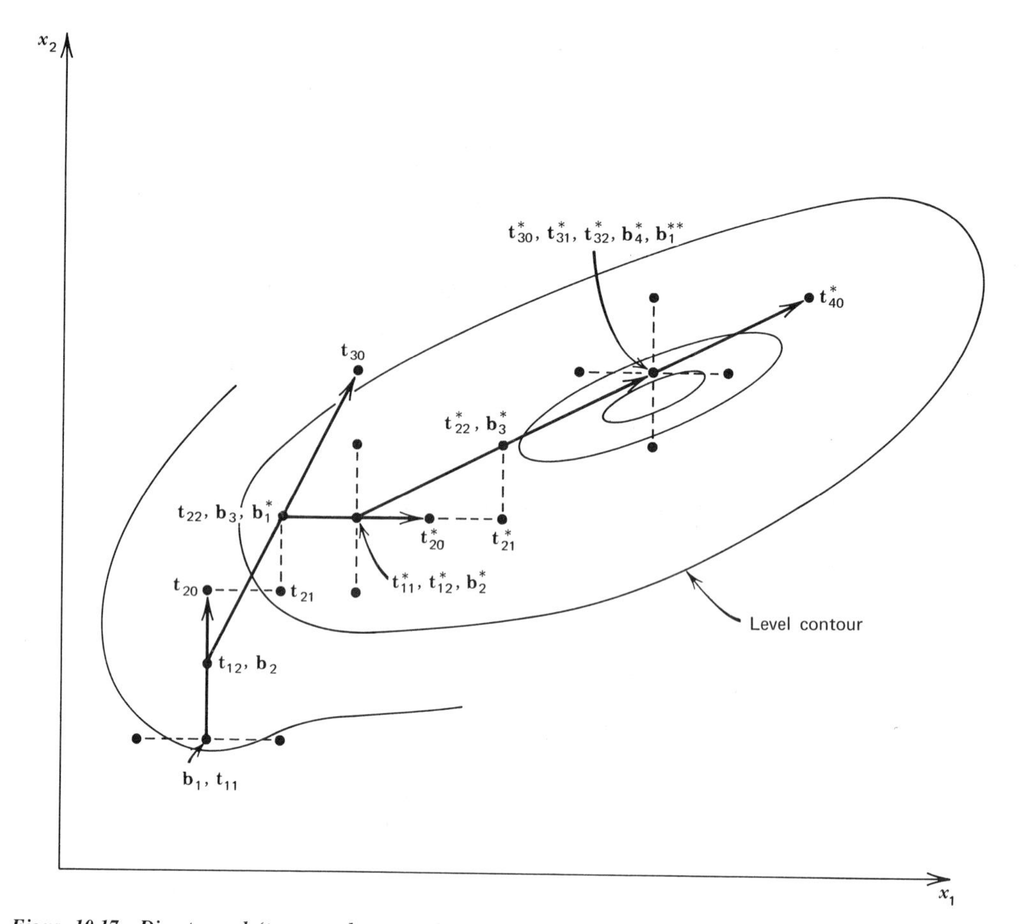

Figure 10.17 Direct search (to stage where search increments are reduced, with pattern moves shown as full lines).

similarly. That is, let $\mathbf{t}_{12}$ equal $\mathbf{t}_{11}+\boldsymbol{\delta}_2$, $\mathbf{t}_{11}-\boldsymbol{\delta}_2$, or $\mathbf{t}_{11}$, depending on which of these three possibilities yields the least functional value.

(4) Continue this process for each of the remaining variables $x_3, x_4, \ldots, x_n$, each time obtaining the jth new temporary point $\mathbf{t}_{1j}$, equal to $\mathbf{t}_{1,j-1}+\boldsymbol{\delta}_j$, $\mathbf{t}_{1,j-1}-\boldsymbol{\delta}_j$, or $\mathbf{t}_{1,j-1}$, depending on which of these three possibilities yields the least functional value. Here, $\boldsymbol{\delta}_j=[0\,0\ldots\delta_j\ldots0\,0]^t$, in which only the jth element is nonzero.

(5) The first exploratory search is now complete, and we define a second base point $\mathbf{b}_2$ equal to $\mathbf{t}_{1n}$.

(6) Next, make the first pattern move by proceeding in the direction $\mathbf{b}_2-\mathbf{b}_1$ along the line from $\mathbf{b}_1$ to $\mathbf{b}_2$ a distance equal to *twice* the distance from $\mathbf{b}_1$ to $\mathbf{b}_2$. Call this point $\mathbf{t}_{20}$, corresponding to the zeroth point for starting the second round; that is,

$$\mathbf{t}_{20}=\mathbf{b}_1+2(\mathbf{b}_2-\mathbf{b}_1)=2\mathbf{b}_2-\mathbf{b}_1. \tag{10.40}$$

(7) Now conduct an exploratory search about the point $\mathbf{t}_{20}$ as was performed with the original base point $\mathbf{b}_1$ (see steps 2 through 5 above). Call the terminal point of this second exploratory search $\mathbf{t}_{2n}$ and define a third base point $\mathbf{b}_3$ as

$$\mathbf{b}_3=\mathbf{t}_{2n}. \tag{10.41}$$

(8) Make the second pattern move in the direction $(\mathbf{b}_3-\mathbf{b}_2)$ to locate a third zeroth temporary point $\mathbf{t}_{30}$ as

$$\mathbf{t}_{30}=\mathbf{b}_2+2(\mathbf{b}_3-\mathbf{b}_2)=2\mathbf{b}_3-\mathbf{b}_2. \tag{10.42}$$

(9) This process of conducting exploratory searches round the point $\mathbf{t}_{k0}$ to yield the kth base point as

$$\mathbf{b}_{k+1} = \mathbf{t}_{kn}, \qquad (10.43)$$

followed by a pattern move to the point $\mathbf{t}_{k+1,0}$ according to

$$\mathbf{t}_{k+1,0} = 2\mathbf{b}_{k+1} - \mathbf{b}_k, \qquad (10.44)$$

continues as long as the functional value at $\mathbf{t}_{k+1,0}$ is smaller than at $\mathbf{b}_{k+1}$.

(10) Should $f(\mathbf{t}_{k+1,0})$ be greater than $f(\mathbf{b}_{k+1})$, then the base point $\mathbf{b}_k$ is dropped from consideration, and a new search is started by designating a new initial base point $\mathbf{b}_1^*$ as

$$\mathbf{b}_1^* = \mathbf{b}_{k+1}, \qquad (10.45)$$

and then returning to step 2.

(11) Should a local exploratory search near the newly designated $\mathbf{b}_1^*$ fail to find a temporary point $\mathbf{t}_{1n}^*$ that is different from $\mathbf{b}_1^*$, that is, should $\mathbf{b}_2^*$ equal $\mathbf{b}_1^*$, then the sizes of the search increments δ_j must be reduced and a new exploratory search initiated by returning to step 2.

(12) When the δ_j have been reduced sufficiently, so that

$$\delta_j < \alpha_j, \qquad j = 1, 2, \ldots, n, \qquad (10.46)$$

where α_j is some small number, then the search is complete and the last designated $\mathbf{b}_1$ becomes the direct-search approximation to the extreme point.

The direct-search method appears to be very efficient for many minimization problems, particularly when ridges are encountered. The directions of successive pattern moves tend to become aligned with the ridge. Experimentally, in such problems, the number of functional evaluations required tends to increase directly with the number of dimensions n, rather than with some higher power of n, as might be expected.

A variation of the above procedure computes the temporary points as follows:

$$\mathbf{t}_{k+1,0} = 2\mathbf{b}_{k+1} - \mathbf{b}_1. \qquad (10.47)$$

That is, all pattern moves are along lines with directions $\mathbf{b}_{k+1} - \mathbf{b}_1$ that pass through $\mathbf{b}_1$. The step sizes in the pattern moves are thus amplified more rapidly than in the standard method.

Note that since the criterion for termination is the failure of a univariate exploratory search using sufficiently small δ_j, the direct-search method can fail in those cases where a univariate search would fail, that is, for surfaces with sharp ridges having certain orientations with respect to the coordinate axes.

There are also many sequential methods that require the evaluation of partial derivatives of $f(x)$, but these are beyond the scope of this text. The interested reader is referred to Beveridge and Schechter [36].

Problems

10.1 Apply tests (10.5) to determine if the functions $f(x) = x^3$ and $f(x) = x^4$ have a local extreme point at the origin. If so, indicate whether the extreme point is a maximum or a minimum point.

10.2 Show that equation (10.9) follows from equations (10.7) and (10.8).

10.3 Develop a general formulation, comparable to equations (10.13) through (10.15), for the m-point equal-interval search methods for m odd. Show that of all the odd-point methods, the three-point method illustrated in Fig. 10.6 is the most efficient, that is, requires the fewest functional evaluations for a specified final interval of uncertainty.

10.4 Develop a general formulation, comparable to equations (10.10) through (10.12), for the m-point equal-interval search methods for m even. Which of the even-point methods is the most efficient? What is its efficiency, relative to the best of the odd-point methods (see Problem 10.3)?

10.5 Comment on the following statement: "If the golden-section method is applied to a function $f(x)$ that is *multimodal* on the starting interval, then the method will at least locate *one* of the local extreme points." (Note: a multimodal function has more than one local extreme point.)

10.6 Show that the golden-section number is exactly $(\sqrt{5}-1)/2$.

10.7 Suppose that an objective function is defined only for a finite number of values of the independent variable (x_0, $x_1, \ldots, x_m$, for example) and that $f(x_0)$, $f(x_1), \ldots, f(x_m)$ are ordered so as to make $f(x_i)$ unimodal over the points $x_0, x_1, \ldots,$ x_m. Without making any assumptions about the spacing between the x_i, devise an analogue of the Fibonacci search to find the x_i that corresponds to the extreme $f(x_i)$.

10.8 A river bed is thought to exhibit cross-stream unimodality (the water always gets deeper as you travel from either shore to the deepest point). If the river is half a mile wide, and you would like to find the deepest point in the river within 30 ft, where would you make depth measurements? How many such measurements would be needed?

10.9 Write a function, named GOLDEN, that implements the golden-section search for finding the minimum of a function. The heading for the function GOLDEN should be

```
GOLDEN (LOWER, UPPER, ALPHA,
                    F, X, NGOLD)
```

The function GOLDEN should find the *minimum* of the function named F, unimodal on the interval LOWER $\leqslant$ X $\leqslant$ UPPER, within an uncertainty interval of length ALPHA. Upon exit, X should contain the value of the independent variable at the minimum and NGOLD should contain the number of functional evaluations required. The value of F at the minimum should be returned as the value of the function GOLDEN.

Test the function GOLDEN by using it to find minimum *and* maximum values of the sine function on intervals where unimodality is assured. The function GOLDEN may also be used in conjunction with Problems 10.12 and 10.14 to 10.16. In such cases, it will be necessary to write a short main program that reads in essential data, calls upon GOLDEN to find the minimum value of an appropriate function F on a specified interval, prints the results, and then returns to read more data.

10.10 Develop one or more techniques, based on the one-dimensional searches discussed in Sections 10.5 to 10.11, for finding a root of the equation $f(x) = 0$.

10.11 A rotating cam has a contour such that the displacement d (in.) of its follower is given by

$$d = \frac{1}{2}(1 + e^{-\theta/2\pi}\sin\theta),$$

where θ $(0 \leqslant \theta \leqslant 2\pi)$ designates the angular position of the cam.

Write a program, based on Problem 10.10, whose input consists of values for d, θ_L and θ_R (lower and upper limits of the starting interval for θ), and α (maximum allowable final interval of uncertainty, radians). The program should then compute and print a value of θ (if such exists) for each value of d. *Suggested test data* are:

d	θ_L	θ_R	α
0.5	2.0	4.0	0.0001
0.9	2.0	4.0	0.0001
0.7	2.0	4.0	0.0001
0.5	0	1.0	0.0001

10.12 Consider the first-order irreversible chemical reaction $A \rightarrow B$, for which the rate constant is k hr^{-1}. A stirred batch reactor is charged at time $t=0$ with V cu ft of reactant solution of concentration a_0 lb moles/cu ft. If the reaction is isothermal, the concentration a of A varies with time t hr according to the differential equation $da/dt = -ka$, so that

$$a = a_0 e^{-kt}.$$

The reaction proceeds for t_r hr and the reactor is then emptied, cleaned, and recharged with fresh reactant, ready for another reaction cycle. This cyclic process is continued indefinitely. If the down time between reaction cycles is t_c hr, show that the yield per unit time (which is the function to be maximized) is given by:

$$y(t_r) = \frac{Va_0(1-e^{-kt_r})}{t_r + t_c}.$$

(a) Use a one-dimensional searching procedure, such as the function GOLDEN developed in Problem 10.9, to find the optimal reaction time t_r for the following situations:

V (cu ft):	10.0	20.0	10.0	10.0	10.0
a_0 (lb moles/cu ft):	0.1	0.1	0.2	0.1	0.1
k (hr^{-1}):	1.0	1.0	1.0	0.5	1.0
t_c (hr):	0.5	0.5	0.5	0.5	1.0

(Note that the problem of maximizing $y(t_r)$ is equivalent to that of minimizing $-y(t_r)$.)

(b) Show that the optimal value of t_r is also the solution of the equation

$$t_r - \ln\,(t_r k + t_c k + 1)/k = 0.$$

Solve this equation by one of the methods discussed in Chapter 5, and compare the results with those found in (a) above.

10.13 Discuss the relative efficiency of solving Problem 10.12 by the methods suggested in (a) and (b) of that problem. One approach is to determine the number of addition, subtraction, multiplication, and division operations required to achieve solutions of comparable accuracy by the two methods.

10.14 After studying Problem 10.10, rephrase Problem 5.28, involving an orbiting satellite, as a minimization problem. Then solve it, using a one-dimensional searching procedure, such as the function GOLDEN developed in Problem 10.9.

10.15 Write a computer program that will indicate how long a car should be kept before being traded. Account as realistically as possible for the various factors, such as mechanical reliability, appearance, etc., that tend to vary with time and hence govern (according to the particular owner's set of values) the net worth of the car at any moment.

10.16 *Introduction.* Consider a small, constant-thrust rocket, of almost constant mass, that is launched vertically upwards from the ground with zero initial velocity ($x=0$ and $v=0$ at $t=0$). Taking into account the drag of the air, it can be shown that the burnout height x_b, the burnout velocity v_b, the peak height x_p (at which the upwards velocity is zero) and coast time t_c (the time required to coast upwards from x_b to x_p) are given by:

$$x_b = \psi \ln [\cosh (t_b\sqrt{n_0 g/\psi})],$$

$$v_b = \sqrt{\psi n_0 g} \tanh (t_b\sqrt{n_0 g/\psi}),$$

$$x_p = x_b + \frac{\psi}{2} \ln \left(1 + \frac{v_b^2}{g\psi}\right),$$

$$t_c = \sqrt{\psi/g} \tan^{-1} (v_b/\sqrt{g\psi}),$$

$$t_p = t_b + t_c.$$

In these equations,

$$n_0 = \frac{T}{W} - 1,$$

$$\psi = \frac{2m}{C_D\rho A},$$

t_b = rocket motor burn time, sec,
T = rocket thrust, newtons,
W = mg = rocket weight, newtons,
m = rocket mass, Kg,
g = gravitational acceleration, 9.81 m/sec^2,
A = frontal area of rocket, = $\pi D^2/4$, m^2,
D = rocket diameter, m,
ρ = air density, 1.225 Kg/m^3 under typical conditions,
t_p = total time to reach x_p from ground, sec,
C_D = drag coefficient, a dimensionless empirical constant.

Problem Statement

Write a program that will accept values for T, t_b, D, and C_D as data, and that will proceed to find the mass m of the rocket that maximizes the peak height x_p. The corresponding values of x_b, v_b, t_c, and t_p should also be computed and printed. A one-dimensional searching procedure, such as the function EXTREM given in Example 10.1, or the function GOLDEN developed in Problem 10.9, should be used.

Suggested Test Data

$D = 0.019$ m, $T = 5$ newtons, with $t_b =$ 1 and 5 sec; $D = 0.05$ m, $T = 50$ newtons, with $t_b = 1$ and 5 sec; $C_D = 0.75$ throughout.

10.17 Consider the n mutually orthogonal vectors $\mathbf{u}_1, \mathbf{u}_2, \ldots, \mathbf{u}_n$ in n-dimensional space. Given a vector $\mathbf{v}_1$, show that a new set of orthogonal vectors $\mathbf{v}_1, \mathbf{v}_2, \ldots, \mathbf{v}_n$ can be generated as follows:

$$\left.\begin{aligned} \mathbf{v}_i &= \mathbf{u}_i + \sum_{j=1}^{i-1} \beta_j \mathbf{v}_j, \\ \text{where} & \\ \beta_j &= -\frac{\mathbf{v}_j^t \mathbf{u}_i}{\mathbf{v}_j^t \mathbf{v}_j}, \end{aligned}\right\} \quad i = 2, 3, \ldots, n.$$

If $\mathbf{u}_1 = [1\,0\,0]^t$, $\mathbf{u}_2 = [0\,1\,0]^t$, $\mathbf{u}_3 = [0\,0\,1]^t$, and $\mathbf{v}_1 = [2\,1\,0]^t$, generate $\mathbf{v}_2$ and $\mathbf{v}_3$ from the above relations; check that $\mathbf{v}_1$, $\mathbf{v}_2$, and $\mathbf{v}_3$ are mutually orthogonal.

10.18 Write a general-purpose function that finds the minimum value of F, a unimodal function of N variables X(1), X(2), . . . , X(N), that is already available with calling sequence F(X). Employ one of the sequential searching techniques discussed in Section 10.15.

For example, suppose that the direct search is to be used, and that the name of the function is DIRECT. A typical call on the function would be

```
FMIN = DIRECT (N, F, X, DELTA,
                        ALPHA, PRINT)
```

The starting coordinates for the search should be stored in the vector X prior to the first call on DIRECT. DELTA(1), . . . , DELTA(N) are the initial step sizes in X(1), . . . , X(N) to be used in performing the local exploratory searches, and ALPHA(1), . . . , ALPHA(N) are the final step sizes to be used in these local searches. PRINT is a logical variable which, when true, indicates that intermediate base points in the search are to be printed. The minimum value of F(X) should be returned as the value of the function DIRECT. The coordinates of the extreme point should be in the X vector on exit.

For applications, see Problems 10.19 and 10.20.

10.19 A long pipeline of diameter D in. is to be used for transporting Q gpm of oil of density ρ lb/cu ft over horizontal terrain. Compressor stations are to be uniformly spaced every L miles along the pipeline; the oil pressure falls from Δp psig at the exit from each station to approximately zero at the inlet to the next station. It may be shown that

$$\Delta p = 1.14\frac{f_M \rho L Q^2}{D^5},$$

in which f_M is the Moody friction factor, and that H, the compressor hp requirement at each station, is given by

$$H = 5.83 \times 10^{-4} Q\Delta p/\eta,$$

where η is the efficiency of pumping. Assume here that D is not restricted to certain standard sizes.

By performing a simplified economic analysis, in which the installed costs C_p and C_c (for the pipeline and compressor, respectively) are amortized over a five-year period, the total annual cost of operation, C_t, is given by

$$C_t = C_f + (C_p + C_c)/5,$$

where C_f represents the fuel and maintenance costs for running the compressors. It is also convenient to normalize C_t to an annual dollar cost per mile of pipeline. The following formulas may be used for computing C_f, C_p, and C_c.

Installed cost of pipeline (\$/ft):	$0.286D^{1.35}(1 + 0.0014\Delta p)$
Installed cost of one compressor (\$):	$75{,}000 + 850\,H^{0.76}$
Fuel and maintenance cost per compressor (\$/hr):	$0.0129\,H + 6.1.$

Note that if f_M, ρ, Q, and η are known, then a choice of the two design variables D and L will serve to determine Δp, H, and hence C_t.

Write a function named COST that returns as its value the total annual cost C_t as described above. Anticipate that a typical reference to the function might be of the form

```
CTOTAL = COST(X)
```

where X(1) and X(2) correspond to the two design variables D and L, respectively. Note that a COMMON declaration will be needed for transmitting other essential information to the function (see, for example, the function CNDNSR in Example 10.1).

Then write a main program that handles input and output and calls on a multidimensional optimization procedure, such as the function DIRECT discussed in Problem 10.18, to find those values of D and L that minimize C_t. After each round of the search, print the current values of C_t, D, and L, so that the path taken from various starting guesses to the final optimum point can be plotted. When the optimum point has been located, also print the corresponding values for the other intermediate variables: Δp, H, C_f, C_p, and C_c.

Suggested Data

$f_M = 0.031$, $\rho = 51.4$ lb/cu ft, $\eta = 0.78$, with several values for Q in the range 100 to 10,000 gpm.

10.20 *Introduction.* The stirred-tank reactor shown in Fig. P10.20 contains liquid at a temperature T_1 °F; it is to be cooled by continuously circulating V gpm of liquid from it through the shell side of shell-and-tube heat exchanger, and returning it to the reactor at T_2 °F. Cooling water enters the tube side of the exchanger at T_3 °F and leaves at T_4 °F. The density ρ, viscosity μ, and the specific heat c_p of the shell-side liquid are given by the following:

$\rho = 55.0 - 0.0187(T - 100)$	lb/cu ft,
$\mu = 1800e^{-0.025T}$	cp,
$c_p = 0.46 + 0.0005(T - 100)$	BTU/lb °F,

in which T is the mean temperature $(T_1 + T_2)/2$.

The shell side heat-transfer coefficient, h BTU/hr sq ft °F, is given by

$$h = 0.041\frac{G^{0.65}}{\mu^{0.27}},$$

where G lb/hr sq ft is the superficial mass velocity across the tubes. (A knowledge of V and G will assist in the actual design of the exchanger.) The overall heat-transfer

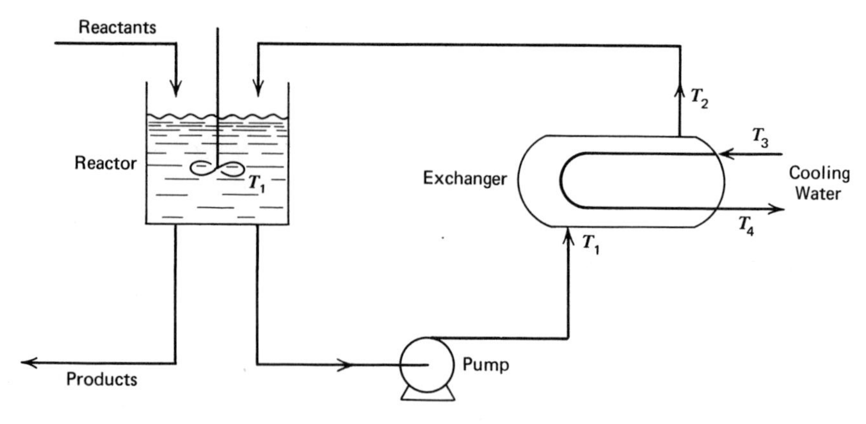

Figure P10.20.

coefficient, U BTU/hr sq ft °F, is to be computed from

$$\frac{1}{U}=\frac{1}{h}+R,$$

where R hr sq ft °F/BTU is the sum of the resistances on the tube side and that due to scale deposit. If Q BTU/hr are removed, the required heat-transfer area A sq ft can be computed from

$$Q=AU\Delta T_m,$$

where ΔT_m is the logarithmic mean temperature difference, corrected for the presence of multiple tube and/or shell passes. Using an exchanger with one shell and two tube passes, ΔT_m is given in terms of the differences $\Delta T_1 = T_1 - T_4$ and $\Delta T_2 = T_2 - T_3$ as:

$$\Delta T_m = \frac{\delta}{\ln\left(\dfrac{\Delta T_1+\Delta T_2+\delta}{\Delta T_1+\Delta T_2-\delta}\right)},$$

where

$$\delta=\sqrt{(T_1-T_2)^2+(T_4-T_3)^2}.$$

The total pressure drop Δp_t in the cooling system can be taken as six times the pressure drop, Δp psi, across one tube bank:

$$\Delta p = 3.33\times 10^{-9}\, fG^2/\rho.$$

Defining a Reynolds number for the shell-side fluid as $Re = 0.019G/\mu$ (with G and μ in the same units as above), the friction factor is given by $f = 190/Re$ for $Re < 340$ and by $f = 1.8/Re^{0.2}$ for $Re \geqslant 340$ (the crossover value is *approximately* 340, and should be computed exactly to avoid creating an artificial discontinuity in the cost function). For a specified pump and motor efficiency η, the rate of consumption of electrical energy, P kw, is given by

$$P = 4.36\times 10^{-4} V\Delta p_t/\eta,$$

from which the corresponding annual consumption of electricity, W kwh, can easily be found.

The installed costs C_h and C_p of the heat exchanger and pump, respectively, are given in dollars by

$$C_h = 220A^{0.596},$$

$$C_p = 264V^{0.582}.$$

If electricity costs C_e \$/kwh, then a very simplified economic analysis gives the following expression for C_t, the total annual dollar cost of installing and operating the proposed system:

$$C_t = WC_e + (C_h + C_p)/5.$$

Problem Statement

Write a function named COST that returns as its value the total annual cost C_t as described above. Anticipate that a typical reference to the function might be of the form

```
CTOTAL = COST (X)
```

where X(1) and X(2) correspond to the two design variables T_2 and G, respectively. Note that a COMMON declaration will be needed for transmitting other essential information to the function (see, for example, the function CNDNSR in Example 10.1).

Write a main program that handles input and output and calls on a multidimensional optimization procedure, such as the function DIRECT discussed in Problem 10.18, to find the minimum of C_t. After each round of the search, print the current values of C_t, T_2, and G, so that the path taken from various starting guesses to the final optimum can be plotted. When the optimum point has been located, also print the corresponding values for the other intermediate variables: ρ, μ, c_p, h, U, A, ΔT_m, V, Δp, Re, f, W, C_h, C_p, and C_t.

Suggested Test Data

$C_e = 0.02$ \$/kwh, $Q = 1.84 \times 10^6$ BTU/hr, $R = 0.004$ hr sq ft °F/BTU, $T_1 = 190$°F, $T_3 = 100$°F, $T_4 = 125$°F, and $\eta = 0.85$, with $T_2 = 125$°F and $G = 20{,}000$ lb/hr sq ft as possible starting guesses.

10.21 Consider the spaceship discussed in Problem 5.41. Use a one-dimensional searching procedure, such as the function EXTREM given in Example 10.1, or the function GOLDEN developed in Problem 10.9, to determine the optimum exhaust velocity that maximizes the payload ratio.

APPENDIX A
Presentation of Computer Examples

The following is an explanation of the various subdivisions appearing in the documentation of each computer example:

1. Problem Statement. Contains a statement of the problem to be solved, and may refer to the main text if appropriate. Occasionally preceded by an introductory section.

2. Method of Solution. Refers to algorithms already developed in the text whenever possible. Also gives notes on any special problems that may occur, and how they will be overcome.

3. Flow Diagram. We have avoided a flow diagram that merely repeats the program, step by step. Rather, the flow diagram is intended to clarify the computational procedure by reproducing only those steps that are essential to the algorithm. Thus, in the final program, there may be several additional steps that are *not* given in the flow diagram. Such steps frequently include: (a) printout of data that have just been read, often for checking purposes, (b) manipulations, such as the floating of an integer when it is to be added to a floating-point number, that may be caused by the limitations of FORTRAN, (c) calculation of intermediate constants used only as a computational convenience, (d) return to the beginning of the program to process another set of data, (e) printing headings and comments, (f) arranging for printout at definite intervals, during an iterative type of calculation, by setting up a counter and checking to see if it is an integer multiple of the specified frequency of printout, etc. The flow-diagram convention is summarized in Appendix B.

4. FORTRAN Implementation. This section serves as the main bridge between the algorithm of the text or flow diagram and the final FORTRAN program. It contains a list of the main variables used in the program, together with their definitions and, where applicable, their algebraic equivalents. These variable names, etc., are listed alphabetically under a heading describing the program in which they first appear, typically in the order: main program, first subroutine or function, second subroutine or function, etc. Variables in a subroutine or function whose meanings are identical with those already listed for an earlier program (the main program or another subroutine or function) are not repeated. All symbolic names that are part of the FORTRAN language —in particular, names of variables, functions, and subroutines—are written in SANSERIF CAPITALS, closely matching the characters used in the program listings and computer outputs.

In addition, we include here any particular programming points such as: (a) descriptions of special subroutines, (b) shifting of subscripts by one in order to accommodate, for example, a_0 in the algebra as A(1) (and not as the illegal A(0)) in the program, etc.

All programs have been run on an IBM 360/67 computer using the IBM level G compiler in conjunction with the Michigan Terminal System (MTS).

5. Program Listing. Contains a printed list of the original FORTRAN-IV program.

6. Data. Contains a printed list of the data cards, if any.

7. Computer Output. In this section, the printed computer output is reproduced. However, if the complete output is rather lengthy, only certain selected portions may be given. In this case, the remaining output is usually summarized somehow. Sometimes, clarifying notes, lines, etc., are added by hand to the printout. Occasionally, the printed output may be cut and reassembled if its clarity is thereby enhanced.

8. Discussion of Results. Contains a short critical discussion of the computed results, including comments on any particular programming difficulties and how they might be overcome.

APPENDIX B
Flow-Diagram Convention

The following boxes, each of characteristic shape, are used for constructing the flow diagrams in this text.

Substitution

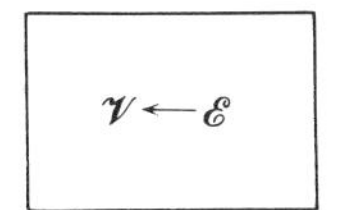

The value of the variable $\mathscr{V}$ is replaced by the value of the expression $\mathscr{E}$.

Label

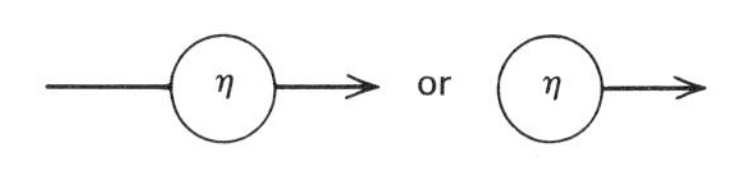

1. When a path leaves from it, the label serves merely as an identification point in the flow diagram; η is usually the same as the statement number in the corresponding program.

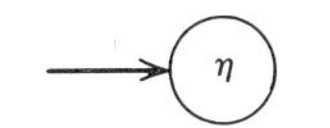

2. When a particular branch *terminates* in a label, control is transferred to the one (and only one) other point in the flow diagram where η occurs as in (1).

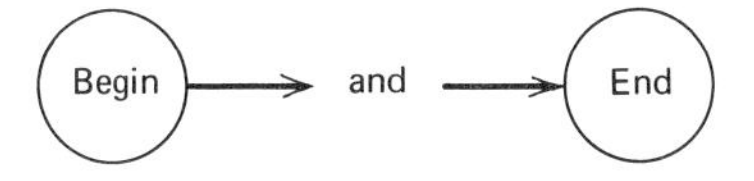

3. These special cases encompass the computations in a main program.

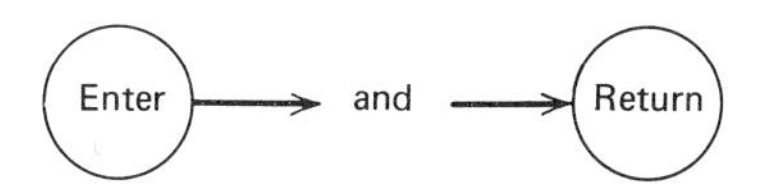

4. Indicate the start and finish of computations in a subroutine subprogram.

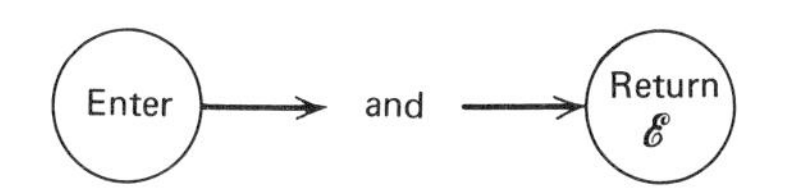

5. Indicate the start and finish of computations in a function subprogram. The value of the expression $\mathscr{E}$ is returned by the function.

Conditional Branching

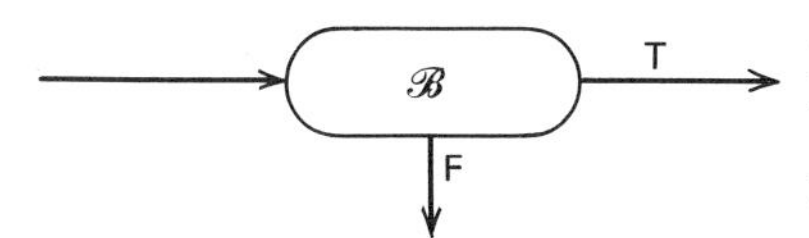

If the logical or Boolean expression $\mathscr{B}$ is true, the branch marked T is followed; otherwise, the branch marked F (false) is followed.

Iteration

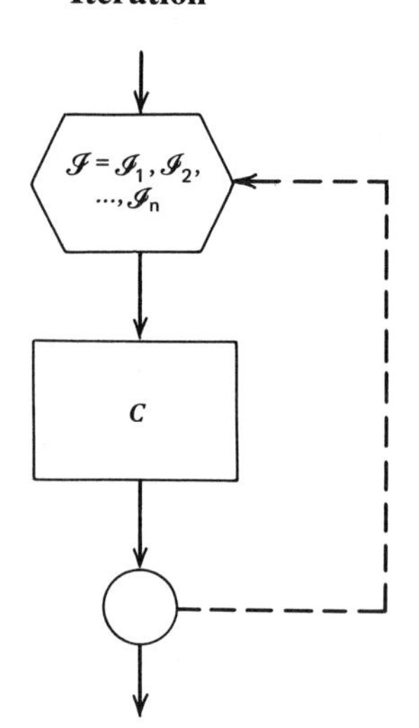

The counter or index $\mathcal{I}$ is incremented uniformly (in steps of $\mathcal{I}_2 - \mathcal{I}_1$) between its initial and final limits $\mathcal{I}_1$ and $\mathcal{I}_n$, respectively. For each such value of $\mathcal{I}$, the sequence of computations C is performed. The broken line emphasizes the return for the next value of $\mathcal{I}$, and the small circle, which is usually inscribed with the statement number of the final statement in the iteration loop in the program, delineates the scope of the iteration.

Input and Output

The values for the variables comprising the list $\mathscr{L}$ are accepted or read from cards.

The values for the variables or expressions comprising the list $\mathscr{L}$ are displayed or printed. A message to be displayed is enclosed in quotation marks.

Occasionally, there are special cases (such as presetting vectors and defining one-line functions) that do not fit conveniently into the above arrangement. However, the intended meaning is readily conveyed by suitable modification of the flow diagram. The flow diagram for a subroutine or function is separate from that for the calling flow diagram, and is started and concluded with the words "Enter" and "Return," as explained above. The arguments for subroutines and functions are also indicated.

APPENDIX C

A Summary of FORTRAN, WATFOR, and WATFIV Statements

All of the FORTRAN, WATFOR, and WATFIV statement types introduced in Chapter 3 are listed below. A prototype for each statement is shown using the following general script references.

Script Reference	*Meaning*
$\mathscr{A}$	Dummy argument for a subprogram.
$\mathscr{A}$†	Calling argument for a subprogram.
$\mathscr{B}$	Logical (Boolean) expression.
c	Arithmetic or logical constant.
$'c'$	Literal (character) constant.
$\mathscr{C}$	Character expression.
$\mathscr{E}$	Arithmetic (real or integer) expression.
$\mathscr{F}$	Subprogram (function or subroutine) name.
$\mathscr{I}$	Integer constant or unsubscripted integer variable.
$\mathscr{E}_{\mathscr{I}}$	Integer expression.
$\mathscr{L}$	A list, usually of variable names.
m	Input/output device number.
$\mathscr{N}$	NAMELIST name or labeled COMMON block name.
$\mathscr{S}$	Executable statement.
$\mathscr{T}$	Type or mode.
$\mathscr{V}$	Arithmetic (real or integer) variable.
$\mathscr{V}$	Logical (Boolean) variable.
$\mathscr{V}_{\mathscr{C}}$	Character variable.
$\mathscr{V}_{\mathscr{I}}$	Integer variable.
η	Statement number or label; an array name in some circumstances.

Appendix C A Summary of FORTRAN, WATFOR, and WATFIV Statements

Section	Page	Executable Statement Type	Prototype	Comments
3.14	74	Arithmetic assignment	$\mathscr{V} = \mathscr{E}$	
			$\mathscr{V}_1 = \mathscr{V}_2 = \mathscr{V}_3 = \cdots = \mathscr{V}_n = \mathscr{E}$	WATFIV only
3.14	75	Logical assignment	$\mathscr{V}_{\mathscr{B}} = \mathscr{B}$	
			$\mathscr{V}_{\mathscr{B}_1} = \mathscr{V}_{\mathscr{B}_2} = \mathscr{V}_{\mathscr{B}_3} = \cdots = \mathscr{V}_{\mathscr{B}_n} = \mathscr{B}$	WATFIV only
3.14	75	Character assignment	$\mathscr{V}_{\mathscr{C}} = \mathscr{C}$	WATFIV only
3.15	76	Unconditional transfer	GO TO η	
3.15	76	Multiple transfer	GO TO $(\eta_1, \eta_2, \ldots, \eta_n), \mathscr{I}$	
3.16	76	Logical conditional	IF $(\mathscr{B})\mathscr{S}$	
3.16	79	Arithmetical conditional	IF $(\mathscr{E})\ \eta_1, \eta_2, \eta_3$	
3.17	80	Iteration	DO $\eta\quad \mathscr{V}_{\mathscr{I}} = \mathscr{I}_1, \mathscr{I}_2, \mathscr{I}_3$	
3.18	82	Continue	CONTINUE	
3.19	83	Formatted input	READ $(m, \eta)\ \mathscr{L}$	
	83		READ $(m, \eta_1,$ END $= \eta_2,$ ERR $= \eta_3)\ \mathscr{L}$	
	87		READ $\eta, \mathscr{L}$	
3.19	84	Formatted output	WRITE $(m, \eta)\ \mathscr{L}$	
	87		PRINT $\eta, \mathscr{L}$	
	87		PUNCH $\eta, \mathscr{L}$	
3.21	100	Format-free input	READ, $\mathscr{L}$	WATFOR and WATFIV only
	102		READ $(m, *)\ \mathscr{L}$	WATFIV only
	102		READ $(m, *,$ END $= \eta_2,$ ERR $= \eta_3)\ \mathscr{L}$	WATFIV only
	103		READ $(m, \mathscr{N})$	FORTRAN and WATFIV only
	103		READ $(m, \mathscr{N},$ END $= \eta_2,$ ERR $= \eta_3)$	FORTRAN and WATFIV only
3.21	102	Format-free output	PRINT, $\mathscr{L}$	WATFOR and WATFIV only
	103		WRITE $(m, *)\ \mathscr{L}$	WATFIV only
	103		WRITE $(m, \mathscr{N})$	FORTRAN and WATFIV only
	103		PUNCH, $\mathscr{L}$	WATFOR and WATFIV only
3.22	107	Stop	STOP	
	107		STOP c	

3.28	121	Core-to-core input		READ ($\mathscr{V}_{\mathscr{C}}$, η) $\mathscr{L}$	WATFIV only
3.28	117	Core-to-core output		WRITE ($\mathscr{V}_{\mathscr{C}}$, η) $\mathscr{L}$	WATFIV only
3.33	141	Call for subroutine execution		CALL $\mathscr{F}(\mathscr{A}_1, \mathscr{A}_2^{\dagger}, \ldots, \mathscr{A}_n)$	
		Nonexecutable Statement Type		Prototype	Comments
3.20	88	Format	η	FORMAT (format specification)	
3.21	103	Namelist		NAMELIST $/\mathscr{N}/$ $\mathscr{L}$	
3.23	108	End of program		END	
3.24	108	Comment	C	message to appear in program listing	
3.25	108	Dimension (array reservation)		DIMENSION $\mathscr{V}$ $(\mathscr{I}_1, \mathscr{I}_2, \ldots, \mathscr{I}_n)$	
3.26	109	Variable initialization		DATA $\mathscr{L}$ $/c_1, c_2, \ldots, c_n/$	
3.27	110	Explicit type		CHARACTER $\mathscr{L}$	WATFIV only
	110			INTEGER $\mathscr{L}$	
	110			LOGICAL $\mathscr{L}$	
	110			REAL $\mathscr{L}$	
	111			DOUBLE PRECISION $\mathscr{L}$	
3.27	112	Implicit type		IMPLICIT $\mathscr{T}*n$ ($\mathscr{L}$)	
3.29	122	Equivalence (alternative naming)		EQUIVALENCE $(\mathscr{V}_1, \mathscr{V}_2, \ldots, \mathscr{V}_n)$	
3.31	136	External (subprogram names as subprogram arguments)		EXTERNAL $\mathscr{L}$	
3.34	142	Common (shared variables)		COMMON $\mathscr{L}$	
	143			COMMON $/\mathscr{N}/$ $\mathscr{L}$	

Appendix C A Summary of FORTRAN, WATFOR and WATFIV Statements (continued)

Section	Page	Function Definition	Prototype	Comments
3.30	124	Heading	$[\mathscr{T}]$ FUNCTION $\mathscr{F}(\mathscr{A}_1, \mathscr{A}_2, \ldots, \mathscr{A}_n)$	
3.30	125	Assign value	$\mathscr{F} = \mathscr{E}$	
	125		$\mathscr{F} = \mathscr{B}$	
	125		$\mathscr{F} = \mathscr{C}$	WATFIV only
3.30	126	Return to calling program	RETURN	
3.30	127	End of function	END	
3.30	133	Additional entry	ENTRY $\mathscr{F}(\mathscr{A}_1, \mathscr{A}_2, \ldots, \mathscr{A}_n)$	
		Subroutine Definition		
3.33	138	Heading	SUBROUTINE $\mathscr{F}(\mathscr{A}_1, \mathscr{A}_2, \ldots, \mathscr{A}_n)$	
3.33	138	Return to calling program	RETURN	
3.33	138	End of subroutine	END	
3.30	133	Additional entry	ENTRY $\mathscr{F}(\mathscr{A}_1, \mathscr{A}_2, \ldots, \mathscr{A}_n)$	
		Statement Function		
3.32	137	Definition	$\mathscr{F}(\mathscr{A}_1, \mathscr{A}_2, \ldots, \mathscr{A}_n) = \mathscr{E}$	
		BLOCK DATA Subprogram		
3.35	144	Heading	BLOCK DATA	
3.35	144	End of subprogram	END	

APPENDIX D

FORTRAN Diagnostic Messages

Compiler Diagnostics. The FORTRAN compiler most often used on the larger IBM 360 series machines, designated level-G FORTRAN by IBM, normally issues its error messages immediately following the offending statement, as described in Section 3.39. Some errors cannot be detected until the entire source program has been examined, however, in which case the error comment will appear at the end of the program. Each error message has the form

n) IEYdddI 'MESSAGE'

A dollar sign ($) is printed under (or close to) the apparent source of the error; if more than one error is detected, more than one dollar sign may appear. Here, n is an integer indicating which dollar sign corresponds to the error comment; the error for the first dollar sign (from left to right in the statement) will have an n of 1, the second an n of 2, etc.

The three-digit integer ddd is a code in the range 001–040 corresponding to the codes in the table shown below, and 'MESSAGE' is a terse message to indicate the nature of the error. A more detailed description of the error follows the entries in the table.

Error Code (ddd)	'MESSAGE' and Explanation of Probable Error.
001	'ILLEGAL TYPE' An expression or assignment statement includes operands of arithmetic and logical type in violation of the syntax, or an argument for one of the FORTRAN functions of Section 3.6 is of improper type (mode).
002	'LABEL' A GO TO statement is followed by an unlabeled statement; this must be a logical error, since the statement following the GO TO cannot be reached.
003	'NAME LENGTH' The name of a variable, labeled COMMON block, namelist, or subprogram has more than six characters; this may indicate that two operands appear without an operator between them.
004	'COMMA' A required comma has been omitted.
005	'ILLEGAL LABEL' Usually indicates that a declaration with a label (for example, a FORMAT declaration) has been referenced in a GO TO, IF, or DO statement.
006	'DUPLICATE LABEL' The same statement label (number) has been used for more than one statement.
007	'ID CONFLICT' A variable or subprogram name appears to have more than one type, or the same variable appears more than once in the dummy argument list of a subprogram.
008	'ALLOCATION' There is some inconsistency in the allocation of storage to variables; for example, the assignment of the same variable to more than one COMMON block.

Error Code (ddd)	'MESSAGE' and explanation of probable error.
009	'ORDER' Statements in the program are not in the required order. This usually means that the declarations appearing at the beginning of the program are in improper sequence; for example, an IMPLICIT statement in a main subprogram is not the first statement.
010	'SIZE' Some constant used in the program is outside the permitted range, or specified lengths for variables are illegal; for example, an attempt to assign six bytes to a real variable.
011	'UNDIMENSIONED' A variable with attached subscripts has not appeared in a DIMENSION declaration.
012	'SUBSCRIPT' The number of subscripts attached to a subscripted variable name does not correspond to the number used in a DIMENSION declaration.
013	'SYNTAX' This is a catchall message that occurs frequently (unfortunately); the FORTRAN translator has been unable to detect the nature of the error. The flagged statement cannot be translated because of some syntactical error.
014	'CONVERT' In a DATA declaration, an assigned constant and its corresponding variable are of different types. This usually indicates that a real variable has been assigned an integer value, or that a double-precision variable has been assigned a single-precision value.
015	'NO END CARD' END declaration is missing.
016	'ILLEGAL STA.' The statement appears in an illegal context; for example, a DO statement in a logical IF statement or an ENTRY statement outside a FUNCTION or SUBROUTINE subprogram.
017	'ILLEGAL STA. WRN.' A RETURN statement appears outside a FUNCTION or SUBROUTINE subprogram.
018	'NUMBER ARG.' A reference to a library subprogram (see Section 3.6) contains an incorrect number of arguments.
019	'FUNCTION ENTRIES UNDEFINED' Usually indicates that the assignment statement $\mathscr{F} = \mathscr{E}$ or $\mathscr{F} = \mathscr{B}$ is missing from a FUNCTION subprogram (see Section 3.30).
020	'COMMON BLOCK name ERRORS' This message appears when variables involved in COMMON and EQUIVALENCE declarations (see Sections 3.29 and 3.34) have been inconsistently specified; 'name' is the name of the COMMON block containing the error.
021	'UNCLOSED DO LOOPS' The statement numbers referenced in one or more DO statements do not appear in the program. The comment, and the associated missing statement labels are printed at the *end* of the program listing. This comment will

Error Code (ddd)	'MESSAGE' and explanation of probable error.
	appear when the terminal statement for a DO loop *is* present, but contains some other error. In particular, note the restrictions on the nature of the terminal statement in a loop described in Section 3.17.
022	'UNDEFINED LABELS' One or more statement numbers have been referenced (in GO TO statements, for example) but do not appear in the program. The message and the missing statement labels are printed at the *end* of the program listing.
023	'EQUIVALENCE ALLOCATION ERRORS' There is a conflict between the variable storage assignments in EQUIVALENCE declarations.
024	'EQUIVALENCE DEFINITION ERRORS' A subscript used with a variable in an EQUIVALENCE declaration is outside the range specified for that variable in a DIMENSION declaration.
025	'DUMMY DIMENSION ERRORS' The variable dimension feature has been used in a subprogram (see Section 3.30), but a variable used to dimension a dummy array is neither a dummy variable nor a COMMON variable.
026	'BLOCK DATA PROGRAM ERRORS' An attempt to preassign values to a variable that does not appear in a labeled COMMON block.
027	'CONTINUATION CARDS DELETED' More than 19 continuation cards have been used in a statement.
032	'NULL PROGRAM' The compiler was unable to find the FORTRAN source program; this usually indicates that the job control cards or FORTRAN program have been arranged improperly in the job deck.
033	'COMMENTS DELETED' More than 30 comment lines appear between two FORTRAN statements. The translator discontinues the listing of all subsequent comment lines until another FORTRAN statement is encountered. Translation of the program will be unaffected.
036	'ILLEGAL LABEL WRN.' A label has been attached to a declaration; this is considered to be an error only if there is an attempt to execute the declaration (see IEY005I). If the END declaration has been labeled (see Sections 3.22 and 3.33) this message will appear.
037	'PREVIOUSLY DIMENSIONED WRN.' An array has been dimensioned more than once. This is not considered to be an error; the dimensions used in the first appearance of the variable are used.
038	'SIZE WRN.' The constant assigned to a variable in a DATA declaration is of improper size, for example, assigning a five-character literal constant to a four-byte variable.
040	'COMMON ERROR IN BLOCK DATA' Usually indicates that the BLOCK DATA subprogram contains references to unlabeled COMMON variables (see Section 3.35).

Other compiler diagnostics with error codes 028, 029, 030, 031, 032, and 035 may be produced when the job is being processed by the standard IBM 360 operating system called OS/360. These errors will normally be generated only when the job control cards contain improper operating system commands; they do not indicate that there is an error in the FORTRAN source program.

Execution Error Messages

A variety of messages can occur as the result of error conditions encountered in execution of an object program produced by the IBM G-Level FORTRAN compiler. All execution error messages have the form

IHCdddI 'MESSAGE'

where ddd is a three-digit error code and 'MESSAGE' is a terse comment (optional in some systems). Most of the error messages deal with input and output or with arguments for standard FORTRAN functions (see Section 3.6) that are outside the allowable number range, and are self-explanatory.

Error Code (ddd)	'MESSAGE'
1	PAUSE
2	STOP
210	PROGRAM INTERRUPT
211	INVALID CHAR IN FORMAT STMT
212	ATTEMPT TO READ/WRITE A TOO LARGE RECORD UNDER FORMAT
213	ATTEMPT TO READ/WRITE A TOO LARGE RECORD NON-FORMATTED
214	ATTEMPT TO WRITE MORE THAN 254 PARTIAL RECORDS IN 1 LOGICAL REC.
215	INVALID CHAR FOR I,E,F,D INPUT FORMAT CODE.
216	ILLEGAL SENSE LIGHT NR IN CALL TO SLITE/SLITET.
217	END OF DATA SENSED.
218	DEVICE ERROR.
220	ILLEGAL DATA SET REFERENCE.
221	NAMEL--INPUT VARIABLE NAME TOO LONG.
222	NAMEL--INPUT VARIABLE NAME NOT IN DICTIONARY.
223	NAMEL--INPUT VARIABLE NAME OR SUBSCRIPT HAS NO DELIMITER.
224	SUBSCRIPT ENCOUNTERED AFTER UNDIMENSIONED INPUT NAME.
225	ILLEGAL CHAR ON INPUT FOR Z-FORMAT CODE.
230	SOURCE ERROR
231	MIXING D-A & SEQUENTIAL STMTS.
232	REC. NR. IN D-A I/O STMT IS <= 0 OR > MAX (SET IN DEFINE FILE).
233	RECORD LENGTH IN DEFINE FILE STMT IS > 32,768 BYTES.
234	DATA SET FOR PRINTING ERROR MESSAGES CAN'T BE A D-A DATA SET.
235	DATA SET ASSIGNED TO D-A SET IS USED FOR SEQUENTIAL DATA SET.
236	READ STMT EXECUTED FOR A DATA SET THAT HAS NOT BEEN CREATED.
241	EXPONENTIATION ERROR - I**J -- I=0 & J<=0 I,J INTEGER.
242	EXPONENTIATION ERROR -- R**J -- R=0 & J<=0 R REAL, J INTEGER.
243	EXPONENTIATION ERROR -- D**J -- D=0 & J<=0 D REAL*8, J INTEGER.
244	EXPONENTIATION ERROR -- R**S -- R=0 & S<=0 R,S REAL.
245	EXPONENTIATION ERROR -- D**P -- D=0 & P<=0 D,P REAL*8.
246	EXPONENTIATION ERROR -- Z**J -- Z=(0,0) & J<=0 Z COMPLEX, J INTEGER.
247	EXPONENTIATION ERROR -- Z**J -- Z=(0,0) & J<=0 Z CMPLX*16, J INT.
251	SQRT ERROR -- ARG <0.
252	EXP ERROR -- ARG > 174.673.
253	ALOG/ALOG10 ERROR -- ARG <= 0.
254	COS/SIN ERROR -- ARG >= .82355E+06
255	ATAN2 ERROR -- BOTH ARGS ZERO.
256	SINH/COSH ERROR -- ARG > 174.673
257	ARSIN/ARCOS ERROR. ARG > 1.
258	TAN/COTAN ERROR. ARG >= .82355E06
259	TAN/COTAN ERROR. ARG TOO CLOSE TO ONE OF SINGULARITIES.
261	DSQRT ERROR -- ARG <0.
262	DEXP ERROR -- ARG > 174.673
263	DLOG/DLOG10 ERROR -- ARG <= 0.
264	DSIN/DCOS ERROR -- ARG >= .35371D16
265	DATAN2 ERROR -- BOTH ARGS ARE ZERO.
266	DSINH/DCOSH ERROR -- ARG >= 174.673
267	DARSIN/DARCOS ERROR -- ARG > 1
268	DTAN/DCOTAN ERROR -- ARG >= .35371E16
269	DTAN/DCOTAN ERROR -- ARG IS TOO CLOSE TO ONE OF SINGULARITIES.
271	CEXP ERROR -- REAL(ARG) > 174.673
272	CEXP ERROR -- IMAG(ARG) >= .82355E06
273	CLOG ERROR -- ARG = (0,0)
274	CSIN/CCOS ERROR -- REAL(ARG) >= .82355E06
281	CDEXP ERROR -- REAL(ARG) > 174.673
282	CDEXP ERROR -- IMAG(ARG) >= .35371E16

Error Code (ddd)	'MESSAGE'
283	CDLOG ERROR -- ARG = (0,0)
284	CDSIN/CDCOS ERROR -- REAL(ARG) >= .35371E16
285	CDSIN/CDCOS ERROR -- IMAG(ARG) > 174.673
290	GAMMA ERROR -- ARG < 2**(-252) OR ARG > 57.5744
291	ALGAMA ERROR -- ARG <= 0 OR ARG > 4.2937E73
300	DGAMMA ERROR -- ARG < 2**(-252) OR ARG > 57.5744
301	DLGAMA ERROR -- ARG <= 0 OR ARG > 4.2937E73
900	ATTEMPT TO OBTAIN SPACE FOR D-A BUFFER FAILED.

The execution error message with code number 210 can refer to any of several causes for interruption of execution. The program interrupt is signalled by the message

IHC2101 PROGRAM INTERRUPT—OLD PSW IS XXXXXXXdXXAAAAAA

The particular condition causing the interrupt is indicated by the code digit or letter d. PSW is a mnemonic for *program status word*. The field AAAAAA normally indicates the address of the machine instruction immediately following the instruction that generated the interrupt condition.

Nonarithmetic interrupts are usually caused by improper alignment of words in the IBM 360 memory (too complex to discuss here) or references to addresses outside the storage area assigned to the object program (for example, having a subscript for a subscripted variable that is outside the dimensioned range for the variable).

Some of the nonarithmetic interrupts that are likely to occur, and possible sources of the interrupts are:

Error Code d	Interrupt type and possible explanation
1	Illegal operation code. This usually means that a data word is being interpreted as an instruction, and is often caused by subscripts outside the range indicated in a DIMENSION statement.
2	Privileged operation exception. Certain instructions in the machine's repertoire are available only to system programs (see Chapter 4). Usually indicates that a data word is being interpreted as an instruction.
4	Protection exception. Certain parts of the machine's memory will be protected against reading, writing, or execution by your program (for example your program must not have access to other user's programs or system programs). Often caused by an attempt to store information outside the reserved boundaries for subscripted variables.
5	Addressing exception. Attempt to use an address outside those physically available in the machine's memory.
6	Specification exception. Usually caused by an improper length specification for an argument to a function or subroutine, for example, using a double-precision argument when a single-precision dummy argument has been specified.

Arithmetic program interrupts are caused by attempted generation of numbers outside the number range for the machine, for example, attempted division by zero. Some of the arithmetic interrupts that are likely to occur with FORTRAN object programs are:

Error Code d	Interrupt type and explanation
8	Integer overflow exception. When high-order significant bits in an integer cannot be retained in the storage word, they are ignored and lost. Processing usually continues, but subsequent results will normally be in error.
9	Integer divide exception. Occurs when a quotient resulting from an integer division would exceed length of the storage word. Usually indicates attempted division by an integer zero. Division is suppressed and computation usually resumes.
C	Exponent overflow exception. Occurs when the exponent of a floating-point (real) result exceeds the maximum size (greater than 63 as an exponent of 16). The result is replaced by a zero and computation usually continues.
D	Exponent underflow exception. Occurs when the exponent of a floating-point (real) result becomes smaller than the minimum size (less than −64 as an exponent of 16). The result is replaced by a zero and computation usually continues.
F	Floating-point divide exception. Attempted division by a floating-point zero. The division is suppressed and computation usually continues. The result is the same as division by 1.

The response of the operating system to these arithmetic program interrupts varies from installation to installation. In many cases, execution of the object program will be stopped permanently after the first interrupt; in others, a fixed number of interrupts will be allowed before permanent suspension of execution.

It should be noted that the error messages shown in this section are for the level-G IBM FORTRAN compiler only. A different set of messages is produced by the level-H IBM FORTRAN compiler, and by FORTRAN compilers for other manufacturers' equipment. Each computing installation will normally make available to its users a listing of diagnostics produced by the compiler in use at that installation.

APPENDIX E
WATFOR Diagnostic Messages

As discussed in Section 3.39, the WATFOR compiler produces a comprehensive list of error and warning messages during both the compilation and execution phases of processing. The following is a complete list of these messages ordered alphabetically by the error code.

```
'ASSIGN STATEMENTS AND VARIABLES'
AS-2    'ATTEMPT TO REDEFINE AN ASSIGNED VARIABLE IN AN ARITHMETIC STATEMENT'
AS-3    'ASSIGNED VARIABLE USED IN AN ARITHMETIC EXPRESSION'
AS-4    'ASSIGNED VARIABLE CANNOT BE HALF WORD INTEGER'
AS-5    'ATTEMPT TO REDEFINE AN ASSIGN VARIABLE IN AN INPUT LIST'

'BLOCK DATA STATEMENTS'
BD-0    'EXECUTABLE STATEMENT IN BLOCK DATA SUBPROGRAMME'
BD-1    'IMPROPER BLOCK DATA STATEMENT'

'CARD FORMAT AND CONTENTS'
CC-0    'COLUMNS 1-5 OF CONTINUATION CARD NOT BLANK'
                PROBABLE CAUSE - STATEMENT PUNCHED TO LEFT OF COLUMN 7
CC-1    'TOO MANY CONTINUATION CARDS (MAXIMUM OF 7)'
CC-2    'INVALID CHARACTER IN FORTRAN STATEMENT '$' INSERTED IN SOURCE LISTING'
CC-3    'FIRST CARD OF A PROGRAMME IS A CONTINUATION CARD'
                PROBABLE CAUSE - STATEMENT PUNCHED TO LEFT OF COLUMN 7
CC-4    'STATEMENT TOO LONG TO COMPILE (SCAN-STACK OVERFLOW)'
CC-5    'BLANK CARD ENCOUNTERED'
CC-6    'KEYPUNCH USED DIFFERS FROM KEYPUNCH SPECIFIED ON JOB CARD'
CC-7    'FIRST CHARACTER OF STATEMENT NOT ALPHABETIC'
CC-8    'INVALID CHARACTER(S) CONCATENTATED WITH FORTRAN KEYWORD'
CC-9    'INVALID CHARACTERS IN COL 1-5. STATEMENT NUMBER IGNORED'
                PROBABLE CAUSE - STATEMENT PUNCHED TO LEFT OF COLUMN 7

'COMMON'
CM-0    'VARIABLE PREVIOUSLY PLACED IN COMMON'
CM-1    'NAME IN COMMON LIST PREVIOUSLY USED AS OTHER THAN VARIABLE'
CM-2    'SUBPROGRAMME PARAMETER APPEARS IN COMMON STATEMENT'
CM-3    'INITIALIZING OF COMMON SHOULD BE DONE IN A BLOCK DATA SUBPROGRAMME'
CM-4    'ILLEGAL USE OF BLOCK NAME'

'FORTRAN CONSTANTS'
CN-0    'MIXED REAL*4,REAL*8 IN COMPLEX CONSTANT'
CN-1    'INTEGER CONSTANT GREATER THAN 2,147,483,647 (2**31-1)'
CN-2    'EXPONENT OVERFLOW OR UNDERFLOW CONVERTING CONSTANT IN SOURCE STATEMENT'
CN-3    'EXPONENT ON REAL CONSTANT GREATER THAN 99'
CN-4    'REAL CONSTANT HAS MORE THAN 16 DIGITS, TRUNCATED TO 16'
CN-5    'INVALID HEXADECIMAL CONSTANT'
CN-6    'ILLEGAL USE OF DECIMAL POINT'
CN-8    'CONSTANT WITH E-TYPE EXPONENT HAS MORE THAN 7 DIGITS, ASSUME D-TYPE'
CN-9    'CONSTANT OR STATEMENT NUMBER GREATER THAN 99999'

'COMPILER ERRORS'
CP-0    'DETECTED IN PHASE RELOC'
CP-1    'DETECTED IN PHASE LINKR'
CP-2    'DUPLICATE PSEUDO STATEMENT NUMBERS'
CP-4    'DETECTED IN PHASE ARITH'
CP-5    'COMPILER INTERRUPT'

'DATA STATEMENT'
DA-0    'REPLICATION FACTOR GREATER THAN 32767, ASSUME 32767'
DA-1    'NON-CONSTANT IN DATA STATEMENT'
DA-2    'MORE VARIABLES THAN CONSTANTS IN DATA STATEMENT'
DA-3    'ATTEMPT TO INITIALIZE A SUBPROGRAMME PARAMETER IN A DATA STATEMENT'
DA-4    'NON-CONSTANT SUBSCRIPTS IN A DATA STATEMENT INVALID IN /360 FORTRAN'
DA-5    '/360 FORTRAN DOES NOT HAVE IMPLIED DO IN DATA STATEMENT'
DA-6    'NON-AGREEMENT BETWEEN TYPE OF VARIABLE AND CONSTANT IN DATA STATEMENT'
DA-7    'MORE CONSTANTS THAN VARIABLES IN DATA STATEMENT'
DA-8    'VARIABLE PREVIOUSLY INITIALIZED. LATEST VALUE USED'
                CHECK COMMON/EQUIVALENCED VARIABLES
DA-9    'INITIALIZING BLANK COMMON NOT ALLOWED IN /360 FORTRAN'
DA-A    'INVALID DELIMITER IN CONSTANT LIST PORTION OF DATA STATEMENT'
DA-B    'TRUNCATION OF LITERAL CONSTANT HAS OCCURRED'
```

WATFOR Error Messages (*continued*)

```
'DIMENSION STATEMENTS'
DM-0   'NO DIMENSIONS SPECIFIED FOR A VARIABLE IN A DIMENSION STATEMENT'
DM-1   'OPTIONAL LENGTH SPECIFICATION IN DIMENSION STATEMENT IS ILLEGAL'
DM-2   'INITIALIZATION IN DIMENSION STATEMENT IS ILLEGAL'
DM-3   'ATTEMPT TO RE-DIMENSION A VARIABLE'
DM-4   'ATTEMPT TO DIMENSION AN INITIALIZED VARIABLE'

'DO LOOPS'
DO-0   'ILLEGAL STATEMENT USED AS OBJECT OF DO'
DO-1   'ILLEGAL TRANSFER INTO THE RANGE OF A DO-LOOP'
DO-2   'OBJECT OF A DO STATEMENT HAS ALREADY APPEARED'
DO-3   'IMPROPERLY NESTED DO-LOOPS'
DO-4   'ATTEMPT TO REDEFINE A DO-LOOP PARAMETER WITHIN RANGE OF LOOP'
DO-5   'INVALID DO-LOOP PARAMETER'
DO-6   'TOO MANY NESTED DO'S (MAXIMUM OF 20)'
DO-7   'DO-PARAMETER IS UNDEFINED OR OUTSIDE RANGE'
DO-8   'THIS DO LOOP WILL TERMINATE AFTER FIRST TIME THROUGH'
DO-9   'ATTEMPT TO REDEFINE A DO-LOOP PARAMETER IN AN INPUT LIST'

'EQUIVALENCE AND/OR COMMON'
EC-0   'TWO EQUIVALENCED VARIABLES APPEAR IN COMMON'
EC-1   'COMMON BLOCK HAS DIFFERENT LENGTH THAN IN A PREVIOUS SUBPROGRAMME'
EC-2   'COMMON AND/OR EQUIVALENCE CAUSES INVALID ALIGNMENT. EXECUTION SLOWED'
           REMEDY - ORDER VARIABLES IN DESCENDING ORDER BY LENGTH
EC-3   'EQUIVALENCE EXTENDS COMMON DOWNWARDS'
EC-7   'COMMON/EQUIVALENCE STATEMENT DOES NOT PRECEDE PREVIOUS USE OF VARIABLE'
EC-8   'VARIABLE USED WITH NON-CONSTANT SUBSCRIPT IN COMMON/EQUIVALENCE LIST'
EC-9   'A NAME SUBSCRIPTED IN AN EQUIVALENCE STATEMENT WAS NOT DIMENSIONED'

'END STATEMENTS'
EN-0   'NO END STATEMENT IN PROGRAMME -- END STATEMENT GENERATED'
EN-1   'END STATEMENT USED AS STOP STATEMENT AT EXECUTION'
EN-2   'IMPROPER END STATEMENT'
EN-3   'FIRST STATEMENT OF SUBPROGRAMME IS END STATEMENT'

'EQUAL SIGNS'
EQ-6   'ILLEGAL QUANTITY ON LEFT OF EQUALS SIGN'
EQ-8   'ILLEGAL USE OF EQUAL SIGN'
EQ-A   'MULTIPLE ASSIGNMENT STATEMENTS NOT IN /360 FORTRAN'

'EQUIVALENCE STATEMENTS'
EV-0   'ATTEMPT TO EQUIVALENCE A VARIABLE TO ITSELF'
EV-1   'ATTEMPT TO EQUIVALENCE A SUBPROGRAMME PARAMETER'
EV-2   'LESS THAN 2 MEMBERS IN AN EQUIVALENCE LIST'
EV-3   'TOO MANY EQUIVALENCE LISTS (MAX = 255)'
EV-4   'PREVIOUSLY EQUIVALENCED VARIABLE RE-EQUIVALENCED INCORRECTLY'

'POWERS AND EXPONENTIATION'
EX-0   'ILLEGAL COMPLEX EXPONENTIATION'
EX-2   'I**J WHERE I=J=0'
EX-3   'I**J WHERE I=0, J.LT.0'
EX-6   '0.0**Y WHERE Y.LE.0.0'
EX-7   '0.0**J WHERE J=0'
EX-8   '0.0**J WHERE J.LT.0'
EX-9   'X**Y WHERE X.LT.0.0, Y.NE.0.0'

'ENTRY STATEMENT'
EY-0   'SUBPROGRAMME NAME IN ENTRY STATEMENT PREVIOUSLY DEFINED'
EY-1   'PREVIOUS DEFINITION OF FUNCTION NAME IN AN ENTRY IS INCORRECT'
EY-2   'USE OF SUBPROGRAMME PARAMETER INCONSISTENT WITH PREVIOUS ENTRY POINT'
EY-3   'ARGUMENT NAME HAS APPEARED IN AN EXECUTABLE STATEMENT'
       BUT WAS NOT A SUBPROGRAMME PARAMETER
EY-4   'ENTRY STATEMENT NOT PERMITTED IN MAIN PROGRAMME'
EY-5   'ENTRY POINT INVALID INSIDE A DO-LOOP'
EY-6   'VARIABLE WAS NOT PREVIOUSLY USED AS A PARAMETER - PARAMETER ASSUMED'

'FORMAT'
  SOME FORMAT ERROR MESSAGES GIVE CHARACTERS IN WHICH ERROR WAS DETECTED
FM-0   'INVALID CHARACTER IN INPUT DATA'
FM-2   'NO STATEMENT NUMBER ON A FORMAT STATEMENT'
FM-5   'FORMAT SPECIFICATION AND DATA TYPE DO NOT MATCH'
FM-6   'INCORRECT SEQUENCE OF CHARACTERS IN INPUT DATA'
FM-7   'NON-TERMINATING FORMAT'

FT-0   'FIRST CHARACTER OF VARIABLE FORMAT NOT A LEFT PARENTHESIS'
FT-1   'INVALID CHARACTER ENCOUNTERED IN FORMAT'
FT-2   'INVALID FORM FOLLOWING A SPECIFICATION'
FT-3   'INVALID FIELD OR GROUP COUNT'
FT-4   'A FIELD OR GROUP COUNT GREATER THAN 255'
FT-5   'NO CLOSING PARENTHESIS ON VARIABLE FORMAT'
FT-6   'NO CLOSING QUOTE IN A HOLLERITH FIELD'
FT-7   'INVALID USE OF COMMA'
FT-8   'INSUFFICIENT SPACE TO COMPILE A FORMAT STATEMENT (SCAN-STACK OVERFLOW)'
FT-9   'INVALID USE OF P SPECIFICATION'
```

WATFOR Error Messages (*continued*)

```
FT-A   'CHARACTER FOLLOWS CLOSING RIGHT PARENTHESIS'
FT-B   'INVALID USE OF PERIOD(.)'
FT-C   'MORE THAN THREE LEVELS OF PARENTHESES'
FT-D   'INVALID CHARACTER BEFORE A RIGHT PARENTHESIS'
FT-E   'MISSING OR ZERO LENGTH HOLLERITH ENCOUNTERED'
FT-F   'NO CLOSING RIGHT PARENTHESIS'

'FUNCTIONS AND SUBROUTINES'
FN-0   'NO ARGUMENTS IN A FUNCTION STATEMENT'
FN-3   'REPEATED ARGUMENT IN SUBPROGRAMME OR STATEMENT FUNCTION DEFINITION'
FN-4   'SUBSCRIPTS ON RIGHT HAND SIDE OF STATEMENT FUNCTION'
            PROBABLE CAUSE - VARIABLE TO LEFT OF = NOT DIMENSIONED
FN-5   'MULTIPLE RETURNS ARE INVALID IN FUNCTION SUBPROGRAMMES'
FN-6   'ILLEGAL LENGTH MODIFIER IN TYPE FUNCTION STATEMENT'
FN-7   'INVALID ARGUMENT IN ARITHMETIC OR LOGICAL STATEMENT FUNCTION'
FN-8   'ARGUMENT OF SUBPROGRAMME IS SAME AS SUBPROGRAMME NAME'

'GO TO STATEMENTS'
GO-0   'STATEMENT TRANSFERS TO ITSELF OR A NON-EXECUTABLE STATEMENT'
GO-1   'INVALID TRANSFER TO THIS STATEMENT'
GO-2   'INDEXED OF COMPUTED 'GOTO' IS NEGATIVE,ZERO OR UNDEFINED'
GO-3   'ERROR IN VARIABLE OF 'GO TO' STATEMENT'
GO-4   'INDEX OF ASSIGNED 'GO TO' IS UNDEFINED OR NOT IN RANGE'

'HOLLERITH CONSTANTS'
HO-0   'ZERO LENGTH SPECIFIED FOR H-TYPE HOLLERITH'
HO-1   'ZERO LENGTH QUOTE-TYPE HOLLERITH'
HO-2   'NO CLOSING QUOTE OR NEXT CARD NOT CONTINUATION CARD'
HO-3   'HOLLERITH CONSTANT SHOULD APPEAR ONLY IN CALL STATEMENT'
HO-4   'UNEXPECTED HOLLERITH OR STATEMENT NUMBER CONSTANT'

'IF STATEMENTS (ARITHMETIC AND LOGICAL)'
IF-0   'STATEMENT INVALID AFTER A LOGICAL IF'
IF-3   'ARITHMETIC OR INVALID EXPRESSION IN LOGICAL IF'
IF-4   'LOGICAL, COMPLEX, OR INVALID EXPRESSION IN ARITHMETIC IF'

'IMPLICIT STATEMENT'
IM-0   'INVALID MODE SPECIFIED IN AN IMPLICIT STATEMENT'
IM-1   'INVALID LENGTH SPECIFIED IN AN IMPLICIT OR TYPE STATEMENT'
IM-2   'ILLEGAL APPEARANCE OF $ IN A CHARACTER RANGE'
IM-3   'IMPROPER ALPHABETIC SEQUENCE IN CHARACTER RANGE'
IM-4   'SPECIFICATION MUST BE SINGLE ALPHABETIC CHARACTER, 1ST CHARACTER USED'
IM-5   'IMPLICIT STATEMENT DOES NOT PRECEDE OTHER SPECIFICATION STATEMENTS'
IM-6   'ATTEMPT TO ESTABLISH THE TYPE OF A CHARACTER MORE THAN ONCE'
IM-7   '/360 FORTRAN ALLOWS ONE IMPLICIT STATEMENT PER PROGRAMME'
IM-8   'INVALID ELEMENT IN IMPLICIT STATEMENT'
IM-9   'INVALID DELIMITER IN IMPLICIT STATEMENT'

'INPUT/OUTPUT'
IO-0   'MISSING COMMA IN I/O LIST OF I/O OR DATA STATEMENT'
IO-2   'STATEMENT NUMBER IN I/O STATEMENT NOT A FORMAT STATEMENT NUMBER'
IO-3   'FORMATTED LINE TOO LONG FOR I/O DEVICE (RECORD LENGTH EXCEEDED)'
IO-6   'VARIABLE FORMAT NOT AN ARRAY NAME'
IO-8   'INVALID ELEMENT IN INPUT LIST OR DATA LIST'
IO-9   'TYPE OF VARIABLE UNIT NOT INTEGER IN I/O STATEMENTS'
IO-A   'HALF-WORD INTEGER VARIABLE USED AS UNIT IN I/O STATEMENTS'
IO-B   'ASSIGNED INTEGER VARIABLE USED AS UNIT IN I/O STATEMENTS'
IO-C   'INVALID ELEMENT IN AN OUTPUT LIST'
IO-D   'MISSING OR INVALID UNIT IN I/O STATEMENT'
IO-E   'MISSING OR INVALID FORMAT IN READ/WRITE STATEMENT'
IO-F   'INVALID DELIMITER IN SPECIFICATION PART OF I/O STATEMENT'
IO-G   'MISSING STATEMENT NUMBER AFTER END= OR ERR='
IO-H   '/360 FORTRAN DOESN'T ALLOW END/ERR RETURNS IN WRITE STATEMENTS'
IO-J   'INVALID DELIMITER IN I/O LIST'
IO-K   'INVALID DELIMITER IN STOP, PAUSE, DATA, OR TAPE CONTROL STATEMENT'

'CONTROL CARDS'
JB-1   'JOB CARD ENCOUNTERED DURING COMPILATION'
JB-2   'INVALID OPTION(S) SPECIFIED ON JOB CARD'
JB-3   'UNEXPECTED CONTROL CARD ENCOUNTERED DURING COMPILATION'

'JOB TERMINATION'
KO-0   'JOB TERMINATED IN EXECUTION BECAUSE OF COMPILE TIME ERROR'
KO-1   'FIXED-POINT DIVISION BY ZERO'
KO-2   'FLOATING-POINT DIVISION BY ZERO'
KO-3   'TOO MANY EXPONENT OVERFLOWS'
KO-4   'TOO MANY EXPONENT UNDERFLOWS'
KO-5   'TOO MANY FIXED-POINT OVERFLOWS'
KO-6   'JOB TIME EXCEEDED'
KO-7   'COMPILER ERROR - INTERRUPTION AT EXECUTION TIME,RETURN TO SYSTEM'
KO-8   'INTEGER IN INPUT DATA IS TOO LARGE (MAXIMUM IS 2147483647)'

'LOGICAL OPERATIONS'
LG-2   '.NOT. USED AS A BINARY OPERATOR'
```

WATFOR Error Messages (*continued*)

```
'LIBRARY ROUTINES'
LI-0   'ARGUMENT OUT OF RANGE DGAMMA OR GAMMA. (1.382E-76 .LT. X .LT. 57.57)'
LI-1   'ABSOLUTE VALUE OF ARGUMENT .GT. 174.673, SINH,COSH,DSINH,DCOSH'
LI-2   'SENSE LIGHT OTHER THAN 0,1,2,3,4 FOR SLITE OR 1,2,3,4 FOR SLITET'
LI-3   'REAL PORTION OF ARGUMENT .GT. 174.673, CEXP OR CDEXP'
LI-4   'ABS(AIMAG(Z)) .GT. 174.673 FOR CSIN, CCOS, CDSIN OR CDCOS OF Z'
LI-5   'ABS(REAL(Z)) .GE. 3.537E15 FOR CSIN, CCOS, CDSIN OR CDCOS OF Z'
LI-6   'ABS(AIMAG(Z)) .GE. 3.537E15 FOR CEXP OR CDEXP OF Z'
LI-7   'ARGUMENT .GT. 174.673, EXP OR DEXP'
LI-8   'ARGUMENT IS ZERO, CLOG, CLOG10, CDLOG OR CDLG10'
LI-9   'ARGUMENT IS NEGATIVE OR ZERO, ALOG, ALOG10, DLOG OR DLOG10'
LI-A   'ABS(X) .GE. 3.537E15 FOR SIN, COS, DSIN OR DCOS OF X'
LI-B   'ABSOLUTE VALUE OF ARGUMENT .GT. 1, FOR ARSIN, ARCOS, DARSIN OR DARCOS'
LI-C   'ARGUMENT IS NEGATIVE, SQRT OR DSQRT'
LI-D   'BOTH ARGUMENTS OF DATAN2 OR ATAN2 ARE ZERO'
LI-E   'ARGUMENT TOO CLOSE TO A SINGULARITY, TAN, COTAN, DTAN OR DCOTAN'
LI-F   'ARGUMENT OUT OF RANGE DLGAMA OR ALGAMA. (0.0  .LT. X .LT. 4.29E73)'
LI-G   'ABSOLUTE VALUE OF ARGUMENT .GE. 3.537E15, TAN, COTAN, DTAN, DCOTAN'
LI-H   'FEWER THAN TWO ARGUMENTS FOR ONE OF MINO, MIN1, AMINO, ETC.'

'MIXED MODE'
MD-2   'RELATIONAL OPERATOR HAS A LOGICAL OPERAND'
MD-3   'RELATIONAL OPERATOR HAS A COMPLEX OPERAND'
MD-4   'MIXED MODE - LOGICAL WITH ARITHMETIC'
MD-6   'WARNING - SUBSCRIPT IS COMPLEX. REAL PART USED'

'MEMORY OVERFLOW'
MO-0   'SYMBOL TABLE OVERFLOWS OBJECT CODE. SOURCE ERROR CHECKING CONTINUES'
MO-1   'INSUFFICIENT MEMORY TO ASSIGN ARRAY STORAGE. JOB ABANDONED'
MO-2   'SYMBOL TABLE OVERFLOWS COMPILER, JOB ABANDONED'
MO-3   'DATA AREA OF SUBPROGRAMME TOO LARGE -- SEGMENT SUBPROGRAMME'
MO-4   'GETMAIN CANNOT PTOVIDE BUFFER FOR WATLIB'

'NAMELIST'
NA-0   'MULTIPLY DEFINED NAMELIST NAME'
NA-1   'NAMELIST NOT IMPLEMENTED'

'PARENTHESES'
PC-0   'UNMATCHED PARENTHESES'
PC-1   'INVALID PARENTHESIS COUNT'

'PAUSE, STOP STATEMENTS'
PS-0   'STOP WITH OPERATOR MESSAGE NOT ALLOWED. SIMPLE STOP ASSUMED'
PS-1   'PAUSE WITH OPERATOR MESSAGE NOT ALLOWED. TREATED AS CONTINUE'

'RETURN STATEMENT'
RE-0   'FIRST CARD OF SUBPROGRAMME IS A RETURN STATEMENT'
RE-1   'RETURN I, WHERE I IS ZERO,NEGATIVE OR TOO LARGE'
RE-2   'MULTIPLE RETURN NOT VALID IN FUNCTION SUBPROGRAMME'
RE-3   'VARIABLE IN MULTIPLE RETURN IS NOT A SIMPLE INTEGER VARIABLE'
RE-4   'MULTIPLE RETURN NOT VALID IN MAIN PROGRAMME'

'ARITHMETIC AND LOGICAL STATEMENT FUNCTIONS'
   PROBABLE CAUSE OF SF ERRORS - VARIABLE ON LEFT OF = WAS NOT DIMENSIONED
SF-1   'PREVIOUSLY REFERENCED STATEMENT NUMBER ON STATEMENT FUNCTION'
SF-2   'STATEMENT FUNCTION IS THE OBJECT OF A LOGICAL IF STATEMENT'
SF-3   'RECURSIVE STATEMENT FUNCTION, NAME APPEARS ON BOTH SIDES OF ='
SF-5   'ILLEGAL USE OF A STATEMENT FUNCTION'

'SUBPROGRAMMES'
SR-0   'MISSING SUBPROGRAMME'
SR-2   'SUBPROGRAMME ASSIGNED DIFFERENT MODES IN DIFFERENT PROGRAMME SEGMENTS'
SR-4   'INVALID TYPE OF ARGUMENT IN SUBPROGRAMME REFERENCE'
SR-5   'SUBPROGRAMME ATTEMPTS TO REDEFINE A CONSTANT,TEMPORARY OR DO PARAMETER'
SR-6   'ATTEMPT TO USE SUBPROGRAMME RECURSIVELY'
SR-7   'WRONG NUMBER OF ARGUMENTS IN SUBPROGRAMME REFERENCE'
SR-8   'SUBPROGRAMME NAME PREVIOUSLY DEFINED -- FIRST REFERENCE USED'
SR-9   'NO MAIN PROGRAMME'
SR-A   'ILLEGAL OR BLANK SUBPROGRAMME NAME'

'SUBSCRIPTS'
SS-0   'ZERO SUBSCRIPT OR DIMENSION NOT ALLOWED'
SS-1   'SUBSCRIPT OUT OF RANGE'
SS-2   'INVALID VARIABLE OR NAME USED FOR DIMENSION'

'STATEMENTS AND STATEMENT NUMBERS'
ST-0   'MISSING STATEMENT NUMBER'
ST-1   'STATEMENT NUMBER GREATER THAN 99999'
ST-3   'MULTIPLY-DEFINED STATEMENT NUMBER'
ST-4   'NO STATEMENT NUMBER ON STATEMENT FOLLOWING TRANSFER STATEMENT'
ST-5   'UNDECODEABLE STATEMENT'
ST-7   'STATEMENT NUMBER SPECIFIED IN A TRANSFER IS A NON-EXECUTABLE STATEMENT'
ST-8   'STATEMENT NUMBER CONSTANT MUST BE IN A CALL STATEMENT'
ST-9   'STATEMENT SPECIFIED IN A TRANSFER STATEMENT IS A FORMAT STATEMENT'
ST-A   'MISSING FORMAT STATEMENT'
```

WATFOR Error Messages (*continued*)

```
'SUBSCRIPTED VARIABLES'
SV-0   'WRONG NUMBER OF SUBSCRIPTS'
SV-1   'ARRAY NAME OR SUBPROGRAMME NAME USED INCORRECTLY WITHOUT LIST'
SV-2   'MORE THAN 7 DIMENSIONS NOT ALLOWED'
SV-3   'DIMENSION TOO LARGE'
SV-4   'VARIABLE WITH VARIABLE DIMENSIONS IS NOT A SUBPROGRAMME PARAMETER'
SV-5   'VARIABLE DIMENSION NEITHER SIMPLE INTEGER VARIABLE NOR S/P PARAMETER'

'SYNTAX ERRORS'
SX-0   'MISSING OPERATOR'
SX-1   'SYNTAX ERROR-SEARCHING FOR SYMBOL,NONE FOUND'
SX-2   'SYNTAX ERROR-SEARCHING FOR CONSTANT,NONE FOUND'
SX-3   'SYNTAX ERROR-SEARCHING FOR SYMBOL OR CONSTANT,NONE FOUND'
SX-4   'SYNTAX ERROR-SEARCHING FOR STATEMENT NUMBER,NONE FOUND'
SX-5   'SYNTAX ERROR-SEARCHING FOR SIMPLE INTEGER VARIABLE,NONE FOUND'
SX-C   'ILLEGAL SEQUENCE OF OPERATORS IN EXPRESSION'
SX-D   'MISSING OPERAND OR OPERATOR'

'I/O OPERATIONS'
UN-0   'CONTROL CARD ENCOUNTERED ON SCARDS DURING EXECUTION'
             PROBABLE CAUSE - MISSING DATA OR IMPROPER FORMAT STATEMENTS
UN-1   'END OF FILE ENCOUNTERED'
UN-2   'I/O ERROR'
UN-4   'REWIND, ENDFILE, BACKSPACE REFERENCES UNIT 5, 6, 7'
UN-5   'ATTEMPT TO READ ON UNIT 5 AFTER IT HAS HAD AN END-OF-FILE'
UN-6   'UNIT NUMBER IS NEGATIVE,ZERO,GREATER THAN 7 OR UNDEFINED'
UN-7   'TOO MANY PAGES'
UN-8   'ATTEMPT TO DO SEQUENTIAL I/O ON A DIRECT ACCESS FILE'
UN-9   'WRITE REFERENCES 5 OR READ REFERENCES 6 OR 7'
UN-C   'I/O ERROR ON WATLIB'

'UNDEFINED VARIABLES'
UV-0   'UNDEFINED VARIABLE - SIMPLE VARIABLE'
UV-1   'UNDEFINED VARIABLE - EQUIVALENCED, COMMONED, OR DUMMY PARAMETER'
UV-2   'UNDEFINED VARIABLE - ARRAY MEMBER'
UV-3   'UNDEFINED VARIABLE - ARRAY NAME WHICH WAS USED AS A DUMMY PARAMETER'
UV-4   'UNDEFINED VARIABLE - SUBPROGRAMME NAME USED AS DUMMY PARAMETER'
UV-5   'UNDEFINED VARIABLE - ARGUMENT OF THE LIBRARY SUBPROGRAMME NAMED'
UV-6   'VARIABLE FORMAT CONTAINS UNDEFINED CHARACTER(S)'

'VARIABLE NAMES'
VA-0   'ATTEMPT TO REDEFINE TYPE OF A VARIABLE NAME'
VA-1   'SUBROUTINE NAME OR COMMON BLOCK NAME USED INCORRECTLY'
VA-2   'NAME LONGER THAN SIX CHARACTERS. TRUNCATED TO SIX'
VA-3   'ATTEMPT TO REDEFINE THE MODE OF A VARIABLE NAME'
VA-4   'ATTEMPT TO REDEFINE THE TYPE OF A VARIABLE NAME'
VA-6   'ILLEGAL USE OF A SUBROUTINE NAME'
VA-8   'ATTEMPT TO USE A PREVIOUSLY DEFINED NAME AS FUNCTION OR ARRAY'
VA-9   'ATTEMPT TO USE A PREVIOUSLY DEFINED NAME AS A STATEMENT FUNCTION'
VA-A   'ATTEMPT TO USE A PREVIOUSLY DEFINED NAME AS A SUBPROGRAMME NAME'
VA-B   'NAME USED AS A COMMON BLOCK PREVIOUSLY USED AS A SUBPROGRAMME NAME'
VA-C   'NAME USED AS SUBPROGRAMME PREVIOUSLY USED AS A COMMON BLOCK NAME'
VA-D   'ILLEGAL DO-PARAMETER,ASSIGNED OR INITIALIZED VARIABLE IN SPECIFICATION'
VA-E   'ATTEMPT TO DIMENSION A CALL-BY-NAME PARAMETER'

'EXTERNAL STATEMENT'
XT-0   'INVALID ELEMENT IN EXTERNAL LIST'
XT-1   'INVALID DELIMITER IN EXTERNAL STATEMENT'
XT-2   'SUBPROGRAMME PREVIOUSLY EXTERNALLED'
```

APPENDIX F
WATFIV Diagnostic Messages

The following is a complete list of diagnostic error and warning messages produced by the WATFIV compiler during both the compilation and execution phases of processing, as discussed in Section 3.39.

```
'ASSEMBLER LANGUAGE SUBPROGRAMS'
AL-0   'MISSING END CARD ON ASSEMBLY LANGUAGE OBJECT DECK'
AL-1   'ENTRY-POINT OR CSECT NAME _ IN AN OBJECT DECK WAS PREVIOUSLY DEFINED.
       FIRST DEFINITION USED'

'BLOCK DATA STATEMENTS'
BD-0   'EXECUTABLE STATEMENTS ARE ILLEGAL IN BLOCK DATA SUBPROGRAMS'
BD-1   'IMPROPER BLOCK DATA STATEMENT. '

'CARD FORMAT AND CONTENTS'
CC-0   'COLUMNS 1-5 OF CONTINUATION CARD NOT BLANK.PROBABLE CAUSE:STATEMENT PUNCHED TO
       LEFT OF COLUMN 7'
CC-1   'LIMIT OF _ CONTINUATION CARDS EXCEEDED'
CC-2   'INVALID CHARACTER IN FORTRAN STATEMENT. A '_' WAS INSERTED IN THE SOURCE LISTING'
CC-3   'FIRST CARD OF A PROGRAM IS A CONTINUATION CARD.PROBABLE CAUSE:STATEMENT PUNCHED TO
       LEFT OF COLUMN 7'
CC-4   'STATEMENT TOO LONG TO COMPILE (SCAN-STACK OVERFLOW)'
CC-5   'BLANK CARD ENCOUNTERED'
CC-6   'KEYPUNCH USED DIFFERS FROM KEYPUNCH SPECIFIED ON JOB CARD'
CC-7   'FIRST CHARACTER OF THE STATEMENT WAS NOT ALPHABETIC'
CC-8   'INVALID CHARACTER(S) ARE CONCATENATED WITH THE FORTRAN KEYWORD. '
CC-9   'INVALID CHARACTERS IN COL 1-5. STATEMENT NUMBER IGNORED.PROBABLE CAUSE:STATEMENT
       PUNCHED TO LEFT OF COLUMN 7'

'COMMON'
CM-0   'VARIABLE _ IS ALREADY IN COMMON'
CM-1   'OTHER COMPILERS MAY NOT ALLOW COMMONED VARIABLE _ TO BE INITIALIZED IN OTHER THAN A
       BLOCK DATA SUBPROGRAM'
CM-2   'ILLEGAL USE OF THE COMMON BLOCK OR NAMELIST NAME _'

'FORTRAN TYPE CONSTANTS'
CN-0   'MIXED REAL*4,REAL*8 IN COMPLEX CONSTANT;REAL*8 ASSUMED FOR BOTH'
CN-1   'AN INTEGER CONSTANT MAY NOT BE GREATER THAN 2,147,483,647 (2**31-1)'
CN-2   'EXPONENT _ ON A REAL CONSTANT IS GREATER THAN 2 DIGITS'
CN-3   'REAL CONSTANT HAS MORE THAN 16 DIGITS. TRUNCATED TO 16'
CN-4   'INVALID HEXADECIMAL CONSTANT.'
CN-5   'ILLEGAL USE OF DECIMAL POINT.'
CN-6   'CONSTANT WITH MORE THAN 7 DIGITS BUT E-TYPE EXPONENT,ASSUMED TO BE REAL*4'
CN-7   'CONSTANT OR STATEMENT NUMBER GREATER THAN 99999. '
CN-8   'EXPONENT OVERFLOW OR UNDERFLOW WHILE CONVERTING A CONSTANT IN SOURCE STATEMENT'

'COMPILER ERRORS'
CP-0   'COMPILER ERROR - LANDR/ARITH'
CP-1   'COMPILER ERROR-LIKELY CAUSE:MORE THAN 255 DO STATEMENTS'
CP-4   'COMPILER ERROR - INTERRUPT AT COMPILE TIME,RETURN TO SYSTEM'

'CHARACTER VARIABLE'
CV-0   'A CHARACTER VARIABLE IS USED WITH A RELATIONAL OPERATOR'
CV-1   'LENGTH OF CHARACTER VALUE ON RIGHT OF EQUAL SIGN EXCEEDS THAT ON LEFT.
       TRUNCATION WILL OCCUR'
CV-2   'UNFORMATTED CORE TO CORE I/O NOT IMPLEMENTED'

'DATA STATEMENT'
DA-0   'REPLICATION FACTOR _ IS ZERO OR GREATER THAN 32767.IT IS ASSUMED TO BE 32767'
DA-1   'MORE VARIABLES THAN CONSTANTS _'
DA-2   'ATTEMPT TO INITIALIZE THE SUBPROGRAM PARAMETER _ IN A DATA STATEMENT'
DA-3   'OTHER COMPILERS MAY NOT ALLOW NON-CONSTANT SUBSCRIPTS IN DATA STATEMENTS'
DA-4   'TYPE OF VARIABLE AND CONSTANT DO NOT AGREE _.(MESSAGE ISSUED ONCE FOR AN ARRAY)'
DA-5   'MORE CONSTANTS THAN VARIABLES _'
DA-6   'A VARIABLE _ WAS PREVIOUSLY INITIALIZED. THE LATEST VALUE IS USED.
       CHECK COMMONED AND EQUIVALENCED VARIABLES'
DA-7   'OTHER COMPILERS MAY NOT ALLOW INITIALIZATION OF BLANK COMMON'
DA-8   'LITERAL CONSTANT _ HAS BEEN TRUNCATED'
DA-9   'OTHER COMPILERS MAY NOT ALLOW IMPLIED DO-LOOPS IN DATA STATEMENTS'

'DEFINE FILE STATEMENTS'
DF-0   'THE UNIT NUMBER IS MISSING.'
DF-1   'THE FORMAT IS INVALID'
```

WATFIV Error Messages (*continued*)

```
DF-2   'ASSOCIATED VARIABLE _ IS NOT A SIMPLE INTEGER VARIABLE'
DF-3   'THE NUMBER OF RECORDS OR RECORD SIZE IS INVALID'

'DIMENSION STATEMENTS'
DM-0   'NO DIMENSIONS WERE SPECIFIED FOR VARIABLE _'
DM-1   'VARIABLE _ HAS ALREADY BEEN DIMENSIONED'
DM-2   'CALL-BY-LOCATION PARAMETER _ CANNOT BE DIMENSIONED'
DM-3   'DECLARED SIZE OF ARRAY _ EXCEEDS SPACE PROVIDED BY CALLING ARGUMENT'

'DO LOOPS'
DO-0   'THIS STATEMENT CANNOT BE THE OBJECT OF A DO-LOOP'
DO-1   'ILLEGAL TRANSFER INTO THE RANGE OF A DO-LOOP FROM LINE _'
DO-2   'THE OBJECT OF THIS DO,STATEMENT _,HAS ALREADY APPEARED'
DO-3   'IMPROPERLY NESTED DO-LOOPS'
DO-4   'ATTEMPT TO REDEFINE THE DO-LOOP PARAMETER _ WITHIN THE RANGE OF THE LOOP'
DO-5   'DO-LOOP PARAMETER IS INVALID'
DO-6   'ILLEGAL TRANSFER TO STATEMENT NUMBER _ WHICH IS INSIDE THE RANGE OF A DO-LOOP'
DO-7   'A DO-LOOP PARAMETER IS UNDEFINED OR OUT OF RANGE._'
DO-8   'BECAUSE OF PARAMETER _,THIS DO-LOOP WILL TERMINATE AFTER THE FIRST TIME THROUGH'
DO-9   'A DO-LOOP PARAMETER MAY NOT BE REDEFINED IN AN INPUT LIST'
DO-A   'OTHER COMPILERS MAY NOT ALLOW THIS STATEMENT TO END A DO-LOOP'

'EQUIVALENCE AND/OR COMMON'
EC-0   '_ HAS BEEN EQUIVALENCED TO A VARIABLE IN COMMON'
EC-1   'COMMON BLOCK _ HAS A DIFFERENT LENGTH THAN WAS SPECIFIED IN A PREVIOUS SUBPROGRAM;
       GREATER LENGTH USED'
EC-2   'COMMON AND/OR EQUIVALENCE CAUSES INVALID ALIGNMENT; EXECUTION SLOWED.
       REMEDY:ORDER VARIABLES BY DECREASING LENGTH'
EC-3   'EQUIVALENCE EXTENDS COMMON DOWNWARDS'
EC-4   'SUBPROGRAM PARAMETER _ APPEARS IN A COMMON OR EQUIVALENCE STATEMENT'
EC-5   'VARIABLE _ WAS USED WITH SUBSCRIPTS IN AN EQUIVALENCE STATEMENT BUT HAS NOT BEEN
       PROPERLY DIMENSIONED'

'END STATEMENTS'
EN-0   'MISSING END STATEMENT;END STATEMENT GENERATED'
EN-1   'AN END STATEMENT WAS USED TO TERMINATE EXECUTION'
EN-2   'AN END STATEMENT CANNOT HAVE A STATEMENT NUMBER. STATEMENT NUMBER IGNORED.'
EN-3   'END STATEMENT NOT PRECEDED BY A TRANSFER'

'EQUAL SIGNS'
EQ-0   'ILLEGAL QUANTITY ON LEFT OF EQUALS SIGN'
EQ-1   'ILLEGAL USE OF EQUAL SIGN.'
EQ-2   'OTHER COMPILERS MAY NOT ALLOW MULTIPLE ASSIGNMENT STATEMENTS'
EQ-3   'MULTIPLE ASSIGNMENT IS NOT IMPLEMENTED FOR CHARACTER VARIABLES'

'EQUIVALENCE STATEMENTS'
EV-0   'ATTEMPT TO EQUIVALENCE VARIABLE _ TO ITSELF'
EV-1   'THERE ARE LESS THAN 2 VARIABLES IN AN EQUIVALENCE LIST'
EV-2   'MULTI-SUBSCRIPTED EQUIVALENCED VARIABLE _ HAS BEEN INCORRECTLY RE-EQUIVALENCED.
       REMEDY:DIMENSION IT FIRST'

'POWERS AND EXPONENTIATION'
EX-0   'ILLEGAL COMPLEX EXPONENTIATION'
EX-1   'I**J WHERE I=J=0'
EX-2   'I**J WHERE I=0, J .LT. 0'
EX-3   '0.0**Y WHERE Y .LE. 0.0'
EX-4   '0.0**J WHERE J=0'
EX-5   '0.0**J WHERE J .LT. 0'
EX-6   'X**Y WHERE X .LT. 0.0,Y .NE. 0.0'

'ENTRY STATEMENT'
EY-0   'ENTRY-POINT NAME _ WAS PREVIOUSLY DEFINED'
EY-1   'PREVIOUS DEFINITION OF FUNCTION NAME IN AN ENTRY IS INCORRECT'
EY-2   'USAGE OF SUBPROGRAM PARAMETER _ IS INCONSISTENT WITH A PREVIOUS ENTRY-POINT'
EY-3   'PARAMETER _ HAS APPEARED IN AN EXECUTABLE STATEMENT BUT WAS NOT A SUBPROGRAM PARAMETER'
EY-4   'ENTRY STATEMENTS INVALID IN MAIN PROGRAM'
EY-5   'ENTRY STATEMENTS INVALID WITHIN RANGE OF DO-LOOP'

'FORMAT'
FM-0   'IMPROPER CHARACTER SEQUENCE OR INVALID CHARACTER IN INPUT DATA'
FM-1   'NO STATEMENT NUMBER ON FORMAT STATEMENT'
FM-2   'FORMAT CODE AND DATA TYPE DO NOT MATCH'
FM-4   'FORMAT PROVIDES NO CONVERSION SPECIFICATION FOR A VALUE IN I/O LIST'
FM-5   'AN INTEGER IN THE INPUT DATA IS TOO LARGE(MAXIMUM = 2,147,483,647 = 2**31-1)'
FM-6   'A REAL NUMBER IN INPUT DATA IS OUT OF MACHINE RANGE (1.E-78,1.E+75)'
FM-7   'FORMAT STATEMENT _ IS UNREFERENCED'
FT-0   'FIRST CHARACTER OF VARIABLE FORMAT NOT A LEFT PARENTHESIS.  '
FT-1   'INVALID CHARACTER ENCOUNTERED IN FORMAT NEAR _'
FT-2   'INVALID FORM FOLLOWING A FORMAT CODE'
FT-3   'INVALID FIELD OR GROUP COUNT NEAR _'
FT-4   'A FIELD OR GROUP COUNT GREATER THAN 255.  '
FT-5   'NO CLOSING PARENTHESIS ON VARIABLE FORMAT'
FT-6   'NO CLOSING QUOTE IN A HOLLERITH FIELD'
FT-7   'INVALID USE OF COMMA.  '
FT-8   'FORMAT STATEMENT TOO LONG TO COMPILE (SCAN STACK OVERFLOW)'
FT-9   'INVALID USE OF P FORMAT CODE.    '
```

WATFIV Error Messages (*continued*)

```
FT-A    'INVALID USE OF PERIOD(.) '
FT-B    'MORE THAN THREE LEVELS OF PARENTHESES'
FT-C    'INVALID CHARACTER BEFORE A RIGHT PARENTHESIS. '
FT-D    'MISSING OR ZERO-LENGTH HOLLERITH ENCOUNTERED'
FT-E    'NO CLOSING RIGHT PARENTHESIS'
FT-F    'CHARACTERS FOLLOW CLOSING RIGHT PARENTHESIS'
FT-G    'WRONG QUOTE USED FOR KEY-PUNCH SPECIFIED'
FT-H    'LENGTH OF HOLLERITH EXCEEDS 255'

'FUNCTIONS AND SUBROUTINES'
FN-1    'PARAMETER _ APPEARS MORE THAN ONCE IN THIS SUBPROGRAM OR STATEMENT
        FUNCTION DEFINITION'
FN-2    'SUBSCRIPTS ON RIGHT-HAND SIDE OF STATEMENT FUNCTION.PROBABLE CAUSE:VARIABLE
        TO LEFT OF = NOT DIMENSIONED'
FN-3    'MULTIPLE RETURNS ARE INVALID IN FUNCTION SUBPROGRAMS'
FN-4    'ILLEGAL LENGTH MODIFIER.'
FN-5    'INVALID PARAMETER. '
FN-6    'PARAMETER _ HAS THE SAME NAME AS THE SUBPROGRAM'

'GO TO STATEMENTS'
GO-0    'THIS STATEMENT COULD TRANSFER TO ITSELF'
GO-1    'THIS STATEMENT TRANSFERS TO _,WHICH IS NON-EXECUTABLE'
GO-2    'ATTEMPT TO DEFINE ASSIGNED GOTO INDEX _ IN AN ARITHMETIC STATEMENT'
GO-3    'ASSIGNED GOTO INDEX _ MAY BE USED ONLY IN ASSIGNED GOTO AND ASSIGN STATEMENTS'
GO-4    'INDEX OF AN ASSIGNED GOTO IS UNDEFINED OR OUT OF RANGE,OR INDEX OF COMPUTED
        GOTO IS UNDEFINED'
GO-5    'ASSIGNED GOTO INDEX _ MAY NOT BE AN INTEGER*2 VARIABLE'

'HOLLERITH CONSTANTS'
HO-0    'ZERO LENGTH SPECIFIED FOR H-TYPE HOLLERITH'
HO-1    'ZERO LENGTH QUOTE-TYPE HOLLERITH'
HO-2    'NO CLOSING QUOTE OR NEXT CARD NOT CONTINUATION CARD'
HO-3    'UNEXPECTED HOLLERITH OR STATEMENT NUMBER CONSTANT'

'IF STATEMENTS (ARITHMETIC AND LOGICAL)'
IF-0    'AN INVALID STATEMENT FOLLOWS THE LOGICAL IF'
IF-1    'ARITHMETIC OR INVALID EXPRESSION IN LOGICAL IF'
IF-2    'LOGICAL, COMPLEX, OR INVALID EXPRESSION IN ARITHMETIC IF'

'IMPLICIT STATEMENT'
IM-0    'THE DATA TYPE IS INVALID'
IM-1    'THE OPTIONAL LENGTH IS INVALID'
IM-3    'IMPROPER ALPHABETIC SEQUENCE IN CHARACTER RANGE.'
IM-4    'A SPECIFICATION IS NOT A SINGLE CHARACTER. THE FIRST CHARACTER IS USED'
IM-5    'IMPLICIT STATEMENT DOES NOT PRECEDE OTHER SPECIFICATION STATEMENTS'
IM-6    'ATTEMPT TO DECLARE THE TYPE OF A CHARACTER MORE THAN ONCE'
IM-7    'ONLY ONE IMPLICIT STATEMENT PER PROGRAM SEGMENT ALLOWED;THIS ONE IGNORED'

'INPUT/OUTPUT'
IO-0    'I/O STATEMENT REFERENCES STATEMENT NUMBERED _ WHICH IS NOT A FORMAT STATEMENT'
IO-1    'VARIABLE FORMAT MUST BE AN ARRAY NAME. '
IO-2    'INVALID ELEMENT IN INPUT LIST OR DATA LIST'
IO-3    'OTHER COMPILERS MAY NOT ALLOW EXPRESSIONS IN OUTPUT LISTS'
IO-4    'ILLEGAL USE OF END= OR ERR= PARAMETERS'
IO-5    '_ IS AN INVALID UNIT NUMBER'
IO-6    '_ IS NOT A VALID FORMAT'
IO-7    'ONLY CONSTANTS,SIMPLE INTEGER*4 VARIABLES AND CHARACTER VARIABLES ARE ALLOWED AS UNI
IO-8    'ATTEMPT TO PERFORM I/O IN A FUNCTION WHICH IS CALLED IN AN OUTPUT STATEMENT'
IO-9    'UNFORMATTED WRITE STATEMENT MUST HAVE A LIST'

'JOB CONTROL CARDS'
JB-0    'CONTROL CARD ENCOUNTERED DURING COMPILATION;PROBABLE CAUSE:MISSING $ CARD'
JB-1    'MIS-PUNCHED JOB OPTION. '

'JOB TERMINATION'
KO-0    'SOURCE ERROR ENCOUNTERED WHILE EXECUTING WITH RUN=FREE'
KO-1    'LIMIT EXCEEDED FOR FIXED-POINT DIVISION BY ZERO'
KO-2    'LIMIT EXCEEDED FOR FLOATING-POINT DIVISION BY ZERO'
KO-3    'EXPONENT OVERFLOW LIMIT EXCEEDED'
KO-4    'EXPONENT UNDERFLOW LIMIT EXCEEDED'
KO-5    'FIXED-POINT OVERFLOW LIMIT EXCEEDED'
KO-6    'JOB-TIME EXCEEDED'
KO-7    'COMPILER ERROR - EXECUTION TIME;RETURN TO SYSTEM'
KO-8    'TRACEBACK ERROR.TRACEBACK TERMINATED'

'LOGICAL OPERATIONS'
LG-0    '.NOT. USED AS A BINARY OPERATOR'

'LIBRARY ROUTINES'
LI-0    'ARGUMENT OF DGAMMA OR GAMMA OUT OF THE RANGE 1.382E-76 .LT. X .LT. 57.57'
LI-1    'ABS(X) .GT. 174.673 FOR SINH,COSH,DSINH OR DCOSH OF X'
LI-2    'SENSE LIGHT OTHER THAN 0,1,2,3,4 FOR SLITE OR 1,2,3,4 FOR SLITET'
LI-3    'REAL(Z) .GT. 174.673 FOR CEXP OR CDEXP OF Z'
LI-4    'ABS(AIMAG(Z)) .GT. 174.673 FOR CSIN, CCOS, CDSIN OR CDCOS OF Z'
```

WATFIV Error Messages (*continued*)

```
LI-5   'ABS(REAL(Z)) .GE. 3.537E15 FOR CSIN, CCOS, CDSIN OR CDCOS OF Z'
LI-6   'ABS(AIMAG(Z)) .GE. 3.537E15 FOR CEXP OR CDEXP OF Z'
LI-7   'X .GT. 174.673 FOR EXP OR DEXP OF X'
LI-8   'ARGUMENT OF CLOG,CLOG10,CDLOG OR CDLOG10 IS ZERO'
LI-9   'ARGUMENT OF ALOG,ALOG10,DLOG OR DLOG10 IS NEGATIVE OR ZERO'
LI-A   'ABS(X) .GE. 3.537E15 FOR SIN, COS, DSIN OR DCOS OF X'
LI-B   'ABS(X) .GT. 1 FOR ARSIN,ARCOS,DARSIN OR DARCOS OF X'
LI-C   'X .LT. 0. FOR SQRT OR DSQRT OF X'
LI-D   'BOTH ARGUMENTS OF DATAN2 OR ATAN2 ARE ZERO'
LI-E   'ARGUMENT TOO CLOSE TO A SINGULARITY OF TAN,COTAN,DTAN OR DCOTAN'
LI-F   'ARGUMENT OF DLGAMA OR ALGAMA OUT OF THE RANGE 0.0 .LT. X .LT. 4.29E73'
LI-G   'ABS(X) .GE. 3.537E15 FOR TAN,COTAN,DTAN,DCOTAN OF X'
LI-H   'LESS THAN TWO ARGUMENTS FOR ONE OF MIN0,MIN1,AMIN0,ETC.'

'MIXED MODE'
MD-0   'RELATIONAL OPERATOR HAS LOGICAL OPERAND'
MD-1   'RELATIONAL OPERATOR HAS COMPLEX OPERAND'
MD-2   'MIXED MODE - LOGICAL OR CHARACTER WITH ARITHMETIC'
MD-3   'OTHER COMPILERS MAY NOT ALLOW SUBSCRIPTS OF TYPE COMPLEX,LOGICAL OR CHARACTER'

'MEMORY OVERFLOW'
MO-0   'INSUFFICIENT MEMORY TO COMPILE THIS PROGRAM. REMAINDER WILL BE ERROR CHECKED ONLY'
MO-1   'INSUFFICIENT MEMORY TO ASSIGN ARRAY STORAGE. JOB ABANDONED'
MO-2   'SYMBOL TABLE EXCEEDS AVAILABLE SPACE. JOB ABANDONED'
MO-3   'DATA AREA OF SUBPROGRAM EXCEEDS 24K;SEGMENT SUBPROGRAM'
MO-4   'INSUFFICIENT MEMORY TO ALLOCATE COMPILER WORK AREA OR WATLIB BUFFER'

'NAMELIST STATEMENTS'
NL-0   'NAMELIST ENTRY _ MUST BE A VARIABLE,BUT NOT A SUBPROGRAM PARAMETER'
NL-1   'NAMELIST NAME _ PREVIOUSLY DEFINED'
NL-2   'VARIABLE NAME _ TOO LONG'
NL-3   'VARIABLE NAME _ NOT FOUND IN NAMELIST'
NL-4   'INVALID SYNTAX IN NAMELIST INPUT.'
NL-6   'VARIABLE _ INCORRECTLY SUBSCRIPTED'
NL-7   'SUBSCRIPT OF _ OUT OF RANGE'

'PARENTHESES'
PC-0   'UNMATCHED PARENTHESIS'
PC-1   'INVALID PARENTHESIS NESTING IN I/O LIST'

'PAUSE, STOP STATEMENTS'
PS-0   'OPERATOR MESSAGES NOT ALLOWED;SIMPLE STOP ASSUMED FOR STOP,CONTINUE ASSUMED FOR PAUSE'

'RETURN STATEMENT'
RE-1   'RETURN I, WHERE I IS OUT OF RANGE OR UNDEFINED;VALUE IS _'
RE-2   'MULTIPLE RETURN NOT VALID IN FUNCTION SUBPROGRAM'
RE-3   '_ IS NOT A SIMPLE INTEGER VARIABLE'
RE-4   'MULTIPLE RETURN NOT VALID IN MAIN PROGRAM'

'ARITHMETIC AND LOGICAL STATEMENT FUNCTIONS'
SF-1   'PREVIOUSLY REFERENCED STATEMENT NUMBER _ APPEARS ON A STATEMENT FUNCTION DEFINITION'
SF-2   'STATEMENT FUNCTION IS THE OBJECT OF A LOGICAL IF STATEMENT'
SF-3   'RECURSIVE STATEMENT FUNCTION DEFINITION:NAME APPEARS ON BOTH SIDES OF EQUAL SIGN.
       LIKELY CAUSE:VARIABLE NOT DIMENSIONED'
SF-4   'A STATEMENT FUNCTION DEFINITION APPEARS AFTER THE FIRST EXECUTABLE STATEMENT'
SF-5   'ILLEGAL USE OF FUNCTION NAME _'

'SUBPROGRAMS'
SR-0   'SUBPROGRAM _ IS MISSING'
SR-1   'SUBPROGRAM _ REDEFINES A CONSTANT,EXPRESSION DO-PARAMETER OR ASSIGNED GOTO INDEX'
SR-2   'SUBPROGRAM _ WAS ASSIGNED DIFFERENT TYPES IN DIFFERENT PROGRAM SEGMENTS'
SR-3   'ATTEMPT TO USE SUBPROGRAM _ RECURSIVELY'
SR-4   'INVALID TYPE OF ARGUMENT IN REFERENCE TO SUBPROGRAM _'
SR-5   'WRONG NUMBER OF ARGUMENTS IN REFERENCE TO SUBPROGRAM _'
SR-6   'SUBPROGRAM _ WAS PREVIOUSLY DEFINED. FIRST DEFINITION IS USED'
SR-7   'NO MAIN PROGRAM'
SR-8   'ILLEGAL OR MISSING SUBPROGRAM NAME'
SR-9   'LIBRARY SUBPROGRAM _ WAS NOT ASSIGNED THE CORRECT TYPE'
SR-A   'METHOD OF ENTERING SUBPROGRAM PRODUCES UNDEFINED VALUE FOR CALL-BY-LOCATION PARAMETER'

'SUBSCRIPTS'
SS-0   'ZERO SUBSCRIPT OR DIMENSION NOT ALLOWED'
SS-1   'SUBSCRIPT NUMBER _'
SS-2   'INVALID SUBSCRIPT FORM'
SS-3   'A SUBSCRIPT OF _ WHICH IS OUT OF RANGE'

'STATEMENTS AND STATEMENT NUMBERS'
ST-0   'MISSING STATEMENT NUMBER _'
ST-1   'STATEMENT NUMBER GREATER THAN 99999.'
ST-2   'STATEMENT NUMBER _ HAS ALREADY BEEN DEFINED'
ST-3   'UNDECODEABLE STATEMENT'
ST-4   'UNNUMBERED EXECUTABLE STATEMENT FOLLOWS A TRANSFER'
ST-5   'STATEMENT NUMBER _ IN A TRANSFER IS A NON-EXECUTABLE STATEMENT'
ST-6   'ONLY CALL STATEMENTS MAY CONTAIN STATEMENT NUMBER ARGUMENTS'
```

WATFIV Error Messages (*continued*)

ST-7 'STATEMENT SPECIFIED IN A TRANSFER STATEMENT _ IS A FORMAT STATEMENT'
ST-8 'MISSING FORMAT STATEMENT _'
ST-9 'SPECIFICATION STATEMENT DOES NOT PRECEDE STATEMENT FUNCTION DEFINITIONS OR EXECUTABLE STATEMENTS
ST-A 'UNREFERENCED STATEMENT _ FOLLOWS A TRANSFER'

'SUBSCRIPTED VARIABLES'
SV-0 'WRONG NUMBER OF SUBSCRIPTS SPECIFIED FOR VARIABLE _'
SV-1 'ARRAY OR SUBPROGRAM NAME _ IS USED INCORRECTLY WITHOUT A LIST'
SV-2 'MORE THAN 7 DIMENSIONS NOT ALLOWED. '
SV-3 'DIMENSION OR SUBSCRIPT IS TOO LARGE (MAXIMUM 10**8-1). '
SV-4 'VARIABLE _,USED WITH VARIABLE DIMENSIONS,IS NOT A SUBPROGRAM PARAMETER'
SV-5 'VARIABLE DIMENSION _ IS NOT ONE OF SIMPLE INTEGER VARIABLE,SUBPROGRAM PARAMETER, IN COMMON'

'SYNTAX ERRORS'
SX-0 'MISSING OPERATOR.'
SX-1 'EXPECTING OPERATOR BUT _ WAS FOUND'
SX-2 'EXPECTING SYMBOL,BUT _ WAS FOUND'
SX-3 'EXPECTING SYMBOL OR OPERATOR,BUT _ WAS FOUND'
SX-4 'EXPECTING CONSTANT,BUT _ WAS FOUND'
SX-5 'EXPECTING SYMBOL OR CONSTANT,BUT _ WAS FOUND'
SX-6 'EXPECTING STATEMENT NUMBER,BUT _ WAS FOUND'
SX-7 'EXPECTING SIMPLE INTEGER VARIABLE,BUT _ WAS FOUND'
SX-8 'EXPECTING SIMPLE INTEGER VARIABLE OR CONSTANT,BUT _ WAS FOUND'
SX-9 'ILLEGAL SEQUENCE OF OPERATORS IN EXPRESSION'
SX-A 'EXPECTING END-OF-STATEMENT,BUT _ WAS FOUND'

'TYPE STATEMENTS'
TY-0 'VARIABLE _ HAS ALREADY BEEN EXPLICITLY TYPED'
TY-1 'LENGTH OF EQUIVALENCED VARIABLE _ MAY NOT BE CHANGED.REMEDY:INTERCHANGE TYPE AND EQUIVALENCE STATEMENTS'

'I/O OPERATIONS'
UN-0 'CONTROL CARD ENCOUNTERED ON UNIT _ AT EXECUTION. PROBABLE CAUSE:MISSING DATA OR INCORRECT FORMAT'
UN-1 'END OF FILE ENCOUNTERED ON UNIT _ (IBM CODE IHC217)'
UN-2 'I/O ERROR ON UNIT _ (IBM CODE IHC218)'
UN-3 'NO DD STATEMENT WAS SUPPLIED (IBM CODE IHC217)'
UN-3 'DEFINE FILE STATEMENT NOT SUPPORTED BY MTS'
UN-4 'REWIND,ENDFILE,BACKSPACE ON UNIT _ REFERENCES READER,PRINTER,OR PUNCH'
UN-5 'ATTEMPT TO READ ON UNIT _ AFTER IT HAS HAD END-OF-FILE'
UN-6 '_ IS NOT A VALID UNIT NUMBER (IBM CODE IHC220)'
UN-7 'PAGE-LIMIT EXCEEDED'
UN-8 'ATTEMPT TO DO DIRECT ACCESS I/O ON A SEQUENTIAL FILE OR VICE VERSA. POSSIBLE MISSING DEFINE FILE STATEMENT (IBM CODE IHC231)'
UN-9 'WRITE REFERENCES UNIT _ OR READ REFERENCES UNITS _ OR _ '
UN-A 'DEFINE FILE REFERENCES A UNIT PREVIOUSLY USED FOR SEQUENTIAL I/O (IBM CODE IHC235)'
UN-B 'RECORD SIZE FOR UNIT EXCEEDS _ OR DIFFERS FROM DD STATEMENT SPECIFICATION (IBM CODES IHC233,IHC237)'
UN-C 'FOR DIRECT ACCESS I/O THE RELATIVE RECORD POSITION IS NEGATIVE, ZERO, OR TOO LARGE (IBM CODE IHC232)'
UN-D 'ON UNIT _,ATTEMPT TO READ MORE INFORMATION THAN LOGICAL RECORD CONTAINS (IBM CODE IHC213)'
UN-E 'FORMATTED LINE EXCEEDS BUFFER LENGTH (IBM CODE IHC212)'
UN-F 'I/O ERROR - SEARCHING LIBRARY DIRECTORY'
UN-G 'I/O ERROR - READING LIBRARY'
UN-H 'ATTEMPT TO DEFINE THE OBJECT ERROR FILE AS A DIRECT ACCESS FILE (IBM CODE IHC234)'
UN-I 'RECFM OTHER THAN V(B) IS SPECIFIED FOR I/O WITHOUT FORMAT CONTROL (IBM CODE IHC214)'
UN-J 'MISSING DD CARD FOR WATLIB;NO LIBRARY ASSUMED'
UN-K 'ATTEMPT TO READ OR WRITE PAST END OF CHARACTER VARIABLE BUFFER'
UN-L 'ATTEMPT TO READ ON AN UNCREATED DIRECT ACCESS FILE (IHC236)'

'UNDEFINED VARIABLES'
UV-0 'VALUE OF _ IS UNDEFINED'
UV-3 'SUBSCRIPT NUMBER _ IS UNDEFINED'
UV-4 'SUBPROGRAM _ IS UNDEFINED'
UV-5 'AN ARGUMENT OF _ IS UNDEFINED'
UV-6 'UNDECODABLE CHARACTERS IN VARIABLE FORMAT'

'VARIABLE NAMES'
VA-0 'NAME _ IS TOO LONG;TRUNCATED TO SIX CHARACTERS'
VA-1 'ATTEMPT TO USE THE ASSIGNED OR INITIALIZED VARIABLE OR DO-PARAMETER _ IN A SPECIFICATION STATEMENT'
VA-2 'ILLEGAL USE OF SUBROUTINE NAME _'
VA-3 'ILLEGAL USE OF VARIABLE NAME _'
VA-4 'ATTEMPT TO USE PREVIOUSLY DEFINED NAME _ AS A FUNCTION OR AN ARRAY'
VA-5 'ATTEMPT TO USE PREVIOUSLY DEFINED NAME _ AS A SUBROUTINE'
VA-6 'ATTEMPT TO USE PREVIOUSLY DEFINED NAME _ AS A SUBPROGRAM'
VA-7 'ATTEMPT TO USE PREVIOUSLY DEFINED NAME _ AS A COMMON BLOCK'
VA-8 'ATTEMPT TO USE FUNCTION NAME _ AS A VARIABLE'
VA-9 'ATTEMPT TO USE PREVIOUSLY DEFINED NAME _ AS A VARIABLE'
VA-A 'ILLEGAL USE OF PREVIOUSLY DEFINED NAME _'

'EXTERNAL STATEMENT'
XT-0 ' _ HAS ALREADY APPEARED IN AN EXTERNAL STATEMENT'

Bibliography

1 B. Carnahan, H. A. Luther, and J. O. Wilkes, *Applied Numerical Methods*, Wiley, New York, 1969.

2 J. M. Kay, *An Introduction to Fluid Mechanics and Heat Transfer*, 2nd ed., Cambridge University Press, London, 1963.

3 L. M. Milne-Thomson, *Theoretical Hydrodynamics*, 4th ed., Macmillan, London, 1960.

4 D. L. Katz et al., *Handbook of Natural Gas Engineering*, McGraw-Hill, New York, 1959.

5 H. S. Carslaw and J. C. Jaeger, *Conduction of Heat in Solids*, 2nd ed., Oxford University Press, London, 1959.

6 W. M. Rohsenow and H. Y. Choi, *Heat, Mass, and Momentum Transfer*, Prentice-Hall, Englewood Cliffs, New Jersey, 1961.

7 S. Timoshenko, *Theory of Elastic Stability*, McGraw-Hill, New York, 1936.

8 J. H. Perry, ed., *Chemical Engineers' Handbook*, 3rd ed., McGraw-Hill, New York, 1950.

9 C. R. Wylie, *Advanced Engineering Mathematics*, McGraw-Hill, New York, 1951.

10 R. Keller, *Basic Tables in Chemistry*, McGraw-Hill, New York, 1967.

11 *Time Magazine*, **89** (12), 51 (1967).

12 *International Critical Tables*, Vol. II, p. 337, McGraw-Hill, New York, 1927.

13 C. D. Hodgman, ed., *Handbook of Chemistry and Physics*, 33rd ed., Chemical Rubber, Cleveland, Ohio.

14 R. M. Nedderman, *Velocity Profiles in Thin Liquid Layers*, Ph.D. Thesis, Dept. of Chemical Engineering, University of Cambridge, 1960.

15 F. A. Jenkins and H. E. White, *Fundamentals of Optics*, 2nd ed., McGraw-Hill, New York, 1951.

16. "Tables of Functions and Zeros of Functions," *Natl. Bur. Stand. Appl. Math. Series*, No. 37, Washington, D.C., 1954.

17 K. G. Denbigh, *The Principles of Chemical Equilibrium*, Cambridge University Press, London, 1957.

18 B. Carnahan, *Radiation-Induced Cracking of Pentanes and Dimethylbutanes*, Ph.D. Thesis, University of Michigan, 1964.

19 P. Henrici, *Discrete Variable Methods in Ordinary Differential Equations*, Wiley, New York, 1962.

20 M. R. Horne and W. Merchant, *The Stability of Frames*, Pergamon Press, New York, 1965.

21 V. L. Streeter, *Fluid Mechanics*, 4th ed., McGraw-Hill, New York, 1966.

22 W. T. Thomson, *Introduction to Space Dynamics*, Wiley, New York, 1961.

23 Merrill Lynch, Pierce, Fenner & Smith Inc., *Annual Report*, New York, 1969.

24 P. E. Gray and C. L. Searle, *Electronic Principles*, Wiley, New York, 1969.

25 R. J. Smith, *Circuits, Devices, and Systems*, Wiley, New York, 1966.

26 I. S. and E. S. Sokolnikoff, *Higher Mathematics for Engineers and Physicists*, 2nd ed., McGraw-Hill, New York, 1941.

27 A. Higdon, E. H. Ohlsen, W. B. Stiles, and J. A. Weese, *Mechanics of Materials*, 2nd ed., Wiley, New York, 1967.

28 J. H. Ahlberg, E. N. Nilson, and J. L. Walsh, *The Theory of Splines and Their Applications*, Academic Press, New York, 1967.

29 B. F. Green, J. E. K. Smith, and L. Klem, "Empirical Tests of an Additive Random Number Generator," *Journal of the A.C.M*, **6**, 527–537 (1959).

30 The Rand Corporation, *A Million Random Digits with 100,000 Normal Deviates*, The Free Press, Glencoe, Illinois, 1955.

31 H. H. Skilling, *Electrical Engineering Circuits*, 2nd ed., Wiley, New York, 1965.

32 G. J. Van Wylen, *Thermodynamics*, Wiley, New York, 1959.

33 R. Aris, *Discrete Dynamic Programming*, Blaisdell, New York, 1964.

34 R. Hooke and T. A. Jeeves, "Direct Search Solution of Numerical and Statistical Problems," *Journal of the A.C.M.*, **8**, 212–229 (1961).

35 H. H. Rosenbrock, "An Automatic Method for Finding the Least Value of a Function," *The Computer Journal*, **3**, 175, (1960).

36 G. S. G. Beveridge and R. S. Schechter, *Optimization Theory and Practice*, McGraw-Hill, New York, 1970.

37 S. I. Gass, *Linear Programming*, 2nd ed., McGraw-Hill, New York, 1964.

38 G. L. Nemhauser, *Introduction to Dynamic Programming*, Wiley, New York, 1966.

39 H. Akima, "A New Method of Interpola-

tion and Smooth Curve Fitting Based on Local Procedures," *Journal of the A.C.M.*, **17**, 589–602, 1970.

40 A. W. Burks, H. H. Goldstine, and J. von Neumann, "Preliminary Discussion of the Logical Design of an Electronic Computing Instrument," a report prepared for the Ordnance Dept., U.S. Army, at the Institute for Advanced Study, Princeton, New Jersey, June, 1946. Reprinted in *Datamation*, **8**, (9 and 10), 1962.

41 Philip and Emily Morrison, *Charles Babbage and his Calculating Engines*, Dover, New York, 1961.

42 L. F. Menabrea, "Sketch of the Analytical Engine Invented by Charles Babbage," Bibliothèque Universelle de Genève, October 1942, No. 82. Translated with notes by Ada Augusta, Countess of Lovelace (a reprint of this article appears in reference 41).

43 Charles Babbage, *The Life of a Philosopher*, Longman, Green, Longman, Roberts, and Green, 1864.

44 Jean E. Sammet, *Programming Languages, History and Fundamentals*, Prentice-Hall, Englewood Cliffs, New Jersey, 1969.

45 "FORTRAN vs Basic FORTRAN," *Communications of the A.C.M.*, **7**, (10), 590–625, 1964.

46 P. Cress, P. Dirksen, and J. W. Graham, "FORTRAN – IV *with* WATFOR and WATFIV," Prentice-Hall, Englewood Cliffs, New Jersey, 1970.

47 E. I. Organick, *A FORTRAN – IV Primer*, Addison-Wesley, Reading, Mass., 1966.

INDEX

Numbers preceded by E refer to computer examples;
numbers preceded by P refer to end-of-chapter problems.